PROGRESS IN
INORGANIC CHEMISTRY

Volume 10

Advisory Board

PROGRESS IN INORGANIC CHEMISTRY

EDITED BY

F. ALBERT COTTON

DEPARTMENT OF CHEMISTRY,
MASSACHUSETTS INSTITUTE
OF TECHNOLOGY, CAMBRIDGE,
MASSACHUSETTS

VOLUME 10

INTERSCIENCE PUBLISHERS   **1968**

a division of JOHN WILEY & SONS   New York · London · Sydney · Toronto

Library of Congress Catalog Card Number 59–13035

SBN 470   17669

Printed in The United States of America

Contents

Covalence and the Orbital Reduction Factor, k, in Magnetochemistry

By M. Gerloch* and J. R. Miller†

Department of Chemistry, The University of Manchester, Manchester, England

* Present address: Department of Chemistry, University College, London.
† Present address: Department of Chemistry, University of Essex, Colchester.

INTRODUCTION

The successful application of quantum mechanics to atomic spectroscopy in the 1920's led inevitably to studies directed to the understanding of similar phenomena in molecules and crystal lattices. It was Bethe (1) who, in 1929, wrote a now classic paper in which the framework of crystal field theory was set out. In this approach, the energy of an ion is taken to be modified by a potential field set up by its neighbors that are assumed to be described by point charges, or simple dipoles. For idealized systems involving high symmetry, Bethe made full use of group theory to establish the nature of crystal field terms arising from a given configuration. His work was put to good use by Van Vleck (2) in the field of magnetism. Van Vleck and also Penney and Schlapp (3) developed the theory of the magnetic susceptibilities of transition metal ions including some particularly fruitful studies of the lanthanides.

Meanwhile Pauling (4), who had been studying the bonding in transition metal complexes, had established his "magnetic criterion for bond type" from which a measurement of the magnetic moment of a complex might lead to a knowledge of the stereochemistry of the ions in it. A subsequent aim of magnetochemistry was to establish something of the character of bonds in complexes. About the same time as the appearance of Pauling's magnetic criterion, Van Vleck (5) showed how the change from a spin-free to a spin-paired situation in iron(III) complexes could be achieved by a smooth increase in the strength of the crystal field set up by the coordinating groups. He did not say, as was asserted by Pauling, that $Fe(CN)_6^{3-}$ ions involved more ionic bonds than FeF_6^{3-}, but merely that the low spin of the cyanide could be explained on the basis of a higher crystal field strength. Detailed calculations of the magnetic anisotropy of potassium ferricyanide by Howard (6) in 1934 based on Van Vleck's model were in fair, and now famous agreement with experiment.

The success of the crystal field approach, particularly in the case of the lanthanide ions, prompted many more studies of the magnetic susceptibilities and anisotropies of complex ions throughout the transition series. It was apparent that the descriptions of the lanthanide energy levels as those of the free ion perturbed only by spin-orbit coupling effects, or of the first-row transition elements perturbed only by a cubic field potential, were essentially correct. But the finer details of magnetic susceptibilities could not be accounted for so simply. The inclusion of spin–orbit coupling effects into strong-field configurations throughout the transition block by Kotani (7) made a considerable advance toward explaining the often large differences between magnetic moments of the 1st-, 2nd-, and 3rd-row elements for a given configuration and their behavior with temperature. Despite these improvements in magnetochemical theory, there remained significant and sometimes large deviations from experiment, most particularly in complexes having a formal orbital triplet ground state. Such ions make an important orbital contribution to the total moment by virtue of their orbital degeneracy—a degeneracy which, in the first row at least, is lifted only a little by the effects of spin–orbit coupling. Experimentally, these ions usually possess somewhat lower moments than expected, suggesting some additional quenching of the orbital contribution. A mechanism is not hard to find, for it was originally invoked by Howard in his calculations on potassium ferricyanide: namely, that distortions from perfect cubic symmetry will reduce the orbital degeneracy of some of the electronic states and cause the magnetic moments to tend to spin-only values.

Figgis (8) has examined in detail the magnetic properties of ions with formal cubic field orbital triplet ground states, perturbed simultaneously by the effects of an axial ligand field distortion and spin–orbit coupling. The energy level scheme involved in such calculations is summarized in Figure 1. The splitting Δ of the cubic triplet state by the low-symmetry field components was not calculated from a knowledge, often unavailable anyway, of the geometric molecular distortion, although its idealized relationship with the cubic field parameters Dq has been discussed recently (9): rather, it was taken as an independent parameter of the system to be

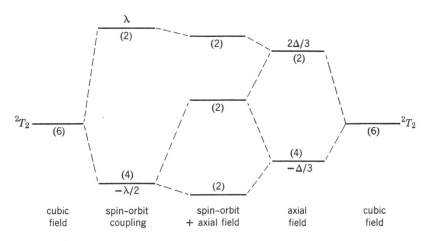

Fig. 1. Perturbation of a $^2T_{2(g)}$ term by spin–orbit coupling and an axial crystal field. Numbers in parentheses are total degeneracies.

adjusted by comparison with experiment. The same is true of λ, the spin–orbit coupling coefficient, which cannot be assumed to have the same value as in the free ion. It proves convenient algebraically to express these variables as v (equal to Δ/λ) and λ instead of Δ and λ. The magnetic moment computation begins with a calculation of the wave functions and energy levels which result from the combined effects of the axial crystal field and spin–orbit coupling acting on the pure metal d wavefunctions of the cubic field triplet. The energy separations between the three Kramers doublets which result lie commonly in the range 50–5000 cm^{-1}. The perturbations induced by the magnetic field—the Zeeman effect—are of the order of 1 cm^{-1}. Accordingly, the Zeeman perturbation calculation is performed sequentially rather than simultaneously; this is a time-saving procedure which involves trivial loss of accuracy. The final calculation (10) of the magnetic susceptibility or moment at any temperature involves a Boltzmann distribution of molecules among the Zeeman levels as expressed in the well-known Van Vleck equation. Figgis has performed these calculations for the cubic field terms $^2T_2$, $^3T_1$, $^4T_1$, and $^5T_2$. For the doublet and quintet terms, the moments may be expressed in terms of the two parameters v and λ and are usually plotted as a family of curves, one for each value of v, of magnetic moment versus $\ell T/\lambda$ (ℓ is the Boltzmann factor and T is the temperature). Some results for the $^2T_2$ term, of octahedral titanium(III) or tetrahedral copper(II) ions, for example, are shown in Figure 2a. These calculations clearly show the sensitivity of the effective moment to the degree of distortion and how the removal of orbital degeneracy with increasing v leads to moments closer to the spin-only value.

This theoretical approach allowed much closer agreement between calculated and observed moments than before, especially for complexes of the first-row transition metals where ligand field perturbations are most significant. Improvement was less marked, and less necessary, in the third row where spin–orbit coupling is relatively much more important. Up to this point, magnetochemistry involved no more than a continuous refinement of the crystal field model as first put forward by Bethe: up to this point, no explicit account had been taken of the undoubted covalence of the bonds in these complexes. Some implicit recognition of the molecular nature of the bonds was inherent in the parameterization of the quantities Dq, Δ, and λ, of course, but in essence the theory was still a crystal field one and it had worked very well.

The inadequacy of the *crystal* field approach was first demonstrated by Owen (11) and Stevens (12) in 1954. Their work grew out of the difficulty of explaining the low ESR g values for NiCl$_6{}^{4-}$ and IrCl$_6{}^{2-}$ ions. In these highly symmetrical ions, the mechanism for quenching some of the

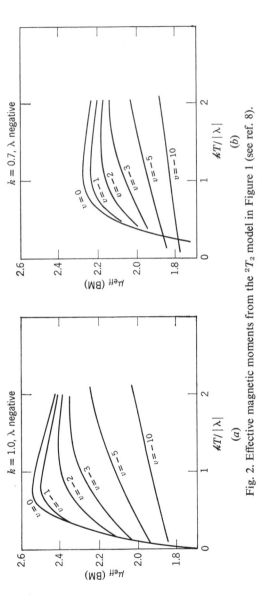

Fig. 2. Effective magnetic moments from the $^2T_2$ model in Figure 1 (see ref. 8).

orbital contribution to the Zeeman effect, discussed above, was not available. The framework of *ligand* field theory, in which the possibility of metal–ligand orbital mixing, that is covalence, was explicitly included, forms the basis of Stevens' work. It was shown that the formation of molecular orbitals, in which the unpaired t_{2g} electrons of the metal may reside partly on the ligands, could lead to a reduced orbital contribution of these electrons to the total angular momentum. Stevens showed that the operation of the magnetic moment operator

$$\mu H = \beta H(\mathbf{L} + 2\mathbf{S})$$

upon wavefunctions with metal d and ligand p character could be represented within a t_{2g} set of orbitals by the retention of pure "crystal field" metal d wavefunctions but with the use of a reduced magnetic moment operator

$$\mu H = \beta H(k\mathbf{L} + 2\mathbf{S})$$

The quantity k may then be regarded as a parameter which describes the "degree of molecular orbital formation." It has been variously called k, the orbital reduction factor, the electron delocalization parameter, the Stevens wandering effect, the covalence factor, and possibly otherwise. That unpaired electron density does exist on the ligands was subsequently demonstrated by Owen (13) for $IrCl_6^{2-}$ from a famous ESR study of the chlorine hyperfine splitting. This has been reviewed by Carrington and Longuet-Higgins (14). A similar study of electron delocalization in some iron-group fluorides has been made by Tinkham (15). Neutron diffraction work on the fluorides of nickel(II) and manganese(II), for example, has also yielded *direct* evidence of the delocalization of electron spin in complexes. The question of covalence in a wide context has been reviewed by Owen and Thornley (16).

The improved agreement between observed magnetic susceptibilities and those calculated from the crystal field model with axial distortion and spin–orbit coupling, discussed above, was usually good but rarely perfect within experimental error. The inclusion of the k factor into the Zeeman operator, in the light of Owen and Stevens' resonance experiments, was an obvious extension to the theory. Figgis performed his calculations with three parameters, v, λ, k: Figure 2b illustrates some of his results for the $^2T_2$ system when $k = 0.7$ and may be compared with those in Figure 2a where $k = 1.0$, equivalent to the "pure" crystal field model. The closer approach to spin-only moments with decreasing k is clearly seen. This three-parameter model has been very successful, allowing, within experimental error, perfect fits with powder susceptibilities, over the temperature range 80–300°K (17–21). For any given complex, best values for v, λ, and k have been found, inviting correlations between geometric distortion (from x-ray structural data) and v, between the free ion spin–orbit coupling coefficient λ_0 and λ in the complex and between λ and k. The last comparison is especially interesting, as one might expect a close connection between the formation of molecular orbitals and the change in the effective value of λ. In fact, k soon came to be regarded as a measure of covalence (22) and delocalization of the metal t_{2g} electrons, so it seemed that ESR and bulk magnetic susceptibility measurements had begun to fulfill one of the original aims of magnetochemistry.

But, as pointed out by Figgis (8), the relationship between observed k values and the electronegativity and π-bonding characteristics of various ligands is quite unclear. Table I presents a selection of magnetic parameters which have been obtained from powder susceptibility measurements. Thus, the smaller k values for the $Fe(CN)_6{}^{3-}$ and $Mn(CN)_6{}^{4-}$ ions compared with $Fe(o\text{-phen})_3{}^{3+}$ and $Fe(dipy)_3{}^{3+}$ are in line with chemical ideas of high π-bonding ability of the cyanide ligands. On the other hand, it is surprising that delocalization is so small in the o-phenanthroline and dipyridyl complexes where π-bonding is again presumed from the low-spin state of the iron. More remarkable still are the low k values for the aquo complexes

TABLE I

Compound or ion	Approximate stereochemistry	k value	Ref.
$K_3Fe(CN)_6$	Oct.	0.8	17
$K_4Mn(CN)_6 \cdot 3H_2O$	Oct.	0.75	17
$Fe(o\text{-phen})_3{}^{3+}$	Oct.	1.0	17
$Fe(dipy)_3{}^{3+}$	Oct.	1.0	17
$Ti(H_2O)_6{}^{3+}$	Oct.	0.8	19
$V(H_2O)_6{}^{3+}$	Oct.	0.8	19
$Cr(H_2O)_6{}^{3+}$	Oct.	0.8	19
$Ni(H_2O)_6{}^{2+}$	Oct.	0.9	18
Cs_2CuCl_4	Tet.	$0.7 \rightarrow 0.5$	23
Cs_3NiCl_5	Tet.	0.95	21
$(Et_4N)_2NiCl_4$	Tet.	1.00	21
$(Et_4N)_2NiBr_4$	Tet.	1.00	21
$(Ph_3P)_2NiCl_2$	Tet.	0.55	21
$(Ph_3P)_2NiBr_2$	Tet.	0.55	21
$(NH_4)_2Fe(SO_4)_2 \cdot 6H_2O$	Oct.	1.00	20
$FeSO_4 \cdot 7H_2O$	Oct.	0.85	20
$Fe(py)_4 \cdot (NCS)_2$	Oct.	1.00	20
$Fe(pyrazine)_2(NCS)_2$	Oct.	0.70	20

of Ti^{3+}, V^{3+}, and Cr^{3+} which involve presumed non-π-bonding water ligands. Again, the k values for the tetrahedral complexes of nickel(II) range very widely: a general observation here is that much lower k values have been observed in tetrahedral molecules than in octahedral, although it is well to remember that few tetrahedral systems with orbital triplet ground states have been available for study, so that such generalizations must be treated cautiously. The nickel compounds, in particular, display greatly reduced spin–orbit coupling coefficients indicative of some relationship between k and λ; this point will be taken up in more detail in the last section of this review.

It is important to emphasize, of course, that the exactness of the fit between theory and experiment obtained for most of these compounds does not necessarily mean the values of the various parameters are right. Such a claim must depend upon the essential correctness of the theoretical model and, in practice, this means upon the neglect of second-order terms, of rhombic symmetry, and so on. Tetrahedral copper(II) complexes offer an interesting example of the dangers of such approximations (23,24). Nevertheless, even though the k values quoted above may be suspect, it is essential at this stage to understand properly the nature of this parameter and try to establish whether there really is any simple relationship between it and other chemical ideas on bonding.

We shall now examine the orbital angular momentum of t_{2g} orbitals in an octahedron and t_2 orbitals in a tetrahedron and investigate the nature of k in these two geometries. We begin with a short summary and explanation of the wavefunctions and operators that will be required.

II. DEFINITIONS

A. Wavefunctions

The angular parts of atomic wavefunctions are conventionally described by spherical harmonics which are discussed in references 25, E57, and B93.* Table II lists those normalized spherical harmonics which we shall require in this review, expressed in cartesian form and phased as described by Condon and Shortley (26).†

---

* References beginning with a letter refer to a textbook, see page 46, and to a specific page in that book.

† A factor for -1 for positive odd values of m_l. We wish to draw particular attention to Professor Ballhausen's footnote in B19.

TABLE II

$$Y_1^{-1} = \sqrt{\frac{3}{8\pi}} \frac{x - iy}{r}$$

$$Y_1^{0} = \sqrt{\frac{3}{4\pi}} \frac{z}{r}$$

$$Y_1^{1} = -\sqrt{\frac{3}{8\pi}} \frac{x + iy}{r}$$

$$Y_2^{-2} = \sqrt{\frac{5}{4\pi}} \sqrt{\frac{3}{8}} \frac{(x - iy)^2}{r^2}$$

$$Y_2^{-1} = \sqrt{\frac{5}{4\pi}} \sqrt{\frac{3}{2}} \frac{z(x - iy)}{r^2}$$

$$Y_2^{0} = \sqrt{\frac{5}{4\pi}} \sqrt{\frac{1}{4}} \frac{3z^2 - r^2}{r^2}$$

$$Y_2^{1} = -\sqrt{\frac{5}{4\pi}} \sqrt{\frac{3}{2}} \frac{z(x + iy)}{r^2}$$

$$Y_2^{2} = \sqrt{\frac{5}{4\pi}} \sqrt{\frac{3}{8}} \frac{(x + iy)^2}{r^2}$$

We may construct the real d and p wavefunctions we shall need as follows:

$$p_z = p_0$$
$$p_x = (1/\sqrt{2})(p_{-1} - p_1)$$
$$p_y = (i/\sqrt{2})(p_{-1} + p_1)$$
$$d_{z^2} = d_0$$
$$d_{x^2 - y^2} = (1/\sqrt{2})(d_{-2} + d_2)$$
$$d_{xz} = (1/\sqrt{2})(d_{-1} - d_1)$$
$$d_{yz} = (i/\sqrt{2})(d_{-1} + d_1)$$
$$d_{xy} = (i/\sqrt{2})(d_{-2} - d_2)$$

where

$$p_m = R_{np} Y_1^m \quad (-1 \le m \le 1) \qquad d_m = R_{nd} Y_2^m \quad (-2 \le m \le 2)$$

in which $R_{np(d)}$ represents the radial part of a $p(d)$ wavefunction of the nth principal quantum shell.

The factor i which appears in some of these combinations is necessary to ensure that the wavefunctions are real.

B. Operators

It is inappropriate here to review the use of operators in quantum mechanics. For the most part, we shall be concerned only with the operator

L_z which is associated with that component of orbital angular momentum about the z axis.

$$L_z = i\frac{h}{2\pi}\frac{\partial}{\partial\phi} = -i\frac{h}{2\pi}\left(x\frac{\partial}{\partial y} - y\frac{\partial}{\partial x}\right)$$

The spherical harmonics are eigenfunctions of L_z, that is they satisfy the eigenvalue equation

$$L_z Y_l^m = (\text{eigenvalue})\, Y_l^m$$

where the eigenvalue is equal to $mh/2\pi$. If we express the orbital angular momentum about the z axis in units* of $h/2\pi$, then we may write,

$$L_z(d_m) = m(d_m)$$

As an example of the operation of L_z, we evaluate

$$
\begin{aligned}
L_z(p_x) &= L_z(1/\sqrt{2})(p_{-1} - p_1)\\
&= (1/\sqrt{2})[(-1)p_{-1} - (1)p_1]\\
&= -(1/\sqrt{2})(p_{-1} + p_1)\\
&= ip_y
\end{aligned}
$$

III. k IN OCTAHEDRAL COMPLEXES

A. Pure Metal Orbitals

Following Stevens' original paper (12) on the orbital reduction factor and Ballhausen's discussion of molecular orbitals (B152), we shall look first at octahedral complexes with reference to the t_{2g} set of orbitals.

We begin by investigating the orbital angular momentum associated with the set of *pure d* orbitals and this is done by evaluating the matrix elements of the t_{2g} set under L_z.† Most are identically zero as may be shown either by the explicit use of L_z as described above or with the aid of group theory. In the subgroup D_{2h} of O_h, L_z transforms as B_{1g}: the matrix elements may then be finite only if the triple direct product contains the totally symmetric representation. For example, for the integral‡

* We do this throughout the review in order to omit the factor $h/2\pi$, merely for convenience.

† (a) Throughout the review, we shall only consider cubic symmetry which makes calculations with L_y or L_x unnecessary in these isotropic environments. (b) We are concerned only with the orbital part of the magnetic moment operator, not the spin.

‡ For bra-ket notation, see Appendix I.

$\langle d_{xy}|\mathbf{L}_z|d_{zx}\rangle$ we look at $B_{1g} \times B_{1g} \times B_{2g}$. This does not contain A_g, so that this integral is zero. In fact there are only two nonzero matrix elements here—$\langle d_{xz}|\mathbf{L}_z|d_{yz}\rangle$ and $\langle d_{yz}|\mathbf{L}_z|d_{zx}\rangle$. In both cases the direct product gives A_g. Explicit use of $\mathbf{L}_z$ gives

$$\langle d_{zx} \mid \mathbf{L}_z \mid d_{yz}\rangle \equiv \left\langle \frac{1}{\sqrt{2}} \cdot (d_{-1} - d_1) \left| \mathbf{L}_z \right| \frac{i}{\sqrt{2}} \cdot (d_{-1} + d_1) \right\rangle$$

$$= \frac{i}{2}[-\langle d_{-1} \mid d_{-1}\rangle + \langle d_{-1} \mid d_1\rangle + \langle d_1 \mid d_{-1}\rangle - \langle d_1 \mid d_1\rangle]$$

$$= \frac{i}{2}[-1 + 0 + 0 - 1]$$

$$= -i$$

Similarly we may show* that

$$\langle d_{yz} \mid \mathbf{L}_z \mid d_{zx}\rangle = +i$$

The matrix of the t_{2g} orbitals under $\mathbf{L}_z$ then appears as,

$$
\begin{array}{c}
 & \begin{array}{ccc} d_{xy} & d_{yz} & d_{zx} \end{array} \\
\begin{array}{c} d_{xy} \\ d_{yz} \\ d_{zx} \end{array} &
\left(\begin{array}{ccc}
0 & 0 & 0 \\
0 & 0 & i \\
0 & -i & 0
\end{array} \right)
\end{array}
$$

leading to the secular determinant

$$
\begin{vmatrix}
0 - \Lambda & 0 & 0 \\
0 & 0 - \Lambda & i \\
0 & -i & 0 - \Lambda
\end{vmatrix} = 0
$$

from which the roots, Λ equal $0, \pm 1$. The substitution of these eigenvalues back into the secular equations will give the c's which define the appropriate linear combinations of the basis set (d_{xy}, d_{yz}, d_{zx}) for which the matrix of $\mathbf{L}_z$ would appear diagonal. We label these linear combinations t_{2g}^{+}, t_{2g}^{0}, t_{2g}^{-} according to their associated eigenvalues being $+1, 0, -1$. This point will be taken up again in a later section. Meanwhile, we conclude that under $\mathbf{L}_z$, the t_{2g} orbitals exhibit z components of orbital angular momentum of $+1, 0,$ and -1, which therefore corresponds to a total orbital angular momentum for the set, of 1 unit (i.e., $h/2\pi$).

* This must follow, anyway, from the Hermitean properties of $\mathbf{L}_z$. (See Appendix I.)

We are now concerned with determining how much angular momentum is associated with a similar set but which contains admixed ligand orbitals.

B. Molecular Orbitals

We must construct suitable linear combinations of metal and ligand orbitals to form molecular orbitals, all of which must transform under the same symmetry representation of the point group, here O_h. It is simple to show (B154) that the six σ bonds in O_h transform as the irreducible representations A_{1g}, E_g, and T_{1u}, from which it follows that no interaction of σ orbitals with the metal T_{2g} set is possible. However, π bonds involving 12 ligand p-orbitals normal to the metal–ligand vectors may be shown to transform as T_{1g}, T_{1u}, T_{2g}, and T_{2u} (C105,B156). The T_{2g} set of ligand p orbitals which may then mix with the metal T_{2g} orbitals are shown in Figure 3 (12,B156).

We wish to evaluate the matrix element between the (xz) and (yz) orbitals. The relevant combinations of ligand orbitals in three dimensions are illustrated in Figure 4.

We prefer here not to follow the conventions of Van Vleck (27) and Ballhausen (B153) but, for simplicity, to assign to all points in space right-handed cartesian axes in parallel orientation. The atomic orbitals represented in Figure 4 by these global coordinates are then,

$$p_x^{(1)} \ p_y^{(1)} \ - p_x^{(2)} \ p_z^{(3)} \ - p_z^{(4)} \ p_z^{(5)} \ - p_z^{(6)}$$

where superscripts label the ligands.

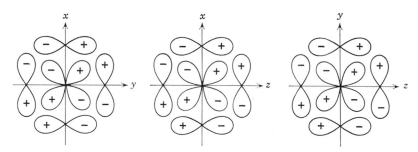

Fig. 3. Atomic orbitals used to construct a t_{2g} set of molecular orbitals in O_h.

Representing the molecular orbital which involves and transforms like d_{zx} by $|zx\rangle$ etc., we may write out the orbitals as:

$$|zx\rangle = N[d_{zx} + \lambda/2(p_x^{(1)} - p_x^{(2)} + p_z^{(3)} - p_z^{(4)})]$$
$$|yz\rangle = N[d_{yz} + \lambda/2(p_y^{(1)} - p_y^{(2)} + p_z^{(5)} - p_z^{(6)})]$$

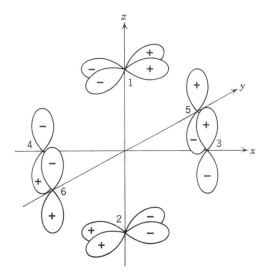

Fig. 4. Three-dimensional view of ligand atomic orbitals which overlap with d_{xz} and d_{yz} metal orbitals in O_h.

[we use $\lambda/2$ as a mixing coefficient rather than λ, to follow Ballhausen (B166)].

The normalizing coefficient, N, is given by

$$1/N^2 = [1 + 4(\lambda^2/4) + 4\lambda(S_{m,p})$$
$$+ \lambda^2/2\,(-\,S_{12} + S_{13} - S_{14} - S_{23} + S_{24} - S_{34})]$$

where $S_{m,p}$ represents the overlap integral between a metal d orbital and a single $p\pi$, and $S_{i,j}$ is the overlap integral between the appropriate p orbitals on ligands i and j, given that in all cases S_{any} is taken to be a positive quantity.

Recognizing that $S_{12} = S_{34}$ and that $S_{13} = S_{14} = S_{23} = S_{24}$ and collecting terms, we get

$$1/N^2 = [1 + \lambda(4S_{m,p}) + \lambda^2(1 + 2S_{13} - S_{12})]$$

C. The Matrix Element $\langle zx \mid \mathbf{L}_z \mid yz \rangle$

Writing this integral out *in extenso*, we have

$$N^2\langle d_{zx} + \lambda/2(p_x{}^{(1)} - p_x{}^{(2)} + p_z{}^{(3)} - p_z{}^{(4)}) \mid \mathbf{L}_z \mid d_{yz}$$
$$+ \lambda/2(p_y{}^{(1)} - p_y{}^{(2)} + p_z{}^{(5)} - p_z{}^{(6)})\rangle$$

Expanding this gives,

Term

$$N^2 \times \begin{cases} \langle d_{zx} \mid \mathbf{L}_z \mid d_{yz}\rangle & 1 \\ \quad + (\lambda/2)[\langle d_{zx} \mid \mathbf{L}_z \mid (p_y^{(1)} - p_y^{(2)} + p_z^{(5)} - p_z^{(6)})\rangle & 2a \\ \qquad + \langle (p_x^{(1)} - p_x^{(2)} + p_z^{(3)} - p_z^{(4)}) \mid \mathbf{L}_z \mid d_{yz}\rangle] & 2b \\ \quad + (\lambda^2/4)\langle (p_x^{(1)} - p_x^{(2)} + p_z^{(3)} - p_z^{(4)}) \mid \mathbf{L}_z \mid (p_y^{(1)} - p_y^{(2)} \\ \qquad + p_z^{(5)} - p_z^{(6)})\rangle & 3 \end{cases}$$

We shall evaluate this term by term.

Term 1: This is simply equal to $-i$

Term 2a: We may utilize the hermitean properties of $\mathbf{L}_z$ thus:

$$\int d_{zx}^* \mathbf{L}_z (p_y^{(1)} - p_y^{(2)} + p_z^{(5)} - p_z^{(6)}) \, d\tau$$

$$= \int (p_y^{(1)} - p_y^{(2)} + p_z^{(5)} - p_z^{(6)}) \mathbf{L}_z^* d_{zx}^* \, d\tau$$

Now $d_{zx}^* = d_{zx}$ (the orbitals were chosen to be real) and

$$\mathbf{L}_z^* = [i(h/2\pi)\,(\partial/\partial\phi)]^* = -\mathbf{L}_z$$

so that

$$\mathbf{L}_z^* d_{zx}^* = -\mathbf{L}_z \mid (1/\sqrt{2})(d_{-1} - d_1)\rangle = -(1/\sqrt{2})(-d_{-1} - d_1)$$
$$= -i d_{yz}$$

Then Term *2a*

$$\approx -i \int (p_y^{(1)} - p_y^{(2)} + p_z^{(5)} - p_z^{(6)}) d_{yz} d\tau$$

$$= -i \cdot 4 S_{m,p}$$

Term 2b: This integral also equals $-i4S_{m,p}$.

Hence Term $2 \equiv -i8\,(\lambda/2)S_{m,p} = -i\lambda(4S_{m,p})$

Term 3: This involves ligand orbitals only and is often approximated. We treat it in Appendix II in some detail, as its importance has hitherto been insufficiently recognized. The result derived there is

$$\text{Term } 3 \equiv -i(\lambda^2/4)(2 - 2S_{12} - 4S_{13})$$

It is important to note that this term involves both possible types of ligand–ligand overlap integral.

The final result for the matrix element $\langle zx \mid \mathbf{L}_z \mid yz\rangle$ is

$$-i \left[\frac{1 + \lambda(4S_{m,p}) + \lambda^2 \tfrac{1}{2}(1 - S_{12} - 2S_{13})}{1 + \lambda(4S_{m,p}) + \lambda^2 \tfrac{1}{2}(2 - 2S_{12} + 4S_{13})} \right]$$

Thus, as a result of forming molecular orbitals, the matrix element $\langle d_{xz} \mid \mathbf{L}_z \mid d_{yz}\rangle$ has been reduced by the factor in square brackets. By using the hermitean properties of $\mathbf{L}_z$, or by repeating the effort of the last few pages, we may show that the matrix element $\langle d_{yz} | \mathbf{L}_z | d_{zx} \rangle$ is reduced by precisely the same factor.

The matrix of the t_{2g} molecular orbitals under $\mathbf{L}_z$ now appears as

$$
\begin{array}{c}
\quad\ xy \ \ yz \ \ zx \\
\begin{array}{c} xy \\ yz \\ zx \end{array}
\left(
\begin{array}{ccc}
0 & 0 & 0 \\
0 & 0 & ik \\
0 & -ik & 0
\end{array}
\right)
\end{array}
$$

That is, we have premultiplied the original matrix for the *pure* metal d orbitals by the scalar* k:

$$
k\left(
\begin{array}{ccc}
0 & 0 & 0 \\
0 & 0 & i \\
0 & -i & 0
\end{array}
\right)
=
\left(
\begin{array}{ccc}
0 & 0 & 0 \\
0 & 0 & ik \\
0 & -ik & 0
\end{array}
\right)
$$

with k equal to the quantity in square brackets above. The eigenvalues are all reduced by k, of course, being k, 0, $-k$. We may conclude that, within the present context, the effects of mixing ligand orbitals into the d orbitals may be completely and accurately accounted for by using pure metal t_{2g} orbitals as a basis but with a "reduced" orbital angular momentum operator, $k\mathbf{L}_z$.

Before discussing the behavior of the expression for k and, indeed, showing that its effect is to *reduce* rather than augment the orbital angular momentum, we note that neglect of ligand–ligand overlap integrals (S_{12} and S_{13}) allows the expression to reduce to that given by Ballhausen (B166) where such overlap integrals were neglected, for simplicity, at an earlier stage in his treatment.

D. Properties of k in Octahedral Complexes

The expression for k above may be rewritten as

$$
k = 1 - \frac{(\lambda^2/2)(1 - S_{12} + 6S_{13})}{1 + \lambda(4S_{m,p}) + (\lambda^2/2)(2 - 2S_{12} + 4S_{13})}
$$

The terms S_{12} and S_{13} represent the overlap integrals between like p orbitals on ligands in *trans* and *cis* positions of the octahedron, respectively. S_{12} is a $p\pi$–$p\pi$ type of overlap while S_{13} may be resolved into both $p\sigma$–$p\sigma$ and $p\pi$–$p\pi$ components, so that we may relate these two integrals. Mulliken et al. (28) have given general expressions for such overlap integrals based on Slater-type atomic orbitals as functions of internuclear distance and

* We consider only isotropic environments here.

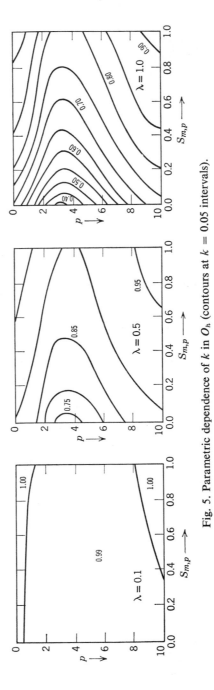

Fig. 5. Parametric dependence of k in O_h (contours at $k = 0.05$ intervals).

effective nuclear charge. For illustration, we may take Mulliken's expressions for $2p$ orbitals in the form

$$S_{12} = (1 + \sqrt{2} \cdot p + \tfrac{4}{5}p^2 + (2\sqrt{2}/15)p^3)e^{-\sqrt{2}\,p}$$
$$S_{13} = (p^2/10)(1 + p + \tfrac{2}{3}p^2)e^{-p}$$

where

$$p = ZR/2a_0$$

in which R is the distance between *adjacent* ligand centers in Å, Z is an effective nuclear charge on the ligand and a_0 is the Bohr radius in Å.

Using these formulas, we have calculated values for k as a function of the metal–ligand mixing coefficient λ, the metal–ligand overlap integral $S_{m,p}$, and the parameter p. The results are presented in the form of a three-dimensional contour map, with representative sections shown in Figure 5. It is immediately apparent from this figure that the minimum values of k occur in a "trough" in the maps which occurs when $p \approx 4$ and this is essentially a result of the behavior with distance of the overlap integrals S_{12} and S_{13}. At first sight it may appear possible for k to be as low as 0.4 or so in octahedral complexes. That this is not so emerges from the following qualitative estimates of values which the parameters λ, $S_{m,p}$, and p may take in practice.

The simplest parameter to fix is p. A typical distance between adjacent ligand donor atoms in the octahedron may be taken as 2.5 Å. Values for the effective nuclear charge may be taken indirectly from Clementi and Riamondi (29), and for carbon, nitrogen, and oxygen, for example, are 3.0, 3.6, and 4.2, respectively. Hence p will be between 7.5 and 10. It is sufficient for the present purposes, however, to recognize that the lower limit of k when p is about 4, will never be achieved in practice. A section of the contour map for constant $p = 8$ is shown in Figure 6. While detailed theoretical or experimental estimates of either λ or $S_{m,p}$ will not be attempted here, we can say that as the metal–ligand mixing coefficient increases, so will the metal–ligand overlap integral, even though the precise relationship between them will depend very much on the particular metal and ligands concerned. Thus, if we vary the metal (say), then as $S_{m,p}$ increases, so will λ. This corresponds to moving along a line in Figure 6, in the general direction of top-left to bottom-right, so that we follow approximately the contours of k. The implication, of course, is that k will alter little along such a series of complexes. While actual values of k for particular octahedral ions must be computed individually, we can say that it is unlikely, in regular octahedral complexes, that the orbital reduction factor will be less than about 0.7; also, experimental values will not be simply

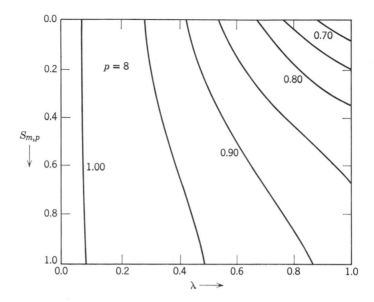

Fig. 6. Variation of k for constant effective nuclear charge and ligand–ligand separation in O_h.

related to any one parameter, so that the apparently similar k values (17, 19) for $Fe(CN)_6^{3-}$ and $Ti(H_2O)_6^{3+}$, for example, do not necessarily imply similar electronegativities or π acceptor properties of the ligands, and so on.

Of course, these arguments are only illustrative, as they must be, when we remember that besides the difficulty of assessing likely values for $S_{m,p}$, λ, and p, these contours for k have been computed for ligands with $2p$ orbitals only. The general forms of the expression for overlap integrals of the $3p$ and other orbitals, and the values of Clementi and Riamondi exponents, indicate somewhat similar behavior of k with ligand donor atoms elsewhere in the periodic table.

IV. k IN TETRAHEDRAL COMPLEXES

A. Ligand Orbital Mixing

We shall now briefly consider the effects upon the orbital reduction factor of ligand orbital mixing in tetrahedral geometries. Here, we may conveniently study the mixing of ligand s orbitals, for as is again simple to

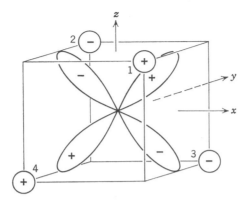

Fig. 7. Atomic orbitals used to construct one component of a t_2 set of molecular orbitals in T_d. This combination of s orbitals transforms as $t_2(xz)$. It does not interact with metal d_{xy} or d_{yz}.

show, four σ bonds in the point group T_d transform as $A_1 + T_2$. Twelve p orbitals on the ligands in a tetrahedron transform as $A_1 + E + T_1 + 2T_2$ and so offer a further possibility for metal ligand mixing: the complications involved in the study of these orbitals are not compensated by any clearer insight for the purpose of this review, and so they will not be considered further. Figure 7 illustrates one component of the T_2 set of ligand s orbitals which may interact with the metal t_{2g} set—in particular, the d_{xz}.

As in the octahedral case, the only nonzero matrix element under $\mathbf{L}_z$ involves the $|zx\rangle$ and $|yz\rangle$ orbitals which we define as,

$$|zx\rangle = N[d_{zx} + \lambda/2(s^{(1)} - s^{(2)} - s^{(3)} + s^{(4)})]$$
$$|yz\rangle = N[d_{yz} + \lambda/2(s^{(1)} - s^{(2)} + s^{(3)} - s^{(4)})]$$

where the normalizing factor N is given by

$$1/N^2 = [1 + 4\lambda(S_{m,s}) + \lambda^2(1 - S_{ss})]$$

in which $S_{m,s}$ represents the overlap integral between a metal d orbital and a single ligand s orbital. As before, we must evaluate the matrix element $\langle zx | \mathbf{L}_z | yz \rangle$ which again involves three different terms of the type, $\langle d | \mathbf{L}_z | d \rangle$, $\langle d | \mathbf{L}_z | ligand \rangle$, and $\langle ligand | \mathbf{L}_z | ligand \rangle$. Evaluation of the first two follows similar lines to those in the octahedral case, while the third is worth commenting upon. The result, derived in Appendix III is $-i\lambda^2\sigma$, where σ resembles a σ-type overlap between an s and p orbital.

The final expression for the matrix element $\langle zx \mid \mathbf{L}_z \mid yz \rangle$ is equal to

$$-i\left[\frac{1 + \lambda(4S_{m,s}) + \lambda^2\sigma}{1 + \lambda(4S_{m,s}) + \lambda^2(1 - S_{ss})}\right]$$

$$\equiv -ik$$

The precise form of the function, σ, will depend upon the size of the tetrahedron and the exact nature of the s orbitals involved. It is sufficient to note for the moment that the general form of this expression for k follows that obtained in the octahedral case and that similar remarks concerning its magnitude will pertain. We shall not investigate this expression for the orbital reduction factor further, for tetrahedral symmetry offers another, possibly more effective, means of reducing orbital angular momentum which we now examine.

As an introduction to this, we include a discussion of the origin and form of the octahedral crystal field potential, $\mathbf{V}_{oct}$.

B. Ligand Field Potentials in the Octahedron

If we represent the Hamiltonian of a free ion as $\mathscr{H}_F$, that for a complex of less than spherical symmetry may be written as

$$\mathscr{H} = \mathscr{H}_F + \mathbf{V}$$

in which $\mathbf{V}$ is the potential provided by the ligands and operates on the basis wavefunctions of the ion as a multiplying function. *Any* function may be expressed (E31) as a linear combination of a complete set of eigenfunctions (involving the same variables and boundary conditions) of *any* operator, and it proves convenient to expand $\mathbf{V}$ as a linear superposition of normalized spherical harmonics. The *complete* set is infinite, involving all harmonics from Y_0^0 to Y_∞^∞. Fortunately, symmetry considerations will always vastly reduce the number of terms necessary in the expansion of $\mathbf{V}$.

The total Hamiltonian $\mathscr{H}$ must transform as the totally symmetric representation of the point group of the molecule; that is, the total energy of the system cannot depend upon any symmetry operation which merely reorients the whole molecule. As $\mathscr{H}_F$ is spherically symmetric and therefore transforms as Γ_1 in any point group, so also must $\mathbf{V}$ transform as Γ_1. This requires, in turn, that each harmonic or group of harmonics of the same order (same l) occurring in the expansion of $\mathbf{V}$ must individually transform as Γ_1. One term may always satisfy this condition, namely Y_0^0, which being dependent only upon the radial term is spherically symmetric. It cannot cause any splitting between orbitals of the same l in the molecule, merely giving rise to a uniform energy shift to all energy levels. Conventionally, this term will be tacitly ignored in the further discussion: in the octahedral point group, for example, it may be included as

$$\mathbf{V} = \mathbf{V}_r + \mathbf{V}_{oct}$$

where $\mathbf{V}_{oct}$ includes all terms except $Y_0{}^0$.

Centrosymmetric crystal fields, like O_h, restrict the terms appearing in $\mathbf{V}$ to those which are even (that is, l even) since in such symmetries the totally symmetric representation, Γ_1, of the point group is even, A_{1g}. The octahedral potential, for example, may then contain terms $Y_0{}^0$, $Y_2{}^m$, $Y_4{}^m$, $Y_6{}^m$. . . . Recognition of the complete symmetry of the rotational subgroup, O, further restricts those terms which may appear in $\mathbf{V}_{oct}$. For example, the complete set of $Y_2{}^m$'s ($|m| \leq 2$) transforms (B60) as a reducible representation Γ under the symmetry operations of O, with the characters

	E	$8C_3'$	$3C_2$	$6C_4$	$6C_2'$
$\chi(\Gamma)$	5	-1	1	-1	1

The irreducible components of Γ are readily shown to be E_g and T_{2g}. As A_{1g} does not appear, harmonics of second order are not included in the expansion of $\mathbf{V}_{oct}$. [The same conclusion may be arrived at by consideration of the potential $\sum_i e^2/r_{ij}$ set up by six point charges of the vertices of a regular octahedron, using the explicit form of the expansion of $1/r_{ij}$ in the spherical harmonics (F13).]

A further restriction on necessary terms in $\mathbf{V}$ is imposed if only d orbitals are considered. d wavefunctions involve the spherical harmonics $Y_2{}^m$ ($|m| \leq 2$). As $Y_2{}^m$ involves radial, θ and ϕ terms, we may write them as,

$$Y_2{}^m \approx R_{n,2}\Theta_2 e^{im\phi}$$

where the maximum and minimum values of m are ± 2. The integral $\int \psi_i^* \mathbf{V} \psi_j d\tau$ will then involve terms like

$$\int R_{n,2}\Theta_2 e^{im\phi} {}^* R_{n,l}\Theta_l e^{ix\phi} R_{n,2}\Theta_2 e^{im'\phi} d\tau$$

and hence

$$\int_0^{2\pi} \exp[i(m + x + m')\phi]d\phi$$

which is zero† unless $(m + m' + x) = 0$. As the maximum and minimum

† Expand $\int_0^{2\pi} e^{in\phi}d\phi$ as $\int_0^{2\pi}(\cos n\phi + i\sin n\phi)d\phi$, etc.

values of $m + m'$ are ± 4, the minimum and maximum values of x must be ∓ 4, so that no harmonics of order (l) greater than 4 will interact with d orbitals. Similarly, no harmonics of order greater than 2 could interact with p orbitals.

The points elaborated in the last two paragraphs have been summarized (B59) by the statements that "if the effects to be described arise from ...d shells... all harmonics of odd order are ineffective, since any matrix representation of an odd-order potential in a basis of orbitals with the same parity vanishes identically," and "when only d electrons are involved, all potentials after the fourth-order terms exert no influence, because the direct product of two d-orbital sets spans no representation of the rotation group of order higher than 4." The final form of the potential set up by a regular octahedron of point charges at a distance a from the nucleus may be arrived at by symmetry considerations above (B61) and is

$$\mathbf{V}_{\text{oct}} = Y_4{}^0 + \sqrt{5/14}\,(Y_4{}^4 + Y_4{}^{-4})$$

C. Ligand Field Potential in the Tetrahedron

The potential set up by a regular tetrahedron of point charges at the same distance from the nucleus is usually represented by $\mathbf{V}_{\text{tet}} = -\frac{4}{9}\,\mathbf{V}_{\text{oct}}$ the minus sign implying the familiar inversion of levels on going from an octahedron to a tetrahedron (30). As pointed out by Ballhausen, for example, the relationship holds good if d orbitals only are considered (B110). It was shown above how the terms in a given potential may depend upon (a) the point-group symmetry and (b) the symmetry of the basis orbitals. Consider the possible inclusion of odd-order harmonics in $\mathbf{V}_{\text{tet}}$. It is readily seen that the set $Y_1{}^m(m = 1, 0, -1)$ transforms like the set of p orbitals which, in T_d, is T_2: these first-order terms cannot, therefore, be included in the totally symmetric potential. Next consider the set $Y_3{}^m$ ($|\,m\,| \leq 3$). Their behavior under the symmetry operations of T_d may be seen by constructing the seven real independent functions from them as a basis set:

$$\psi_1 = -(i/\sqrt{2})(Y_3{}^3 + Y_3{}^{-3})$$
$$\psi_2 = (1/\sqrt{2})(Y_3{}^3 - Y_3{}^{-3})$$
$$\psi_3 = (1/\sqrt{2})(Y_3{}^2 + Y_3{}^{-2})$$
$$\psi_4 = -(i/\sqrt{2})(Y_3{}^2 - Y_3{}^{-2})$$
$$\psi_5 = -(i/\sqrt{2})(Y_3{}^1 + Y_3{}^{-1})$$
$$\psi_6 = (1/\sqrt{2})(Y_3{}^1 - Y_3{}^{-1})$$
$$\psi_7 = Y_3{}^0$$

In T_d these real orbitals transform as follows:

	E	C_3	C_2	S_4	σ_d
ψ_1	ψ_1	$\sqrt{15}/4\,\psi_6 - 1/4\,\psi_2$	$-\psi_1$	ψ_2	ψ_2
ψ_2	ψ_2	$-\sqrt{5}/8\,\psi_7 + \sqrt{3}/8\,\psi_3$	$-\psi_2$	$-\psi_1$	ψ_1
ψ_3	ψ_3	$-\sqrt{10}/4\,\psi_5 + \sqrt{6}/4\,\psi_1$	ψ_3	ψ_3	$-\psi_3$
ψ_4	ψ_4	ψ_4	ψ_4	ψ_4	ψ_4
ψ_5	ψ_5	$-1/4\,\psi_6 - \sqrt{15}/4\,\psi_2$	$-\psi_5$	$-\psi_6$	$-\psi_6$
ψ_6	ψ_6	$\sqrt{3}/8\,\psi_7 + \sqrt{5}/8\,\psi_3$	$-\psi_6$	ψ_5	$-\psi_5$
ψ_7	ψ_7	$\sqrt{5}/8\,\psi_1 + \sqrt{3}/8\,\psi_5$	ψ_7	$-\psi_7$	ψ_7

Clearly, ψ_4, and only ψ_4, transforms in T_d as A_1. Using the table of spherical harmonics given earlier we may write:

$$-i/\sqrt{2}(Y_3{}^2 - Y_3{}^{-2}) \approx -(i/\sqrt{2})[z(x + iy)^2 - z(x - iy)^2]$$
$$\approx (i/\sqrt{2})[z(x^2 + 2ixy - y^2) - z(x^2 - 2ixy - y^2)]$$
$$\approx -(i/\sqrt{2})[4izxy] \approx xyz$$

Thus, we see that the tetrahedral potential should be written (for orbitals no higher than d), as

$$\mathbf{V}_{\text{tet}} = -\tfrac{4}{9}\mathbf{V}_{\text{oct}} + c'i(Y_3{}^2 - Y_3{}^{-2})$$

or

$$\mathbf{V}_{\text{tet}} = -\tfrac{4}{9}\mathbf{V}_{\text{oct}} + cxyz$$

That this "extra" xyz term is absent in O_h is clear from the oddness of this function in a group in which parity is defined.

Again we may consider integrals of type (a), (b), and (c) below in the tetrahedron. Parity is not defined and so no conclusions which rest on the oddness or evenness of the orbitals may be made.

$$\int \psi_d^* \mid xyz \mid \psi_d d\tau \qquad\qquad\qquad \text{(a)}$$

$$\int \psi_p^* \mid xyz \mid \psi_p d\tau \qquad\qquad\qquad \text{(b)}$$

$$\int \psi_p^* \mid xyz \mid \psi_d d\tau \qquad\qquad\qquad \text{(c)}$$

In T_d, the p orbitals transform as T_2 while the d orbitals transform as $T_2(d_\varepsilon)$ and $E(d_\gamma)$. The diagonal matrix elements of type (a) and (b) involve the triple direct products, $E \times A_1 \times E$, $T_2 \times A_1 \times T_2$, and $T_2 \times A_1 \times T_2$, each containing A_1. For integrals of type (c), $\int \psi_p^* \mid xyz \mid \psi_{d_\varepsilon} d\tau$, transform as

$T_2 \times A_1 \times T_2 = A_1$ and may be finite, while $\int \psi_p^* \mid xyz \mid \psi_{d_\gamma} d\tau$ transforms as $T_2 \times A_1 \times E \neq A_1$ and is identically zero, That is, the component xyz of $\mathbf{V}_{\text{tet}}$ mixes p and d_ε but not p and d_γ.

The amount of mixing of $4p$ into $3d$ may be approached by first-order perturbation theory. In the latter case, we may write (B111,E96),

$$\psi(3d_\varepsilon) = \psi^\circ(3d_\varepsilon) + \sum \frac{\int \psi_{3d_\varepsilon}^{0*} \mid xyz \mid \psi_{4p}{}^0 d\tau}{E_d{}^0 - E_p{}^0} \cdot \psi_{4p}{}^0$$

Using a point charge model (31) it is found that the admixed portion of $4p$ orbitals is typically about 8%. Further quantitative comments about this are made later. The effects of the various terms in $\mathbf{V}_{\text{tet}}$ are represented in Figure 8.

D. The Crystal Field Wavefunctions in the Tetrahedron

We have shown how the $4p$ orbital may mix, under $\mathbf{V}_{\text{tet}}$, into the $3d_\varepsilon$ set. The resulting set is still orbitally triply degenerate, of course, and for a subsequent discussion of the effects of the orbital angular momentum operator, we must separate these degenerate wavefunctions into three orthogonal (i.e., independent) functions. In general, an n-fold degeneracy may be expressed in an infinite number of ways so that apparently any three linear combinations of the three $3d_\varepsilon$ plus the admixed $4p$ functions

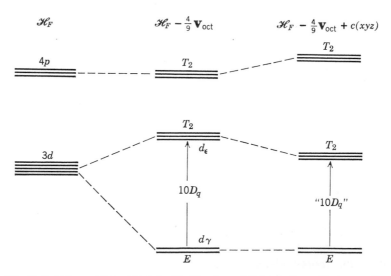

Fig. 8. Splitting of $3d$ orbitals in T_d. The $4p$ and $3d_\varepsilon$ sets "repel" one another.

will do, provided that they are made orthogonal. While this is strictly correct in T_d symmetry in which the transformation properties of these orbitals under the group of symmetry operations involves the set *as a whole* (i.e., T_2), no indication of which linear combination to take is immediately obvious; nor, in general, will such arbitrary linear combinations usefully serve as zeroth-order wavefunctions for many typical distortions from exact tetrahedral symmetry. We may arrive at suitable linear combinations in several ways.

1. Let us evaluate the matrix element connecting the $4p$ and $3d_\varepsilon$ wavefunctions. Dropping the principal quantum number, we may write this as

$$\int (p_x + p_y + p_z)^* xyz(d_{xy} + d_{xz} + d_{yz})d\tau$$

which may be resolved into 9 components, for example:

$$\int p_x^*(xyz)d_{xy}d\tau$$

$$\int p_x^*(xyzd)_{yz}d\tau \qquad \text{etc.}$$

To illustrate the properties of these two integrals, by way of example, we may substitute the forms of p_x, d_{xy}, and d_{yz} given in the tables above. We omit all numerical constants and also the radial parts of the p and d wavefunctions.

$$\int p_x^*(xyz)d_{xy}d\tau$$

$$\approx \int \left(\frac{x - iy}{r} - \frac{x + iy}{r} \right) xyz \, i \left[\frac{(x - iy)^2}{r^2} - \frac{(x + iy)^2}{r^2} \right] d\tau$$

$$\approx \int \frac{x}{r} xyz \frac{xy}{r^2} \, d\tau$$

$$\approx \int \frac{x^3yz}{r^3} \, d\tau \tag{d}$$

Compare:

$$\int p_x^*(xyz) \, d_{yz} \, d\tau \approx \int \frac{x}{r} xyz \, i \left[\frac{z(x - iy)}{r^2} + \frac{-z(x + iy)}{r^2} \right] d\tau$$

$$\approx \int \frac{x}{r} xyz \frac{zy}{r^2} \, d\tau$$

$$\approx \int \frac{(xyz)^2}{r^3} \, d\tau \tag{e}$$

It is simply checked that the three integrals of the type $\int p_i^* \, xyz \, d_{jk} \, d\tau$ all behave as (e) while all others ($\int p_i^* \, xyz \, d_{ij} \, d\tau$) behave as, or similar to (d). Integrals of the type (e) will clearly transform as Γ_1 while those of type (d) will not.

Earlier we proved the general statement that $\int (4p \text{ set})^* xyz (3d \text{ set}) d\tau$ was finite because the triple direct product transformed as A_1. Now we have shown that, in particular, only the matrix elements $\int p_i^* \, xyz \, d_{jk} d\tau$ are nonzero. The appropriate linear combinations are, therefore:

$$Nd_{xy} + np_z$$
$$Nd_{xz} + np_y$$
$$Nd_{yz} + np_x$$

2. These linear combinations may be arrived at another way, Consider a subgroup of T_d in which all three d (or p) orbitals transform separately — D_2.

D_2	E	$C_2(x)$	$C_2(y)$	$C_3(x)$		
A	1	1	1	1		x^2, y^2, z^2
B_1	1	1	-1	-1	z, R_z	xy
B_2	1	-1	1	-1	y, R_y	xz
B_3	1	-1	-1	1	x, R_x	yz

From this it is immediately clear that z transforms as xy, y as xz, and x as yz, leading once more to the linear combinations above.

3. Each p orbital points in a different (i.e., independent) direction which we indicate by a single subscript in the symbols p_x, p_y, p_z. But there are other ways of defining a given direction. In three-dimensional space, we may do so by describing the plane perpendicular to the required direction: this is simply the reverse of the usual definition of a plane by its normal. Thus the orbital d_{yz} defines the same direction as does p_x and so the linear combination of these two wavefunctions above appears only too reasonable. This line of argument depends upon the possible choice of, say, the $d_{x^2-y^2}$ orbital having been removed, of course; and in any case it is implicit in the other two approaches, particularly 2.

We have devoted considerable space to the linear combinations for d–p mixing in the tetrahedron, for, as will become clear, they are of central importance to the nature of k in this symmetry.

E. Matrix Elements of L_z for $d–p$ Hybrid Orbitals

We have shown how the T_2 set of d orbitals in the tetrahedron may mix under the ligand field operator with a T_2 set of metal p orbitals by an amount which depends on the exact form of the tetrahedral environment and on the $d–p$ energy gap. Taking the wavefunctions,

$$| xy \rangle = Nd_{xy} + np_z \tag{1}$$

$$| yz \rangle = Nd_{yz} + np_x \tag{2}$$

$$| zx \rangle = Nd_{zx} + np_y \tag{3}$$

as basis, we wish to calculate matrix elements under L_z. As before, we make use of the usual group-theoretical rule that such integrals are zero unless the representation products $\Gamma_{(xy)} \times \Gamma_{(L_z)} \times \Gamma_{(yz)}$ etc., contain the fully symmetric representation of the group. Again, we may use the subgroup D_2 in which the orbitals (1), (2), and (3) above belong to the representations B_3, B_1, and B_2, respectively, while L_z transforms as B_3. Now, for a triple product $B_i \times B_j \times B_k$ to contain A_1 in D_2 it is easy to show that i, j, and k must be different. Thus, of the nine matrix elements, only two are nonzero, namely,

$$\langle zx \,|\, L_z \,|\, yz \rangle \quad \text{and} \quad \langle yz \,|\, L_z \,|\, zx \rangle$$

These will again be related by the Hermitean properties of L_z. We may now evaluate these integrals:

$$\langle zx \,|\, L_z \,|\, yz \rangle \equiv \int (zx)^* L_z (yz) d\tau$$

$$= \int (Nd_{zx} + np_y)^* L_z (Nd_{yz} + np_x) \cdot d\tau$$

$$= N^2 \int d_{zx}^* L_z d_{yz} \cdot d\tau + n^2 \int p_y^* L_z p_x \cdot d\tau$$

$$+ Nn \left[\int d_{zx}^* L_z p_x \cdot d\tau + \int p_y^* L_z \cdot d_{yz} \cdot d\tau \right]$$

Now,

$$\int d_{zx}^* L_z d_{yz} \cdot d\tau = -i$$

and

$$\int p_y^* L_z p_x \cdot d\tau = \int p_y^* \cdot i p_y \cdot d\tau$$

$$= +i$$

while the other integrals vanish because of d–p orthogonality. Therefore:

$$\langle zx \mid \mathbf{L}_z \mid yz \rangle = -i(N^2 - n^2)$$

and the secular determinant is

$$
\begin{array}{c|ccc}
 & xy & yz & zx \\
\hline
xy & 0 - \Lambda & 0 & 0 \\
yz & 0 & 0 - \Lambda & i(N^2 - n^2) \\
zx & 0 & -i(N^2 - n^2) & 0 - \Lambda
\end{array} = 0
$$

giving the eigenvalues under $\mathbf{L}_z$ as $(N^2 - n^2)$, 0, $-(N^2 - n^2)$. The reduction of the eigenvalues may be allowed for by using $k\mathbf{L}$ on pure d orbitals as was done for the octahedral case, but now with k equal to $(N^2 - n^2)$.

It is important to notice, however, that, although the extent of d–p mixing may be small, its effect may be relatively large. For, not only is k reduced from unity by virtue of the "diluting" effect ($N^2 < 1$) of the p orbitals, but also because the orbital contribution from the p orbitals themselves *opposes* that from the d functions, as is evidenced by the minus sign in the expression for k. This mechanism is essentially a "crystal field" phenomenon, in contrast to the previous orbital reduction processes which involved ligand orbital mixing and may be termed the "ligand field" type.

F. A Pictorial View of the Orbital Reduction by Configurational Mixing

The opposing orbital contribution of metal d and p orbitals may be illustrated qualitatively by the model shown in Figure 9. The possession of orbital angular momentum about the z axis by the pair of orbitals d_{xz}, d_{yz} is often demonstrated (M425,F261) by the observation that these two *degenerate* orbitals may be transformed into one another by rotation about the z axis. Such a rotation cannot transform the d_{z^2} orbital into the $d_{x^2 - y^2}$ and this pair of orbitals is accordingly devoid of angular momentum—being a "nonmagnetic" doublet. Now consider the "mixed" orbitals:

$$d_{xz} + p_y \tag{1}$$

$$d_{yz} + p_x \tag{2}$$

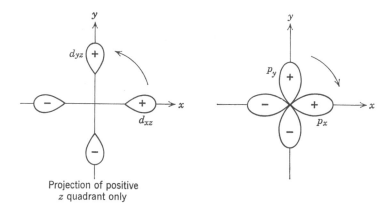

Projection of positive
z quadrant only

Fig. 9. Opposed angular momenta of d_ε and p orbitals in the t_2 set.

(from which mixing coefficients are omitted here, for simplicity). The four components are drawn looking down the positive z direction onto the xy plane (Fig. 9).

The correctly defined p and d orbitals have positive angular parts along the positive axes or in the positive–positive quadrant, respectively. Let us rotate from "mixed" orbital (1) to orbital (2): this means from d_{xz} to d_{yz} and *at the same time* from p_y to p_x. As indicated by the curved arrows in Figure 9, these rotations must be made in opposite directions; that is, the orbital angular momentum contribution from the p orbitals opposes that from the d orbitals.

The argument depends ultimately upon the appropriate choice of d–p orbital combinations which is why so much space has been devoted to that particular aspect. This pictorial approach does not lend itself to any extrapolation: it must always be realized that the operation $\mathbf{L}_z$ does not actually rotate one orbital into another but that the angular momentum is associated with the *pair* of orbitals. The diagonalized wavefunctions correspond to $t_{2g}{}^+$, $t_{2g}{}^0$, and $t_{2g}{}^-$ and, for orbitals in this complex form, a simple pictorial description is not possible.

G. Isomorphism of the d_ε- and p-Orbital Sets

The opposing contributions to the total orbital angular momentum of the d_ε- and p-orbital sets may be deduced also from a brief discussion of the isomorphism of the properties of these functions.

The operations of the three angular momentum operators, L_z, L_x, and L_y upon the p-orbital set may be written as (B89):

$$L_z \begin{pmatrix} p_1 \\ p_0 \\ p_{-1} \end{pmatrix} = \begin{pmatrix} 1 & 0 & 0 \\ 0 & 0 & 0 \\ 0 & 0 & -1 \end{pmatrix} \begin{pmatrix} p_1 \\ p_0 \\ p_{-1} \end{pmatrix}$$

$$L_x \begin{pmatrix} p_1 \\ p_0 \\ p_{-1} \end{pmatrix} = \frac{1}{\sqrt{2}} \begin{pmatrix} 0 & 1 & 0 \\ 1 & 0 & 1 \\ 0 & 1 & 0 \end{pmatrix} \begin{pmatrix} p_1 \\ p_0 \\ p_{-1} \end{pmatrix}$$

$$L_y \begin{pmatrix} p_1 \\ p_0 \\ p_{-1} \end{pmatrix} = \frac{i}{\sqrt{2}} \begin{pmatrix} 0 & 1 & 0 \\ -1 & 0 & 1 \\ 0 & -1 & 0 \end{pmatrix} \begin{pmatrix} p_1 \\ p_0 \\ p_{-1} \end{pmatrix}$$

Let us now construct a similar set of d_ε orbitals and investigate the same operations on them. By "similar," we mean here that the relationship within the d_ε set is to be analogous to that within the p set; so we begin by examining the relationships within the p set. These are characterized by the operation of the raising and lowering operators (E42). For our present purposes we need to define these operators merely as,

$$L_+(\psi m_l) \propto \psi m_{l-1}$$
$$L_-(\psi m_l) \propto \psi m_{l+1}$$

Thus,

$$L_-(p_1) \rightarrow p_0$$
$$L_-(p_0) \rightarrow p_{-1}$$
$$[L_-(p_{-1}) \rightarrow 0]$$

An isomorphous behavior of the d_ε set of orbitals may be achieved by choosing the functions:

$$\phi_1 = d_{-1}$$
$$\phi_0 = (1/\sqrt{2})(d_{-2} - d_2)$$
$$\phi_{-1} = -d_1$$

Only this choice leads to the correct relationships between the d_ε orbitals. Thus,

$$L_-(\phi_1) \equiv L_-(d_{-1}) \rightarrow d_{-2}, \quad \text{(part of } \phi_0)$$
$$L_-(\phi_0) \equiv L_-[(1/\sqrt{2})(d_{-2} - d_2)] \rightarrow 0 - d_1, \quad (\phi_{-1})$$

The operations of $\mathbf{L}_z$, $\mathbf{L}_x$ and $\mathbf{L}_y$ on these functions then appear as:

$$\mathbf{L}_z\begin{pmatrix}\phi_1\\\phi_0\\\phi_{-1}\end{pmatrix} = -\begin{pmatrix}1&0&0\\0&0&0\\0&0&-1\end{pmatrix}\begin{pmatrix}\phi_1\\\phi_0\\\phi_{-1}\end{pmatrix}$$

$$\mathbf{L}_x\begin{pmatrix}\phi_1\\\phi_0\\\phi_{-1}\end{pmatrix} = -\frac{1}{\sqrt{2}}\begin{pmatrix}0&1&0\\1&0&1\\0&1&0\end{pmatrix}\begin{pmatrix}\phi_1\\\phi_0\\\phi_{-1}\end{pmatrix}$$

$$\mathbf{L}_y\begin{pmatrix}\phi_1\\\phi_0\\\phi_{-1}\end{pmatrix} = -\frac{i}{\sqrt{2}}\begin{pmatrix}0&1&0\\-1&0&1\\0&-1&0\end{pmatrix}\begin{pmatrix}\phi_1\\\phi_0\\\phi_{-1}\end{pmatrix}$$

The matrices describing the operations $\mathbf{L}_z$, $\mathbf{L}_x$, and $\mathbf{L}_y$ on the d orbital set are now the negatives of those for the p set. Once again we have the result that the orbital contribution from the p and d_ε sets are opposed.

The isomorphism of the p- and d_ε-orbital sets is only one special case of general isomorphous relationship between sets transforming in similar ways. The signs of the effective quantum numbers associated with orbital angular momentum are intimately bound up in this property and related to the parity of the sets. Griffith has discussed the phenomenon in general terms and has also shown (32), for example, how it is possible to treat a T_{1g} term in octahedral symmetry with an effective quantum number $L = -1$ and has derived a modified formula for the Landé splitting factor, g.

H. Quantitative Aspects of the Orbital Reduction for d–p Hybrids

If we represent the d–p hybrid wavefunctions by

$$| zx \rangle = Nd_{zx} + np_y \qquad \text{etc.}$$

then, for normalization we have $N^2 + n^2 = 1$, so that k, the orbital reduction factor, is given by $1 - 2n^2$. Values for k as a function of the mixing coefficients N and n are given in the table accompanying Figure 10. Clearly, at first k is very insensitive to small admixtures of p character into the d orbitals but this situation changes rapidly so that when there is equal d and p character, the opposition of angular momentum is absolute and k is zero. With more p character, k is not negative, but rather we should then regard d character as being mixed into p.

Now, as mentioned earlier, point charge calculations have indicated values for n around -0.08 and so it would appear that the mechanism we

n	N	k
0.0	1.000	1.00
0.1	0.995	0.98
0.2	0.980	0.92
0.3	0.954	0.82
0.4	0.915	0.68
0.5	0.865	0.50
0.6	0.800	0.28
0.7	0.713	0.02
$1/\sqrt{2}$	$1/\sqrt{2}$	0.00

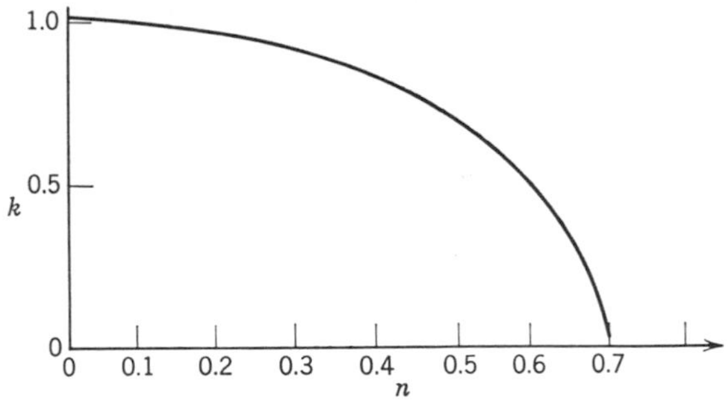

Figure 10

have discussed at great length causes only a trivial orbital reduction. However, molecular orbitals and ligand field theory again show that this is not so. There are grounds for believing that d–p mixing in tetrahedral complexes may be much more extensive as the following discussion shows.

I. A More Complete Model for the Tetrahedron

It is not really valid to separate the effects of metal–ligand mixing from those of metal d–metal p mixing, or configurational interaction, in the tetrahedron. We have done this in the previous paragraphs merely to illustrate the nature of the two effects. If we restrict ourselves to metal d and p orbitals and some ligand orbitals (X) unspecified we require the molecular orbitals,

$$| zx \rangle = N d_{zx} + n p_y + l X_y$$
$$| yz \rangle = N d_{yz} + n p_x + l X_x$$
$$| xy \rangle = N d_{xy} + n p_z + l X_z$$

The evaluation of the matrix element $\langle zx \mid \mathbf{L}_z \mid yz \rangle$ would then involve six types of integrals, $viz.$:

$$\langle d \parallel d, \rangle \ \langle p \parallel p \rangle, \ \langle X \parallel X \rangle, \ \langle d \parallel p \rangle, \ \langle d \parallel X \rangle, \ \langle p \parallel X \rangle$$

Of these, only $\langle d \parallel p \rangle$, involving the orthogonal metal wavefunctions, will be identically zero. We shall not treat the complete system in general, merely noting the increased complexity of the problem in the tetrahedron relative to the octahedron.

Let us represent the Hamiltonian for the tetrahedron by $\mathscr{H}$ but, for the sake of illustration, omit that term responsible for d–p interaction: that is, the matrix element $\langle d \mid \mathscr{H} \mid p \rangle$ is set as zero. The linear combination of a basis set of orbitals which includes metal d and p functions and ligand χ functions may be found by diagonalization of the matrix of these basis orbitals under $\mathscr{H}$. We can write this matrix as

$$
\begin{array}{c}
\begin{array}{ccc} d & \chi & p \end{array} \\
\begin{array}{c} d \\ \chi \\ p \end{array}
\begin{pmatrix} \alpha & \delta & 0 \\ \delta & \beta & \varepsilon \\ 0 & \varepsilon & \gamma \end{pmatrix}
\end{array}
$$

in which δ, for example, represents the integral $\int d^* \mathscr{H} \chi d\tau$. This matrix will not factorize or "block off" so that the resulting eigenfunctions will contain something of all three d, χ, and p components.

The conclusion to be drawn from this is that even though the direct mixing of metal d and p orbitals may be very small, some indirect mixing will occur by virtue of the fact that the ligand functions χ may overlap and mix with *both* the metal d and p wavefunction. To quote a single experimental result: on the basis of ESR studies of tetrahedral $CuCl_4^{2-}$ ions, Sharnoff (25) has deduced the values: $N \sim 0.86$, $n \sim -0.34$, $l \sim -0.42$. In general, we may infer that metal d–p mixing will be enhanced by the presence of the ligand functions; that estimates for d–p mixing are underestimated in a crystal field model compared with those derived from a ligand field one; and, therefore, any evaluation of k in tetrahedral molecules should not neglect configurational interaction.

V. CONCLUSION

The orbital reduction factor as used in the magnetochemistry of transition metal ions is a device which makes *implicit* recognition of ligand field theory and configurational interactions while allowing the calculation of orbital angular momenta within the framework of crystal field pure atomic d orbitals: it represents the effects of modifying metal d wavefunctions by substituting for the angular momentum operator $\mathbf{L}$, the reduced operator $k\mathbf{L}$.

In this review we have investigated the orbital angular momentum associated with molecular orbitals transforming as T_{2g} in an octahedron or as T_2 in a tetrahedron. In both geometries, such orbitals involve linear combinations of metal and ligand atomic orbitals, and give rise to a "ligand field" orbital reduction effect, while in the tetrahedral case only, there is the additional "crystal field" factor of the admixture of other metal orbitals of different parity.

The parameter k has often been referred to as an "electron delocalization factor" and has been used to account for the time spent by unpaired t_{2g} electrons on the ligands where, it was assumed, they made little contribution to the total orbital angular momentum. Our discussion of k, above, has shown that this description is not justified, for the reduction of orbital angular momentum resulting from the time spent by these electrons away from the central metal is offset to a greater or lesser degree by two factors. One results from a term in the expression for k which involves the metal–ligand overlap integral, while the other describes how t_{2g} electrons on the ligand atoms may contribute to the total orbital angular momentum by an amount dependent upon the overlap integrals between the ligand orbitals themselves. The importance of this last term has previously been underestimated, for it is largely responsible for setting a rather high limit for the minimum value of k in octahedral complexes.

Much lower values for k appear possible in the tetrahedral stereochemistry for, as a result of the lack of a center of symmetry, metal p orbitals (and, in principle, other odd metal wavefunctions) may mix into the ground state molecular orbitals. The contribution to the total orbital angular momentum of odd and even parity metal wavefunctions are opposed, so that any admixture of p orbitals into the d may cause a reduction in k, much greater than expected merely from the "dilution" of the d wavefunctions. The ability of the ligand orbitals in the tetrahedron to mix with *both* the metal d and p orbitals will enhance such d–p interaction and hence the reduction of k. In general, this "crystal field" mechanism might be expected to operate in other geometries lacking an inversion center.

The general conclusions show, on the one hand, why much lower values of the orbital reduction factor may be expected in tetrahedral complexes than in octahedral ones and, on the other, why simple correlations between π-acceptor properties of ligands and k do not appear to work. It should now be clear, however, that the use of this parameter to allow such excellent agreement between theory and experiment in magnetochemistry is quite valid, even though the ultimate aim of its correlation in detail with other chemical facts has yet to be realized. This, of course, will require many more experimental as well as theoretical studies in the subject: it is surprising perhaps in view of the long history of magnetochemistry that relatively few systematic attempts to measure the powder, and especially the crystal susceptibilities of series of compounds, have been made. As the theoretical models become more sophisticated by including low symmetry fields, ligand orbitals, more excited states on the central metal, configurational interactions, and so on, the number of parameters required to define the system increases. Quite often it is necessary to assume more or less arbitrary values for, or relationships between, many of these parameters in order to handle the theoretical model. This is why it is so desirable to perform experiments with single crystals as well as powder, over as wide a temperature range as possible and by several complementary techniques. The original concept of the k factor came out of ESR work, of course: the often very accurate g values obtained by the resonance method may be used to complement bulk susceptibility measurements. These techniques have also been used (33,34) to estimate the extent of any magnetic coupling between structurally isolated paramagnetic ions. It is clearly important not to ascribe to distortion, molecular orbital formation, or whatever, effects due to antiferromagnetism or other cooperative phenomena. The energies and something of the nature of higher-lying states in complexes can be studied optically: low-temperature, single-crystal, polarized, infrared, visible, and ultraviolet spectroscopy may all help to "fix" the values of some of the parameters required in the magnetic model. One example of this is the parameter (33,34) A, related to the mixing of excited $^3T_1$ (P) or $^4T_1$ (P) states into ground $^3T_1$ (F) or $^4T_1$ (F) states, which may be determined from spectral studies. Mössbauer spectroscopy (35) has given some information about lower lying energy bands, complementary to both microwave and optical regions.

There are, of course, many aspects of the orbital reduction factor which we have not yet discussed. For example, as originally introduced by Stevens, k was concerned with the π-type overlap of ligand p orbitals in octahedral complexes with the metal t_{2g} d-orbital set; this k factor has been called (16) $k_{\pi\pi}$. This is to distinguish it from $k_{\pi\sigma}$ which describes the reduced matrix elements of the orbital angular momentum operator between wavefunctions of the t_{2g} set, overlapping with p–π orbitals, and those of the higher lying e_g set (d_{z^2} and $d_{x^2-y^2}$) which overlap with ligand p–σ orbitals. Alternatively, consider symmetries other than O_h and T_d. Earlier, we discussed the question of axial distortion from pure cubic symmetry; we did so in order to show that despite such refinements in magnetochemical theory, good fits with experiment still required the parameter k. However, the discussions on k were made in the context of the pure cubic fields so as to bring out, as clearly as possible, the nature of orbital reduction. If we superimpose an axial distortion, however, the resultant symmetries no longer span representations of order three: the orbital triplets split into doublets and singlets. In consequence, the coefficients in the molecular orbitals of the doublet are not the same as in the singlet. In the tetrahedral case, for example, we might write for our ground state wavefunctions

$$\left.\begin{array}{l} N_1 d_{yz} + n_1 p_x + l_1 \chi_x \\ N_1 d_{zx} + n_1 p_y + l_1 \chi_y \end{array}\right\} \text{orbital doublet}$$

$$N_2 d_{xy} + n_2 p_z + l_2 \chi_z \quad \text{orbital singlet}$$

If we insist on using pure metal d wavefunctions with suitably reduced orbital angular momentum operators, we shall then require one k for the doublet and another k for matrix elements connecting the doublet and singlet (the orbital singlet, having zero orbital angular momentum, will not require any k value). We shall also require a set of k's connecting these two levels with the various components of the excited e_g levels, or derivatives thereof. In short, such systems are complicated. One thing is clear, it is *not* satisfactory to recognize low symmetry on the one hand by using the parameter Δ or v, but to neglect it on the other by using a single k. The *anisotropy* in k is, in fact, just another way of defining the presence of noncubic terms in the ligand field potential. Recent measurements of the magnetic anisotropies of crystals of some tetrahedral copper(II) complexes (23,24) have revealed considerable anisotropy in k values.

We have made reference to a $k_{\pi\sigma}$ parameter to describe effects in matrix elements between ground $t_{2(g)}$ and excited $e_{(g)}$ levels. It might appear that such effects should be very small, for the second-order Zeeman effect which involves matrix elements of the magnetic moment operator, μ, between nondegenerate levels is of the order of 0.01–0.1 cm^{-1}. There is, however, a more significant perturbation which may mix excited states into the ground state and which involves the angular momentum operator; namely, spin–orbit coupling.

A. Spin–Orbit Coupling

The spin–orbit perturbation is taken prior to the magnetic or Zeeman effect in magnetic calculations. This is because it may cause energy shifts of between tens and thousands of wavenumbers rather than units. Its effect is not only to split up ground state degeneracies but also to mix excited state wavefunctions into the ground state: these admixtures are by no means trivial. The operator for spin–orbit coupling is often represented by

$$\lambda \mathbf{L S}$$

where λ, the spin–orbit coupling coefficient, is a constant for a free-ion term. The rest of the operator requires the multiplication of the spin and orbital angular momenta. In our examination of k we saw how the formation of molecular orbitals and the effects of configurational interaction could be allowed for by replacing the operator $\mathbf{L}$ by $k\mathbf{L}$. The same arguments hold for the spin–orbit coupling operator which should accordingly be taken as

$$\lambda k \mathbf{LS}$$

with the same remarks about matrix elements within a set and those between sets being applicable. This is why we expect some sort of relationship between the values of k and λ derived from magnetic measurements. The values of λ in Table III were calculated using the operator λLS and so the greatly reduced effects of spin–orbit coupling often observed in tetrahedral nickel complexes, for example, are not to be attributed completely to reduced spin–orbit coupling *coefficients*, but rather to reduced orbital angular momentum *operators*. It is largely the spin–orbit coupling energy which changes on complex ion formation, rather than the spin–orbit coefficient. For precisely the same reasons as discussed above, the λ values derived from the use of the operator $\lambda \mathbf{LS}$ may be expected to be anisotropic in noncubic environments; again, the work (23) on tetrahedral copper(II) crystals, cited earlier, demonstrates this effect. Strictly speaking, the three-parameter model using v, k, λ is logically inconsistent and, no doubt, its successful use to date partly reflects the averaged nature of the information obtainable from powdered samples.

It is important to recognize that the coefficient λ, even as derived with the operator $\lambda k \mathbf{LS}$, will differ from the free-ion value and be

TABLE III

Compound	Stereochem.	k	λ/λ_0	Ref.
Cs_3NiCl_5	Tet.	0.95	0.48	21
$(Et_4N)_2NiCl_4$	Tet.	1.00	0.62	21
$(Et_4N)_2NiBr_4$	Tet.	1.00	0.41	21
$(Ph_3P)_2NiCl_2$	Tet.	0.55	0.41	21
$(Ph_3P)_2NiBr_2$	Tet.	0.55	0.41	21
$(Ph_3P)_2NiI_2$	Tet.	0.55	0.45	21
		0.70	0.66	
$(NH_4)_2Fe(SO_4)_26H_2O$	Oct.	1.00	0.90	20
$Fe(py)_4(NCS)_2$	Oct.	1.00	0.65	20
$Fe(pyrazine)_2(NCS)_2$	Oct.	0.70	0.70	20
			0.80	
$K_3Fe(CN)_6$	Oct.	0.8	0.9	17
$K_4Mn(CN)_6 \cdot 3H_2O$	Oct.	0.75	0.67	17

anisotropic in noncubic symmetries. This is not too well understood quantitatively. On the one hand it may be argued that the difference from the free-ion value λ_0 should be small in the complex since λ is a function of the distance of electrons from the nucleus to the inverse cube power and so the perturbation of ligands at a distance of about 2 Å is unlikely to be felt. On the other hand, λ is also a function of the effective nuclear charge to the fourth power and must therefore be very sensitive to the ligand field. Then again the metal electrons spend some time near the ligand nuclei which may accordingly directly affect the value of λ. It has been generally observed that λ values in complexes are smaller than in the free ion, but there appears to be no fundamental reason why the reverse should not be so. In general, then, we may expect the magnitude and anisotropy of the ratio λ/λ_0 to reflect the magnitude and anisotropy of k but not to be strictly related to them.

It is just these correlations between k, spin–orbit coupling, v, distortion, and the rest which still require much detailed investigation. The complexity of polyparameter theoretical models such as those we have discussed means that such investigations must be broadly based with respect to experimental conditions and techniques and that, in the first instance at least, they must be empirical. In this area of magnetochemistry, where complex coordination number and stereochemistry are a beginning rather than an end, a great deal of detailed and accurate experimental data are still required.

VI. APPENDIX I.   The Secular Determinant

The calculation of magnetic susceptibilities makes extensive use of an approximate method called perturbation theory (E92,H197). Although our need of it in the present review is minimal, it is appropriate to discuss, without proof, some of the basic concepts—these may also be developed by other methods.

We begin with a degenerate set of wavefunctions ϕ of some Hamiltonian operator $\mathscr{H}^0$: for example, this might be an orbital triplet ground state resulting from a cubic crystal field potential acting upon a transition metal ion. Under the action of some perturbing influence—a magnetic field, perhaps—the degeneracy may be removed. We wish to know the energies, E of the new levels and also the new wavefunctions ψ: we must express these in terms of what we know initially; namely, the original Hamiltonian $\mathscr{H}^0$, the original set of wavefunctions ϕ, and the perturbing Hamiltonian $\mathscr{H}$. The situation may be represented pictorially as in Figure 11.

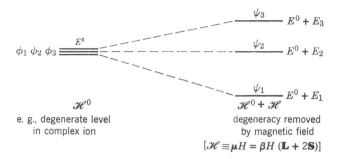

Fig. 11. Perturbation of a degenerate set of levels by a magnetic field.

To a first approximation, the new wavefunctions ψ are simply linear combinations of the old ones, ϕ. If we represent such combinations by

$$\psi = c_1\phi_1 + c_2\phi_2 + c_3\phi_3$$

it can be shown that the coefficients c_1, c_2, c_3 are given by solving three simultaneous equations, called the secular equations, which take the form:

$$[\langle \phi_1 \mid \mathcal{H} \mid \phi_1 \rangle - E]c_1 + \langle \phi_1 \mid \mathcal{H} \mid \phi_2 \rangle c_2 + \langle \phi_1 \mid \mathcal{H} \mid \phi_3 \rangle c_3 = 0$$
$$\langle \phi_2 \mid \mathcal{H} \mid \phi_1 \rangle c_1 + [\langle \phi_2 \mid \mathcal{H} \mid \phi_2 \rangle - E]c_2 + \langle \phi_2 \mid \mathcal{H} \mid \phi_3 \rangle c_3 = 0$$
$$\langle \phi_3 \mid \mathcal{H} \mid \phi_1 \rangle c_1 + \langle \phi_3 \mid \mathcal{H} \mid \phi_2 \rangle c_2 + [\langle \phi_3 \mid \mathcal{H} \mid \phi_3 \rangle - E]c_3 = 0$$

where E is one of the shifts in energy caused by the perturbation (see Fig. 3) and the quantities $\langle \phi_i \mid \mathcal{H} \mid \phi_j \rangle$ represent the integrals, $\int \phi_i^* \mathcal{H} \phi_j d\tau$.

The equations cannot be solved immediately because possible values for E are not yet known. However, if the three equations are independent then the determinant of the coefficients of the variables, i.e., of the c's, must be zero. That is,

$$\begin{vmatrix} \langle \phi_1 \mid \mathcal{H} \mid \phi_1 \rangle - E & \langle \phi_1 \mid \mathcal{H} \mid \phi_2 \rangle & \langle \phi_1 \mid \mathcal{H} \mid \phi_3 \rangle \\ \langle \phi_2 \mid \mathcal{H} \mid \phi_1 \rangle & \langle \phi_2 \mid \mathcal{H} \mid \phi_2 \rangle - E & \langle \phi_2 \mid \mathcal{H} \mid \phi_3 \rangle \\ \langle \phi_3 \mid \mathcal{H} \mid \phi_1 \rangle & \langle \phi_3 \mid \mathcal{H} \mid \phi_2 \rangle & \langle \phi_3 \mid \mathcal{H} \mid \phi_3 \rangle - E \end{vmatrix} = 0$$

The determinant is called the secular determinant. The integrals $\langle \phi_i \mid \mathcal{H} \mid \phi_j \rangle$ involve *known* quantities only and may be evaluated directly: they are called matrix elements. We refer to the matrix of the basis functions ϕ under the operator $\mathcal{H}$. Knowing the values of all the matrix elements, evaluation of the determinantal equation above, here a cubic in E, gives three roots for E. These are E_1, E_2, E_3, as represented in Figure 11. For each root, one at a time, we may substitute back into the secular equations and obtain values for the c's; this also requires the use of the normalizing condition, $\sum c_i^2 = 1$. We then have the new wavefunctions by a process which simultaneously gave the new energies. The new energies, $E^0 + E_i$, are called first-order perturbation energies while the new wavefunctions, which are merely linear combinations of the old, are called zeroth-order wavefunctions, Had we chosen, by accident perhaps, the new wavefunctions ϕ in the beginning, instead of the set ϕ, as basis, the matrix under $\mathscr{H}$ would have appeared diagonal:

$$\begin{pmatrix} \langle \psi_1 \mid \mathscr{H} \mid \psi_1 \rangle & 0 & 0 \\ 0 & \langle \psi_2 \mid \mathscr{H} \mid \psi_2 \rangle & 0 \\ 0 & 0 & \langle \psi_3 \mid \mathscr{H} \mid \psi_3 \rangle \end{pmatrix}$$

so that we would deduce that the functions ψ were already the correct zeroth-order wavefunction under the perturbing Hamiltonian $\mathscr{H}$. The process of arriving at zeroth-order wavefunctions from some basis set (ϕ) is often called "diagonalization" of the matrix of ϕ. It is important to remember, in what went before, that the energies E, or, in general, the roots of the secular determinant are the same whichever set of linear combinations of ϕ we take as our basis.

Finally, because the roots of any quantum mechanical determinant are observables (or part thereof), they must be real. In general, this means that the operators we use must be Hermitean (E27): this property is defined in two ways:

(a) that the integral $\int \phi^* \mathscr{H} \psi d\tau = \int \psi \mathscr{H}^* \phi^* d\tau$

(b) that nondiagonal matrix elements are the complex conjugates of the elements related by reflection in the leading diagonal.

VII. APPENDIX II. Matrix Elements of the Type $\langle \text{ligand} \mid \mathbf{L}_z \mid \text{ligand}' \rangle$ in O_h

When evaluating these integrals, it must be remembered that $\mathbf{L}_z$ lies along the *molecular z* axis which is often different in position from the given ligand z axis. To calculate $\mathbf{L}_z \mid p_y^{(1)} - p_y^{(2)} + p_z^{(5)} - p_z^{(6)} \rangle$, we may proceed in one of two ways: either by referring each ligand orbital to the central coordinate framework, or by transforming $\mathbf{L}_z$ into the local ligand framework for each part. The latter method is found to be particularly convenient and simple here.

The z axes (see Fig. 4) for ligands (1) and (2) are coincident with the molecular z axis, so that we readily obtain,

$$L_z^{(1)} \equiv L_z^{(2)} \equiv L_z^{(M)}$$

where superscripts label ligands or metal. Then,

$$L_z^{(M)} \mid p_y^{(1)}\rangle = -ip_x^{(1)}$$

and

$$L_z^{(M)} \mid p_y^{(2)}\rangle = -ip_x^{(2)}$$

But $L_z^{(5)}$ involves a displacement of $+a$ in the y direction (where a is the metal–ligand distance). Now, from above, we have:

$$L_z^{(M)} = -i\frac{h}{2\pi}\left(x\frac{\partial}{\partial y} - y\frac{\partial}{\partial x}\right)^{(M)}$$

so that,

$$
\begin{aligned}
L_z^{(5)} &= -i\frac{h}{2\pi}\left(x\frac{\partial}{\partial y} - (y+a)\frac{\partial}{\partial x}\right)^{(M)} \\
&= -i\frac{h}{2\pi}\left(x\frac{\partial}{\partial y} - y\frac{\partial}{\partial x}\right)^{(M)} + i\frac{h}{2\pi}a\frac{\partial}{\partial x}^{(5)} = L_z^{(M)} + i\frac{h}{2\pi}a\frac{\partial}{\partial x}^{(5)}
\end{aligned}
$$

from which

$$L_z^{(M)} \mid p_z^{(5)}\rangle = 0 + i\frac{h}{2\pi}a\frac{\partial}{\partial x}(p_z^{(5)})$$

Similarly,

$$L_z^{(6)} = -i\frac{h}{2\pi}\left(x\frac{\partial}{\partial y} - (y-a)\frac{\partial}{\partial x}\right)^{(M)}$$

giving

$$L_z^{(M)} \mid p_z^{(6)}\rangle = 0 - i\frac{h}{2\pi}a\frac{\partial}{\partial x}(p_z^{(6)})$$

We shall now investigate the function

$$i\frac{\partial}{\partial x}(p_z)$$

A. The Function $i(\partial/\partial x)(p_z)$ for the Second Principal Quantum Shell

The normalized hydrogenlike $2p_z$ wavefunction may be written as:

$$\psi_{2p_z} = \frac{1}{4\sqrt{2\pi}}\left(\frac{Z}{a_0}\right)^{3/2}\rho e^{-\rho/2}\cos\theta \qquad (E89)$$

where

$$\rho = (Z/a_0)r$$

i.e.,

$$\psi_{2p_z} \approx \left(\frac{Z}{a_0}\right) \exp\left[-\frac{Zr}{2a_0}\right] z$$

So that

$$i\frac{\partial}{\partial x}(\psi_{2p_z}) \approx i\left(\frac{Z}{a_0}\right) z \frac{\partial}{\partial x}\left(\exp\left[-\frac{Zr}{2a_0}\right]\right)$$

Now,

$$\frac{\partial}{\partial x}\left(\exp\left[-\frac{Zr}{2a_0}\right]\right) = -\frac{Z}{2a_0}\exp\left[-\frac{Zr}{2a_0}\right]\frac{x}{r} = -\frac{\rho}{2}e^{-\rho/2}\frac{x}{r}$$

and

$$i\frac{\partial}{\partial x}(\psi_{2p_z}) \approx -i\left(\frac{Z}{a_0}\right)\frac{\rho}{2}e^{-\rho/2}\frac{xz}{r}$$

$$\approx -i\frac{\rho^2}{2}e^{-\rho/2}\frac{xz}{r^2}$$

That is, the result of operating on $p_z^{(5)}$ with $\mathbf{L}_z$ centered on the molecule is to give a totally imaginary function with the angular properties of a d_{xz} function centered on the ligand, but with the radial part unchanged

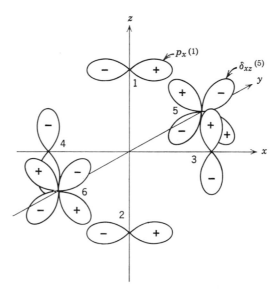

Fig. 12. Functions derived from the $\mathbf{L}_z$ operation in O_h.

and the whole function being magnified by a, the metal–ligand distance.*
We shall temporarily represent this process by:

$$L_z^{(M)}p_z^{(5)} = i\delta_{xz}^{(5)} \quad \text{etc.}$$

Term *3* above (see page 14) may now be written as,

$$\frac{\lambda^2}{4}\int (p_x^{(1)} - p_x^{(2)} + p_z^{(3)} - p_z^{(4)})^* \times -i(p_x^{(1)} - p_x^{(2)} + \delta_{xz}^{(5)} + \delta_{xz}^{(6)})d\tau$$

We may represent the functions in this integral as in Figure 12.
By inspection, we have:

$$\int p_x^{(1)*}p_x^{(1)}d\tau = \int p_x^{(2)*}p_x^{(2)}d\tau = 1$$

$$\int p_x^{(1)*}p_x^{(2)}d\tau = S_{12}$$

$$\int p_x^{(1)*}p_z^{(3)}d\tau = -\int p_z^{(4)*}p_x^{(1)}d\tau = \int p_z^{(4)*}p_x^{(2)}d\tau$$

$$= -\int p_z^{(3)*}p_x^{(2)}d\tau = S_{13}$$

$$\int p_x^{(1)*}\delta_{xz}^{(5)}d\tau = \int p_x^{(1)*}\delta_{xz}^{(6)}d\tau = -\int p_x^{(2)*}\delta_{xz}^{(5)}d\tau$$

$$= -\int p_x^{(2)*}\delta_{xz}^{(6)}d\tau$$

$$= \int p_z^{(3)*}\delta_{xz}^{(5)}d\tau = \int p_z^{(3)*}\delta_{xz}^{(6)}d\tau$$

$$= -\int p_z^{(4)*}\delta_{xz}^{(5)}d\tau = -\int p_z^{(4)*}\delta_{xz}^{(6)}d\tau$$

$$= \sigma_{15} \quad \text{(say)}$$

Hence, Term *3* becomes,

$$-i(\lambda^2/4)(2 - 2S_{12} + 4S_{13} + 8\sigma_{15})$$

The integral σ_{15}, which resembles an overlap integral, may be simply
replaced. Thus,

$$\int p_x^{(1)*}L_z p_z^{(5)}d\tau = \int p_z^{(5)}L_z^* p_x^{(1)}d\tau$$

(using the Hermitean properties of L_z again)

* As $a \to 0$, $L_z^{(M)}(p_z^{(L)}) \to 0$ as expected.

Remembering that p_x, etc., are real, this gives,

$$\int p_x^{(1)*} \mathbf{L}_z p_z^{(5)} d\tau (= i\sigma_{15})$$

$$= -\int p_z^{(5)*} \mathbf{L}_z p_x^{(1)} d\tau$$

$$= -i \int p_z^{(5)*} p_y^{(1)} d\tau$$

$$= -iS_{13}$$

That is

$$\sigma_{15} = -S_{13} \quad !$$

giving the final result for Term 3 as

$$-i(\lambda^2/4)(2 - 2S_{12} - 4S_{13})$$

VIII. APPENDIX III. Matrix Elements of the Type $\langle \text{ligand} \mid \mathbf{L}_z \mid \text{ligand}' \rangle$ in T_d

We must again consider the translation $\mathbf{L}_z^{(M)}$ to $\mathbf{L}_z^{(\text{ligand})}$. If we represent the edge of the cube in which the tetrahedron is inscribed (Fig. 7) as $2a$, then $\mathbf{L}_z^{(1)}$ involves a translation of $+a$ in the x direction and $-a$ in the y direction, whence:

$$\mathbf{L}_z^{(M)} \mid s^{(1)} \rangle = -i \frac{h}{2\pi} \left[(x + a) \frac{\partial}{\partial y} - (y - a) \frac{\partial}{\partial x} \right] \mid s^{(1)} \rangle$$

$$= \mathbf{L}_z^{(1)} \mid s^{(1)} \rangle - i \frac{h}{2\pi} a \left(\frac{\partial}{\partial y} + \frac{\partial}{\partial x} \right) \mid s^{(1)} \rangle$$

$$= 0 - i \frac{h}{2\pi} a \left(\frac{\partial}{\partial y} + \frac{\partial}{\partial x} \right) \mid s^{(1)} \rangle$$

Similarly:

$$\mathbf{L}_z^{(M)} \mid s^{(2)} \rangle = +i \frac{h}{2\pi} a \left(\frac{\partial}{\partial y} + \frac{\partial}{\partial x} \right) \mid s^{(2)} \rangle$$

$$\mathbf{L}_z^{(M)} \mid s^{(3)} \rangle = -i \frac{h}{2\pi} a \left(\frac{\partial}{\partial y} - \frac{\partial}{\partial x} \right) \mid s^{(3)} \rangle$$

$$\mathbf{L}_z^{(M)} \mid s^{(4)} \rangle = +i \frac{h}{2\pi} a \left(\frac{\partial}{\partial y} - \frac{\partial}{\partial x} \right) \mid s^{(4)} \rangle$$

Analogous to the function $\partial/\partial x \, (p_z)$ discussed in Appendix I, the functions

$\partial/\partial y(s)$ and $\partial/\partial x(s)$ have angular properties similar to $-p_y$ and $-p_x$ orbitals, respectively. We may write,

$$i\frac{h}{2\pi}a\frac{\partial}{\partial x}(s) \equiv -i\Pi_x$$

$$i\frac{h}{2\pi}a\frac{\partial}{\partial y}(s) \equiv -i\Pi_y$$

(where the Π functions are defined to be positive along the positive axes). Therefore, the $\langle$ligand $\mid \mathbf{L}_z \mid$ ligand$'\rangle$ term reduces to

$$\frac{\lambda^2}{4}i\,\langle s^{(1)} - s^{(2)} - s^{(3)} + s^{(4)} \mid \Pi_y^{(1)} + \Pi_x^{(1)}$$

$$+ \Pi_y^{(2)} + \Pi_x^{(2)} + \Pi_y^{(3)} - \Pi_x^{(3)} + \Pi_y^{(4)} - \Pi_x^{(4)}\rangle$$

Some of these functions are represented in Figure 13.

By inspection we find that,

$$\langle s^{(j)} \mid \Pi_x^{(j)}\rangle = \langle s^{(j)} \mid \Pi_y^{(j)}\rangle = 0$$

$$\langle s^{(1)} \mid \Pi_y^{(4)}\rangle = \langle s^{(1)} \mid \Pi_x^{(3)}\rangle = 0 \quad \text{etc.}$$

$$\langle s^{(1)} \mid \Pi_x^{(2)}\rangle = -\langle s^{(1)} \mid \Pi_y^{(2)}\rangle = \sigma \quad \text{(say), etc.}$$

From this it is simple to show that the integral

$$\langle\text{ligand} \mid \mathbf{L}_z \mid \text{ligand}'\rangle \quad \text{is} \quad -i\sigma$$

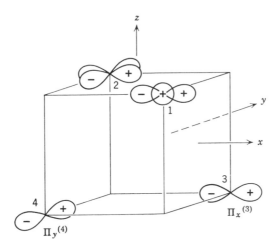

Fig. 13. Functions derived from the $\mathbf{L}_z$ operation in T_d.

Acknowledgment

We wish to thank Professor J. Lewis for his interest and encouragement throughout this work.

References

Textbooks referred to by page numbers

B C. J. Ballhausen, *Introduction to Ligand Field Theory*, McGraw-Hill, New York, 1962.

C F. A. Cotton, *Chemical Applications of Group Theory*, Interscience, New York, 1963.

E H. Eyring, J. Walter, and G. E. Kimball, *Quantum Chemistry*, Wiley, New York, 1944.

F B. N. Figgis, *Introduction to Ligand Fields*, Interscience, New York, 1966.

H H. F. Hameka, *Introduction to Quantum Theory*, Harper and Row, New York, 1967.

M J. Lewis and R. Wilkins, Eds., *Modern Coordination Chemistry*, Interscience, New York, 1960.

1. H. A. Bethe, *Ann. Physik*, *3*, [5], 133 (1929).
2. J. H. Van Vleck, *Theory of Electric and Magnetic Susceptibilities*, Oxford University Press, Oxford, 1932.
3. W. G. Penney and R. Schlapp, *Phys. Rev.*, *42*, 666 (1932).
4. L. Pauling, *The Nature of the Chemical Bond*, Cornell University Press, Ithaca, New York, 1948.
5. J. H. Van Vleck, *J. Chem. Phys.*, *3*, 807 (1935).
6. J. Howard, *J. Chem. Phys.*, *3*, 811 (1935).
7. M. Kotani, *J. Phys. Soc. Japan*, *4*, 293 (1949).
8. B. N. Figgis, *Trans. Faraday Soc.*, *57* (2), 198 (1961).
9. B. N. Figgis, *J. Chem. Soc.*, *1965*, 4887.
10. B. N. Figgis and J. Lewis, in *Progress in Inorganic Chemistry*, Vol. 6, F. A. Cotton, Ed., Interscience, New York, 1964.
11. J. H. E. Griffiths, J. Owen, and I. M. Ward, *Proc. Roy. Soc. A*, *219*, 526 (1953).
12. K. W. H. Stevens, *Proc. Roy. Soc. A*, *219*, 542 (1953).
13. J. H. E. Griffiths and J. Owen, *Proc. Roy. Soc. A*, *226*, 96 (1954).
14. A. Carrington and H. C. Longuet-Higgins, *Quart. Rev.*, *1960*, 427.
15. M. Tinkham, *Proc. Roy. Soc. A*, *236*, 535 and 549 (1956).
16. J. Owen and J. H. M. Thornley, *Rep. Prog. Phys.*, *29*, 676 (1966).
17. B. N. Figgis, *Trans. Faraday Soc.*, *57* (2), 204 (1961).
18. J. Owen, *Discussions Faraday Soc.*, *19*, 127 (1955).
19. B. N. Figgis, J. Lewis, and F. E. Mabbs, *J. Chem. Soc.*, *1960*, 2480.
20. B. N. Figgis, J. Lewis, F. E. Mabbs, and G. A. Webb, *J. Chem. Soc. A*, *1967*, 442.
21. B. N. Figgis, J. Lewis, F. E. Mabbs and G. A. Webb, *J. Chem. Soc. A*, *1966*, 1411.
22. L. E. Orgel, *Chemistry of Transition Metal Ions*, Methuen, London, 1960.
23. M. Gerloch, to be published in *J. Chem. Soc. A*.
24. B. N. Figgis, M. Gerloch, J. Lewis, and R. Slade, to be published in *J. Chem. Soc. A*.

25. M. Sharnoff, *J. Chem. Phys.*, *42*, 3, 383 (1965).
26. E. U. Condon and G. H. Shortley, *Theory of Atomic Spectra*, Cambridge University Press, Cambridge, 1957.
27. J. H. Van Vleck, *J. Chem. Phys.*, *3*, 803 (1935).
28. R. S. Mulliken, C. A. Rieke, D. Orloff, and H. Orloff, *J. Chem. Phys.*, *17*, 1248 (1949).
29. F. Clementi and D. L. Riamondi, *J. Chem, Phys.*, *38*, 2686 (1963).
30. C. J. Gorter, *Phys. Rev.*, *42*, 437 (1932).
31. C. J. Ballhausen and A. D. Liehr, *Mol. Spectry.*, *2*, 342, 1958; *4*, 190 (1960).
32. J. S. Griffith, *The Theory of Transition-Metal Ions*, Cambridge University Press, London, 1961.
33. D. Baker, M. Gerloch, and J. Lewis, to be published.
34. J. Owen, *Discussions Faraday Soc.*, *26*, 53 (1958).
35. M. Weissbluth, in *Structure and Bonding*, Vol. 2, Springer Verlag, New York, 1967.

Metal 1,2-Dithiolene and Related Complexes

By J. A. McCleverty

University of Sheffield,
Sheffield, England

I. INTRODUCTION

During the last ten years, there has been an enormous upsurge of interest in the chemistry of transition metal sulfur complexes. This interest has manifested itself in the general area of novel complex synthesis, in the formation of esoteric organometallic molecules, and in the field of "biological-inorganic" chemistry. There have been several important and useful reviews of the different aspects of metal sulfur chemistry, among which those by Livingstone (1a), Harris (1b), Jorgensen (2), and Gray (3) are particularly worthy of mention.

The particular subject of this review, metal 1,2-dithiolene and related complexes, is by no means as novel as the title may suggest. There was considerable interest in the analytical uses of dithiols or dithiolates in the mid 1930's. However, the attention which the chemistry of metal complexes of unsaturated 1,2-dithiols has received has risen rapidly within the past six or seven years.

In this introduction, it is my purpose to outline the early development of this field in an "historical" sense, and then to indicate how the early preoccupation with sulfur chemistry has given way to a much broader interest in the reactions of complexes containing not only sulfur but also nitrogen and oxygen donor atoms. The remainder of this article is divided into six sections, the first of which, II, is concerned with the syntheses and chemistry of the 1,2-dithiol or dithietene ligands, themselves. In the following two sections, III and IV, the characterization and investigations of bis- and tris-substituted dithiolene metal complexes and their analogs are discussed, and in Section V, dithiolene complexes containing two, three, and four electron donors (that is, CO, phosphines, NO, dipyridyl, and diphosphines) are described. Section VI is concerned with organometallic dithiolenes while the final section, VII, contains some discussion of new developments.

In the mid and late 1930's, R. E. D. Clarke and co-workers studied the reactions of toluene-3,4-dithiol and 1-chlorobenzene-3,4-dithiol with metal halides including those of zinc, cadmium, mercury, and tin. They discovered (4,5) that these dithiols readily formed complexes and, with tin, species of the type [Sn(dithiol)$_2$] were isolated. It was soon found that these dithiols were particularly effective in the sequestration of tin. Work on the analytical aspects of toluene-3,4-dithiol seems to have been carried on during the 1940's and early 1950's (6,7) but not until 1957 did any serious work on the characterization of metal complexes of unsaturated 1,2-dithiols begin. Then, Bähr and Schleitzer in Munich noted (8) that a ligand which they themselves had prepared, Na$_2$S$_2$C$_2$(CN)$_2$ (1), readily formed complexes with palladium; they formulated these as Pd(NH$_3$)$_2$-S$_2$C$_2$(CN)$_2$ and Na$_2$[Pd(S$_2$C$_2$(CN)$_2$)$_2$], and they further observed that the latter was apparently oxidized by iodine, although they were unable to identify the reaction product. In 1959, Stevancevic and Drazic (9) described the formation of complexes of nickel, cobalt, and palladium containing quinoxaline-2,3-dithiol (2) (QDTH$_2$). These were prepared in alkaline solution and appeared to be of the type [M(QDT)$_2$]$^{-2}$, although

(1)

(2)

cobalt also appeared to form a tris-substituted complex. Ayres and Janota (10) also observed that palladium formed a complex with QDT and reported that polymeric materials could also be obtained. In 1960, Gilbert and Sandell (11) of the University of Minnesota reinvestigated the reports that toluene-3,4-dithiol formed complexes with molybdates in acid solution and they described the isolation and characterization of a diamagnetic green species, $Mo(S_2C_6H_3Me)_3$, believed to contain Mo(VI), and also a red species of indefinite composition thought to contain Mo(IV); this was the first report of an authentic tris-substituted dithiolene complex.

In August 1962, G. N. Schrauzer and V. P. Mayweg, of the University of Munich, reported (12) that they had prepared an unusual nickel complex, $NiS_4C_4Ph_4$, in the somewhat esoteric reaction between $Ni(CO)_4$, sulfur, and diphenylacetylene. Apparently, their purpose in attempting such a reaction was to investigate the role of transition metal catalysts in the formation of thioaromatics, particularly thiophenes, from acetylenes and sulfur, and also to investigate the stability of 1,2-stilbene dithiol and its analogs. This new nickel complex was intensely colored (green), diamagnetic, and volatile and was quite unlike any other nickel complex known at that time. It was found that in solution in pyridine or piperidine, the green color was gradually discharged and was replaced by a red–brown color, the final solution being paramagnetic. Almost concurrently (in September), H. B. Gray and his colleagues at Columbia University published an account (13) of the formation of metal complexes of $Na_2S_2C_2(CN)_2$, of the type $[MS_4C_4(CN)_4]^{-2}$ where M = Cu, Ni, Pd, and Co. In June of the following year (1963), King reported (14) that bis-perfluoromethyldithietene reacted with molybdenum hexacarbonyl, forming a tris-substituted complex $[MoS_6C_6(CF_3)_6]$. However, perhaps the most significant publications in that year came from Davison, Edelstein, Maki, and Holm at Harvard University, and from Gray and co-workers. The more important publication is clearly that of the Harvard workers (15) who pointed out that there was only a two-electron difference between the neutral complex of Schrauzer, $NiS_4C_4Ph_4$, and the dianionic species of Gray, $[NiS_4C_4(CN)_4]^{-2}$ (considering the valence electrons only and ignoring the ligand substituents) and so it should be possible to effect an oxidation of the dianion to the neutral species, or a reduction of the neutral compound to the dianionic complex, via an intermediate monoanionic complex. They discovered that this was possible and isolated several different *paramagnetic* monoanions, $[Ni-S_4]^{-1}$. Another, and most valuable, contribution was the demonstration that the one-electron transfer reactions linking the dianions, monoanions, and neutral species could be detected polarographically or voltammetrically. Gray and colleagues (16) found that the most stable complexes of nickel, cobalt, and iron containing toluene-3,4-dithiol were monoanions, the first having a spin–doublet ground state and the other two reputedly spin–quartet ground states. Later in 1963, Schrauzer and co-workers (17) reported the formation of a large number of metal complexes of dialkyl and diaryl "α,β-dithioketones" $[MS_4C_4R_4]^0$.

During 1964, further work on the planar species $[M—S_4]^z$ was published, but the most important developments were in the study of six-coordinate complexes which, like their planar counterparts, also underwent electron transfer reactions (18–20). Perhaps the most exciting development in the tris-substituted systems was the discovery that the neutral species, $[M—S_6]^0$, were trigonal prisms. The first complex to be recognized as a trigonal prism was $[ReS_6C_6Ph_6]^0$ whose structure was elucidated by Eisenberg and Ibers (21) in 1965. Two other complexes were subsequently investigated—$[MoS_6C_6H_6]^0$ by Smith, Schrauzer, and colleagues (22), and $[VS_6C_6Ph_6]^0$ by Eisenberg, Gray and co-workers (23). As a consequence of these structural studies several molecular orbital treatments of the trigonal prismatic complexes were devised. There was some speculation about the structures of the charged tris-substituted species but it was not until mid-1967 that a crystal structural determination of $[Me_4N]_2[VS_6C_6(CN)_6]$ was reported (24). It was shown that this complex had a structure best described (24) as a compromise between a trigonally distorted octahedron and a trigonal prism.

Early in 1966, Balch and Holm discovered (25) that many of the charged cobalt and iron dithiolenes previously thought to be monoanionic were, in fact, dimeric $[M—S_4]_2^{-2}$. The dimeric unit had already been detected in $[CoS_4C_4(CF_3)_4]_2^0$, and Enemark and Lipscomb (26) had shown that the two halves of the molecule were connected by Co—S bridges and not through a metal–metal bond. Davison, Shawl, and Howe also reported (27) that the "monoanionic" cobalt and iron dicyano– and bis-perfluoromethyl–dithiolenes were dimeric in certain weakly polar solvents. It was observed (25) that many of the dimeric units themselves underwent discrete one-electron transfer reactions and some of the monoanionic dimers, $[M—S_4]_2^{-1}$, were isolated.

The majority of the work described in this review is concerned with the electron transfer reactions of complexes having the basic $[M—S_4]^z$ or $[M—S_6]^z$ coordination units. The existence of a three-membered series of planar complexes,

$$[M—S_4]^0 \longleftrightarrow [M—S_4]^{-1} \longleftrightarrow [M—S_4]^{-2}$$

has been recognized and the nature of the electronic structure of these species has received much attention. However, in 1965 interest in the general scope of electron transfer reactions was extended (28) to include complexes having the coordination units $[M—N_4]^z$ $[M—O_4]^z$, $[M—N_2S_2]^z$, and $[M—O_2S_2]^z$. It was suggested by Balch, Holm, and Röhrscheid (28) that if such species having these coordination units could be prepared, a five-membered series related by one-electron transfer reactions should, in principle, exist:

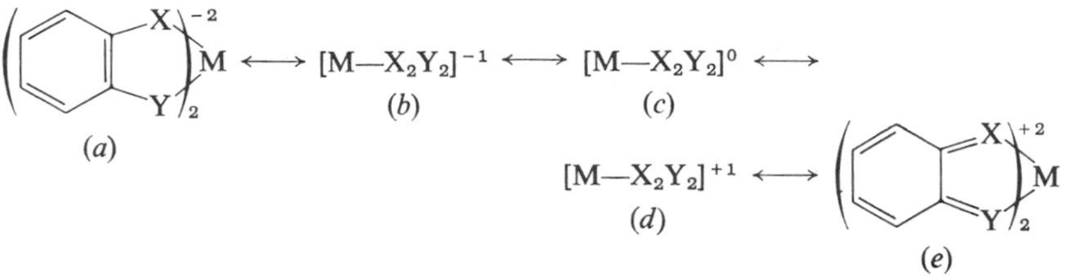

The terminal members of the series were expected to be defined as metal complexes of the potentially stable dianions $[C_6H_4XY]^{-2}$ (a) and the "quinonoidal" species $[C_6H_4XY]^0$ (e). Although $[C_6H_4(NH)_2]^{-2}$ is unknown, $[C_6H_4(NH)_2]^0$, 1,2-diimino-3,5-cyclohexadiene, has been obtained (29) in solution; both $[C_6H_4O_2]^{-2}$ and o-benzoquinone are known in addition to several ring-substituted analogs. However, the dithia-o-quinones, thia-o-benzoquinones, and thia-o-benzoquinonimines are unknown.

Feigl and Fürth in 1927 had shown (30) that Ni^{2+} ions reacted with o-phenylenediamine in concentrated ammonia with oxygen giving a blue complex which they formulated as $NiC_{12}H_{12}N_4$ having the structure 3.

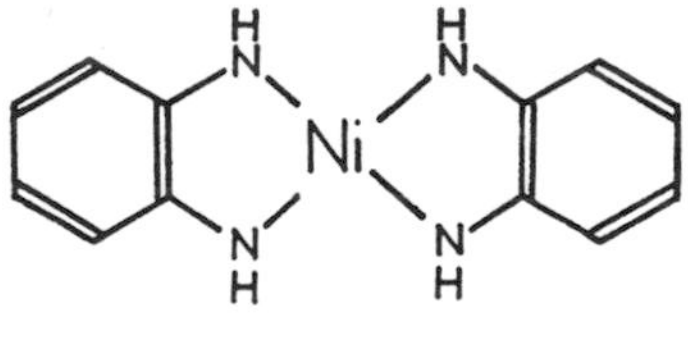

(3)

Bardodej subsequently questioned this proposed structure (31) and suggested that the complex was better represented as **4**. Nyholm (32) also

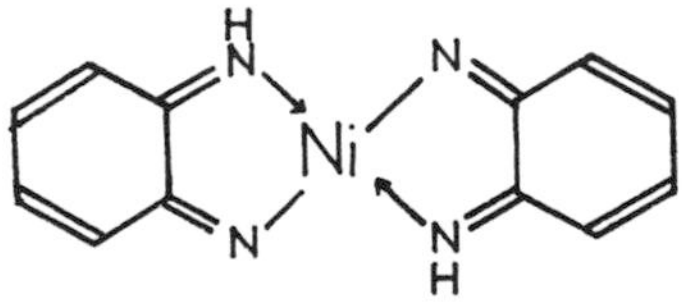

(4)

investigated this complex and considered such alternative structures as **5**, but after a study of the infrared spectrum of the complex, he concluded that amino groups did not contribute to the structure of Feigl and Fürth's material. Holm and co-workers (28,33), and very shortly thereafter Gray, Stiefel, Billig, and Waters (34), reinvestigated the nickel complex and discovered that it underwent facile electron transfer reactions. All of the five-membered series, *a–e* were detected (33) using this complex and also the

(5)

related palladium, platinum, and cobalt systems. Mass spectral studies of the nickel complex revealed a parent ion peak at m/e values corresponding to $NiC_{12}H_{12}N_4^+$, thereby eliminating Bardodej's proposed structure (**4**). A further peak at m/e values corresponding to $MC_6H_6N_2^+$ was detected in all of the complexes, thereby indicating very strongly that these complexes had structure **3** in which the metal ion was complexed by symmetrical ligands. Further confirmation of the structure of the nickel complex was obtained by its direct synthesis from 1,2-diimino-3,5-cyclohexadiene and $Ni(CO)_4$ in ether (28,33).

The complete electron transfer series for complexes represented as $[M—O_4]^z$ would consist of the five members *a–e* in which *a* would be described as a complex of a divalent metal ion with the catecholate dianion and *e* as a complex of a divalent metal ion with *o*-benzoquinone. The intermediate members of the series could be less well defined although this ligand system is one of the few in which distinct stages of oxidation—dianion, semiquinone, and quinone—have been detected. Many bis-catecholate complexes containing cobalt, nickel, copper, zinc, and cadmium were previously known (35) before interest developed in their potential electron transfer properties, although many of these had very unusual formulations. However, it was established by Eaton (36) that metal-complexed semiquinone systems did exist, and also that some metal complexes of substituted *o*-quinones could be prepared (37–39). Röhrscheid, Balch, and Holm (40) were able to prepare discrete salts of the dianions, $[M(O_2C_6H_4)_2]^{-2}$ and $[M(O_2C_6Cl_4)_2]^{-2}$ and to demonstrate that these complexes could be oxidized polarographically and voltammetrically.

However, the general existence of the complete five-membered series a–e was not established.

Hieber and Brück (41) observed that the yellow–brown polymeric Ni(II) complex of o-aminothiophenol could apparently be oxidized by air in basic solutions giving a blue product which they formulated as $\{Ni[S(NH_2)-C_6H_4]_2\}_2O_2$ (6). A reinvestigation of this compound (28,33,34,44) led to its

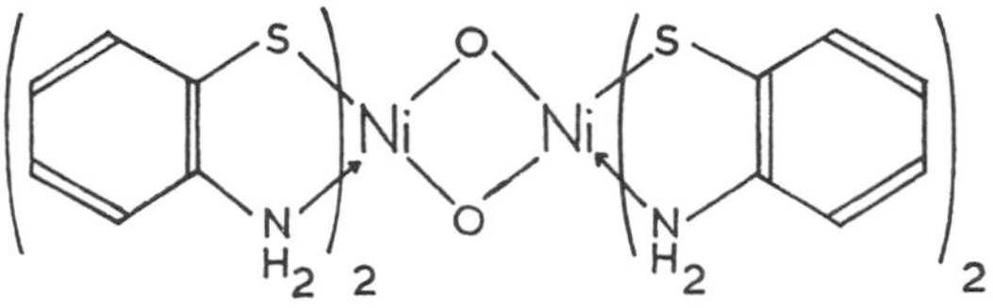

(6)

reformulation as $\{Ni[S(NH)C_6H_4]_2\}^0$ (7), a nickel complex of the unknown thia-o-benzoquinoneimine. This complex underwent electron transfer reactions of the type already discovered in the $[M-S_4]^z$ series, although only three members, a, b, and c, of the five-membered series were detected.

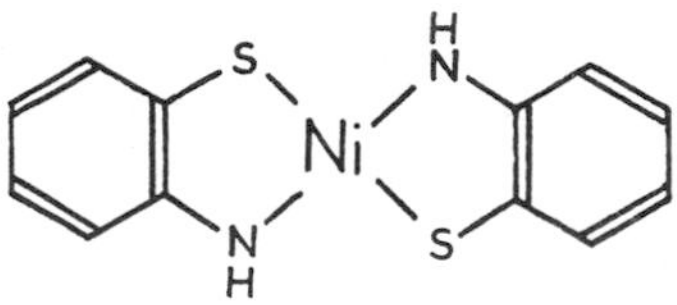

(7)

Other complexes, such as the blue oxidized nickel complex of thiobenzoyl–hydrazide, $\{Ni[SC(Ph)N,NH]_2\}^0$ (8) discovered by Jensen and Miguel (43), were also found (44) to undergo electron transfer reactions. Metal complexes of the monothiocatecholate dianion, $[C_6H_4OS]^{-2}$, behaved similarly (28).

Parallel with the developments in the complex chemistry of the bis-planar and tris-trigonal prismatic systems, work on the reactions with the

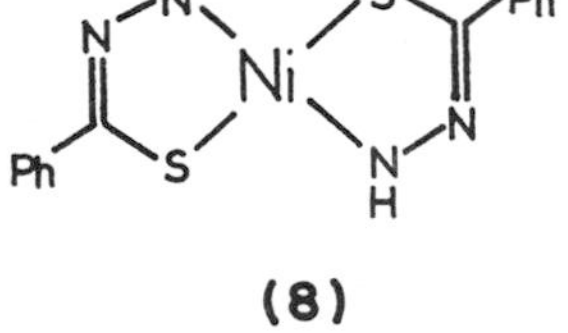

(8)

dithiolate and dithietene ligands with organometallic compounds such as carbonyls, nitrosyls, and π-cyclopentadienyls progressed. One of the major contributions in this area has been that of King who synthesized many different types of π-cyclopentadienyl bis-perfluoromethyl–dithiolenes (45,46). Some of these have been shown to undergo electron transfer reactions (47,48), although the nature of the oxidized or reduced species has not been elucidated. The chemistry of nitrosyl cobalt, iron, and manganese complexes of the type $[M(NO)—S_4]^z$ has recently been investigated (49–51) and all of these complexes could be voltammetrically oxidized and reduced.

A. NOMENCLATURE

Throughout this review, the term "1,2-dithiolene" or "dithiolene" will be used to denote metal complexes of the general types $[M—S_4]_n^z$, $[M—S_6]^z$, and $[M(L)_x—S_{2,4}]^z$ (where L may be any other ligand such as π-C_5H_5, NO, PPh_3, or O), e.g., **9**, **10**, or **11**, which undergo, or are capable

(9)   **(10)**   **(11)**

of undergoing, facile (and usually reversible) electron transfer reactions. This terminology is simple and does not imply any particular ligand structure or valence formalism for a particular value of z. This does not represent any criticism, actual or implied, of the proposals for the electronic structures of these complexes. It might be argued that the most reduced species of an electron transfer series, e.g., $[NiS_4C_4R_4]^{-2}$ or $[CrS_6C_6R_6]^{-3}$, could be described as containing a di- or trivalent metal ion complexed with two or three dianionic dithiolates, $[^-SC(R){=}C(R)S^-]$, and this would be strictly correct. However, these reduced species are regarded as members of a general class of molecules which are termed generically as "metal-1,2-dithiolene complexes." By a logical extension of this nomenclature to dithiolene analogs, complexes containing the coordination units $[M—N_4]^z$, $[M—N_2S_2]^z$, $[M—O_2S_2]^z$, and $[M—O_4]^z$ would be referred to as metal complexes of 1,2-diiminolenes, 1,2-iminothiolenes, 1,2-oxythiolenes, and 1,2-dioxylenes, respectively. However, in some cases the use of trivial names or standard organic nomenclature [as in the metal complexes of glyoxal bis(2-mercaptoanil)] is easier and more appropriate. Whereas such complexes as $[Ni(S_2C_6H_3Me)_3]^z$ would be referred to as nickel bis-toluene-3,4-dithiolenes, $[Zn(S_2C_6H_3Me)_2]^{-2}$, which cannot undergo electron transfer reactions of the type described in this review, must be referred to as the zinc bis-toluene-3,4-dithiolate dianion.

Schrauzer has proposed another nomenclature (184) for the oxidized complexes such as $[Ni—S_4]^{0,-1}$ or $[M—S_6]^{0,-1}$. These are referred to as "dithienes" and this terminology is strictly reserved for those species which are regarded as having "delocalized ground states." Those most reduced complexes are normally referred to as "dithiolates."

II. THE 1,2-DITHIOL LIGANDS AND THEIR DERIVATIVES

The *cis*-1,2-dicyanoethylene-1,2-dithiolate dianion (also known as the dimercaptomaleonitrile ion or maleonitriledithiolate ion, MNT^{-2}) was formed when sodium dithiocyanoformate dimerized in water over 24 hr (52) or in refluxing chloroform (52,53) during 8 hr. The best yields were obtained (53) by the latter process which was also advantageous in that the product was ready for instant use and did not require further purification.

$$2[S_2C \cdot CN]^- \longrightarrow [cis\text{-}S_2C_2(CN)_2]^{-2} + 2S$$

Partial isomerization of this salt to the *trans* form (disodium dimercapto-fumaronitrile) occurred under UV irradiation of the *cis* form, the final distribution of isomers being 1:1 *cis/trans*.

Methylation of the *cis* salt readily occurred (52,54), affording $Na[MeS_2-C_2(CN)_2]$ and $Me_2S_2C_2(CN)_2$, the latter of which formed complexes with titanium and vanadium (55). Oxidation (54) of the *cis* salt with a number of one-electron oxidizing agents afforded the *cis–cis* disulfide (12), which

(12)

itself isomerized to the *trans–trans* form in dimethoxyethane in 2 hr. Oxidation of *cis*-$Na_2S_2C_2(CN)_2$ with thionyl chloride in dimethoxyethane at 0° afforded the tetracyanodithiin (13),

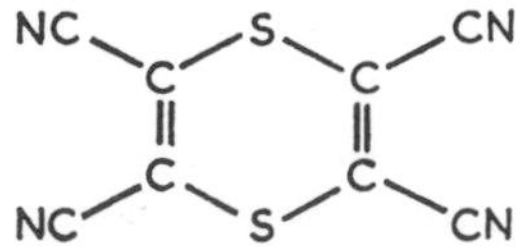

(13)

Two polarographic oxidation waves were detected (54) in aqueous solution (pH ~ 5) at +0.08 and +0.43 V, respectively (vs. SCE); other similar voltammetric results are described in Section III-C. The *trans–trans* disulfide was also polarographically oxidized at +0.40 V. These data were interpreted as in the scheme shown here

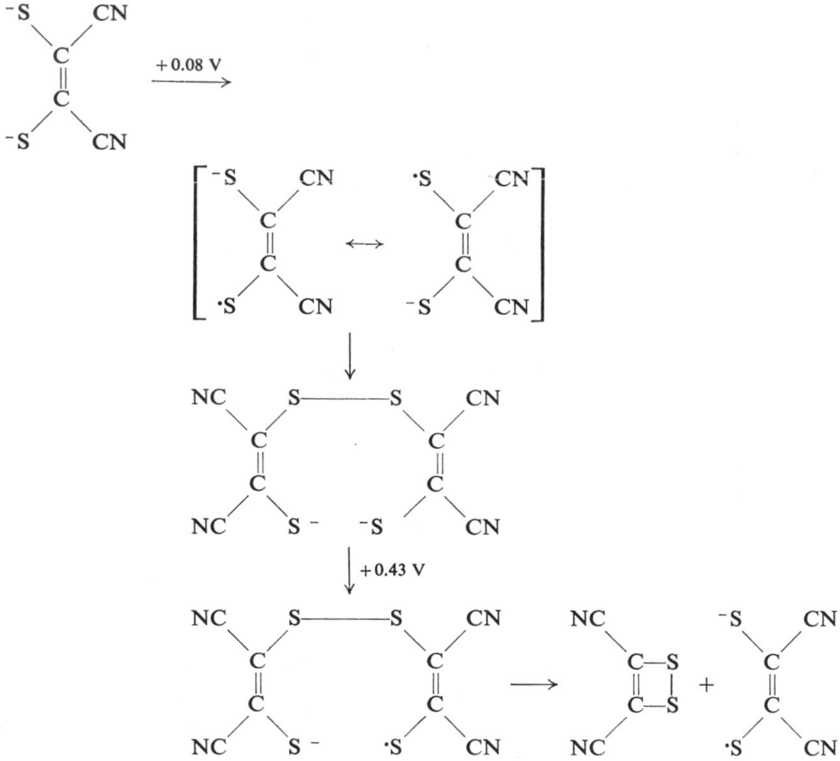

There was other evidence for the transient existence of the dithietene, $S_2C_2(CN)_2$, and theoretical calculations (56) showed that this S—S-bonded four-membered ring form would be more stable that the corresponding α,β-dithioketone.

·Attempts to prepare ethylene-*cis*-dithiol directly from glyoxal (57)

were unsuccessful, but the compound and its sodium salt were readily
obtained (58) by treating cis-dichloroethylene with benzyl chloride and
thiourea, thereby forming the dibenzyl thioether of the required dithiol.
Cleavage of the benzyl groups was conveniently effected using sodium in
liquid ammonia, and the dithiol was obtained by acidification of this with
HCl. Because of the inherent instability both of the disodium salt and the
dithiol, it was found practical to store the ethylene dithiol species as a
diacetyl derivative (59); the dithiolate was easily recovered from this by
treatment with two equivalents of methanolic KOH. Polarographic oxidation in two apparently irreversible one-electron steps has been detected
(60) (Sec. III-C).

Bis-perfluoromethyldithietene, $S_2C_2(CF_3)_2$, was formed (61) in the
direct reaction between hexafluorobut-2-yne and boiling sulfur. A vile-
smelling liquid, it was stored below 0° to prevent polymerization. Traces of
diethylamine caused dimerization to $[S_2C_2(CF_3)_2]_2$ and it was shown that
the dithietene was unstable with respect to this dimer at room temperature
but that the monomer was the most stable form at 200°. Reduction of the
dithietene to $[S_2C_2(CF_3)_2]^{-2}$ was achieved (62) by prior formation of the
mercury(II) polymer, $[HgS_2C_2(CF_3)_2]_n$, and its treatment with an aqueous
solution containing a lanthanide ion and the appropriate organic cation.
The expected dithiin has been prepared (61).

Useful derivatives of the dialkyl- and diarylethylene dithiols, which
are themselves very unstable with respect to the corresponding dithiins,
were prepared (63) by refluxing together the appropriate acyloin or benzoin with P_4S_{10} in xylene or dioxan. The major product of this reaction was
the dithiophosphoric ester of the desired dithiol,

Although the amber resins obtained from these reactions approximated
the above constitution closely, there was infrared evidence for traces of
phosphate (oxygen) groups. The amounts of oxygen incorporated in the
resin were small and did not justify the exclusion of air from the preparations. The metal complexes of these dithiols were obtained by treatment of
metal salts with the dithioester in alcohol, acetone, or water. Prolonged
contact between the diphenyl dithioester and Ni^{2+} salts resulted in the
formation, and isolation, of the novel complex **14** which, in part, confirmed the proposed constitution of the resins. The synthesis of mono-

(14)

substituted dithioesters $[RHC_2S_2PS_2]_2$ was achieved by using α-haloketones rather than aldehydes, $RCO \cdot CHO$, or alcohols, $RCO \cdot CH_2OH$.

Toluene-3,4-dithiol and quinoxaline-2,3-dithiol are available commercially, but the other arene dithiols must be prepared, usually from o-dibromo-substituted benzene derivatives. Treatment (64,65) of the o-dibromo arenes with copper ethyl- or n-butyl-mercaptide resulted in the formation of the o-dithioesters, from which the disodium salt of the arene dithiol was obtained by cleavage of the alkyl groups with sodium in liquid ammonia or THF. Thus, the disodium salts of benzene-1,2-, o-xylene-4,5-, and tetramethylbenzene-1,2-dithiols were prepared, and the dithiols were liberated by acidification of these salts. Tetrachlorobenzene-1,2-dithiol was obtained (65) by refluxing hexachlorobenzene with iron powder and sodium sulfide in DMF. On treatment of this mixture with base, the iron complex, $[Fe(S_2C_6Cl_4)_2]_n{}^0$, was precipitated (50) and the dithiol was liberated from this by its treatment with zinc oxide.

Monothiocatechol has been prepared by several methods (66). o-Aminophenol was converted to the O-benzyl ether, diazotized, and treated with potassium ethyl xanthate; the diazo-xanthate was pyrolyzed yielding the O-benzyl ether derivative of the desired thiol which was itself generated by cleavage of the benzyl group. The monothiol was also obtained by heating sodium phenoxide and sulfur at 200° followed by reduction of the resultant 2:2'-dihydroxydiphenyldisulfide. Tetrachlorocatechol, C_6Cl_4-$(OH)_2$, was produced (67) from pyrocatechol (o-dihydroxybenzene) by chlorination, and the corresponding o-quinones are available commercially.

It was not found necessary to isolate glyoxal bis(2-mercaptoanil) prior to complex formation since the zinc, cadmium, and nickel complexes 15 were readily obtained (44,68) from a mixture of the metal acetate, glyoxal, and o-aminothiophenol in the appropriate quantities. However, the

(15)

reaction between glyoxal and *o*-aminothiophenol afforded 2,2′-bis-benzthiazoline (68), and the dimethyl and diphenyl 2,2-′substituted derivatives were obtained using diacetyl or benzil in place of glyoxal. Reaction of these thiazolines with metal salts generally afforded the metal-Schiff base complex. Biacetylbis(thiobenzoylhydrazone) (16) and its benzoylhydrazone

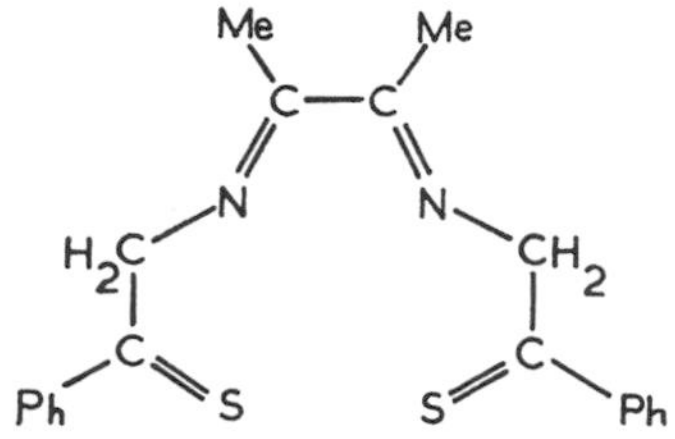

(16)

analog were obtained (44,70) by treatment of thiobenzoylhydrazone with the appropriate α-diketones, and the thiobenzoylhydrazine (17) itself was synthesized (71) by treatment of $PhCS_2CH_2CO_2H$ with hydrazine.

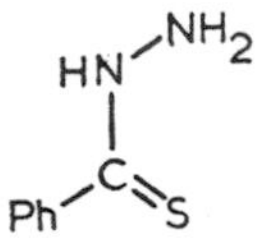

(17)

III. BIS-1,2-DITHIOLENE AND RELATED METAL COMPLEXES

A. Syntheses and General Characteristics

To discuss the syntheses and characteristics of each and every one of the various types of planar and dimeric 1,2-dithiolene complexes and their analogs would be tedious and unnecessary. It is fortunate that this descriptive chemistry can be divided into nine parts, the first five of which describe generalized syntheses using certain classes of ligands, the next three reactions carried out using the preformed complexes, and the last the general characteristics of these complexes. The known planar and dimeric species are listed in Table I.

1. Complexes Derived from Dianionic Dithiolates

As described in Section II, the ligands most readily available as dianionic dithiolates are $Na_2S_2C_2(CN)_2$ and $Na_2S_2C_2H_2$ or its potassium salt. Treatment of the divalent metal salts of zinc, copper, the nickel group, cobalt, rhodium, and manganese with $Na_2S_2C_2(CN)_2$ in aqueous or alcoholic solutions resulted in the formation of dianionic species, $[MS_4C_4(CN)_4]^{-2}$. Addition of the dicyanoethylene dithiolate to aqueous solutions containing salts of Ag(I) and Au(III) afforded the insoluble and light-sensitive $Ag_2S_2C_2(CN)_2$, or $[AgS_4C_4(CN)_4]^{-3}$, depending on the relative concentrations of ligand and metal ion (72), and $[AuS_4C_4(CN)_4]^{-1}$, respectively.

The product of the reaction between Fe(II) or Fe(III) salts and $Na_2S_2C_2(CN)_2$ was dependent on the type of cation used as precipitant (73). If tetraalkyl ammonium salts were used, the product was invariably the dimeric dianion, $[FeS_4C_4(CN)_4]_2^{-2}$, also referred to in early papers as the monoanion, $[FeS_4C_4(CN)_4]^{-1}$ (see Sec. III-B and III-F). If the precipitant was a tetraaryl phosphonium or arsonium salt, however the product was usually the tris-substituted complex, $[FeS_6C_6(CN)_6]^{-2}$ (Sec. IV-A), although, under carefully controlled conditions, $[Ph_4P]_2$ $[FeS_4C_4(CN)_4]_2$ could be isolated.

Treatment of $Re_2Cl_8^{-2}$ with $Na_2S_2C_2(CN)_2$ resulted in the formation of dimeric complex $[ReS_4C_4(CN)_4]_2^{-2}$ (84).

When $Na_2S_2C_2H_2$ was used as the complexing agent, the charged species normally produced were the monoanions, $[MS_4C_4H_4]^{-1}$, where M = Co, Pd, or Cu. An exception to this occurred when Ni(II) salts were treated with the dithiolate and a black amorphous powder with empirical formula $[NiS_4C_4H_4]_n$ was formed (57); the purple neutral complex $[NiS_4C_4H_4]^0$ could be extracted from this powder by ether extraction. Similar polymers were apparently formed (59) in reaction of the dithiolate with iron or manganese salts; the iron-containing polymer yielded $[Fe(PBu_3)S_4C_4H_4]^0$ (Sec. V-C) when extracted with tri-n-butyl phosphine.

2. Complexes Derived from Dithietenes

The neutral bis-perfluoromethyldithiolenes, $[NiS_4C^2(CF_3)_4]^0$ and $[MS_4C_4(CF_3)_4]_2^0$, where M = Co or Fe, were obtained from the direct reaction between the appropriate metal carbonyls and the dithietene, $S_2C_2(CF_3)_2$. If a suitable carbonyl was unavailable or unknown, phosphine complexes such as $Pt(PPh_3)_4$ or $[PPh_3CuI]_4$ could be used, and in this way several mono- and dianionic bis-perfluoromethyl–dithiolene complexes of Cu, Au, and the nickel group were obtained.

Treatment of mercuric salts with $[S_2C_2(CF_3)_2]^{-2}$ (62) resulted in the formation of the polymeric $[HgS_2C_2(CF_3)_2]_n$.

TABLE I

Bis-Substituted 1,2-Dithiolene Complexes and their Analogs

Complex	z	cation/anion	Color	mp,°C	μ_{eff}, BM	Synthesis[a]	Refs.[b]
$ZnS_4C_4(CN)_4$	-2	Bu_4N^+	Yellow	156–158	Dia.	1	97
$Zn(S_2C_6H_3Me)_2$	-2	$Quin^+$	Yellow		Dia.	1	4
$Zn(gma)$	0	—	Blue–black (red)[c]		Dia.	5	68,44
$Zn—N_2S_2(cis)$[d]	0	—	Orange		Dia.	5	44
$Zn—N_2O_2(cis)$[d]	0	—	Yellow		Dia.	5	44
$Zn(O_2C_6Cl_4)_2$	-2	Pr_4N^+	White	222 d	Dia.	3	40
$Zn(O_2C_6H_4)_2$	-2	Ph_4As^+	Yellow–orange	213–214 d	Dia.	3	40
$Cd(gma)$	0	—	Black	290 d	Dia.	3,6	37,40
$Cd—N_2S_2(cis)$[d]	0	—	Blue		Dia.	5	44,68
$Cd(O_2C_6Cl_4)_2$	-2	Pr_4N^+	Red		Dia.	5	44
$Cd(O_2C_6H_4)_2$	-2	Et_4N^+	Yellow	207 d	Dia.	3	40
$CuS_4C_4(CN)_4$	-2	Bu_4N^+	Red–brown	112 d	1.78	1	97
$CuS_4C_4(CF_3)_4$	-2	Ph_4As^+	Dark red	146–147	Dia.	1,6	98
	-1	Ph_4As^+	Brick red	198–200		2	27
	-1	Ph_4As^+	Olive–green	175–176		2	27
$Cu(S_2C_6Cl_4)_2$	-1	Bu_4N^+	Green	192	Dia.	2	65
$Cu(S_2C_6H_4)_2$	-1	Bu_4N^+	Green	178	Dia.	3	65
$Cu(S_2C_6H_3Me)_2$	-2	Ph_3MeAs^+	Orange–brown	92–95 d	1.78	3,7	110
	-1	Ph_3MeAs^+	Green	146–148	Dia.	3	110
$Cu(S_2C_6H_2Me_2)_2$	-1	Bu_4N^+	Green	152–153	Dia.	3	110
$Cu(S_2C_6Me_4)_2$	-1	Bu_4N^+	Green	215	Dia.	3	110
$CuS_4C_4H_4$	-1	Et_4N^+	Green	242	Dia.	3	110
$Cu—N_2S_2(cis)$[d]	-1	—	Pale brown	136	Dia.	1	59
$Cu—N_2O_2(cis)$[d]	0	—	Green–brown		Dia.	5	44,69
$Cu(O_2C_6Cl_4)_2$	0	—	Olive–green		Dia.	5	44
$Cu(O_2C_6H_4)_2$	-2	Pr_4N^+	Olive–green	198 d	1.87	3	40
	-2	Pr_4N^+	Green	191–192 d	1.87	3	40

Compound	Charge	Cation	Colour	M.p. (°C)	μ_{eff} (B.M.)/Dia.	Prepn.	Ref.
AuS₄C₄(CN)₄	-2	Bu₄N⁺	Green		1.85	1,7	111
	-1	Et₄N⁺	Red-brown	179–181	Dia.	1	98
AuS₄C₄(CF₃)₄	-1	Ph₄As⁺	Pale green	179	Dia.	2	27
Au(S₂C₆Cl₄)₂	-1	Bu₄N⁺	Pale green	152–153	Dia.	3	65
Au(S₂C₆H₃Me)₂	-1	Bu₄N⁺	Green		Dia.	3	110
NiS₄C₄(CN)₄	-2	Et₄N⁺	Brick red	143–145	Dia.	1	97
	-1	Bu₄N⁺	Red	171–173	Dia.	1	97
	-1	Et₄N⁺	Dark brown–red		1.02	1,6	93,98
			Dark brown–red		1.83[e]	1,6	93
NiS₄C₄(CF₃)₄	-2	Ph₃MeP⁺	Yellow	241–251 d	1.15	2,7	98
	-2	Et₄N⁺	Yellow	234–235	Dia.	2,7	98
	-1	Ph₄As⁺	Green–brown	119–120	1.85	2,7	98
	-1	Et₄N⁺	Green–brown	174–175	1.82	2,7	98
	0	—	Blue–violet	134–135	Dia.	2	98
Ni(S₂C₆Cl₄)₂	-1	Bu₄N⁺	Dark green	194	1.89	3	65
Ni(S₂C₆H₄)₂	-1	Bu₄N⁺	Dark green	173	1.83	3	65
Ni(S₂C₆H₃Me)₂	-2	Ph₃MeAs⁺	Light red	131–134	Dia.	3,7	110
	-1	Bu₄N⁺	Green	152–153	1.79	3	110
	-1	Ph₃MeAs⁺	Green		1.89	2	110
Ni(S₂C₆H₂Me₂)₂	-1	Bu₄N⁺	Dark green	65	1.83	3	65
Ni(S₂C₆Me₄)₂	-1	Bu₄N⁺	Green	57	1.82	3	65
NiS₄C₄(p-ClC₆H₄)₄	0	—	Green	325 d	Dia.	4	79
NiS₄C₄Ph₄	-2	N₂H₅⁺	Orange	140 d	1.86	4,7	79
	-1	Ph₄As⁺	Purple–red	254–256 d	1.82	4,7	98
	-1	Et₄N⁺	Purple–red		Dia.	4,7	98
NiS₄C₄H₂Ph₂(cis)	0	—	Green	293 d	Dia.	4	79
NiS₄C₄H₄	0	—	Green	174–175	Dia.	4	79
	-1	Et₄N⁺	Dark brown	176	1.72	1,7	57,59
	0	—	Purple	68 d	Dia.	1	57
NiS₄C₄(p-MeC₆H₄)₄	0	—	Green	337 d	Dia.	4	79
NiS₄C₄(p-MeOC₆H₄)₄	0	—	Green	329 d	Dia.	4	79

(continued)

TABLE I (continued)

Complex	z	cation/anion	Color	mp,°C	μ_{eff},BM	Synthesis[a]	Refs.[b]
$NiS_4C_4Me_2Ph_2(cis)$	0	—	Blue	152	Dia.	4	60
$NiS_4C_4Me_4$	0	—	Blue	256 d	Dia.	4	79
$NiS_4C_4Et_4$	0	—	Blue	112	Dia.	4	79
$NiS_4C_4(i\text{-}Pr)_4$	0	—	Blue	52–54 d	Dia.	4	79
$\{Ni[(NH)_2C_6H_4]_2\}$	0	—	Blue		Dia.	3	30,33,34
$\{Ni[4\text{-}i\text{-}PrC_6H_4(NH)_2]_2\}$	+1	I^-	Grey–green		1.18	3,6	33
	+2	I^-	Blue		Dia.	3,6	33
$\{Ni[(MeCNPh)_2]_2\}$	0	—	Violet		Dia.	3	33
	−2	I	Red-orange		3.02	(5)	33
		NO_3^-	Red-orange		3.12	(5)	33
$Ni[S(NH)C_6H_4]_2$	0	—	Blue		Dia.	3	34,41,44
$Ni[SC(Ph)N \cdot NH]_2$	0	—	Blue		Dia.	3	43,44
$Ni(H_2\,gma)$	−1	Bu_4N^+	Green		Para.	5,7	34,78
$Ni(gma)$	0	—	Dark red		Dia.	5	77,78
	−1	Bu_4N^+	Red–brown		Para.	5,7	78
$Ni\text{—}N_2S_2(cis)$[d]	0	—	Red–brown		Dia.	5	44,69
$Ni\text{—}N_2O_2(cis)$[d]	0	—	Red		Dia.	5	44
$Ni(OSC_6H_4)_2$	−2	—	Brown		Dia.	3	28,44
$Ni(O_2C_6Cl_4)_2$	−2	Pr_4N^+	Dark brown	219 d	Dia	3	40
	"—1"	Pr_4N^+	Blue–green	150 d	3.80	(8)	40
$Ni(O_2C_6H_4)_2$	0	—	Brown	188 d	3.49	3	40
	−2	Pr_4N^+	Green–black		Dia.	(3)	40
$Ni(O_2Phen)_2$[f]	0	—	Red		2.70	3	40
	−2	Pr_4N^+			Dia.	(3)	40
$Ni[3,5\text{-}(tBu)_2C_6H_2O_2]_2$	0	—	Deep green	155–158	3.92	3	40
$PdS_4C_4(CN)_4$	−2	Bu_4N^+	Green		Dia.	1	97
	−1	Et_4N^+	Dark red	275 d	Dia.	1,6	93,98

$PdS_4C_4(CN)_4$		Bu_4N^+	Dark red	190–191	Dia.	1,6	98
$PdS_4C_4(CF_3)_4$	−2	Ph_4As^+	Pale green	248–251	Dia.	2,7	99
	−1	Ph_4As^+	Red–brown	245–247 d	1.73	2	99
$PdS_4C_4Ph_4$	−2	$N_2H_5^+$	Orange	132 d		4,7	79
	0	—	Blue		Dia.	4	79
$PdS_4C_4H_4$	−1	Et_4N^+	Brown–violet	136	1.61	1	59
$PdS_4C_4(p\text{-}MeOC_6H_4)_4$	0	—		294 d	Dia.	4	60
$Pd[(NH)_2C_6H_4]_2$	0	—	Blue		Dia.	3	33
	+1	I^-	Green			3,6	33
$PtS_4C_4(CN)_4$	−2	Bu_4N^+	Red	164–167	Dia.	1	97
	−1	Et_4N^+	Black	288 d	1.05	1,6	93,98
$PtS_4C_4(CF_3)_4$	−2	Ph_4As^+	Golden–yellow	240–242	Dia.	2,7	99
	−1	Ph_4As^+	Red	169–169	1.73	2,7	99
		Ph_4P^+	Red	174–175	1.79	2,7	99
	0	—	Purple	174–175	Dia.	2	99
$Pr(S_2C_6H_3Me)_2$	−2	Ph_4As^+	Orange		Dia.	3,7	110
	−0	Bu_4N^+	Green–brown	160–161	1.77	3	110
$PtS_4C_4Ph_4$	0	—	Red	310	Dia.	4	79
$PtS_4C_4(p\text{-}ClC_6H_4)_4$	0	—	Red	340	Dia.	4	79
$PtS_4C_4(p\text{-}MeOC_6H_4)_4$	0	—	Red	341	Dia.	4	60
$PtS_4C_4Me_2Ph_2$(*cis*)	0	—		350	Dia.	4	60
$PtS_4C_4Me_4$	0	—		360 d	Dia.	4	79
$Pt[(NH)_2C_6H_4]_2$	0	—	Blue		Dia.	3	33
	+1	I^-	Green			3,6	33
$CoS_4C_4(CN)_4$	−2	Ph_4As^+	Red		2.16	1,7	13
$[CoS_4C_4(CN)_4]_2$	−2	Bu_4N^+	Green–brown	198–199	Dia.	1,6	98
		Ph_3MeP^+	Green–brown	230	Dia.	1,6	93
$CoS_4C_4(CF_3)_4$	−2	Et_4N^+	Orange	245 d	2.06	2,7	99
$[CoS_4C_4(CF_3)_4]_2$	−2	Et_4N^+	Green–brown	268–271	Dia.	2,7	99
	−1	Et_4N^+				2,7	25
	0	—	Purple–brown	189–190	Dia.	2	99
$[Co(S_2C_6Cl_4)_2]_2$[g]	−2	Bu_4N^+	Blue–black	214	Dia.	3	65

(continued)

TABLE I (continued)

Complex	z	cation/anion	Color	mp,°C	μ_{eff},BM	Synthesis[a]	Refs.[b]
$Co(S_2C_6H_4)_2$	−2	Bu_4N^+	Blue	177	3.27	3	65
$Co(S_2C_6H_3Me)_2$	−1	Bu_4N^+	Blue			3	16,110
		Ph_3MeAs^+	Blue		3.27	3	16,110
$Co(S_2C_6H_2Me_2)_2$	−1	Bu_4N^+	Blue	220	2.34	3	65
$Co(S_2C_6Me_4)_2$	−1	Bu_4N^+	Blue	253	3.23	3	65
$[CoS_4C_4Ph_4]_2$	−2	Et_4N^+	Green	198 d	Dia.	4,7	96
		Ph_4As^+	Green	98 d	Dia.	4,7	96
	0	—	Black	360 d	Dia.	4	96
$CoS_4C_4H_4$	−1	Et_4N^+	Green–black	160	2.96	1	59
		Ph_4As^+	Green–black	185	2.70	1	59
$[CoS_4C_4(p\text{-}MeC_6H_4)_4]_2$	0	—	Black	294 d	Dia.	4	96
$[CoS_4C_4(p\text{-}MeOC_6H_4)_2]_2$	0	—	Black	275 d	Dia.	4	96
$Co(gma)$	0	—	Black			5	76
$\{Co[(NH)_2C_6H_4]_2\}$	0		Violet		2.20	3,6	33
	−1	I^-	Blue		Dia.	3	33
$Co(O_2C_6Cl_4)_2$	−2	Pr_4N^+	Red	224 d	4.96	3	40
$RhS_4C_4(CN)_4$	−2	Bu_4N^+	Red(green)[h]		1.91	1	122
$[FeS_4C_4(CN)_4]_2$	−2	Et_4N^+	Dark red–brown	280	1.62	1	93
		Ph_4P^+	Dark red–brown			1	73
$[FeS_4C_4(CF_3)_4]_2$	−2	Et_4N^+	Green	290–295 d	1.83	2,7	99
		Ph_4As^+	Green	251–253	3.98[i]	2,7	99
	−1	Et_4N^+			1.39	2,7,6	25
$[Fe(S_2C_6Cl_4)_2]_n$	0	—	Blue–black	189–190	Dia.	2	99
	−2	Bu_4N^+	Purple–red			3,7	50,65
	0	—	Blue–black			(3)	50,65
$[Fe(S_2C_6H_3Me)_2]_2$	−2	Bu_4N^+	Purple–red	232–234	2,07	3	16,112
	−1	Bu_4N^+				3,6	50

Complex	Charge	Cation	Color	Dec. (°C)	Magnetic	Method	Ref.
$[FeS_4C_4Ph_4]_2$	-2	Et_4N^+	Green	214 d	—	4,7	96
		$N_2H_5^+$	Red[j]		—	4,7	50
		Ph_4As^+	Green	112 d	Dia.	4,7	96
$[FeS_4C_4H_4]_n$	0	—	Black	295 d	Dia.	4	96
$[FeS_4C_4(p\text{-}MeC_6H_4)_4]_2$	0	—	Black			1	59,96
$[FeS_4C_4(p\text{-}MeOC_6H_4)_4]_2$	0	—	Black	224 d	Dia.	4	96
$[Fe_2S_2(S_2C_2Ph_2)_2]$	-2	Ph_4As^+	Black	272 d	Dia.	4	96
			Green	174–179			96
	0	—	Violet	360	Dia.		96
$[Fe_2S_2(S_2C_2(p\text{-}MeC_6H_4)_2)_2]$	0	—	Violet		Dia.		96
$[Fe_2S_2(S_2C_2(p\text{-}MeOC_6H_4)_2)_2]$	0	—	Violet		Dia.		96
Fe(gma)	0	—	Black				76
$MnS_4C_4(CN)_4$	-2	Ph_4P^+	Red–brown		Para.	1	73
$[ReS_4C_4(CN)_4]_2$	-2	Bu_4N^+	Black		Dia.	1	84

[a] The numbers refer to the general methods described in Section III-A.

[b] The references refer to the most important paper concerning the syntheses and characterization of the complexes.

[c] The red form is apparently polymeric and precipitates from the blue DMF solutions.

[d] The ligands are biacetylbis(thiobenzoylhydrazone) and biacetylbis(bensoylhydrazone).

[e] Magnetically dilute sample.

[f] Phenanthrene-9,10-diol.

[g] Complex dimeric only in the solid state but dissociated in solution.

[h] Complex is red in solution but green in solid state.

[i] Magnetic moment when dissociated in polar solvents.

[j] Complex may contain coordinated hydrazine.

3. *Complexes Derived from Arene-1,2-dithiols and Their Analogs*

The monoanionic complexes of copper, the nickel group, and cobalt, and the dianionic dimeric complexes of iron were obtained by treatment of the divalent metal salts with the disodium salts of benzene-1,2-dithiol, and its 4-methyl (TDT), 4,5-dimethyl, and 3,4,5,6-tetramethyl derivatives, and tetrachlorobenzene-1,2-dithiol in ethanol. If the concentration of sodium in ethanol was very high, dianionic complexes of copper, nickel, and platinum could be prepared. $[Fe(S_2C_6Cl_4)_2]_n^0$ was the product of the reaction between iron powder, hexachlorobenzene, sodium sulfide, and DMF, as described in Section II, and dissolution of this compound in acetone/DMF caused the generation of the dimeric dianion.

Dianionic metal complexes of quinoxaline-2,3-dithiol (2) containing copper, nickel, palladium, and cobalt were synthesized (9,74,75) by treating the divalent metal salts with the ligand in the presence of base.

It was thought that the dimeric oxygen-bridged complex (6) was the product (41) of the aerial oxidation of polymeric $\{Ni[S(NH_2)C_6H_4]_2\}_n$ in strong base. However, a reinvestigation of this compound (28,34,44) led to its reformulation as $\{Ni[S(NH)C_6H_4]_2\}^0$; its composition was confirmed by mass spectrometry (34,44). In the presence of base, monothiocatechol formed complexes of the type $[M(OSC_6H_4)_2]^{-2}$ with divalent copper, nickel, palladium, and cobalt (28).

The dianionic species, $[M—O_4]^{-2}$, derived from the ligand catechol, $C_6H_4(OH)_2$, tetrachlorocatechol, $C_6Cl_4(OH)_2$, and phenanthrene-9,10-diol, were prepared (40) by treating Zn(II), Cd(II), Cu(II), Ni(II), and Co(II) salts with these ligands in alkaline solutions containing n-Pr_4NOH. The neutral complexes, $[Ni(O_2C_6Cl_4)_2]^0$ and $[Ni(O_2C_6H_2-3,5-(t-Bu)_2)_2]^0$, were formed by treating $Ni(CO)_4$ with the appropriate o-quinones in cold hydrocarbon solvents. The dication, $[M—O_4]^{+2}$, where the ligand was o-phenanthrene-9,10-dione and M = Zn or Ni, were obtained by addition of the dione to the metal salts. However, it seems that the zinc complex is the only authentic bis-substituted species (it may be tetrahedral) since the nickel compound has a distorted octahedral structure where the counteranions functioned as fifth and sixth ligands.

Suspension of divalent nickel, palladium, platinum, and cobalt salts in aqueous concentrated ammonia containing o-phenylenediamine, and passage of air or oxygen through these suspensions, caused the formation (30,31,33) of the diimino complexes, $\{M[(NH)_2C_6H_4]_2\}^0$, whose compositions were confirmed by mass spectrometry. The nickel complex was also obtained (33) by allowing cyclohexadienediimine to react with $Ni(CO)_4$ in cold ether.

4. Complexes Derived from Disubstituted Ethylene Dithiophosphoric Esters

The neutral complexes $[MS_4C_4R_4]^0$ (where M is a nickel group metal) and $[MS_4C_4R_4]_2^0$ (where M = Co or Fe) were synthesized (63) by addition of a suitable metal compound, e.g., acetate, halide, acetylacetonate, or carbonyl, solubilized in water, ethanol, or benzene, to the appropriate thiophosphoric ester (see Sec. II) in hot dioxan or xylene. Optimum yields were obtained if the added metal compound contained the metal ion in its lowest feasible oxidation state, since the use of metal ions in high oxidation states necessitated the use of larger amounts of the thio-ester, which functioned as the reducing as well as the complexing agent. Neutral complexes of the nickel group containing many different alkyl or aryl groups have been described (Table I), but the only cobalt and iron complexes which have been prepared are bis-diphenyl-1,2-dithiolenes and their toluoyl analogs.

$[Fe_2S_2(S_2C_2Ph_2)_2]^0$ was obtained (96) by thermal decomposition of $[FeS_4C_4Ph_4]_2^0$ in toluene under pressure, or by refluxing $Fe_3(CO)_{12}$, sulfur, and $PhC{\equiv}CPh$ together in toluene.

5. Complexes Derived from Schiff Bases and Related Species

Complexes of the Schiff-base glyoxal bis(2-mercaptoanil), gma, were prepared (44,68) either by allowing the aqueous Ni^{2+}, Zn^{2+}, or Cd^{2+} ions to react with the ligand bis-benzthiozaline (which is derived from gma) or in the self-condensation reaction between two equivalents of o-amino-thiophenol and one equivalent each of glyoxal (or any other α,β-diketone) and the metal ion in ethanol. The iron and cobalt complexes (76) were conveniently prepared by refluxing the metal carbonyls with bis-benzthiazoline in methanol. The reactions of metal ions with gma and its analogs is summarized in Figure 1.

Reaction of $Ni(CO)_4$ with bis-acetylbisanil (**18**) provided the neutral complex $[Ni(MeCNPh)_2]^0$ and treatment of NiI_2 with the same ligand in

(18)

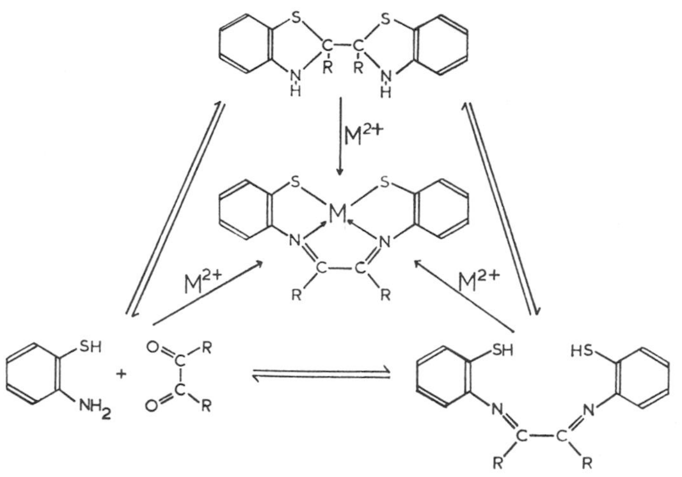

Fig. 1. The formation of glyoxal bis(2-mercaptoanil) and its analogs, and their reaction with metal ions.

acetone afforded the diiodide, $[Ni(MeCNPh)_2]I_2$. The zinc, cadmium, copper, and nickel group complexes of biacetylbis(thiobenzoylhydrazone) (**16**) were prepared from solutions of the ligand and the metal acetates in ethanol (44,69), and $[Ni(SC(Ph)N \cdot NH)_2]^0$ from Ni^{2+} salts and the hydrazide (43) in aqueous ammoniacal ethanol.

6. Oxidation of Preformed Dithiolene Complexes and Their Analogs

Oxidation of the preformed dithiolene complexes whose methods of formation have just been described may be achieved by using a number of different oxidizing agents. The actual choice of reagent depends upon the value of the half-wave potentials for the one- or two-electron transfer reactions which are required. The polarographic and voltammetric methods by which these data were acquired are discussed in Section III-C. The commonest oxidizing agents were air and iodine, but in certain special cases, notably the formation of $[CuS_4C_4(CN)_4]^{-1}$ or $[PdS_4C_4(CN)_4]^{-1}$ from their respective dianions, or of $[FeS_4C_4(CF_3)_4]_2^{-1}$ from $[FeS_4C_4(CF_3)_4]_2^{-2}$, use of a more powerful one-electron oxidizing agent was essential, and $[NiS_4C_4(CF_3)_4]^0$ or $[FeS_4C_4(CF_3)_4]_2^0$ was found to be suitable. Oxidation of the dianion $[Zn(O_2C_6H_4)_2]^{-2}$ to the neutral species was effected by using persulfate ion in aqueous solution. Treatment of $[Cd(O_2C_6Cl_4)_2]^{-2}$ with $[NiS_4C_4(CF_3)_4]^0$ (40) did afford a radical species whose existence was demonstrated by ESR spectral methods, but it was not established whether this species was in fact $[Cd—O_4]^{-1}$. By refluxing a mixture of $[Pr_4N]_2[Ni(O_2C_6Cl_4)_2]$ and $[Ni(O_2C_6Cl_4)_2]^0$ in dichloromethane, a complex described (40) as $[Pr_4N][Ni(O_2C_6Cl_4)_2]$ was obtained. On the basis of the lack of redox properties of this compound, and its magnetic and spectral dissimilarities when compared with authentic $[Ni—S_4]^{-1}$ species, it was suggested that it was a charge-transfer complex, $[Pr_4N]_2[Ni(O_2C_6Cl_4)_2] \cdot [Ni(O_2C_6Cl_4)_2]$.

Chemical oxidation of the complexes obtained from o-phenylene-diamine, $[M—N_4]^0$, to the corresponding monocations was achieved using stoichiometric amounts of iodine. Treatment of the nickel complex and its 4-isopropyl analog with larger amounts of iodine led to the formation of compounds of empirical formulas $[Ni—N_4]I_2$, but evidence gleaned from electrochemical studies suggested that these should be regarded as $[Ni—N_4]$ $I_3 \cdot [Ni—N_4]I$.

7. Reduction of Preformed Dithiolene Complexes and Their Analogs

The choice of reducing agent was made on the basis of the half-wave potentials for the desired reaction:

$$[M—X_4]^z + ne^- \rightleftharpoons [M—X_4]^{z+ne}$$

Common reducing agents included o- and p-phenylenediamine, hydrazine, borohydride ion, zinc in pyridine, amalgams, and alkali metal alkoxides. In a few instances, notably that of $[NiS_4C_4(CF_3)_4]^0$ and the neutral dimeric cobalt and iron bis-perfluoromethyl–dithiolenes, weakly basic solvents such as ketones or alcohols effected one-electron reductions.

The use of borohydride ion as a reducing agent for simple dithiolene complexes was usually satisfactory if the electrochemical data indicated that its use was desirable. However, in one case, its use has been the source of some embarrassment and confusion. The reduction of $[Ni(gma)]^0$ to the monoanion was reported independently by two groups of workers; in one case the one-electron reduction was effected using sodium amalgam in THF (44,77), and in the other, BH_4^- in THF (34). The species formed in each case was isolated as the Bu_4N^+ salt and each apparently analyzed correctly for the 1:1 electrolyte $[Bu_4N][Ni(gma)]$. However, one compound was brown (34) and the other green (44,77) and there were serious discrepancies in the g values obtained from their ESR spectra in solution at room temperature (Sec. III-F). A very thorough reexamination (78) of this system has resolved this problem. If the reduction of $[Ni(gma)]^0$ was carried out using borohydride ion in THF, a green solution was formed which became grey–black on addition of Bu_4N^+ salts in ethanol. Admission of traces of oxygen caused this solution to become bright green and the crystalline complex $[Bu_4N][Ni(H_2 gma)]$ was obtained from this solution. Using BD_4^-, the partially deuterated species, $[Ni(HD gma)]^{-1}$ was produced. If $[Ni(gma)]^0$ was reduced by sodium amalgam in THF under strictly anaerobic conditions, the deep red–brown monoanion, $[Ni(gma)]^{-1}$, which was spectrally similar to the species produced by electrolytic reduction of the neutral compound, was formed; it was isolated as the brown crystalline complex $[Bu_4N][Ni(gma)]$. The mechanism of formation, and possible structures, of these two different species is further discussed in Section III-F.

Hydrazine as a reducing agent also had disadvantages. In some cases, such as the reduction of $[NiS_4C_4Ph_4]^0$ (79) and $[FeS_4C_4Ph_4]_2^0$ and its cobalt analog (171), hydrazine was retained in the reduced complex after it had been isolated. It seems fairly certain that in $[N_2H_5]_2[NiS_4C_4Ph_4]\cdot N_2H_4$, the residual reductant was hydrogen-bonded to the cation and, indeed, if N,N'-dimethylhydrazine was used as reducing agent, the simple 1:1 complex, $[N_2H_3Me_2]_2[NiS_4C_4Ph_4]$, was isolated. The reduction products of the iron and cobalt diphenyl–dithiolenes, $[N_2H_5][MS_4C_4Ph_4]\cdot N_2H_4$ are, however, less well characterized, but it is thought that the extra hydrazine functions as a fifth ligand giving the molecules pyramidal symmetry: such behavior is common among iron and cobalt dithiolenes which have been dissolved in donor solvents (Sec. V-C).

8. Ligand Exchange Reactions

The use of dithiolene complexes as oxidizing agents raised the interesting problem as to whether in such reactions ligand exchange or "scrambling" took place, i.e.,

$$[MS_4C_4R_4']^z + [MS_4C_4R_4'']^z \longrightarrow 2[MS_4C_4R_2'R_2'']^z$$

It was found (80) that these reactions were best investigated polarographically and are discussed in more detail in Section III-C. However, by refluxing a neutral complex (say with R') and a dianionic species (R''), or two equivalent monoanions, ligand exchange could be effected at different rates depending on the solvent used, the reflux temperature, and the nature of R' and R''. The product of the reactions was mainly the mixed ligand complex as a monoanion, but there were often traces of the precursor complexes, also as monoanions.

9. General Characteristics

It seems clear that all the copper and nickel group dithiolenes, $[M—S_4]^z$, and the cobalt and rhodium dianions, $[M'—S_4]^{-2}$, are monomeric and planar. The zinc group dithiolates are, for the most part, mono-meric but are probably tetrahedral except where the ligand (or ligands) imposes some stereochemical restraint as in $[Zn(gma)]^{0,-1}$. The oxidized iron and cobalt dithiolenes, that is, the neutral and "monoanionic" species, appear to be dimeric $[M—S_4]_2^z$ ($z = 0$ or -2) with the exception of $[Co(S_2C_6H_nMe_{4-n})_2]^{-1}$, $[Co(S_2C_6Cl_4)_2]^{-1}$ (see Sec. III-B), and $[CoS_4C_4H_4]^{-1}$. The rhenium complex, $[ReS_4C_4(CN)_4]_2^{-2}$, although apparently similar to the cobalt and iron analogs, may have a different molecular structure (see Sec. III-B).

The spin ground states of the various types of complexes discussed in this section are summarized in Table II.

TABLE II

Spin Ground States for Monomeric and Dimeric Bis-Substituted Dithiolene Complexes and their Analogs

| $[M—L_4]^z$ | Ligand, L^a | Spin ground states for species | | | | |
		$z = -2$	$z = -1$	$z = 0$	$z = +1$	$z = +2$
Zn—S$_4$		0	—	—	—	—
Zn—N$_2$S$_2$	gma	0	½	0	—	—
Cd—N$_2$S$_2$	gma	0	½	0	—	—
Cu—S$_4$		½	0	—	—	—
Cu—O$_4$		½	0	—	—	—
Cu—N$_2$S$_2$		(½)	0	½	(0)	—
Ag—S$_4$	S$_2$C$_2$(CN)$_2$	½	0	—	—	—
Au—S$_4$	S$_2$C$_2$(CN)$_2$	½	0	—	—	—
M—S$_4$^{b}		0	½	0	(½)^c	—
M—O$_4$^{b}		0	½	(³⁄₂)	?	—
M—N$_4$^{b}		0	½	0	½	0
M—O$_2$S$_2$^{b}		0	½	—	—	—
M—N$_2$S$_2$^{b}		0	½	0	(½)	0
Co—S$_4$		½	1	—	—	—
[Co—S$_4$]$_2$		0^d	½	0	—	—
Co—N$_4$		—	—	½	0	—
Rh—S$_4$		½	—	—	—	—
[Fe—S$_4$]$_2$		½^{e,f}	½^g	0	—	—

^{a} All ligands implicit except where limit stated.
^{b} M = Ni, Pd, and Pt.
^{c} $[NiS_4C_4Ph_4]^{+1}$ detected voltammetrically.
^{d} $S = 1$ when complex dissociated.
^{e} $S = ³⁄₂$ when complex dissociated.
^{f} Per Fe.
^{g} Per dimeric molecule.

Many of the iron complexes have been investigated by Mössbauer spectroscopy (83). An x-ray structural investigation of "$[Bu_4N][FeS_4C_4$-$(CN)_4]$" (94) revealed that the anion was dimeric (Sec. III-B), each iron being surrounded by five sulfur atoms in a distorted square-pyramidal arrangement. The Mössbauer parameters (Table III) for this compound, and for several similar complexes, were consistent with a dimeric formulation for the anion in the solid state. The spectra of the neutral complexes, $[FeS_4C_4Ph_4]_2{}^0$ and $[Fe(S_2C_6Cl_4)_2]_n{}^0$ were similar to those of the dimeric dianions.

The neutral nickel-group dithiolenes were volatile and exhibited characteristic mass spectral cracking patterns (81,82). With the exception of $[MS_4C_4(CF_3)_4]^0$, the principle fragmentation pattern could be summarized as follows:

$$[MS_4C_4R_4]^+ \longrightarrow [MS_4C_2R_2]^+ \longrightarrow [MS_2C_2R_2]^+ \longrightarrow$$
$$[MSC_2R_2]^+ \longrightarrow \text{complex fragmentation}$$

The bis-perfluoromethyl-dithiolenes fragmented in a slightly different way:

$$[MS_4C_4(CF_3)_4]^+ \longrightarrow [MC_8F_{11}S_4]^+ \text{ and}$$
$$[MC_6F_6S_2]^+ \longrightarrow \text{complex fragmentation}$$

TABLE III

Mössbauer Data Obtained from Iron Bis- and Tris-1,2-dithiolene Complexes

Complex	Temp., °K	Isomer shift, mm/sec	Quadrupole splitting, mm/sec
$[Et_4N]_2[FeS_4C_4(CN)_4]_2$	77	0.59	2.76
	295	0.50	2.81
$[Et_4N]_2[FeS_4C_4(CF_3)_4]_2$	77	0.59	2.50
	295	0.49	2.50
$[Et_4N]_2[FeS_4C_4Ph_4]_2$	77	0.61	2.37
	295	0.53	2.45
$[Bu_4N]_2[Fe(S_2C_6H_3Me)_2]_2$	77	0.60	2.95
	295	0.54	2.99
$[Bu_4N]_2[Fe(S_2C_6Cl_4)_2]_2$	77	0.58	3.02
	295	0.49	3.03
$[FeS_4C_4Ph_4]_2$	77	0.51	2.01
	295	0.42	2.05
$[Fe(S_2C_6Cl_4)_2]_n$	77	0.58	3.15
	295	0.49	3.15
$[Ph_4P]_2[FeS_6C_6(CN)_6]$	77	0.50	1.57
	295	0.42	1.59
$[Ph_4P]_3FeS_6C_6(CN)_6]$	77	0.65	1.69

The remaining fragmentation processes were very complex, but in the nickel compound, species having m/e corresponding to $[NiC_4F_5S_2]^+$, $[NiC_3F_3-S_2]^+$, and $[NiS_2]^+$ were detected. The mass spectrum of $[FeS_4C_4(CF_3)_4]_2^0$ was even more complicated, but provided the first direct evidence that this molecule was dimeric in the vapor phase. The principal ions detected were

$$[Fe_2C_{16}F_{24}S_8]^+ \longrightarrow [Fe_2C_{16}F_{23}S_8]^+ \quad \text{and}$$
$$[FeC_{12}F_{18}S_6]^+ \longrightarrow \text{complex fragmentation}$$

Other ions observed had m/e corresponding to $[Fe_2C_{12}F_{17}S_6]^+$, $[FeC_8-F_{12}S_4]^+$, and $[FeC_8F_{11}S_4]^+$.

B. Structural Data

The structures of ten bis-1,2-dithiolene complexes or their analogs have been analyzed by single-crystal x-ray diffraction techniques. Seven of these have been reported fully and the remaining four are at various stages of refinement. Of these ten complexes, eight are genuine four-coordinate systems with slight (but not chemically significant) distortions from planarity, and three are dimeric species containing five-coordinate metal atoms. The available structural parameters are given in Tables IV and V.

TABLE IV

Bond Lengths in Planar and Dimeric 1,2-Dithiolenes

Complex	Cation	M—S[a]	M—S′[b]	"C=C"	C—S[c]
$[CuS_4C_4(CN)_4]^{-1}$ [d]	Bu_4N^+	2.17 ± 0.01		1.31 ± 0.02	1.72 ± 0.02
$[NiS_4C_4(CN)_4]^{-2}$ [e]	Me_4N^+	2.165 ± 0.005		1.33 ± 0.02	1.75 ± 0.01
$[NiS_4C_4(CN)_4]^{-1}$ [f]	Ph_3MeP^+	2.146 ± 0.001	3.591	1.356 ± 0.007	1.714 ± 0.004
$[NiS_4C_4Ph_4]^0$ [g]	—	2.101 ± 0.002	4.6	1.37 ± 0.014	1.71 ± 0.01
$[CoS_4C_4(CN)_4]^{-2}$ [h]	Bu_4N^+	2.16 ± 0.005		1.34 ± 0.010	1.723 ± 0.015
$[CoS_4C_4(CF_3)_4]_2^0$ [i]	—	2.161 ± 0.016	2.38	1.393 ± 0.006	1.694 ± 0.016
$[FeS_4C_4(CN)_4]_2^{-2}$ [j]	Bu_4N^+	2.23 ± 0.01	2.46 ± 0.01	1.39 ± 0.03	1.73 ± 0.03

[a] Basal bond lengths, Å.
[b] Apical bond lengths.
[c] Ring carbon and sulfur atoms.
[d] Cu····Cu′ 4.026 ± 0.005, 4.431 ± 0.006 Å; C≡N 1.15 ± 0.02 Å.
[e] Ni····Ni′ 8.04 Å; C≡N 1.13 Å.
[f] Ni····Ni′ 4.40, 4.45 Å, C≡N 1.140 ± 0.005 Å.
[g] C—C(Ph) 1.38 Å.
[h] Co····Co′ 9.81 Å; C≡N 1.15 ± 0.01 Å.
[i] Co····Co′ 2.781 ± 0.006 Å.

TABLE V

Bond Angles in Planar and Dimeric Metal 1,2-Dithiolenes

Bond angle	$[Cu]^{-1}$	$[Ni]^{-2}$	$[Ni]^{-1}$	$[Ni]^0$	$[Co]^{-2}$	$[Co]_2^0$	$[Fe]_2^{-2}$
S—M—S (intraring)	92^a	92		90	91	90	90
S—M—S (extraring)	88	—		—	89	86	88
M—S—C (intraring)	102	104		107	104	106	104
S—C=C	122	121		118	121	119	121
S—C—R	116	118		118	117	116	118

^{a} In degrees.

The symmetry of the planar dithiolene anions and molecules is essentially D_{2h}, and in $[Me_4N]_2[NiS_4C_4(CN)_4]$ (86), $[Bu_4N]_2[CoS_4C_4(CN)_4]$ (87), $[Bu_4N]$ $[Co(S_2C_6H_3Me)_2]$ (88), and $[NiS_4C_4Ph_4]^0$ (89) the metal atoms are well separated from each other in the crystal lattices.

Few details are yet known about the structure of the borohydride reduction product of biacetyl bis(2-mercaptoanil) nickel (88), but it is clear that the CH_3—C—CH_3 angles are consistent with a tetrahedral configuration around at least one of the carbon atoms in the bridge, thus lending further support to the proposal that hydride attack of $[Ni(gma)]^0$ occurs somewhere on the axomethine bridge. The structure of a copper complex (19) has recently been described (90) in which the molecule, which contains the basic cis-$[Cu—N_2S_2]$ coordination unit, is essentially planar.

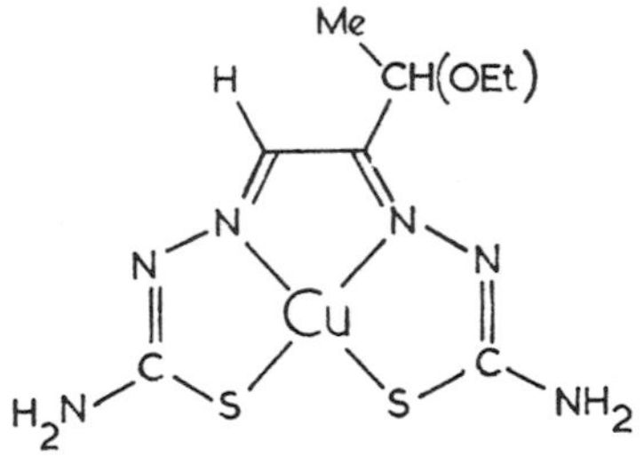

(19)

The crystal structures of $[Bu_4N][CuS_4C_4(CN)_4]$ (91) and its nickel analog, $[Ph_3MeP][NiS_4C_4(CN)_4]$ (92), have certain similarities. In both, the arrangements of anions and cations are virtually independent and the former are stacked in column and associated in pairs, whereas the latter form either columns running through the crystal (Ni) or lattice works about

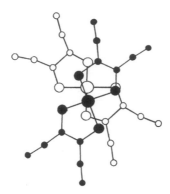

Fig. 2. The stacking of the anions in $[Bu_4N][CuS_4C_4(CN)_4]$.

the anions (Cu). There is a difference, however, in the mode of pairwise association. In the copper crystal, there are two stacks of anions per unit cell and the four copper atoms in each stack are almost exactly above each other in the a direction (the crystal is monoclinic) (Fig. 2); the anions are rotated in such a way as to give the stack mm symmetry in the bc direction. The distances between each copper atom are 4.03 and 4.43 Å. In $[Ph_3MeP]$ $[NiS_4C_4(CN)_4]$, the pairwise association does not occur directly between pairs of nickel atoms, but involves two Ni····S interactions (Fig. 3) between adjacent pairs of anions, and this effectively renders each nickel atom five coordinate with respect to sulfur. It seems likely that antiferromagnetic coupling of these pairs of nickel atoms via this Ni····S−−−Ni interaction is responsible for unusually low magnetic moments (93) of this complex and of its palladium and platinum analogs.

From the structural determinations of $[Bu_4N]_2[Co(S_2C_6Cl_4)_2]_2$ (88), $[CoS_4C_4(CF_3)_4]_2^0$ (26), and $[BuN]_2[FeS_4C_4(CN)_4]^2$ (94), it was established that all three complexes contain two sulfur-bonded five-coordinate

Fig. 3. The stacking of the anions in $[Ph_3MeP][NiS_4C_4(CN)_4]$.

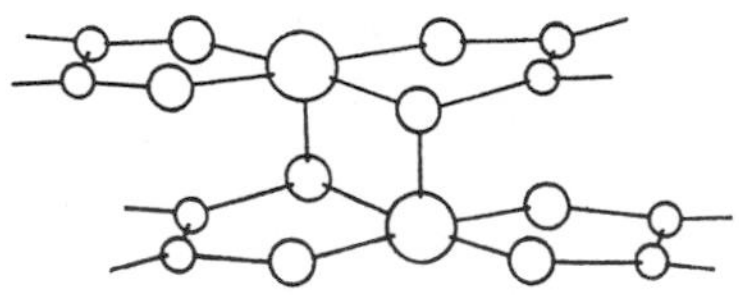

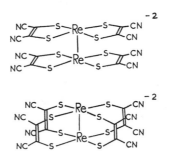

Fig. 4. The idealized structure of the dimeric dithiolenes.

metal atoms which have an essentially square-pyramidal configuration, with each pyramid sharing a common edge (Fig. 4). No detailed reports of the cobalt anionic species are yet available, but it may be seen that in the iron complex the apical Fe—S bond lengths are only 0.2 Å longer than the basal Fe—S distances. In the neutral cobalt compound, with which the analogous iron complex, $[FeS_4C_4(CF_3)_4]_2^0$, is isomorphous and probably isostructural (25,95,96), the difference between apical and basal Co—S distances is 0.2 Å and each cobalt atom is displaced by about 0.37 Å out of the basal plane toward the apical S atoms.

It was observed in Section III-A-9 that $[ReS_4C_4(CN)_4]_2^{-2}$ (84) may be structurally different from the dimeric cobalt and iron dithiolenes. This could arise by virtue of the synthesis of the complex from $Re_2Cl_8^{-2}$ where there is a direct, and short, Re—Re bond. Two possible structures are shown in Figure 5.

Of particular interest is the variation of bond lengths with overall charge, z, on the complexes, $[M—S_4]^z$, and a useful comparison can be made in the nickel series where all charge types, $z = 0, -1$, and -2, have been structurally investigated. The "ethylenic" bond distances decrease and the Ni—S and S—C(sp^2) distances increase as the charge increases, but the total variations are small, the "C=C" range being 1.33–1.37 Å in the nickel series, and 1.31–1.39 Å overall. The Ni—S and S—C(sp^2) dis-

Fig. 5. Two possible structures for $[ReS_4C_4(CN)_4]_2^{-2}$.

tances vary from 2.00 to 2.17 and 1.69 to 1.75 Å, respectively. If it is assumed (92) that the S atoms approach sp^2 hybridization and the nickel radius is that of Ni^{2+} (in square-planar compounds), then the S(sp^2)—C(sp^2) distance is calculated to be 1.77 Å and the Ni—S(sp^2) and Ni—S(sp^3) bond lengths would be 2.24 and 2.20 Å, respectively. A comparison of these calculated values with those actually observed (Table IV) would seem to imply a degree of multiple bonding between the sulfur and nickel atoms, and between the sulfur and carbon (sp^2) atoms. Certainly, the closest approach to the limits imposed by the calculated values occurs when $z = -2$ and so the description of this complex as a Ni(II) derivative of the dithiolate ligands is reasonable (57,79,97). The assumption that the radius of the nickel atom is always that of Ni^{2+} regardless of the charge on the basic [Ni—S$_4$]^z coordination unit is naïve but, for want of a better yardstick, must serve as a fair approximation to the truth. Thus it would seem that as the nickel dithiolene complexes are progressively oxidized, the sulfur ligand gradually loses its "dithiolate" character and becomes more "dithioketonic." This would seem to be the logical conclusion to draw from the contraction of the S—C(sp^2), and expansion of the "C=C" bond lengths as z goes from -2 through -1 to 0. Therefore, the neutral species could be regarded, to a first approximation, as a resonance structure of the two forms **20a** and **20b**.

(20a)   **(20b)**

However, it is perhaps more pictorially useful to employ the "delocalized dotted line" approach, namely, **21**, since this avoids the representation of awkward valence bond forms in the monoanionic systems and does indicate some degree of multiplicity in the metal–sulfur, sulfur–carbon, and carbon–carbon bonds.

There exists some confusion about the early reports of the isomorphism (98) of cobalt, nickel, and copper complexes, [Bu$_4$N][MS$_4$C$_4$(CN)$_4$].

(21)

It is now known from solution measurements (27) that the cobalt complex is dimeric and a careful examination (27) of its unit cell and space group requirements has revealed that it has the same systematic absences as the copper complex, but that the intensities of the x-ray spectral lines are significantly different.

C. Polarographic and Voltammetric Studies

The synthetic chemistry of the bis-1,2-dithiolenes, and their stability towards redox reagents, was considerably systematized by a study (99,100) of their half-wave potentials for the one-electron transfer reactions which most of them undergo. For reversible reactions, these potentials may be related to standard electrode potentials and can be used similarly. The majority of complexes were insoluble in water and so all polarographic and voltammetric studies were carried out in solvents such as acetonitrile, dichloromethane, acetone, DMF, or DMSO.

The results of the various studies are collected and summarized in Tables VI, VII, VIII, IX, X, XI, XII, and XIII. It is regrettable that there has been little or no standardization of these results—standardization, that is, with respect to solvent and reference cell. Therefore, in order to make effective comparisons of the data, the $E_{1/2}$ values have been arbitrarily converted to a standard acetonitrile–SCE scale, and the converted values are shown in the Tables in italics. It must be appreciated that these converted values are not accurate, but at least they do serve as an indication of the value of the expected experimental half-wave potential.

Certain generalizations about the synthetic and chemical behavior of the planar and dimeric 1,2-dithiolenes have been made (100). These are as follows :

1. Reduced species in couples less positive than ca. 0.00 V are susceptible to aerial oxidation in solution, whereas all reduced species in couples more positive than this value are air-stable.

2. Oxidized species in couples more positive than ca. +0.20 V are unstable to reduction by weakly basic solvents such as ketones or alcohols.

3. Oxidized species in couples within the approximate range +0.20– −0.12 V can be reduced by stronger bases such as aromatic amines (e.g., *o*- or *p*-phenylenediamines).

4. Oxidized species in couples more negative than ca. −0.10 V are readily reduced by strong reducing agents such as hydrazine, sodium amalgam, zinc in pyridine, or alkali metals in ethanol.

5. Reduced forms in couples less positive than ca. +0.40 V can be oxidized by iodine. Otherwise, one-electron oxidizing agents such as $[NiS_4C_4(CF_3)_4]^0$ and its analogs must be used.

Several points arise from these generalizations. Firstly, it must be emphasized that the limits quoted are only approximate. Secondly, in synthetic oxidations or reductions, care must be exercised in the choice of oxidizing or reducing agent. In some cases, it is necessary to employ $[NiS_4C_4(CF_3)_4]^0$ as the agent, and contact time between the reactants must be kept as short as possible in order to avoid ligand exchange reactions of the type described in Section III-A-8. Certain reducing agents, such as hydrazine, can function as coordinating ligands (171) and thereby upset the course of the reducing reaction. It is important to realize that the reductions of oxidized species in couples more negative than ca. -0.95 V, or the oxidations of reduced species in couples more positive than ca. $+0.95$ V, are not possible by normal chemical methods, since reagents capable of effecting these electron transfer reactions invariably destroy the complex.

The half-wave potentials obtained from the bis-perfluoromethyl–dithiolenes for the couple,

$$[M\text{—}S_4]_n{}^0 + ne^- \rightleftharpoons [M\text{—}S_4]_n{}^{-n}$$

where M $=$ Ni, Co, or Pt, are considerably more positive than those for the oxidation of the tetracyanoethylene or tetracyanoquinone radical anions (102). Tetracyanoethylene and tetracyanoquinone are regarded as being among the strongest organic π acids and react rapidly with alkali metals with aliphatic and aromatic amines forming these radical anions. Thus, the comparably greater activity of the bis-perfluoromethyl–dithiolenes as one-electron oxidizing agents may be appreciated.

As described previously, $[NiS_4C_4(CF_3)_4]^0$ has been used as a one-electron oxidizing agent for the formation of $[CuS_4C_4(CN)_4]^{-1}$ and $[PdS_4C_4(CN)_4]^{-1}$. The reaction with this compound may be visualized as

$$[MS_4C_4(CN)_4]^{-2} + [NiS_4C_4(CF_3)_4]^0 \longrightarrow [MS_4C_4(CN)_4]^{-1} + [NiS_4C_4(CF_3)_4]^{-1}$$

and, generally, the products of these and similar reactions could be separated by fractional crystallization. The use of this type of reaction presupposed that very little or no ligand exchange took place. However, recent studies (80) have shown that reactions of this type between neutral and dianionic species, or between pairs of monoanionic dithiolenes, did result in the formation of mixed-ligand species:

$$\left.\begin{array}{l}[NiS_4C_4R_4]^{-2} + [NiS_4C_4(CF_3)_4]^0 \\ [NiS_4C_4R_4]^{-1} + [NiS_4C_4(CF_3)_4]^{-1}\end{array}\right\} \longrightarrow 2[NiS_4C_4(CF_3)_2R_2]^{-1}$$

and voltammetry has proved to be a particularly convenient technique in detecting the occurrence and extent of the ligand reorganization. Initially,

TABLE VI

Polarographic and Voltammetric Data Obtained from Bis-1,2-dithiolenes

Couple: $[ML_2]^{-1} + e^- \rightleftharpoons [ML_2]^{-2}$ (volts)

M	Ligands, L_2	DMSO[a]	CH_3CN[a]	CH_2Cl_2[b]	DMF[c]	DMF[d]
Cu	$S_4C_4(CN)_4$		+0.34		−0.201	+0.37
	$S_4C_4(CF_3)_4$		−0.01			
	$(S_2C_6Cl_4)_2$		−0.29[e]		−0.752	−0.25[f]
	$(S_2C_6H_4)_2$		−0.68		−1.14	−0.65
	$(S_2C_6H_3Me)_2$		−0.69		−1.15	−0.66
	$(S_2C_6H_2Me_2)_2$		−0.75		−1.21	−0.72
	$(S_2C_6Me_4)_2$		−0.95		−1.41	−0.92
	$S_4C_4H_4$	−0.74				
Au	$S_4C_4(CN)_4$		−0.419			
	$S_4C_4(CF_3)_4$		—[g]			
	$(S_2C_6Cl_4)_2$		−1.21		−1.67	
Ni	$S_4C_4(CN)_4$	+0.226	+0.23[h]	+0.30	−0.218	+0.259
	$S_4C_4(CN)_2(CF_3)_2$		+0.06[i]			
	$S_4C_4(CF_3)_4$		−0.12			−0.088
	$(S_2C_6Cl_4)_2$		−0.07		−0.532	−0.04
	$S_4C_4(CN)_2Ph_2$		−0.35[j]			
	$S_4C_4(CF_3)_2Ph_2$		−0.55			
	$(S_2C_6H_4)_2$		−0.59		−1.05	−0.56
	$(S_2C_6H_3Me)_2$	−0.52	−0.58	−0.40	−1.07	−0.55
	$(S_2C_6H_2Me_2)_2$		−0.68		−1.14	−0.65
	$(S_2C_6Me_4)_2$		−0.78		−1.24	−0.75
	$S_4C_4(p\text{-}ClC_6H_4)_4$		−0.79			−0.757
	$S_4C_4Ph_4$		−0.82			−0.881
	$S_4C_4Ph_2H_2$		−0.91			−0.879
	$S_4C_4H_4$	−0.84	−0.95	−0.72		−0.921
	$S_4C_4(p\text{-}MeC_6H_4)_4$		−0.99			−0.960
	$S_4C_4Ph_2Me_2$		−1.02			−0.988
	$S_4C_4(p\text{-}MeOC_6H_4)_4$		−0.98			−0.945
	$S_4C_4Me_4$		−1.14			−1.114
	$S_4C_4Et_4$		−1.17			−1.138
	$S_4C_4(n\text{-}Pr)_4$		−1.18			−1.154
	$S_4C_4(i\text{-}Pr)_4$		−1.23			−1.204
Pd	$S_4C_4(CN)_4$	+0.440	+0.46		−0.001	+0.473
	$S_4C_4(CF_3)_4$		+0.081			+0.114
	$S_4C_4Ph_4$		−0.51			−0.636
	$S_4C_4H_4$	−0.645	−0.75			−0.718
	$S_4C_4(p\text{-}MeOC_6H_4)_4$		−0.75			−0.720
	$S_4C_4Me_4$		−0.90			−0.870

(continued)

TABLE VI (*continued*)

M	Ligands, L_2	DMSO[a]	CH$_3$CN[a]	CH$_2$Cl$_2$[b]	DMF[d]	DMF
Pt	$S_4C_4(CN)_4$	+0.21	+0.21		−0.231	*+0.243*
	$S_4C_4(CF_3)_4$		−0.267			*−0.234*
	$(S_2C_6H_3Me)_2$		*−0.59*		−1.05	*−0.56*
	$S_4C_4Ph_4$		*−0.806*			−0.844
	$S_4C_4(p\text{-}MeC_6H_4)_4$		*−0.93*			−0.900
	$S_4C_4(p\text{-}MeOC_6H_4)_4$		*−0.95*			−0.919
	$S_4C_4Me_4$		*−1.10*			−1.069
Co	$S_4C_4(CN)_4$	−0.05	+0.05		−0.384	
	$S_4C_4(CF_3)_4$		−0.24	−0.20		
	$(S_2C_6Cl_4)_2$		*−0.39*		−0.847	
	$(S_2C_6H_4)_2$		*−0.92*		−1.38	
	$(S_2C_6H_3Me)_2$		*−0.95*		−1.41	
	$(S_2C_6H_2Me_2)_2$		*−1.00*		−1.46	
	$(S_2C_6Me_4)_2$		*−1.11*		−1.57	
	$S_4C_4H_4$	−0.93				
Fe	$S_4C_4(CN)_4$		*−0.40*		−0.862	
	$S_4C_4(CF_3)_4$		−0.44	−0.62	*−1.08*	
	$(S_2C_6H_3Me)_2$		*−1.00*		−1.46	

[a] Versus standard calomel electrode with DME or rotating Pt electrodes.
[b] Versus Ag/AgI electrode (40).
[c] Versus Ag/AgClO$_4$ electrode.
[d] Versus Ag/AgCl electrode
[e] All estimated values given in italics.
[f] All estimated values given in italics.
[g] No wave found in range 0 to −1.5 V.
[h] In CH$_2$Cl vs. SCE, $E_{1/2}$ = +0.12 V.
[i] In CH$_2$Cl$_2$ vs. SCE, $E_{1/2}$ = −0.01 V.
[j] In CH$_2$Cl$_2$ vs. SCE, $E_{1/2}$ = −0.38 V.

the characteristic voltammograms of the reactants in the chosen solvent (dichloromethane or acetonitrile) were observed but (gradually) intermediate waves due to the mixed-ligand species appeared which grew at the expense of the initial waves (Table IX). The mixed-ligand complexes were formed to a much greater extent than that expected for random reorganization of the ligands, and it was established that the rate of reorganization was both temperature and solvent dependent (Table X). It was possible, therefore, to establish the conditions under which ligand exchanges proceeded to near completion so that isolation of the complexes could be effected. The half-wave potentials (Table IX) for the reactions

$$[MLL']^0 + e^- \rightleftharpoons [MLL']^{-1} \quad \text{and} \quad [MLL']^{-1} + e^- \rightleftharpoons [MLL']^{-2}$$

TABLE VII

Polarographic and Voltammetric Data Obtained from Bis-1,2-dithiolenes

Couple: $[ML_2]^0 + e^- \rightleftharpoons [ML_2]^{-1}$ (volts)

M	Ligands, L_2	DMSO[a]	CH_3CN[a]	CH_2Cl_2[b]	DMF[c]
Ni	$S_4C_4(CN)_4$		+1.02	+1.38	+1.049
	$S_4C_4(CF_3)_4$	+0.997	+0.92[d]		+1.030
	$S_4C_4(CN)_2Ph_2$		+0.57		
	$S_4C_4(CF_3)_2Ph_2$		+0.43		
	$(S_2C_6H_3Me)_2$	+0.45		+0.52	
	$S_4C_4(p\text{-}ClC_6H_4)_4$		+0.19		+0.218
	$S_4C_4Ph_4$		+0.12[e]		+0.134
	$S_4C_4H_4$	+0.16	+0.09		+0.120
	$S_4C_4H_2Ph_2$		+0.09		+0.115
	$S_4C_4(p\text{-}MeC_6H_4)_4$		+0.05		+0.083
	$S_4C_4Me_2Ph_2$		0.00		+0.025
	$S_4C_4(p\text{-}MeOC_6H_4)_4$		0.00		+0.035
	$S_4C_4Me_4$		-0.14		-0.107
	$S_4C_4Et_4$		-0.15		-0.119
	$S_4C_4(n\text{-}Pr)_4$		-0.15		-0.121
	$S_4C_4(i\text{-}Pr)_4$		-0.18		-0.151
Pd	$S_4C_4(CN)_4$		—[f]		
	$S_4C_4(CF_3)_4$		+0.963		+0.996
	$S_4C_4Ph_4$		+0.15		+0.182
	$S_4C_4H_4$		+0.14		+0.165
	$S_4C_4(p\text{-}MeOC_6H_4)_4$		+0.06		+0.086
	$S_4C_4Me_4$		-0.09		-0.006
Pt	$S_4C_4(CN)_4$		—[f]		
	$S_4C_4(CF_3)_4$		+0.819		+0.853
	$S_4C_4Ph_4$		+0.06		+0.090
	$S_4C_4(p\text{-}MeC_6H_4)_4$		+0.01		+0.043
	$S_4C_4(p\text{-}MeOC_6H_4)_4$		-0.03		-0.004
	$S_4C_4Me_4$		-0.16		-0.133

[a] Versus standard calomel electrode with DME or RPE; estimated values in italics.

[b] Versus Ag/AgI electrode (40).

[c] Versus Ag/AgCl electrode.

[d] In CH_2Cl_2 vs. SCE, $E_{1/2} = +0.80$ V.

[e] In CH_2Cl_2 vs. SCE, $E_{1/2} = 0.01$ V; a further oxidation wave also detected.

[f] Not detected.

TABLE VIII

Arene 1,2-Dithiolene Complexes

Couple: $[ML_2]^{-1} + e^- \rightleftharpoons [ML_2]^{-2}$ (volts)

M	Ligand	CH₃CN[a]	DMSO[a]	Acetone[a]	CH₂Cl₂[b]	DMF[c]
Ni	$C_6H_4S_2$	-0.59				-1.05
	$C_6H_4(NH)S$	-1.04	-1.04			-0.720
	C_6H_4OS	-0.42	-0.42			
	$C_6H_4(NH)_2$	-1.59	-1.59	-1.43		
	$C_6H_4O_2$	-0.38			-0.06	
	$C_6Cl_4S_2$	-0.07				-0.532
	$C_6Cl_4O_2$	$+0.38$			-0.48	
	$C_6Me_4S_2$	-0.78				-1.24
	$C_6H_2Me_2S_2$	-0.68				-1.14
	$C_6H_3MeS_2$	-0.61				-1.07
Cu	$C_6H_4S_2$	-0.68				-1.14
	$C_6H_4O_2$	-0.03			$+0.07$	
	$C_6Cl_4S_2$	-0.29				-0.752
	$C_6Cl_4O_2$	-0.50			$+0.60$	
	$C_6He_4S_2$	-0.95				-1.41
	$C_6H_2Me_2S_2$	-0.75				-1.21
	$C_6H_3MeS_2$	-0.69				-1.15
B	$C_6H_4O_2$				$+1.10$	
Mg	$C_6H_4O_2$				$+1.27^d$	

[a] Versus SCE with n-Pr₄NClO₄ as supporting electrolyte.
[b] Versus Ag/AgI electrode, with n-Bu₄NPF₆ as supporting electrolyte.
[c] Versus Ag/AgClO₄.
[d] Two-electron step.

were nearly midway between those for the symmetrical complexes $[ML_2]^z$ and $[ML_2']^z$.

The existence of discrete dimeric complex ions which could undergo one-electron transfer reactions has been demonstrated beautifully (25) by voltammetry and polarography (Table XI). Both the cobalt and iron complexes, $[MS_4C_4(CF_3)_4]_2^0$, revealed two reversible one-electron reduction waves and one reversible two-electron reduction wave corresponding to the following process:

$$[M—S_4]_2^0 \underset{-e^-}{\overset{+e^-}{\rightleftharpoons}} [M—S_4]_2^{-1} \underset{-e^-}{\overset{+e^-}{\rightleftharpoons}} [M—S_4]_2^{-2} \underset{-2e^-}{\overset{+2e^-}{\rightleftharpoons}} [M—S_4]^{-2}$$

TABLE IX

Voltammetric Data for Mixed-Ligand Dithiolene Complexes

Couple: $[MLL']^z + e^- \rightleftharpoons [MLL']^{z-1}$

Ligands		$z = -2 \rightarrow -1$		$z = -1 \rightarrow 0$	
L	L'	CH_3CN^a	$CH_2Cl_2^b$	CH_3CN^a	$CH_2Cl_2^b$
$S_2C_2Ph_2$	$S_2C_2Ph_2$	-0.82^c	-0.83	$+0.12$	-0.01
$S_2C_2Ph_2$	$S_2C_2(CF_3)_2$	-0.55	-0.55	$+0.43$	$+0.35$
$S_2C_2Ph_2$	$S_2C_2(CN)_2$	-0.35	-0.38	$+0.57$	$+0.43$
$S_2C_2(CF_3)_2$	$S_2C_2(CF_3)_2$	-0.12	-0.22	$+0.92$	$+0.80$
$S_2C_2(CF_3)_2$	$S_2C_2(CN)_2$	$+0.06^d$	-0.01		
$S_2C_2(CN)_2$	$S_2C_2(CN)_2$	$+0.23$	$+0.12$		

^{a} Measured with a RPE using SCE as reference cell.
^{b} Measured with DME using SCE as reference cell.
^{c} In volts.
^{d} Measured with DME.

The dimeric dianions both exhibited two oxidation waves of equal diffusion current and a reduction wave of twice this diffusion current which was consistent with, and supported, the interpretation of the behavior of the neutral species. Similar electrode processes were observed in the voltammograms of $[FeS_4C_4(CN)_4]_2^{-2}$, $[Fe(S_2C_6H_3Me)_2]_2^{-2}$, and

TABLE X

Ligand Exchange Reactions in Solution

Mixed species	Solvent	Equilibrium time^a	% Mixed species^b
$[Ni(S_2C_2Ph_2)(S_2C_2(CF_3)_2)]^{-1}$	CH_3CN	13 days at 25°	92
	CH_2Cl_2	6 days at 40°	77
$[Ni(S_2C_2Ph_2)(S_2C_2(CN)_2)]^{-1}$	CH_3CN	4 hr at 82°	88
	CH_2Cl_2	6 days at 40°	81
$[Ni(S_2C_2(CF_3)_2)(S_2C_2(CN)_2)]^{-1}$	CH_2Cl_2	37 days at 40°	74

^{a} Time required for voltammograms of a solution containing equimolar amounts (ca. $10^{-3}M$) of parent compounds to show no further change.
^{b} Equilibrium values determined from voltammograms recorded at 25°; estimated error $+ 3\%$

TABLE XI

Voltammetric Data Obtained from Dimeric Dithiolene Complexes[a]

Metal	Ligand	$[ML_2]^{-2} \rightleftharpoons [ML_2]_2^{-2}$	$[ML_2]_2^{-2} \rightleftharpoons [ML_2]_2^{-1}$	$[ML_2]_2^{-1} \rightleftharpoons [ML_2]_2^{0}$
Co[b]	$S_2C_2(CN)_2$	$(+0.07)^c$	$+1.08$	
	$S_2C_2(CF_3)_2$	-0.20	$+0.50$	$+1.24$
Fe[b]	$S_2C_2(CN)_2$	$(-0.40)^c$	$+1.16$	
	$S_2C_2(CF_3)_2$	-0.62	$+0.62$	$+1.23$
	$S_2C_6H_3Me$	$(-1.00)^c$	$+0.17$	$+0.48$

[a] In CH_2Cl_2 vs. Ag/AgI (40) using RPE; in volts.
[b] Other complexes known to be dimeric in solution include $[FeS_4C_4Ph_4]_2^z$, $[CoS_4C_4Ph_4]_2^z$, and $[Fe(S_2C_6Cl_4)_2]_2^z$ ($z = 0$ or -2), although no voltammetric data to support this has been published.
[c] Data obtained in CH_3CN vs. SCE.

$[CoS_4C_4(CN)_4]_2^{-2}$. In the dicyano-dithiolene complexes, one oxidation wave corresponding to the step,

$$[M—S_4]_2^{-2} \rightleftharpoons [M—S_4]_2^{-1} + e^-$$

was observed but no further oxidation waves were detected; the iron toluene-3,4-dithiolene complex exhibited two oxidation waves corresponding to the formation of the dimeric monoanion and neutral species.

The polarographic study of $[Fe_2S_2(S_2C_2Ph_2)_2]^0$ (96) indicated the existence of four charged species, namely $[Fe_2S_2(S_2C_2Ph_2)_2]^z$, $z = -1$, -2, -3, and -4, and the dianionic complex was obtained by hydrazine reduction of the neutral compound.

The preceding discussion has dealt exclusively with complexes having the $[M—S_4]^z$ coordination unit. As a logical extension of the electro-chemical studies of planar complexes, systems containing the coordination units $[M—O_4]$ (40), $[M—N_4]$ (33), $[M—N_2S_2]$ (34,44), $[M—N_2O_2]$ (44), and $[M—O_2S_2]$ (28) were investigated (Table XII). It was established that in complexes containing these donor atom sets, electron transfer reactions occurred fairly generally and the oxidative stabilities of complexes of the same general composition and charge type had a marked dependence on the nature of the donor atoms (this is referred to in more detail later). The investigations revealed that in a metal–ligand system, $[M(XYC_6H_4)_2]^z$, a five-membered series could be generated where $z = 0$, ± 1, or ± 2, i.e.,

$$[M—X_2Y_2]^{-2} \longleftrightarrow [M—X_2Y_2]^{-1} \longleftrightarrow [M—X_2Y_2]^0 \longleftrightarrow [M—X_2Y_2]^{+1}$$
$$\longleftrightarrow [M—X_2Y_2]^{+2}$$

TABLE XII

Polarographic and Voltammetric Data from Dithiolene Analogs and Related Complexes

$$[ML_2]^z + e^- \rightleftharpoons [ML_2]^{z-1}$$

M	Ligands	Solvent	Cell[a]	Couple, in volts				
				$-2 \rightarrow -1$	$-1 \rightarrow 0$	$0 \rightarrow +1$	$+1 \rightarrow +2$	Othe
Zn	cis-N_2S_2[b]	DMSO	a	-1.44	-1.08			
		CH_3CN	a		-1.10			
		Ch_2Cl_2	b		-1.31			
	gma	DMSO	a	-1.15	-0.75			
Cd	cis-N_2S_2	DMSO	a	-1.55	-1.13			
	gma	DMSO	a	-1.33	-0.79			
Cu	cis-N_2S_2	DMSO	a		-0.15			
		CH_3CN	a		-0.23			
		CH_2Cl_2	b		-0.09	$+1.15$		
	cis-N_2O_2[c]	CH_2Cl_2	b					ca. -0.4
	$(O_2C_6H_4)_2$	CH_2Cl_2	b	$+0.07$				
	$(O_2C_6Cl_4)_2$	CH_2Cl_2	b	$+0.60$				
Ni	cis-N_2S_2	DMSO	a	-1.24	-0.53			
		CH_3CN	a	-1.00	-0.55			
		CH_2Cl_2	b	-1.26	-0.48			
	cis-N_2O_2	DMSO	a	-1.67	-0.92			
		CH_3CN	a		-1.00			
		CH_2Cl_2	b		-0.86			
	gma	DMSO	a	-1.05	-0.30			
		DMF	c	-1.605	-0.823			
	H_2gma	DMF	c	-1.591	-0.758			
	$[S(NH)C_6H_4]_2$	DMSO	a	-1.04	-0.19			
		DMF	c	-1.573	-0.72			
		CH_2Cl_2	b	-0.93	-0.03	$+1.05$		
	$[SC(Ph)NNH]_2$	DMSO	a	-1.13	-0.14			
	$(OSC_6H_4)_2$	DMSO	a	-0.42	$+0.38$			
	$[(NH)_2C_6H_4]_2$	DMSO	a	-1.59	-0.88			$+0.2$
		DMF	c	-2.075	-1.404			
		acetone	a	-1.43	-0.89	$+0.14$	$+0.73$	
	$\{4\text{-}i\text{-}Pr[(NH)_2\text{-}C_6H_4]_2\}_2$[f]	acetone	a	-1.51	-0.98	$+0.09$	$+0.81$	
		CH_2Cl_2	b	-1.4	-0.77	$+0.44$	$+1.32$	
	$[(MeCNPh)_2]_2$	CH_3CN	a	-1.80	$-1.6-$		-0.61[g]	
	$(O_2C_6H_4)_2$	CH_3CN	a	-0.38				
		CH_2Cl_2	b	-0.06				
		DMSO	a		$+0.17$[g]	$+0.78$		
	$(O_2C_6Cl_4)_2$	CH_2Cl_2	b	$+0.48$				$+0.1$

(continu

TABLE XII (*continued*)

| M | Ligands | Solvent | Cell[a] | Couple, in volts | | | | |
				$-2 \rightarrow -1$	$-1 \rightarrow 0$	$0 \rightarrow +1$	$+1 \rightarrow +2$	Other
i	$(O_2phenan)_2{}^i$	CH_2Cl_2	b	-0.38				-0.65^h
d	$[(NH)_2C_6H_4]_2$	DMSO	a	-1.44	-0.89	$+0.10$	$+0.78$	
t	$[(NH)_2C_6H_4]_2$	DMSO	a	-1.70	-1.01	$+2.03$	$+0.78$	
o	$[(NH)_2C_6H_4]_2$	DMSO	a	-1.83	-0.80	-0.17		$+0.61^j$
	$(O_2C_6H_4)_2$	CH_2Cl_2	b	$+1.10$				
Ig	$(O_2C_6H_4)_2$	CH_2Cl_2	b			$+1.27^g$		

Voltammetric and polarographic oxidation waves for the $[M-O_4]^{-2}$ series in acetonitrile (Table VIII and XII) were generally indicative of irreversible electrode processes but a slight improvement in wave profile could be made by using dichloromethane and a special reference cell (40); even so, the oxidation waves were not completely reversible. It was not established unequivocally that only one electron was involved in these oxidations, but by comparison with $[NiS_4C_4(CN)_4]^{-2}$ and $[Ni(S_2C_6H_3-Me)_2]^{-2}$, the diffusion currents of which for the known one-electron oxidation were very similar to those in the $[M-O_4]^{-2}$ species, this seemed very likely. Despite the observation of electrochemical oxidation behavior, attempts to prepare authentic $[Ni(O_2C_6Cl_4)_2]^{-1}$ from the corresponding dianion were unsuccessful; the compound formally described as $[Pr_4N]$-$[Ni(O_2C_6Cl_4)_2]$ exhibited no redox behavior in dichloromethane. The neutral species, $[M-O_4]^0$ showed no well-defined redox behavior in dichloromethane and the only information concerning $[M-O_4]^{+1}$ species was obtained from the voltammograms of $[Ni(O_2C_6Cl_4)_2]^{-2}$ in DMSO where waves were observed at $+0.17$ V (a two-electron step) and at $+0.78$ V (a one-electron step). The former appeared to correspond to the oxidation of the dianion to the neutral compound and the latter to the step:

$$[Ni-O_4]^0 + e^- \rightleftharpoons [Ni-O_4]^{+1}$$

The latter wave was not observed in any other solvent and a corresponding wave was not detected in the voltammograms of other $[Ni-O_4]^z$ species.

In the $[M-N_4]^z$ series, the existence of the complete five-membered system, $z = -2$ to $+2$, was confirmed (Table XII) (33). For each of the o-phenylenediimine complexes of the nickel group, $\{M[(NH)_2C_6H_4]_2\}^0$, oxidation or reduction waves were observed which linked the $[M-N_4]^{-2}$ and $[M-N_4]^{+2}$ species by one-electron steps, and no other waves were detected in the voltammograms and polarograms of these complexes; a similar observation was made in the related $\{Ni[(MeCNPh)_2]_2\}^z$ series. Fortunately, the synthesis of the monoiodides, $[M-N_4]I$, was possible, and the voltammograms of these were essentially similar to those of the neutral compounds. However, the half-wave reduction potentials obtained from the nickel monocations were very solvent dependent and this seemed to be related to the instability of the $[Ni-N_4]^{+1}$ species in solution. The oxidation waves of these monocations were very close to, or coincident with, the oxidation of iodide to iodine.

The voltammetric examination of $\{Co[(NH)_2C_6H_4]_2\}^0$ and the corresponding monoiodide provided clear evidence for an incomplete series where $z = 0, \pm 1$ and -2. Unambiguous evidence for $[Co-N_4]^{+2}$ was lacking since the number of electrons transferred in the oxidation wave at $+0.61$ V was not directly ascertained.

The voltammograms of $\{Ni[(MeCNPh)_2]_2\}^{+2}$, either as dinitrate or diiodide, consisted of a two-electron reduction wave (in acetonitrile), corresponding to the step $z = +2 \rightarrow 0$, followed by two one-electron reduction waves, corresponding to the steps $z = 0 \rightarrow -1$ and $z = -1 \rightarrow -2$.

Because of the very negative half-wave potentials for the processes $z = 0 \rightarrow -1$ and $z = -1 \rightarrow -2$ in the $[M-N_4]^z$ series, isolation of the mono- or dianions was not attempted. However, the authenticity of the monoanions was assured by the observation of ESR signals in solutions containing the paramagnetic nickel group complexes generated by controlled potential electrolysis of $[M-N_4]^0$; these were very similar to those obtained from the analogous $[M-S_4]^{-1}$ species. Iodine proved to be the most useful reagent for oxidation of the neutral complexes to the monocations, and all evidence strongly supported the similarity between the electrochemically generated and chemically synthesized complexes. Halogen oxidation of $[Ni—N_4]^0$ complexes afforded partially character- ized dicationic species. Voltammetric reduction waves corresponding to the process $z = +2 \rightarrow +1$ were probably obscured by waves arising from the reduction of halogen (in a tri- or polyhalide counteranion). The ex- istence of dications of Pd and Pt was clearly indicated electrochemically, but the potentials for their formation from $[M—N_4]^0$ were too positive to permit their formation and isolation by chemical means.

In the $[M—N_2S_2]^z$ system (44), complexes containing the ligand o-aminothiophenol, glyoxal bis(2-mercaptoanil), thiobenzoylhydrazine, and biacetylbis(thiobenzoylhydrazone) have been investigated (Table XII). ${Ni[S(NH)C_6H_4]_2}^0$ displayed two one-electron reduction waves and one very anodic one-electron oxidation wave, corresponding to the complexes having $z = \pm 1$ and -2; the most oxidized member of the five-membered series ($z = +2$) was not detected. $[Ni(gma)]^0$ could be voltammetrically reduced in two one-electron steps, corresponding to the formation of mono- and dianions. The zinc and cadmium analogs behaved similarly, but the potentials were much more cathodic than those of $[Ni(gma)]^0$. Because of the extremely negative potentials, no attempt was made to isolate $[M(gma)]^{-1}$, M = Zn or Cd, but synthesis of $[Ni(gma)]^{-1}$ has already been discussed in Section III-A-5. The nickel complex of thio- benzoylhydrazine bears a formal resemblance to ${Ni[S(NH)C_6H_4]_2}^0$ in the same way that the nickel complex of biacetylbis(thiobenzoylhydrazone) is related to $[Ni(gma)]^0$. Accordingly, the reduction potentials of $[Ni(SC(Ph)- N \cdot NH)_2]^0$ are very similar to ${Ni[S(NH)C_6H_4]_2}^0$. However, the reduction potentials of the biacetylbis(thiobenzoylhydrazone) complex are con- siderably more negative than those of $[Ni(gma)]^0$. Zinc, cadmium, and copper complexes of biacetylbis(thiobenzoylhydrazone) were also investi- gated electrochemically and evidence for mono- and dianionic zinc and cadmium species was obtained; the copper complex was reduced and oxidized in one-electron steps, but despite the ease with which the reduc- tion could take place, no monoanion was isolated. For comparison pur- poses, nickel, copper, and zinc complexes of biacetylbenzoylhydrazone, that is, complexes having the basic coordination unit $[M—N_2O_2]$, were studied voltammetrically. It was discovered that only the nickel complex underwent electron transfer reactions, being reduced in two one-electron steps. The potentials at which these reductions occurred were very much more cathodic than those of the sulfur analog.

Little data has yet been presented for electron transfer behavior in $[M—O_2S_2]^z$ systems. However, there have been brief reports (28) of redox behavior in monothiocatecholate complexes of copper, nickel, palladium, and cobalt, and, apparently, $[Ni(OSC_6H_4)_2]^{-1}$ was formed by aerial oxidation of the corresponding dianion.

Two significant points arise from a perusal of the foregoing discussion. The first of these, the use of half-wave potentials in the design of synthetic procedures, has already been discussed and the facility with which this kind of data can be obtained and applied cannot be overemphasized. The second of these is that the half-wave potentials are highly dependent on the nature of the ligand, that is, the nature of R in $S_2C_2R_2$ and $S_2C_6H_{4-n}R$, and the donor atom set, XY in C_6H_4XY, and the nature of the central metal atom in any given ligand system.

Vlček has observed (103) that the effects exerted by given substituents in ligands (complexed to various metal ions) on the oxidation or reduction potentials of these complexes were parallel but not proportional and proposed that the couples of the metal complexes should behave with respect to substitution on the ligand in the same way as the free ligand couples would behave. This is an eminently reasonable postulate which perhaps can be illustrated by drawing an analogy with the redox behavior of aromatic hydrocarbons. It has been demonstrated (104) that the values of the half-wave oxidation (or reduction) potentials for a series of aromatic hydrocarbons differing only in ring substituents depend on the nature of the substituents and, more precisely, on the relative energy of the orbital (or orbitals) involved in the electron transfer process. These remarks must apply to free ligands of the type under discussion and, indeed, the redox potentials of these do vary with the substituent or "donor atom" set (40,60) (Table XIII). On complexing with the metal ions, the relative energies of the ligand orbitals will be modified by mixing with metal orbitals of the appropriate symmetry and so the orbital (or orbitals) involved in the electron transfer reactions must, almost by definition, be comprised of both ligand and metal character. Therefore, a substituent, donor atom set, or metal ion effect on $E_{1/2}$ values need not be surprising, and an attenuation of these half-wave potentials of the ligands, after complexing, is to be expected.

In the simple 1,2-dithiolenes, the decreasing oxidative stability of the dianions, $[M—S_4]^{-2}$, that is, the order of increasingly negative potentials for the reaction

$$[M—S_4]^{-1} + e^- \rightleftharpoons [M—S_4]^{-2}$$

was in the order

$$R = CN > CF_3 > Ph > H > Me > Et$$

In other words, the dianions were the most unstable to oxidation when R was an electron-releasing group, and a very comprehensive study

TABLE XIII

Polarographic and Voltammetric Data Obtained from the Free Ligands

Couple: $[L]^z + e^- \rightleftharpoons [L]^{z-1}$ (volts)

Ligand	CH$_3$CN[a]		CH$_2$Cl$_2$[b]		DMF[c]	
	$-2 \rightarrow -1$	$-1 \rightarrow 0$	$-2 \rightarrow -1$	$-1 \rightarrow 0$	$-2 \rightarrow -1$	$-1 \rightarrow 0$
9,10-Phenanthroquinone	-1.21	-0.65	-1.3	-0.51		
1,2-Benzoquinone	-0.90	-0.31	(-0.15)	(-0.74)		
o-chloranil	-0.61	$+0.14$	-0.70	$+0.30$		
[S$_2$C$_2$(CN)$_2$]$^{-2}$					-0.31[d]	0.00[d]
[S$_2$C$_2$H$_2$]$^{-2}$					-0.81[d]	-0.21[d]
[MeCNPh]$_2$	-1.82[e]					

[a] Versus SCE using DME.
[b] Versus Ag/AgI using RPE (40).
[c] Versus Ag/AgCl using RPE.
[d] Highly irreversible waves.
[e] Two-electron reduction wave.

of the voltammetric and polarographic behavior (60) of nickel group complexes revealed a linear relationship between $E_{1/2}$ and an inductive substituent constant (Taft's σ^* constant). Unfortunately, many of the free ligands could not be investigated polarographically but those which were (60) (Table XIII) showed the expected parallel substituent effects on $E_{1/2}$. Similar behavior was observed in the substituted arene-1,2-dithiolene complexes (65) where the monoanionic species were stabilized with respect to the dianionic complexes in proportion to the number of electron-releasing groups around the benzene ring. Thus, the order of decreasing potentials for the reaction

$$[M-S_4]^{-1} + e^- \rightleftharpoons [M-S_4]^{-2}$$

was

$$C_6Cl_4S_2 > C_6H_4S_2 > C_6H_3MeS_2 > C_6H_2Me_2S_2 > C_6Me_4S_2$$

The half-wave potentials of the tetrachloro- and tetramethyl-substituted complexes differed by about 0.7 V and the introduction of only one methyl group onto the ring was sufficient to stabilize the monoanion by about 0.05 V. The potentials of the couples of the tetrachlorobenzene–dithiolene complexes were approximately similar to those of the analogous bis-perfluoromethyl–dithiolenes and the overall order of stability of reduced complexes versus their oxidized species would seem to be

$$CN > CF_3 > C_6Cl_4S_2 > C_6H_4S_2 > C_6H_2Me_2S_2 > C_6Me_4S_2 > p\text{-}C_6H_4Cl$$
$$> Ph > H > p\text{-}C_6H_4Me > p\text{-}C_6H_4OMe > Me > Et > n\text{-}Pr$$

In the $[M—O_4]^z$ series of complexes (40), the half-wave potentials of the apparent one-electron oxidations of the copper and nickel complexes could be compared directly with the free ligand couples (Table XIII). Holm and co-workers were therefore able to arrange the species in order of increasing ease of oxidation, and the potentials shown below are differences between those potentials for one-electron oxidations of the indicated species:

−0.53

$[Cu(O_2C_6H_4)_2]^{-2}$ $[Cu(O_2C_6Cl_4)_2]^{-2}$

−0.32 −0.54

$[Ni(phenan)_2]^{-2}$ $[Ni(O_2C_6H_4)_2]^{-2}$ $[Ni(O_2C_6Cl_4)_2]^{-2}$

−0.6 −1.4

$[phenan]^{-2}$ $[C_6H_4O_2]^{-2}$ $[C_6Cl_4O_2]^{-2}$

−0.36 −0.45

$[phenan]^{-1}$ $[C_6H_4O_2]^{-1}$ $[C_6Cl_4O_2]^{-1}$

The most appropriate comparisons are between the oxidation potentials for the ligand and complex dianions, and it is seen that the order of oxidative stability is the same for both the ligands and the complexes although the latter potentials are diminished with respect to the former.

In the particular case of the copper complexes $[Cu—O_4]^z$, the potentials for the catecholate and tetrachlorocatecholate systems were markedly different. ESR results indicated that the metal ion could be correctly described as Cu(II) (in $[Cu—O_4]^{-2}$) in a $^2B_{1g}$ ground state, and the covalence parameter, α^2, relating to the molecular orbital of the odd electron, $[\phi(\sigma^*, b_{1g})]$, in D_{2h} symmetry, was estimated and found to be almost the same for each complex. Accepting the σ^* electron as the one involved in the oxidation process, it followed that only substituent changes on the ligands were responsible for the changes in $E_{1/2}$. These, and the preceding results, are in close accord with Vlček's proposals.

It is clear that the nature of X and Y in the complex system $[M(XYC_6H_4)_2]^z$ has a profound effect on the half-wave potentials of a given couple for the same metal. Because of the close geometrical and electronic similarity of the complexes for a given charge type, meaningful comparisons of these potentials could be made. The orders of oxidative stability were

$$[Ni—N_4] > [Ni—N_2S_2] > [Ni—S_4] > [Ni—O_2S_2] > [Ni—O_2N_2]$$

The $[Ni—O_4]$ systems have been excluded because of the uncertain nature of the monoanionic species. This result can be qualitatively understood in the light of Vlček's proposals assuming that they can be extended to include variations in the donor atom sets as well as in the ring substituents. Unfortunately, it is not possible to compare these data with the corresponding ligand couples since the quinonediimines are too unstable to permit measurements and the dithio-*o*-quinones and the thia-*o*-benzoquinoneimines are unknown. The required potentials could be obtained from oxidation of the dianions, $[C_6H_4XY]^{-2}$, but of these, only the catechol and tetrachlorocatechol derivatives had been characterized satisfactorily.

Finally, it has been proposed (33) on the basis of the results and previous discussion, that free ligands which are most stable in the anionic form tend to stabilize the dianionic and monoanionic (and possibly neutral) members of the five-membered electron transfer series (see Sec. I), whereas free ligands which are most stable in the neutral (or oxidized) form tend to stabilize the neutral, monocationic, and dicationic members of the series.

D. Infrared and NMR Spectral Studies

Three characteristic absorptions, ν_1, ν_2, and ν_3 (Table XIV), excluding those due to ligand substituents R in $S_2C_2R_2$ and the cations in charged species, were observed in the infrared spectra of bis-dimethyl and bis-diphenyl–dithiolene complexes of the nickel group (79), cobalt, and iron (96) between 3000 and 600 cm^{-1}. ν_1 and ν_2 were described as the perturbed "C=C" and "C=S" stretching frequencies, respectively, while ν_3, which was strongly substituent dependent, was assigned to the stretching vibrations of the $R—C\overset{\diagup S}{\underset{\diagdown C}{}}$ group. Metal–sulfur "stretching" frequencies ν_4 and ν_5) were also reported, and occurred in a region comparable with those in $[Pt(S_2C_2O_2)_2]^{-2}$ (105), namely 490–300 cm^{-1}. The "C=C" stretching frequencies of a number of planar and dimeric bis-perfluoromethyl–dithiolenes have also been reported (99).

The value of ν_1 was found to be dependent, to a small degree, on the electron-attracting ability of the ligand substituent R, decreasing in the order $CF_3 >$ Ph $>$ Me, and, furthermore, ν_1 also decreased in the order Ni $>$ Pt $>$ Pd. One of the most striking features about ν_1 was its variation with the overall charge, z, on the complexes. Thus, when the overall charge decreased (i.e., when the complexes were oxidized), ν_1 also decreased, and, incidentally, ν_2 decreased but less markedly. Two interpretations of these effects have been offered, one based on the premise (99) that the

TABLE XIV

Infrared Stretching Frequencies in Planar and Dimeric Dithiolene Complexes

Complex	Cation	Infrared frequencies, cm^{-1}				
		ν_1	ν_2	ν_3	ν_4	ν_5
$[NiS_4C_4Me_4]^0$ [a]		1333	914	558	435	333
$[PtS_4C_4Me_4]^0$ [a]		1324	908	563	405	310
$[NiS_4C_4Ph_4]^0$ [a]		1359	1136	882	408	354
$[NiS_4C_4Ph_4]^{-1}$ [a]		1428	1168	869		
$[NiS_4C_4Ph_4]^{-2}$ [a]	$N_2H_5^+$	1481				
$[PdS_4C_4Ph_4]^0$ [a]		1342	1136	884	401	352
$[PtS_4C_4Ph_4]^0$ [a]		1351	1139	877	403	373
$[Mi_4(S_2C_2Ph_2)_4]$ [a,b]		1474				345
$[CoS_4C_4Ph_4]_2^0$ [a]		1370	1170,1152	902,871		
$[CoS_4C_4Ph_4]_2^{-2}$ [a]	Ph_4As^+		1145	860		
$[FeS_4C_4Ph_4]_2^0$ [a]		1378	1176,1151	910,982		
$[FeS_4C_4Ph_4]_2^{-2}$ [a]	Et_4N^+		1175	910		
$[Fe_2S_2(S_2C_2Ph_2)_2]^0$ [a]		1400	1163	887		
$[NiS_4C_4(CF_3)_4]^0$ [c]		1422				
$[NiS_4C_4(CF_3)_4]^{-1}$ [c]	Et_4N^+	1502				
$[NiS_4C_4(CF_3)_4]^{-2}$ [c]	Et_4N^+	1534				
$[PdS_4C_4(CF_3)_4]^{-1}$ [c]	Ph_4As^+	1513				
$[PtS_4C_4(CF_3)_4]^0$ [c]		1422				
$[PtS_4C_4(CF_3)_4]^{-1}$ [c]	Ph_4As^+	1493				
$[PtS_4C_4(CF_3)_4]^{-2}$ [c]	Ph_4As^+	1515				
$[CoS_4C_4(CF_3)_4]_2^0$ [c]		1466				
$[CoS_4C_4(CF_3)_4]_2^{-2}$ [c]	Et_4N^+	1502				
$[FeS_4C_4(CF_3)_4]_2^0$ [c]		1515				
$[FeS_4C_4(CF_3)_4]_2^{-2}$ [c]	Et_4N^+	1538				

[a] KBr disk.
[b] CsBr disk.
[c] Fluorolube mulls.

metal ion was being oxidized when z decreased from -2 to O, and the other (57) on the belief that oxidation caused removal of electrons from molecular orbitals which were predominantly ligand in character. In the absence of other data, it would be hard to decide which of these views was correct, if, indeed, either one was entirely so. If the oxidation state of the metal was increasing on oxidation, then donation *from* the ligand *to* the metal would be expected, and ν_1 should decrease and ν_2 increase; the ease with which this could occur would have to depend to some extent on the electron-releasing or attracting ability of R, as has been observed. However, if the view of Schrauzer and Mayweg is accepted, namely that the electrons are being withdrawn from predominantly ligand molecular orbitals and that the ligands themselves are being progressively oxidized from dithiolates to dithioketones, the most oxidized species $[MS_4C_4R_4]^0$ (20) (p. 81) being, in effect, the half-way house (i.e., a resonance hybrid between dithiolate and dithioketone), then the data is also compatible. Certainly, such x-ray data that is available demonstrates quite clearly that on oxidation, there is a shortening of the C—S bond and a lengthening of the "ethylenic" bond with which the infrared data is entirely consistent.

The ^{19}F NMR spectra (99) of several planar bis-perfluoromethyl–dithiolene complexes have been recorded (Table XV). Of particular interest were the results obtained from the iron complexes which exhibited solvent-dependent magnetic moments. In acetone, it was observed that the complex had a spin-doublet ground state whereas in DMF it had a spin-quartet ground state and it is now known that this behavior is entirely consistent with the dissociation of the dimer $[FeS_4C_4(CF_3)_4]_2^{-2}$ into a monomer in the more strongly polar solvent. The conditions for observation of hyperfine contact-interaction shifts in $[FeS_4C_4(CF_3)_4]_2^{-2}$ were favorable, since the failure to observe ESR signals from acetone solutions of this complex implied that its spin-lattice relaxation time was short compared with that in $[NiS_4C_4(CF_3)_4]^{-1}$ where an ESR signal had been detected at room temperature. The total contact shift for $[FeS_4C_4(CF_3)_4]_2^{-2}$ was estimated to be

TABLE XV

^{19}F NMR Data Obtained from Planar and Dimeric
Bis-perfluoromethyl–Dithiolene Complexes

Complex	Solvent[a]	—[b], cps
$(CF_3)_2C_2S_2$	CH_2Cl_2[c]	-24
$[NiS_4C_4(CF_3)_4]^0$	CH_2Cl_2	-238
$[NiS_4C_4(CF_3)_4]^{-2}$	CH_2Cl_2	-337
$[PtS_4C_4(CF_3)_4]^0$	DMF[d]	-346
$[CoS_4C_4(CF_3)_4]_2^{-2}$	Acetone	-469
$[FeS_4C_4(CF_3)_4]_2^{-2}$	Acetone	-759 ± 50[e]
$[FeS_4C_4(CF_3)_4]^{-1}$	DMF	-1100 ± 50[e]

[a] Data measured at 22°.

[b] Measured relative to $PhCF_3$ as internal reference;
± 1 cps unless otherwise stated.

[c] 2% hydrazine added to prevent oxidation.

[d] In CH_2Cl_2, -247 cps.

[e] Line width at half-height 180 cps.

-460 cps and for the monomer ($S = \frac{3}{2}$), -800 cps. Although the observation of considerable anisotropy in the g tensor in related nickel complexes with $S = \frac{1}{2}$ suggested that pseudo-contact interactions might be important in the dimer, the isotropic pseudo-contact shift was calculated to be ca. 30 cps, an order of magnitude smaller than the observed shift which was believed to be due mainly to a Fermi contact interaction. The order of magnitude of the spin density on the "ethylene" ring carbon atom, ρ_C, which was calculated to be ca. $+0.003$, was considered to be too low to be consistent with the presence of an unpaired spin in a highly delocalized π-molecular orbital, but could be explained in terms of spin polarization of the ligand π orbitals by the unpaired spin. However, since these calculations were carried out assuming that the low-spin form of this complex was planar, whereas in fact it is a dimer, the conclusions reached must be regarded with caution.

A single proton NMR resonance, at $\tau = 3.2$, occurred in $Na_2S_2C_2H_2$ (57) and this was shifted to $\tau = 0.8$ (CS_2) on complexing in $[NiS_4C_4H_4]^0$. These data were regarded as evidence for the more "aromatic" nature of the protons in the metal complex when compared with the free dithiolate ligand.

E. ESR Spectral and Magnetic Properties of Planar and Dimeric 1,2-Dithiolenes

In this section, the ESR spectral and magnetic properties of the dithiolene complexes are discussed and classified under metals or metal groups. No paramagnetic zinc-group complexes containing the unit $[M-S_4]^{-1}$ have been discovered. However, both zinc and cadmium glyoxal bis(2-mercaptoanil) complexes (44) and their biacetylbis(thiobenzoylhydrazone) analogs (44), were reduced chemically or by controlled potential electrolysis to form paramagnetic monoanions, $[M-N_2S_2]^{-1}$. These anions displayed ESR signals in solution and their g tensors, which were isotropic in frozen solutions, were very similar to each other (Table XVI) and close to the free electron value ($g = 2.0023$). These results were much like those found in aromatic free radicals and were consistent with other work on closed-shell cation-stabilized radical anion systems such as, for example, Eaton's studies (36) of semiquinone complexes of zinc and cadmium.

The single isotropic resonance due to the reduced blue zinc complex, $[Zn(gma)]^{-1}$, was accompanied by satellites attributed to zero-field splitting in a molecular triplet state, and a half-filled $\Delta M = 2$ transition was also observed. These results were thought to be consistent with a polymeric structure for these complexes, in which the metal atoms were six coordinate.

Eaton has shown (36) that aerial oxidation of solutions containing Mg^{2+}, Zn^{2+}, or Cd^{2+} ions and catechol produced a species formulated as $[M(O_2C_6H_4)(H_2O)_n]^{+1}$ in which the semiquinone was coordinated to the metal. In the investigations of the $[M{-}O_4]^{+1}$ systems, Holm and his colleagues (40) attempted to oxidize the dianions, $[M{-}O_4]^{-2}$, with one-electron oxidizing agents, although clear voltammetric evidence for the existence of the monoanions was not available. These attempted oxidations were accompanied by color changes and the observation of strong ESR signals due to the reduced oxidant. However, an additional, but much weaker, signal at about $g = 2.002$ was also evident but no proton hyperfine splittings similar to those reported by Eaton were observed. The oxidation of $[Cd(O_2C_6Cl_4)_2]^{-2}$ by $[NiS_4C_4(CF_3)_4]^0$ produced a radical with $g = 2.002$ and an apparent satellite splitting of 7.6 G possibly due to [111,113]Cd $(I = 1/2, 25.2\%$ abundance), but it was not established unequivocally that the resonances were due to the anticipated monoanion. Aerial oxidation of the zinc group complexes, $[M{-}O_4]^{-2}$, did indeed produce radicals, but these were identified by ESR studies as o-benzoquinone and its further oxidation product (36).

The copper dianions, $[Cu{-}S_4]^{-2}$ and $[Cu{-}O_4]^{-2}$, were paramagnetic $(S = \frac{1}{2})$ and had solid-state magnetic moments consistent with one unpaired electron; the oxidized species, $[Cu{-}S_4]^{-1}$, were diamagnetic.

All dianionic copper complexes exhibited characteristic ESR spectra (Table XVII) consisting of a quartet due to hyperfine splittings caused by [63]Cu $(I = \frac{3}{2})$; additional structure only observed on the high-field $(+\frac{3}{2})$ line was caused by hyperfine splittings due to [65]Cu $(I = \frac{3}{2})$.

ESR spectral measurements in a variety of solvents showed that $[CuS_4C_4(CN)_4]^{-2}$ was not axially perturbed by donor molecules and results obtained in methanol glasses (97) at 77°K and in magnetically

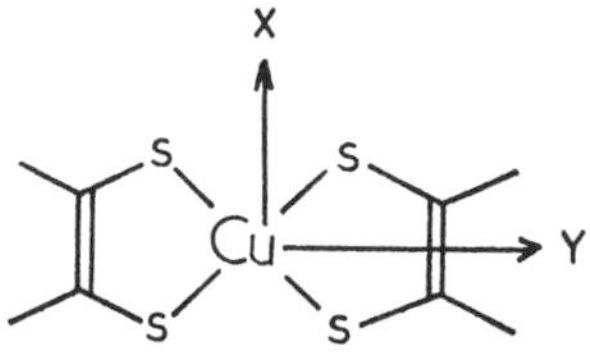

Fig. 6. The coordinate axes assumed in ESR spectral studies of $[CuS_4C_4(CN)_4]^{-2}$.

TABLE XVI

ESR Data Obtained from Monoanionic and Monocationic Planar Dithiolenes
(including the zinc and nickel groups)

Metal	Ligand	Solvent	Solution, g	g_1	g_2	g_3
Zn	gma[a,l]	DMF/CHCl$_3$	2.0027		2.0027[c]	
	gma[d]	2-MeTHF	2.0033		2.0033[c]	
	cis-N$_2$S$_2$[e,d]	2-MeTHF	2.0023		2.0023[c]	
Cd	gma[a,d]	2-MeTHF	2.0024		2.0024[c]	
	cis-N$_2$S$_2$[e,d]	2-MeTHF	2.0015		2.0015[c]	
Ni	S$_2$C$_2$H$_2$	pyr/CHCl$_3$	2.056 $\pm$ 0.001	1.996	2.039	2.126
	S$_2$C$_2$(CF$_3$)$_2$	DMF/CHCl$_3$	2.0618 $\pm$ 0.0004	1.996	2.043	2.137
	S$_2$C$_2$(CN)$_2$	DMF/HCl$_3$	2.0633 $\pm$ 0.004	1.996	2.043	2.140
	S$_2$C$_2$(CN)$_2$	Single-crystal[f]		1.998[f]	2.042[f]	2.160[f]
	S$_2$C$_2$Ph$_2$	DMSO	2.0568 $\pm$ 0.0003[g]			
	S$_2$C$_6$H$_3$Me	DMF/CHCl$_3$	2.082	2.016	2.046	2.183
	SOC$_6$H$_4$	DMF/CHCl$_3$	2.083	2.017	2.036	2.191
	S(NH)C$_6$H$_4$[d]	2-MeTHF	2.0533	2.005	2.028	2.126
	SC(Ph)N·NH[d]	2-MeTHF	2.0435	2.006	2.025	2.094
	H$_2$gma[h]	DMF/CHCl	2.051	2.009	2.027	2.119
	gma[a,l]	DMSO/CHCl$_3$	(2.003)	1.978	2.005	2.026
	gma[l]	DMF	2.0041			
	gma[d]	2-MeTHF	(2.004)	1.979	2.006	2.028
	gma[j]	DMSO/CHCl$_3$	2.0041	1.980	2.004	2.025
	gma[j]	DMF/CHCl$_3$	2.0042	1.975	2.005	2.026
	cis-N$_2$S$_2$[e,l]	CHCCN	1.9933			
	cis-N$_2$S$_2$[d]	2-MeTHF	1.9979			
	cis-N$_2$O$_2$[k,l]	CH$_3$CN	2.0006			
	cis-N$_2$O$_2$[d]	2-MeTHF	2.0009		1.997[l]	
	(NH)$_2$C$_6$H$_4$	DMF/acetone	2.031	1.990	2.006	2.102
	(NH)$_2$C$_6$H$_4$	DMSO	2.034			
Pd	S$_2$C$_2$(CN)$_2$	DMF/CHCl$_3$	2.0238 $\pm$ 0.0002[m]	1.956	2.056	2.065
	S$_2$C$_2$(CF$_3$)$_2$	DMF/CHCl$_3$	2.0238 $\pm$ 0.0002[m]	1.955	2.049	2.065
	S$_2$C$_6$H$_3$Me	DMF/CHCl$_3$	2.022			
	(NH)$_2$C$_6$H$_4$	DMF/acetone	2.006	1.946	2.008	2.062
Pt	S$_2$C$_2$(CN)$_2$	DMF/CHCl$_3$	2.042 $\pm$ 0.001[n]	1.825	2.067	2.221
	S$_2$C$_2$(CF$_3$)$_2$	DMF/CHCl$_3$	2.039 $\pm$ 0.001[n]	1.823	2.074	2.210
	S$_2$C$_6$H$_3$Me	CMF/CHCl$_3$		1.810	2.059	2.229
	S$_2$C$_6$H$_3$Me	Acetone	2.031			
	(NH)$_2$C$_6$H$_4$	DMF/acetone	1.988[p]	1.759	1.979	2.217

(continued)

TABLE XVI (*continued*)

Metal	Ligand	Solvent	Solution, g	ESR data Glass or crystal g_1	g_2	g_3
Ni	$(NH)_2C_6H_4{}^b$	$DMSO^q$	1.997			
	$(NH)_2C_6H_4{}^b$	$Solid^q$	2.000			
Pd	$(NH)_2C_6H_4{}^b$	$DMSO^q$; DMF	1.996			
	$(NH)_2C_6H_4{}^b$	$DMSO^l$	1.997			
	$(NH)_2C_6H_4{}^b$	$Solid^q$	2.000^c			
Pt	$(NH)_2C_6H_4{}^b$	$DMSO^q$; $DMSO^l$	1.982			
	$(NH)_2C_6H_4{}^b$	$Solid^q$	1.985	1.940^r	2.009^s	

  [a] Glyoxalbis(2-mercaptoaniol); all complexes monoanions unless labeled with superscript b.
  [b] Monocations.
  [c] No measurable anisotropy in glass, 70°K.
  [d] Generated by reduction using Na/Hg.
  [e] Biacetylbis(thiobenzoylhydrazone).
  [f] Magnetic dilution in $[Bu_4N][CuS_4C_4(CN)_4]$; $g_1 \equiv g_z$; $g_2 \equiv g_y$; $g_3 \equiv g_x$.
  [g] Isotopically enriched with ^{61}Ni, $I = \frac{1}{2}$; $\langle A \rangle = 5.4 \pm 1$ G.
  [h] Reduction product of $[Ni(gma)]^0$ using $BH_4{}^-$.
  [i] Controlled potential electrolysis.
  [j] As $[Bu_4N]^+$ salt.
  [k] Biacetylbis(benzoylhydrazone).
  [l] Anisotropy not resolvable; measurement made at center of absorption.
  [m] Hyperfine splitting due to ^{105}Pd ($I = \frac{5}{2}$); $\langle A \rangle = 7.7 \pm 0.3$ G.
  [n] Hyperfine splitting due to ^{195}Pt ($I = \frac{1}{2}$); $\langle A \rangle = 82 \pm 7$ G.
  [p] Hyperfine splitting due to ^{195}Pt; $\langle A \rangle = 145$, $\langle A_2 \rangle = 200$, and $\langle A_3 \rangle = 68$ G.
  [q] Measured as I^- salt.
  [r] $g_{\parallel}$.
  [s] $g_\perp$.

diluted single crystals (106) revealed that the g factor had a twofold anisotropy. From the single-crystal studies, it was possible to make the assignments $g_{\parallel} = g_{zz}$ and $g_\perp = g_{xx}, g_{yy}$ (the coordinate axes as shown in Fig. 6), and from Table XVII it is seen that $g_{zz} > g_{xx}, g_{yy}$ and $|A_{zz}| > |A_{xx}|$, $|A_{yy}|$. These results were entirely typical for Cu(II) complexes containing the basic $[Cu—S_4]$ coordination unit and indicated that the ligand field in the complexes was axially symmetric. They were also consistent with an electronic configuration d^9 where, in molecular orbital terms, the unpaired electron was in a σ^* (b_{1g}) orbital of d_{xy} symmetry (D_{2h} symmetry assumed) and they further indicated that there was considerable σ covalence in the

TABLE XVII

ESR Data from Copper Dithiolene and Analogous Complexes

Ligand	Solvent, temp.[a]	$\langle g \rangle$	$\langle A \rangle$[b,c]	$g_{\parallel}$	$g_{\perp}$	$A_{\parallel}$[c]	$A_{\perp}$[c]
$S_2C_2(CN)_2$[d]	MeOH (300)	2.0458	75.99[e]				
	MeOH (77)	(2.044)		2.082	2.024	158.9[e]	43.3[e]
	DMF (300)	2.0458	75.63[e]				
	Pyridine (300)	2.0471	75.16[e]				
	Single crystal[f]			2.086	2.026	162	39
					2.023		
$S_2C_2O_2$[d]	MeOH (300)	2.0427	81.26[e]				
	MeOH (77)	(2.044)		2.083	2.024	176	51
	DMF (300)	2.0440	81.34[e]				
S_2CNEt_2[g]	DMF (300)	2.0445	81.87[e]				
N_2S_2[h,g]	2MeTHF	2.062	81.8[i]				
N_2O_2[j,g]	2-MeTHF	2.109	71.9[k]			168[k]	
$Acac(O_4)$[l,g]	Glass	(2.112)	68	2.264	2.036	145.5	29
	Single crystal	(2.124)	66	2.266	2.053	160	19.5
$O_2C_6H_4$[d]	DMF–CHCl$_3$ (77)	2.118	82	2.233	2.058	205	—
$O_2C_6Cl_4$[d]	DMF–CHCl$_3$ (77)	2.224	79	2.254	2.056	200	—

[a] Measured in °K.

[b] ^{63}Cu ($I = \frac{3}{2}$) hyperfine splitting.

[c] In $cm^{-1} \times 10^4$, unless otherwise indicated.

[d] Charge, z, on complex -2.

[e] In gauss.

[f] Using $[n\text{-}Bu_4N]_2[NiS_4C_4(CN)_4]$ as host crystal.

[g] Charge on complex 0.

[h] Biacetylbis(thiobenzoylhydrazone).

[i] ^{14}N hyperfine splitting = 13.8[c].

[j] Biacetylbis(benzoylhydrazone).

[k] Isotropic ^{14}N hyperfine splitting = 11.3[c]; $\langle A_{\parallel} \rangle = 13.7$[c].

[l] Acetylacetone.

Cu—S bonds. Similar results had been obtained from more detailed molecular orbital calculations (107,108) performed in $[Cu(S_2CNR_2)_2]$ systems.

The ESR data obtained from $[Cu(O_2C_6H_4)_2]^{-2}$ and its tetrachloro analog (40) were similar to those obtained from $Cu(acac)_2$ (109). It is very likely that the electronic structures of the catechol derivatives are similar to the β-diketonate although the degree of σ covalence of the Cu—O bonds appears to be greater in the latter. A comparison of the data obtained from the $[Cu—S_4]^{-2}$ and $[Cu—O_4]^{-2}$ species showed that $g_{o\perp} > g_{s\perp}$, which reflected the reduction in orbital contributions to the magnetic moment in the $[Cu-S_4]$ complexes in the parallel direction; in other words, delocalization of the σ^* electron occurred in both types of complexes, but was greater in the tetrasulfur systems.

The ESR spectra of copper(II) complexes of biacetylbis(thiobenzoyl-hydrazone) and biacetylbis(benzoylhydrazone) were entirely typical (44) for those of the complex types $[Cu-N_2S_2]$ and $[Cu-N_2O_2]$ and both $^{63,65}Cu$ and ^{14}N hyperfine splittings were clearly resolved.

In acetonitrile, $[AgS_4C_4(CN)_4]^{-2}$ was identified (110) by its ESR spectrum (Table XVIII) which consisted of a doublet arising from nuclear hyperfine splittings due to $^{107,109}Ag$ $(I = \frac{1}{2})$. This spectrum was very similar to that obtained by Petterson and Vänngard (108) from benzene solutions believed to contain $Ag(S_2CNEt_2)_2$.

The spectrum of $[AuS_4C_4(CN)_4]^{-2}$ (111) (Table XVIII), which was a four-line multiplet due to ^{197}Au $(I = \frac{3}{2})$ hyperfine interactions, was some-what similar to that of $Au(S_2CNEt_2)_2$ (108,112). However, the g and A values of these formally isoelectronic species were markedly different, in contrast to their copper and silver analogs. There may be structural differences between the two gold complexes leading to differences in their electronic configurations, but in the absence of definitive structural information, any interpretation of the data must be speculative. It has been suggested that both complexes contain Au(II), but there are few examples of complexes containing divalent gold (113). The magnetic moment of $[Bu_4N]_2[AuS_4C_4(CN)_4]$ is certainly consistent with a doublet ground state and there is no evidence of antiferromagnetic interactions of the type found in salts of $[NiS_4C_4(CN)_4]^{-1}$ and its platinum-group metal analogs (93). The tetrabutyl ammonium salt is isomorphous with the analogous platinum complex and, therefore, this gold compound is probably planar. No

TABLE XVIII

ESR Data from Copper, Silver, and Gold Sulfur Complexes

	Cu		Ag		Au	
Parameters	S_2CNEt_2	$S_2C_2(CN)_2$	S_2CNEt_2	$S_2C_2(CN)_2$	S_2CNEt_2	$S_2C_2(CN)_2$
$\langle g \rangle$	2.0445[a]	2.0458[a]	2.019[b]	2.019[c]	2.040[d]	2.009[e]
$\langle A \rangle$	81.87[f]	75.63[f]	29[g]	31.4[g]	29[h]	41.7[h]

[a] Measured in DMF.   [b] Measured in benzene.   [c] Measured in acetonitrile.
[d] Measured in benzene.   [e] Measured in DMF:diglyme.   [f] ^{63}Cu, $I = \frac{3}{2}$.
[g] $^{107,108}Ag$, $I = \frac{1}{2}$.   [h] ^{197}Au, $I = \frac{3}{2}$.

comparable information about the dithiocarbamate is available but it seems pertinent to observe that the crystal structure of $[Cu(S_2CNEt_2)_2]$ (114) clearly shows that there is a pairwise association involving a $Cu\cdots S$ interaction not unlike that in $[Ph_3MeP][NiS_4C_4(CN)_4]$, thereby rendering each copper atom five coordinate with respect to sulfur; five-coordinate gold complexes are not unknown (115).

The monoanions of the nickel group $[M—S_4]^{-1}$, $[M—N_2S_2]^{-1}$, $[M—O_2S_2]^{-1}$, and $[M—N_4]^{-1}$, are paramagnetic $(S = \frac{1}{2})$ and display ESR signals in solution (Table XVI). Metal hyperfine satellite splittings were normally observed in complexes containing Pt or Pd, and in one case, that of $[NiS_4C_4Ph_4]^{-1}$ enriched with ^{61}Ni $(I = \frac{1}{2})$, $\langle A \rangle$ was found to be 4.5 G (98). The range of g values for the $[M—S_4]^{-1}$ complexes, the largest group in this series, was

$[Ni—S_4]^{-1}$, $\langle g \rangle$ between 2.057 and 2.082
$[Pd—S_4]^{-1}$, $\langle g \rangle$ between 2.022 and 2.024
$[Pt—S_4]^{-1}$, $\langle g \rangle$ between 2.031 and 2.042

If the $[M—N_4]^{-1}$ and $[M—O_2S_2]^{-1}$ species were included in this, then the lower limit for the isotropic g values in the nickel complexes was 2.03, in the palladium complexes 2.01, and in the platinum complexes ca. 2.00. The g factors of these monoanions were also characterized by a threefold anisotropy.

The analysis of the single-crystal spectrum of $[NiS_4C_4(CN)_4]^{-1}$ (106), magnetically diluted in $[Bu_4N][CuS_4C_4(CN)_4]$, enabled the assignment of g_1 as g_{zz}, g_2 as g_{yy}, and g_3 as g_{xx}. It should be noted that the largest anisotropic g values were those of g_{xx}, that there was a significant difference between g_{xx} and g_{yy}, and that the smallest anisotropic g values were those of g_{zz}. These results indicated that the effective ligand field was of rhombic and not axial symmetry.

Three-fold anisotropies have been observed in frozen solutions containing sulfur or sulfur-containing radicals and in some cases anisotropic g values as high as 2.066 were observed (116). In radicals where it was extremely probable that the odd electron was localized on the sulfur atom, the g factor has been observed within the range 2.03–2.01. Thus, it was concluded (44) that the larger g value anisotropies and isotropic g values of these metal complex monoanions were indications of appreciable admixture of metal d orbitals in the wave function of the odd electron.

The ESR spectra of the reduced nickel glyoxal bis(2-mercaptoanil) are of some interest. It will be recalled from Section III-A-7 that the nature of the reduced species depended on the method by which it was produced. Thus, reduction using borohydride ion (34) in THF afforded green solutions having isotropic g values of 2.048. Precipitation with Bu$_4$NBr afforded a dark green complex, [Bu$_4$N][Ni(H$_2$gma)], whose polycrystalline ESR spectrum gave $\langle g \rangle = 2.050 \pm 0.005$ and, in various polar solvents, g values close to 2.051 but never less than 2.048. However, the sodium amalgam-reduced complex (77) showed, in 2-methyl-THF, a g value of 2.0042, identical with the value obtained from electrochemically produced samples of [Ni(gma)]$^{-1}$. The course of the borohydride reduction of [Ni(gma)]^0 could be followed (77) by monitoring the ESR spectrum at various stages during the reaction. On addition of BH$_4^-$ to the THF solution containing [Ni(gma)]^0, a green color was generated and two signals, at $\langle g \rangle = 2.004$ and 2.048, were observed. These decayed in intensity with time until, in 30 min, they had vanished. The signals were never strong, arising from only 6% of the total nickel-containing material in the solution. On addition of Bu$_4$NBr in ethanol, the green color was discharged, a gas was evolved, and a grayish-black solution, which exhibited no ESR signals, was formed. When a small amount of oxygen was passed through this solution, an intense green color was produced and a strong resonance (arising from 81% of the nickel-containing material in solution was observed at $\langle g \rangle = 2.047$. Evaporation and dilution of this solution with n-hexane afforded [Bu$_4$N][Ni(H$_2$gma)]. This complex was very air-sensitive and if further oxygen was passed into the intense green solutions described above, the resonance line as $\langle g \rangle = 2.047$) was observed to decrease and a new line at $\langle g \rangle = 2.004$ appeared. Although this new species was not characterized, it was clear that it was not Ni(gma)]$^{-1}$. Using borodeuteride ion as reductant, and analysis of the reduced product revealed that only one deuterium atom was incorporated in the molecule. A possible reaction mechanism is summarized in Figure 7.

The g values, isotropic and anisotropic, of [Ni(H$_2$gma)]$^{-1}$ were very similar to those obtained from Ni[S(NH)C$_6$H$_4$]$_2^{-1}$. Those of [Ni(gma)]$^{-1}$, however, were markedly smaller than either of these, and the anisotropies of the monoanions formed by the reduction of nickel complexes of 16 and its benzoylhydrazone analog were barely observable. Thus, it would appear that there is a subtle electronic difference between those complexes containing conjugated tetradentate cyclic Schiff bases and those where there is no conjugation between the two (bidentate) ligands. This point is discussed further in Section III-G.

The monoanions [M—N$_4$]$^{-1}$ exhibited (33) g tensor anisotropies of magnitudes comparable with those observed in their [M—S$_4$]$^{-1}$ analogs and the order of (g_3–g_1), Pt > Ni > Pd, was more or less preserved. The platinum complex displayed ^{195}Pt hyperfine splittings but in no cases were ^{14}N hyperfine interactions resolved. The monocationic complexes,

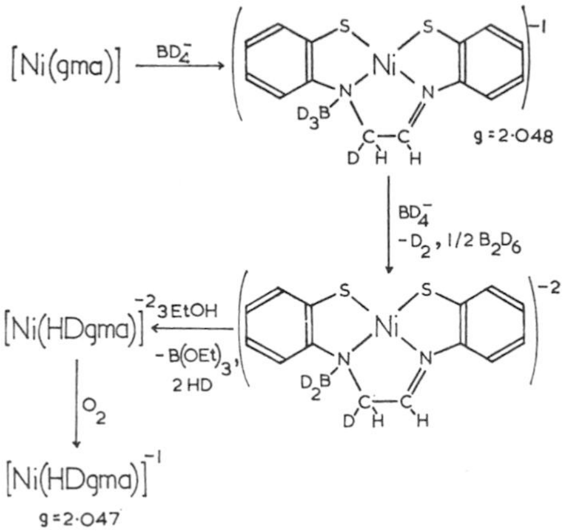

Fig. 7. A possible mechanism for the formation of [Ni(HDgma)]$^{-1}$.

[M—N$_4$]$^{+1}$, had g values very close to the free electron value, and this seemed to suggest that the odd electron was localized in an orbital of predominantly ligand character. However, the g factor of the platinum monoiodide (measured as a microcrystalline solid) was resolved into a twofold anisotropy but again, no ^{195}Pt or ^{14}N hyperfine splittings were detected.

[Ni(OSC$_6$H$_4$)$_2$]$^{-1}$ had ESR spectral properties very similar (28) to those of other [Ni—S$_4$]$^{-1}$ systems, particularly [Ni(S$_2$C$_6$H$_3$Me)$_2$]$^{-1}$.

Controlled potential electrolysis of the [Ni—O$_4$]$^{-2}$ species (40) at potentials appropriate to the generation of the monoanions failed to produce detectable ESR signals, but it was conceivable that these monoanions, if formed, might have had triplet ground states and sufficiently short spin-lattice relaxation times that their spectra were not detected at room temperature.

With the exception of the dicyano–dithiolenes and "[Pr$_4$N][Ni(O$_2$-C$_6$Cl$_4$)$_2$]," the monoanionic nickel-group complexes exhibited full magnetic moments consistent with doublet ground states both in solution and in the solid state. The dicyano–dithiolenes behaved normally in solution but in the solid state had moments considerably less than those expected for one unpaired electron (93) (Table I). The agreement between the observed and calculated temperature dependences of the magnetic moments strongly indicated that the system involved a singlet ground state with a thermally accessible triplet state. Interactions *within* the molecules were excluded by the observation of normal Curie-Weiss behavior of magnetically diluted [Ph$_3$MeP][NiS$_4$C$_4$(CN)$_4$] and it was concluded that inter-actions occurred between pairs of metal ions through the sulfur atoms of adjacent ligands. This suggestion was subsequently confirmed by the single-crystal x-ray analysis of the nickel complex (92) (Sec. III-B).

Although the neutral species of the nickel group, [M—S$_4$]°, [M—N$_4$]°, and [M—N$_2$S$_2$]°, were diamagnetic, those complexes derived from o-quinones and catechol, [M—O$_4$]° (40), were paramagnetic. They had magnetic moments ranging from 2.7 to 3.9 BM, results indicating a distinct lack of structural similarity to their sulfur and nitrogen donor-atom analogs. It was suggested that the complexes might have π-quinonoidal structures similar to those proposed by Schrauzer (117) for some nickel p-quinone derivatives, e.g., [Ni(π-duroquinone)$_2$], (Fig. 8) and [Ni(diole-fin)(π-duroquinone)]. However, it was noted that some p-quinones of higher electron affinity than duroquinone could oxidize Ni(CO)$_4$ to Ni^{2+} salts and that o-quinones did not give products structurally similar to those of duroquinone. In general, o-quinones are more powerful oxidizing agents than their para analogs, and so perhaps the most realistic descrip-tion of the [M—O$_4$]° complexes was that they were probably polymeric species containing six-coordinate Ni(II).

The monocations, [M—N$_4$]$^{+1}$, were generally not obtained in suf-ficient amounts for bulk susceptibility measurements to be made. However, a moment of 1.18 BM was recorded for {Ni[(NH)$_2$C$_6$H$_4$]$_3$}I, which was low, spin–spin interactions of the type displayed by [NiS$_4$C$_4$(CN)$_4$]$^{-1}$ salts in the solid state being suggested as a possible reason for the decrease in the moment. The dications, [M—N$_4$]$^{+2}$, as iodides or nitrates, had moments consistent with triplet ground states and, therefore, the com-plexes were presumed to have distorted octahedral stereochemistries.

Early claims (13) that [CoS$_4$C$_4$(CN)$_4$]$^{-2}$ was high spin ($S = \frac{3}{2}$) (the Bu$_4$N$^+$ salt was reported to have $\mu_{eff} = 3.92$ BM) were the subject of some controversy. However, it was later established (118), and accepted (88), that this planar anion was low spin ($S = \frac{1}{2}$), and ESR spectral results

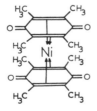

Fig. 8. The structure of [Ni(π-duroquinone)$_2$].

(Table XIX) were consistent with this. The solid-state magnetic moments of a large number of salts of this anion varied from 1.80 to 2.20 BM depending on cation, and in solution from 2.09 to 2.17 BM. Low temperature (ca. 77°K) moments were, in general, lower than their room temperature counterparts, a phenomenon often observed in other low-spin Co(II) complexes (119). Room temperature ESR spectra of this complex ion consisted of a single broad line and no hyperfine interaction due to ^{59}Co ($I = \frac{7}{2}$) was resolved; the isotropic g value was 2.25 ± 0.01. In magnetically diluted single crystals (106), a three-fold anisotropy of the g factor was observed and the assignments $g_1 = g_{zz}, g_2 = g_{yy}$, and $g_3 = g_{xx}$

TABLE XIX

Magnetic Moment and ESR Spectral Data from Dianionic Cobalt and Rhodium Dithiolenes

Complex	Cation	Phase, temp.	μ_{eff},BM	$\langle g \rangle$	Glass or solid spectra		
					g_1	g_2	g_3
$[CoS_4C_4(CN)_4]^{-2}$	Ph_4As^+	Solid	2.17				
	Et_4N^+	Solid	2.08				
		Solid (77 °K)	1.87				
	$n\text{-}Bu_4N^+$	Solid	2.16				
		Solid (77°K)	1.93				
		Acetone	2.17	2.255[a]			
		DMSO	2.14				
		Single crystal[c]			1.977[b]	2.025	2.798
$[CoS_4C_4(CF_3)_4]^{-2}$	Et_4N^+	Solid	2.06				
		$CHCl_3$/DMF			2.04[d]	2.71	
$[CoS_4C_4Ph_4]^{-2}$		Diglyme	2.33				
$[RhS_4C_4(CN)_4]^{-2}$	$n\text{-}Bu_4N^+$	Solid	1.91		1.950[c]	2.015	2.35
		Single crystal[c]			1.936[e]	2.019	2.447
$[Co\{(NH)_2C_6H_4\}_2]^0$		Solid	2.20				
		DMF		2.24[f]			
$[Co(O_2C_6Cl_4)_2]^{-2}$	$n\text{-}Pr_4N^+$	Solid	4.96				
		CH_2Cl_2	4.94				

[a] Broad line, peak-to-peak width ~220 G.

[b] $g_1 \equiv g_{zz}$; $g_2 \equiv g_{xx}$; $g_3 \equiv g_{yy}$. Anisotropic hyperfine splittings, $|a_1| = 23 \pm 1$, $|a_2| = 28 \pm 1$ and $|a_3| = 50 \pm 1$ cm^{-1} × 10$_4$.

[c] $[n\text{-}Bu_4N]_2[NiS_4C_4(CN)_4]$ as host.

[d] Estimated error ±0.02, due to broad line.

[e] $g_1 \equiv g_{zz}$; $g_2 \equiv g_{yy}$; $g_3 \equiv g_{xx}$. Anisotropic hyperfine splittings, $|a_1| = <4$, $|a_2| = 7.5 \pm 0.1$, and $|a_3| = <4$ cm^{-1} × 10^4.

[f] Peak-to-peak width ~ 260 G; uncertainty ± 0.02 due to broad line.

were made. It was found that $A_{xx} > A_{yy} > A_{zz}$ but that $g_{xx} > g_{yy} > g_{zz}$. The ESR and magnetic properties of $[Bu_4N]_2[CoS_4C_4(CF_3)_4]$ were very similar.

The magnetic properties of the cobalt "monoanionic" species, $[Co—S_4]^{-1}$, presented, at first sight, a rather confusing picture (Table XX). However, a certain degree of rationalization was introduced by the discovery that many of these apparently monoanionic complexes were dimeric dianions, $[Co—S_4]_2^{-2}$. This was established by the x-ray crystallographic examination of $[Bu_4N]_2[Co(S_2C_6Cl_4)_2]_2$ (88), voltammetric studies of cobalt dithiolenes (25), and by conductivity measurements on the dicyano- and bis-perfluoromethyl–dithiolenes (27). Complexes which were dimeric in the solid state and in noncoordinating solvents were diamagnetic. In DMSO, the tetrachlorobenzene– and dicyano–dithiolene complexes had moments of 2.37 and 2.81 BM, respectively, and this may indicate that dissociation of the dimers occurs, the resultant monomers having spin-triplet ground states. Divergence of behavior occurred in cyclohexanone where the tetrachlorobenzene–dithiolene was paramagnetic and the dicyano–dithiolene diamagnetic. However, it seems that the tetrachloro complex dissociates in all solvents regardless of their polarity. The known monomeric species, that is all arene–dithiolenes except the tetrachloro species in the solid state, were high-spin ($S = 1$) in the solid state and in solution.

Oxidation of the dimeric cobalt dianions produced paramagnetic

TABLE XX

Magnetic Moments of Monoanionic, Monomeric, and Dianionic Dimeric Cobalt Dithiolenes

	Ligand						
Phase	$S_2C_2(CN)_2$	$S_2C_2(CF_3)_2$	$S_2C_6Cl_4$	$S_2C_6H_4$	$S_2C_6H_3Me$	$S_2C_6H_2Me$	$S_2C_6Me_4$
lid	Dia.[a]	Dia.[a]	Dia.[a]	3.27	3.18	3.24	3.23
clo-hexanone	Dia.[a]	—[b]	3.14	—	3.29	—	—
HF	—	—	3.18	—	—	—	—
MSO	2.81	—	2.37	—	3.39	—	—
MF	—	—	—	—	3.40	—	—
ridine	Dia.[c]	Dia.[c]	Dia.	—	3.40	—	—

[a] Complex dimeric.
[b] Not measured.
[c] Five-coordinate adduct isolated.

monoanions, $[Co—S_4]_2^{-1}$ (25) and dimeric, diamagnetic neutral complexes, $[Co—S_4]_2^0$. Of the former, $[Et_4N][CoS_4C_4(CF_3)_4]_2$ had a magnetic moment consistent with that for one unpaired electron (Table XXI) and exhibited normal Curie–Weiss behavior although the Weiss constant was ca. 25°.

The p-phenylenediimine complex $Co[(NH)_2C_6H_4]_2^0$, and the corresponding iodide (33), were paramagnetic ($\mu_{eff} = 2.20$ BM) and diamagnetic, respectively (Table XIX). The former had $\langle g \rangle = 2.24$ in DMF, with a peak-to-peak width of about 260 G; neither ^{59}Co nor ^{14}N hyperfine splittings were detected.

$[Pr_4N]_2[Co(O_2C_6Cl_4)_2]$ had a magnetic moment (40) of 4.96 BM in the solid state (Table XIX), a value which was very similar to those obtained from $[Co(acac)_2]_4$ and $[Co(acac)_2(H_2O)_2]$ (120) [both of which contained octahedrally coordinated Co(II) (121)] and differed considerably from the moments expected for planar or tetrahedral species. Thus, it seemed

TABLE XXI

Magnetic and ESR Data from Dimeric 1,2-Dithiolene Complexes of Iron and Cobalt

Complex	Cation	μ_{eff},BM[a]	θ[b]	$\langle g \rangle$
$[CoS_4C_4(CN)_4]_2^{-2}$		Dia.		
$[CoS_4C_4(CF_3)_4]_2^{-2}$		Dia.		
$[CoS_4C_4(CF_3)_4]_2^{-1}$	Et_4N^+	1.91[c]	ca. 25°	2.043
$[CoS_4C_4(CF_3)_4]_2^0$		Dia.		
$[CoS_4C_4Ph_4]_2^0$		Dia.		
$[Co(S_2C_6Cl_4)_2]_2^{-2}$	Bu_4N^+	Dia.		
$[FeS_4C_4(CN)_4]_2^{-2}$	Et_4N^+	1.62[f]		
$[FeS_4C_4(CF_3)_4]_2^{-2}$	Et_4N^+	1.48[f,d]		
$[FeS_4C_4(CF_3)_4]_2^{-1}$	Et_4N^+	2.14[d]		
$[FeS_4C_4(CF_3)_4]_2^0$		—[e]		
$[Fe(S_2C_6H_3Me)_2]_2^{-2}$	Bu_4N^+	2.07[f]		
$[Fe(S_2C_6H_3Me)_2]_2^{-1}$	Et_4N^+	3.48[d]		
$[FeS_4C_4Ph_4]_2^0$		Dia.		

[a] Measured at 298°K in the solid state; all data per dimer unit, except where indicated.[f]

[b] Curie-Weiss constant.

[c] Complex obeys Curie-Weiss behavior.

[d] Moments decrease with decreasing temperature and do not obey Curie–Weiss behavior.

[e] Moment unknown, but complex presumed diamagnetic.

[f] Moments determined assuming complex monomeric.

very likely that this complex was polymeric and probably contained distorted octahedral Co(II).

The planar rhodium complex $[Bu_4N]_2[RhS_4C_4(CN)_4]$ (122) had a magnetic moment of 1.81 BM (Table XIX), consistent with a spin–doublet ground state. The g tensor exhibited threefold anisotropy, and in magnetically diluted single crystals, the assignments $g_1 \equiv g_{zz}$, $g_2 \equiv g_{yy}$, and $g_3 \equiv g_{xx}$ were made. The ordering of the g- and A-values was $g_{xx} > g_{yy} > g_{zz}$ and $A_{yy} > A_{xx} \sim A_{zz}$, which was similar to the results obtained from the analogous cobalt complex.

The crystal structure of $[Bu_4N]_2[FeS_4C_4(CN)_4]_2$ (94), and voltammetric studies of many iron dithiolenes (25), have established that many, if not all, "monoanionic" iron complexes should be reformulated as dimeric dianions. An early report (16) that the dicyano–dithiolene was high-spin $(S = \frac{3}{2})$ in the solid state was refuted by two different groups of workers (75,93). In fact, the magnetic moments (Table XXII) of these complexes in the solid state are consistent with two unpaired electrons, one each per iron atom. These dimeric units are almost certainly completely dissociated in polar solvents and, indeed, stable five-coordinate pyridine and phosphine adducts have been isolated. These dissociated complexes, and the five-coordinate pyridine adducts (Sec. V-C) are high-spin $(S = \frac{3}{2})$.

The dimeric monoanions, $[FeS_4C_4(CF_3)_4]_2^{-1}$ and $[Fe(S_2C_6H_3Me)_2]_2^{-1}$ (25), were paramagnetic with moments of 2.14 and 2.48 BM, respectively; the former exhibited a broad ESR signal with $\langle g \rangle = 2.043$. The magnetic

TABLE XXII
Magnetic Moments of Monoanionic and Dianionic
Dimeric Iron Dithiolenes

Phase	Ligand		
	$S_2C_2(CF_3)_2$	$S_2C_2(CF_3)_2$	$S_2C_6H_3Me$
Solid	1.89[a]	1.48[a]	2.04[a]
Cyclohexanone	3.75	—	—
Acetone	—	1.83[a]	—
DMSO	4.09	—	3.87
DMF	4.19	3.98	3.77
Pyridine	4.19[b]	—	3.77[b]

[a] Complex dimeric.
[b] Five-coordinate adduct isolated.

moments (Table XXI) decreased with decreasing temperature but deviated considerably from normal Curie-Weiss behavior. The neutral dimers were diamagnetic.

F. The Electronic Spectra of Bis-Substituted 1,2-Dithiolene Complexes

The most characteristic features of the neutral and monoanionic, and some dianionic, species of the type $[M—S_4]_n{}^z$ were the intense absorptions occurring at low energy in the visible and near infrared regions. It was quite obvious that these bands were not of "d–d" type but had their origin in some form of charge transfer. In general, the weak ligand-field bands were obscured by, or appeared as shoulders on, these intense transitions but in the dianions of the nickel and copper groups, the strong charge transfer bands were absent and the d–d transitions could be clearly identified. The details of the spectra, together with some of the proposed assignments of the transitions, are given in Table XXIII.

The strong low-energy absorptions in $[NiS_4C_4R_4]^{0,\,-1}$ were assigned by Schrauzer and Mayweg (57) as a $\pi \rightarrow \pi$ charge transfer transition, and the remaining bands were described as of $M = \pi$, or $n_{=s} \rightarrow M$ or π type. It was found that the $\pi \rightarrow \pi$ charge transfer band was very sensitive to the inductive effect of the R group in these oxidized dithiolenes. In the dianions, however, the intense transitions were absent, thus permitting the observation of the d–d transitions which were assigned as $d_{x^2-y^2} \rightarrow d_{xy}$ and shifted in the expected manner when Ni was replaced by Pd or Pt.

The assignments of the spectral transitions in $[NiS_4C_4(CN)_4]^{-2}$ by Gray and his colleagues (123) were broadly similar to those of Schrauzer and Mayweg, but differences occurred in the description of the low-energy charge transfer bands. It was observed that the two suggested $L(\pi) \rightarrow M$ transitions appearing at high frequencies in $[MS_4C_4(CN)_4]^{-2}$, $M = Pd$ or Pt, were in the order Pd > Pt, similar to that found in the corresponding planar tetrahalides (124), and that the energies of the assigned $M \rightarrow L(\pi)$ transitions were in precisely the order, Ni > Pt > Pd, that they were in the corresponding tetracyano complexes (124). In $[NiS_4C_4(CN)_4]^{-1}$, the low-energy intense band was assigned to a $L(\pi) \rightarrow M$ transition, and the remaining bands in the same way as the dicyano–dithiolene dianion.

The assignment of the spectral features of complexes containing these dithiolene ligands and other metals were carried out (65,110) using an extended and simplified molecular orbital scheme which is described in the next section (Sec. III-G). The separation between the σ^* and π_1^* levels (Fig. 10) which corresponded to $\Delta_1(x^2 - y^2 \rightarrow xy)$ in square-planar halides

TABLE XXIII

Electronic Spectra of Some Planar and Dimeric Dithiolene Complexes and their Analogs

Complex	Solvent	Maxima (intensity), cm^{-1}	Assignments[a] (1)	(2)
Dianions				
$[CuS_4C_4(CN)_4]^{-2}$		8,300 (94)		
$[AuS_4C_4(CN)_4]^{-2}$		13,750 (368)		
$[NiS_4C_4(CN)_4]^{-2}$	CH_3CN	11,690 (30)	$d \rightarrow d$	$\pi_1 \rightarrow \sigma^*$
		17,500 (570 sh)	$n_{=S} \rightarrow M$	$\pi_2 \rightarrow \sigma^*$
		19,250 (1,250)	$n_{=S} \rightarrow M$	$M \rightarrow L(\pi)$
		21,000 (3,800)		$M \rightarrow L(\pi)$
		26,400 (6,600)		$L(\pi) \rightarrow M$
		31,300 (30,000)		$L(\pi) \rightarrow L(\pi^*)$
		37,000 (50,000)		$L(\sigma) \rightarrow M$
$[NiS_4C_4(CF_3)_4]^{-2}$	N_2H_4	12,900 (sh)	$d \rightarrow d$	
		17,000 (sh)	$n_{=S} \rightarrow M$	
		26,312	$n_{=S} \rightarrow M$	
$[NiS_4C_4Me_4]^{-2}$	N_2H_4	11,110	$d \rightarrow d$	
		12,739	$d \rightarrow d$	
		18,520	$n_{=S} \rightarrow M$	
		24,691	$n_{=S} \rightarrow M$	
$[NiS_4C_4H_4]^{-2}$	N_2H_4	11,765	$d \rightarrow d$	
		12,987	$d \rightarrow d$	
		18,182	$n_{=S} \rightarrow M$	
		23,756	$n_{=S} \rightarrow M$	
$[NiS_4C_4Ph_4]^{-2}$	N_2H_4	12,579	$d \rightarrow d$	
		23,529	$n_{=S} \rightarrow M$	
$[Ni(S_2C_6H_3Me)_2]^{-2}$		14,540		$\pi_1 \rightarrow \sigma^*$
		19,510		$\pi_2 \rightarrow \sigma^*$
$[PdS_4C_4(CN)_4]^{-2}$	CH_3CN	15,700 (64)	$d \rightarrow d$	$d \rightarrow d$
		22,700 (5,700)	$n_{=S} \rightarrow M$	$L(\pi) \rightarrow M$
		25,800 (2,840 sh)		$M \rightarrow L(\pi)$
		30,800 (20,200 sh)		$L(\pi) \rightarrow L(\pi^*)$
		33,900 (47,000)		$L(\sigma) \rightarrow M$
		37,800 (45,000)		
		42,800 (34,000)		
$[PdS_4C_4(CF_3)_4]^{-2}$	CH_2Cl_2	16,950 (40)	$d \rightarrow d$	
$[PdS_4C_4Ph_4]^{-2}$	N_2H_4	16,340 (33)	$d \rightarrow d$	
		24,270 (6,590)	$n_{=S} \rightarrow M$	
$[PtS_4C_4(CN)_4]^{-2}$	CH_3CN	14,410 (49 sh)		$d \rightarrow d$
		15,650 (56 sh)		$d \rightarrow d$
		18,500 (1,220 sh)		$d \rightarrow d$
		21,100 (3,470)		$M \rightarrow L(\pi)$
		29,700 (15,600)		$L(\pi) \rightarrow L(*)$

(*continued*)

TABLE XXIII (continued)

Complex	Solvent	Maxima (intensity), cm^{-1}	Assignments[a] (1)	(2)
$[PtS_4C_4(CN)_4]^{-2}$		32,300 (13,400)		$L(\pi) \rightarrow M$
		38,500 (17,000 sh)		
		43,800 (43,500)		$L(\sigma) \rightarrow M$
$[PtS_4C_4Me_4]^{-2}$	N_2H_4	18,018 (7)	$d \rightarrow d$	
		20,833 (51)	$d \rightarrow d$	
$[PtS_4C_4Ph_4)^{-2}$	N_2H_4	19,608 (4,915)	$d \rightarrow d$	
$[CoS_4C_4(CN)_4]^{-2}$	CH_3CN	21,700 (2,200)		$L(\pi) \rightarrow \pi_1^*$
$[RhS_4C_4(CN)_4]^{-2}$	DMF	15,800 (4,000)		$L(\pi) \rightarrow \pi_1^*$
Monoanions				
$[CuS_4C_4(CN)_4]^{-1}$		6,400 (337)		$\pi_1^* \rightarrow \sigma^*$
		13,000 (110)		$\pi_2^* \rightarrow \sigma^*$
$[Cu(S_2C_6Cl_4)_2]^{-1}$	CH_2Cl_2[b]	8,875 (200)		
		17,200 (300)		
		24,600 (44,000)		
$[Cu(S_2C_6H_4)_2]^{-1}$	CH_2Cl_2[b]	8,300 (300)		
		15,900 (310)		
		25,400 (30,000)		
$[Cu(S_2C_6H_3Me)_2]^{-1}$	CH_2Cl_2[b]	8,090 (382)		$\pi_1^* \rightarrow \sigma^*$
		15,600 (318)		$\pi_2^* \rightarrow \sigma^*$
$[Cu(S_2C_6Me_4)_2]^{-1}$	CH_2Cl_2[b]	8,425 (260)		
		15,550 (350)		
		24,500 (30,000)		
$[AuS_4C_4(CN)_4]^{-1}$	CH_2Cl_2	13,400 (44)		$\pi_1^* \rightarrow \sigma^*$
		21,700 (112)		$\pi_2^* \rightarrow \sigma^*$
$[Au(S_2C_6H_3Me)_2]^{-1}$	CH_2Cl_2[b]	15,500 (87)		$\pi_1^* \rightarrow \sigma^*$
		23,530 (195)		$\pi_2^* \rightarrow \sigma^*$
$[NiS_4C_4(CN)_4]^{-1}$	CH_3CN	8,330 (329)	$\pi \rightarrow \pi(f)$[c]	$\pi_2^* \rightarrow \pi_1^*$
		11,790 (8,000)	$\pi \rightarrow \pi$	$L(\pi) \rightarrow \pi_1^*$
		16,700 (500)	$\pi \rightarrow M(f)$[c]	$d \rightarrow d$
		18,350 (2,500)	$n_{=S} \rightarrow M$	$d \rightarrow d$
		20,800 (13,650)	$\pi \rightarrow \pi$ $=$	$L(\sigma) \rightarrow M$
		27,300 (42,350)	$M \rightarrow \pi$ $\rightarrow$	$L(\pi) \rightarrow M$
		31,800 (42,300)	$\pi \rightarrow \pi$ $-$	$L(\pi) \rightarrow L(\pi^*)$
		36,600 (23,900)		$L(\sigma) \rightarrow M$
		43,500		
$[NiS_4C_4H_4]^{-1}$	Pyridine	7,246 (54)	$M \rightarrow \pi(f)$[c]	
		8,695 (323)	$M \rightarrow \pi(f)$[c]	
		11,495 (2,374)	$\pi \rightarrow \pi$	
		15,748 (85)	$\pi \rightarrow M(f)$[c]	
		19,120 (833)	$n_{=S} \rightarrow \pi$	
		19,685 (956)	$n_{=S} \rightarrow M(f)$[c]	
		20,284 (787)	$\pi \rightarrow \pi$	

(continued)

TABLE XXIII (*continued*)

Complex	Solvent	Maxima (intensity), cm^{-1}	Assignments[a] (1)	(2)
$(NiS_4C_4H_4)^{-1}$		22,989 (360)	$n_{=S} \rightarrow M(f)^c$	
		23,529 (512)		
		28,571 (2,720)	$M \rightarrow \pi$	
		33,898 (7,516)	$M \rightarrow \pi$	
$[NiS_4C_4(CF_3)_4]^{-1}$	Acetone	12,400 (6,480)		$L(\pi) \rightarrow \pi_1{}^*$
$[NiS_4C_4Ph_4]^{-1}$	THF	11,100 (7,950)	$\pi \rightarrow \pi$	
		15,600 (400)	$\pi \rightarrow M(f)^c$	
		17,200 (720)	$n_{=S} \rightarrow \pi$	
		19,400 (1,570)	$n_{=S} \rightarrow M$	
		20,800 (1,900)	$\pi \rightarrow \pi$	
		28,000 (13,000)	$M \rightarrow \pi$	
		31,400 (38,400)	$\pi \rightarrow \pi$	
$[Ni(S_2C_6Cl_4)_2]^{-1}$	CH_2Cl_2	7,150 (250)		
		11,300 (15,700)		
		19,600 (280)		
	CH_3CN	6,900 (220)		
		8,400 (90)		
	Pyridine	6,670 (175)		
		7,950 (72)		
	Mineral oil	7,550		
$[Ni(S_2C_6H_4)_2]^{-1}$	$CH_2Cl_2{}^b$	7,220 (200)		
		11,350 (13,200)		
		19,800 (340)		
$[Ni(S_2C_6H_3Me)_2]^{-1}$	CH_3CN	7,280 (240)		$\pi_2{}^* \rightarrow \pi_1{}^*$
		11,230 (16,270)		$L(\pi) \rightarrow \pi_1{}^*$
$[Ni(S_2C_6Me_4)_2]^{-1}$	$CH_2Cl_2{}^b$	7,550 (350)		
		10,800 (16,400)		
		19,400 (240)		
$[PdS_4C_4(CN)_4]^{-1}$	CH_2Cl_2	9,000 (13,800)		$L(\pi) \rightarrow \pi_1{}^*$
$[PtS_4C_4(CN)_4]^{-1}$	Acetone	11,700 (11,700)		$L(\pi) \rightarrow \pi_1{}^*$
$[Pt(S_2C_6H_3Me)_2]^{-1}$	CH_3CN	11,200 (18,000)		$L(\pi) \rightarrow \pi_1{}^*$
$[CoS_4C_4(CN)_4]_2{}^{-2}$	Acetone	12,800 (3,700)		
		14,000 (1,200)		
$[Co(S_2C_6Cl_4)_2]^{-1}$	CH_2Cl_2	5,375 (50)		
		9,530 (100)		
		15,000 (12,300)		
	CH_3CN	5,320 (50)		
		9,620 (128)		
		15,100 (12,400)		
	DMSO	6,750 (140)		
		7,270 (148)		
		9,350 (185)		
		15,050 (11,700)		

(*continued*)

TABLE XXIII (*continued*)

Complex	Solvent	Maxima (intensity), cm^{-1}	Assignments[a] (1)	(2)
$[Co(S_2C_6Cl_4)_2]^{-1}$	Pyridine	7,950 (14)		
		16,250 (1,300)		
$[Co(S_2C_6Cl_4)_2]^{-1}$	Mineral oil	5,270		
		6,670		
		11,100		
		14,500		
$[Co(S_2C_6H_4)_2]^{-1}$	CH_2Cl_2[b]	5,710 (25)		
		9,540 (45)		
		15,200 (12,100)		
$[Co(S_2C_6H_3Me)_2]^{-1}$	CH_3CN	15,200 (16,600)		$L(\pi) \to \pi_2^*$
$[Co(S_2C_6Me_4)_2]^{-1}$	CH_2Cl_2[b]	5,820 (40)		
		9,440 (75)		
		14,900 (13,000)		
$[Fe(S_2C_6H_3Me)_2]_2^{-2}$	CH_2Cl_2	17,800		
	CH_3CN	19,050		
	DMSO	20,350		
	DMF	20,100		
	DMSO	20,350		
	DMF	20,100		
Neutral species				
$[NiS_4C_4H_4]^0$	CH_3OH	11,857 (1,000)	$M \to \pi(f)$[c]	
		13,870 (7,700)	$\pi \to \pi$	
		18,150 (11,400)	$n_{=s} \to \pi$	
		19,360 (weak)	$n_{=s} \to M(f)$?	
		22,750	$n_{=s} \to \pi$	
		24,000–	$M \to \pi$	
		30,600	$n_{=s} \to M(f)$[c]	
			$n_{=s} \to \pi(f)$[c]	
$[NiS_4C_4Me_4]^0$	$CHCl_3$	12,920 (28,180)		
		17,570 (2,344)		
		25,000 (2,455)		
		27,030 (2,620)		
		32,890 (37,150)		
		35,970 (32,360)		
$[NiS_4C_4(CF_3)_4]^0$	*n*-Pentane	13,990 (12,300)		
		18,080 (1,905)		
		24,390 (1,995)		
		33,900 (24,550)		
		35,710 (20,420)		
$[NiS_4C_4Ph_4]^0$	$CHCl_3$	11,550 (30,900)		
		16,610 (2,188)		
		23,980 (5,129)		

(*continued*)

TABLE XXIII (*continued*)

Complex	Solvent	Maxima (intensity), cm^{-1}	Assignments[a]	
			(1)	(2)
$[NiS_4C_4Ph_4]^0$		26,530 (12,020)		
		31,650 (51,290)		
		37,040 (37,150)		
$\{Ni[(NH)_2C_6H_4]_2\}^0$	DMSO	12,658 (54,900)		
		15,385 (5,640)		
		17,857 (2,140 sh)		
		19,231 (1,560)		
		23,810 (1,260 sh)		
		25,000 (1,160)		
		29,851 (4,210)		
		32,258 (4,670)		
		36,232 (9,040)		
$\{Pd[(NH)_2C_6H_4]_2\}^0$	DMSO	12,821 (48,200)		
		16,103 (1,700)		
		17,065 (1,280)		
		19,569 (1,130)		
		21,834 (924)		
		28,571 (3,870)		
		32,154 (5,520)		
		36,232 (7,430)		
$\{Pt[(NH)_2C_6H_4]_2\}^0$	DMSO	12,788 (3,310)		
		13,333 (8,950 sh)		
		14,065 (96,700)		
		14,389 (69,900)		
		14,706 (52,200 sh)		
		15,385 (12,900 sh)		
		16,750 (32,270)		
		17,699 (3,710)		
		18,349 (3,180)		
		19,048 (2,420)		
		20,964 (1,680)		
		27,397 (1,150)		
		30,488 (4,390)		
$\{Co[(NH)_2C_6H_4]_2\}^0$	DMSO	8,811 (3,840)		
		13,106 (11,700)		
		17,007 (17,700)		
		23,810 (2,730)		
		29,674 (4,640)		

[a] Assignments according to (1) G. N. Schrauzer and V. P. Mayweg, (57) and (2) Gray et al. (65,110,123).
[b] Also in pyridine and CH_3CN.
[c] Forbidden band.

and cyanides (124), was estimated for the dianionic dicyano–dithiolenes and it was found that $\Delta_1(Pt) > \Delta_1(Pd) > \Delta_1(Ni)$ and that for the mono-anionic copper and gold species, $\Delta_1(Au) > \Delta_1(Cu)$. However, it was discovered that $\Delta_1[Ni-S_4]^{-2} > \Delta_1[Cu-S_4]^{-1}$ and $\Delta_1[Pt-S_4]^{-2} > \Delta_1[Au-S_4]^{-1}$, and it was reasoned that the abnormally small value of Δ_1 for the copper and gold monoanions was consistent with the predominantly ligand-based character of the levels π_1^* and π_2^*. Since the σ^* orbital was mainly metal in character, the transitions $\pi_1^* \rightarrow \sigma^*$ and $\pi_2^* \rightarrow \sigma^*$ were of $L(\pi) \rightarrow M$ type and were expected to shift to lower energies on moving from the nickel group dianions to the copper group monoanions.

Solvent perturbations were not observed in the spectra of the mono- and dianions of the copper- and nickel-group dithiolenes with the exception of $[Ni(S_2C_6Cl_4)_2]^{-1}$ (Table XXIII). However, such interactions were quite pronounced (65) in the spectra of the cobalt and iron complexes, particularly in the dimeric dianion series. Only the spectra of the cobalt arene–dithiolenes monoanions were unaffected by donor solvents, and it seemed likely that dissociation of the dimers in polar solvents was responsible for the observed shifts in the principal spectral bands (Table XXIII). Indeed, the spectra of many of the five-coordinate adducts of cobalt and iron dithiolenes (101) were identical to the spectra of the dissociated dimeric dianions in donor solvents or solutions containing donor molecules.

Schrauzer and Mayweg (57) devised a spectrochemical series of increasing ligand field strength for the sulfur ligands based on the observed values of their assigned d–d transitions in $[NiS_4C_4R_4]^{-2}$. That order was

$$R = CN < H < p\text{-}ClC_6H_4 \sim Ph \sim p\text{-}MeC_6H_4$$
$$\sim p\text{-}MeOC_6H_4 < Et < n\text{-}Pr \sim i\text{-}Pr < Me < CF_3$$

The ordering reflected the inductive properties of R and it will be recalled that the polarographic half-wave potentials (Sec. III-C) and infrared stretching frequencies, ν_1 and ν_2 (Sec. III-D) were also dependent on this effect. Only when $R = CF_3$ was there a serious divergence from the expected ordering.

Gray and his colleagues (3,110,123) placed some of the dithiolate ligands in a spectrochemical series relative to more conventional ligands. Data obtained from the spectra of some "normal" Pd(II) complexes indicated that the order was

$$S_2C_2(CN)_2^{-2} < Br^- < Cl^- < S_2C_2(CF_3)_2^{-2} < -SCN^- < S_2P(OEt)_2^-$$
$$< S_2CNR_2^- < H_2O \sim O_2C_2O_2^{-2} < NH_3 < NH_2CH_2CH_2NH_2 < CN^-$$

A comparison of Δ_1 values obtained from different planar Ni(II) complexes containing the [Ni–S_4] coordination unit provided the series, in order of decreasing Δ_1,

$$S_2C_2(CN)_2{}^{-2} < S_2C_2(CF_3)_2{}^{-2} < S_2C_6H_3Me^{-2} < S_2P(OEt)_2{}^- < S_2COEt^-$$
$$< S_2CNR_2{}^- < SCH_2CH_2S^{-2} \sim S_2C_2O_2{}^{-2}$$

It was suggested that the large variation in ligand-field strength of the ligands in the nickel complexes might be due to a large variation in the ligand character of the two most important orbitals, $\pi_1{}^*$ and σ^*. In every case, however, it was found that the dithiolates were weak-field ligands.

The electronic spectrum of the [Cu—O_4]$^{-2}$ complexes (40) were very similar to those of Cu(acac)$_2$ but indicated that the catechol ligands exerted a weaker ligand field than did the β-diketonate (125); a similar result was obtained from the [Ni—O_4]$^{-2}$ series (126). The spectra of [Co(O$_2$C$_6$Cl$_4$)$_2$]$^{-2}$ were very similar to those of [Co(acac)$_2$]$_4$ and [Co-(acac)$_2$(H$_2$O)$_2$] (120) but markedly different from [Co(dipivaloylmethane)$_2$] (127) which was known to contain a planar [Co—O_4] coordination unit. Thus, it was established that this catechol complex contained a distorted octahedral [Co—O_6] coordination unit. Assuming that the spectral assignments in [Ni(O$_2$C$_6$H$_4$)$_2$]$^{-2}$ and [Ni(S$_2$C$_6$H$_3$Me)$_2$]$^{-2}$ were correct, Δ_1 was found to be very similar in the two different species, and this suggested that the two ligands occupied similar positions in the spectrochemical series for Ni(II) species.

G. Electronic Structures of the Planar and Dimeric Dithiolenes and Their Analogs

There have been three major attempts to devise unambiguous descriptions of the electronic structures of the planar bis-1,2-dithiolenes. Two of these used simple, modified Hückel and Wolfsberg–Helmholtz theories and applied the resultant molecular orbital schemes to interpreting the spectral and chemical properties of this class of complexes as a whole. The third approach was developed from a detailed examination and interpretation of the ESR spectra of a number of complexes having spin–doublet ground states.

Schrauzer and Mayweg (57) considered three possible structures for molecules of the general formula [MS$_4$C$_4$(R$_4$)]^0 (where M was specifically Ni and R was H). These structures are represented by **22a, 22b,** and **22c.** **22a** was dismissed because it was thought most unlikely that zerovalent d^8 metal atoms would be stabilized by α,β-dithioketones, the most likely structure being, in that situation, tetrahedral [which was not the case in view

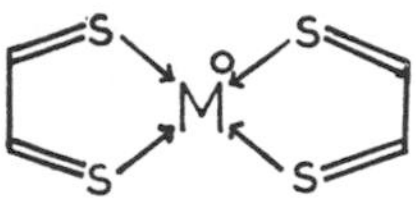

(22a) 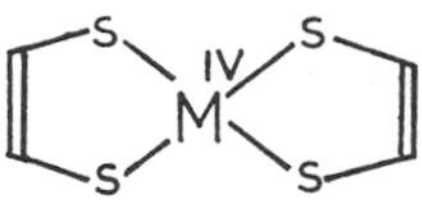   **(22b)** 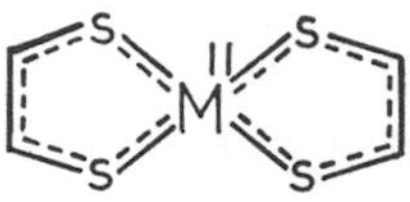   **(22c)**

of the known crystal structure of $[NiS_4C_4Ph_4]^0$ (89)], and **22b** was thought
to be unreasonable in view of the rather unlikely stabilization of such high
oxidations states [Ni(IV)] by highly polarizable sulfur donor atoms. Thus,
these authors were left with **22c** which was described as containing a
divalent central metal ion complexed by two spin-paired monoanions. An
alternative representation, in Kekule resonance forms, was given by **20a**
and **20b** (Sec. III-B).

The essential feature of the electronic structure of the neutral molecules was believed to lie in the behavior of the lowest unoccupied ligand
π molecular orbitals and their interaction with the central metal ion
orbitals. These two ligand orbitals arose from a combination of the two
originally degenerate (the lowest being empty but weakly antibonding)
orbitals in the hypothetical α,β-dithioketones. The B_{1u} combination of these
two ligand orbitals (in D_{2h} symmetry) became stabilized on interaction with
the metal np_z orbital whereas the B_{2g} combination became weakly antibonding on interaction with the metal nd_{xz} orbital. The calculated ground
state for $[MS_4C_4R_4]^0$ was

$$\cdots\cdot(2b_{1u})^2(2b_{2g})^2(3a_g)^2(2b_{3g})^2(4a_g)^2(3b_{2g})^0(3b_{1g})^0$$

(using the coordinate axes as shown in Fig. 9). The three uppermost orbitals
shown above all consisted of both metal and ligand character, but, according to the calculations, the $4a_g$ level was predominantly metal ($nd_{x^2-y^2}$
symmetry), the $3b_{2g}$ ligand (nd_{xz} symmetry) and the $3b_{1g}$ metal (nd_{xy} symmetry) in character, respectively; the metal character of the $3b_{2g}$ level
was calculated to be 18.32%. This configuration was fully represented by
structure **23c**.

On reduction of $[MS_4C_4R_4]^0$, the $3b_{2g}$ orbital became half-filled in the

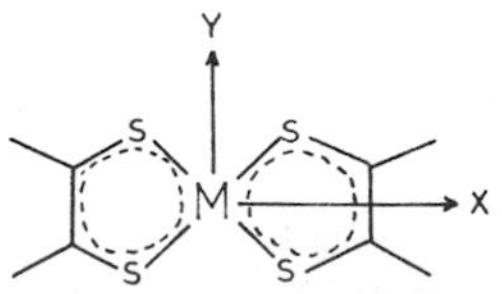

Fig. 9. The coordinate axes used in calculations of the ground state of $[MS_4C_4R_4]$.

monoanions and totally filled in the dianions. The complete filling of this level had the effect of conferring true dithiolate character on the sulfur ligands so that the dianionic complexes were correctly represented as containing divalent metal ions complexed by two dianionic dithiolate ligands. The neutral and monoanionic species were also regarded as containing divalent metal ions although they were complexed by oxidized dithiolate ligands (i.e., ligands having considerable dithioketonic character). The monoanions could be regarded (when M was a nickel group metal), in valence bond resonance forms, as **23**.

(23)

The effect of reduction of $[MS_4C_4R_4]^0$ on the bond lengths of the ligands has been described in Sec. III-B. Thus, the "ethylenic" and $S-C(sp^2)$ bond lengths decreased and increased, respectively, as charge was added to the basic $[M-S_4]$ coordination unit. It was believed that the proposed molecular orbital scheme was consistent with these facts, since filling of the $3b_{2g}$ level resulted in the conversion of the ligands from a partially dithioketonic form to a totally dithiolate form. As a consequence of the structural changes, the infrared stretching frequencies ν_1 and ν_2 (Sec. III-D) increased and decreased, respectively.

The predominance of ligand (π-molecular orbital) character in the $3b_{2g}$ level would suggest that alterations in the energy of the *ligand* orbital would have direct and measurable consequences on the values of the reduction potentials for the reactions

$$[M-S_4]^0 + e^- \rightleftharpoons [M-S_4]^{-1} \quad \text{and} \quad [M-S_4]^{-1} + e^- \rightleftharpoons [M-S_4]^{-2}$$

The absolute energy of the ligand π orbitals depends partly on the inductive effect of the substituents on the ligand (and, incidentally, this applies to ring substituents in arene-1,2-dithiolates *and* donor atoms sets in the 1,2-disubstituted benzenes) and, therefore, $+I$ substituents would cause an increase in the absolute energy of these ligands orbitals and, therefore, in the absolute energy of the $3b_{2g}$ level, whereas $-I$ substituents would cause a decrease in the energy of this level. The dependence of $E_{1/2}$ values on the substituents was discussed in Sec. III-C, and it may be recalled that these potentials could be linearly correlated with Taft's inductive substituent constant, σ^* (128). These data were thought to be one of the most compelling pieces of evidence in favor of the proposed molecular orbital scheme. However, it has already been stated that whatever molecular orbital was involved in the electron transfer reaction, in this scheme or in any other, a substituent effect on the values of the redox potentials would be expected provided, of course, that the orbital involved contained significant metal *and* ligand character.

The single occupancy of the $3b_{2g}$ level in the nickel group monoanions was believed to be consistent with the observed threefold anisotropy of the g tensor (Sec. III-E). The observation of ^{61}Ni hyperfine splittings in the ESR spectra of $[^{61}NiS_4C_4Ph_4]^{-1}$ (98) would be expected in view of the calculated one-fifth metal character of this orbital.

The electronic spectra of the neutral and monoanionic nickel group species, and of the neutral dimeric dithiolene complexes of iron and cobalt, were characterized by very intense absorptions at low energy in the visible region. These strong absorptions were entirely absent from the spectra of the dianions, indicating that their electronic structure was distinctly different from those of the more oxidized species. The principal absorption at low energy in the neutral nickel group complexes was assigned to the transition $2b_{1u} \rightarrow 3b_{2g}$ (or, in the notation used in Table XXIII, $\pi \rightarrow \pi$ or $L(\pi) \rightarrow \pi_1^*$). The remaining bands are described as of $n_{=S} \rightarrow M$, $n_{=S} \rightarrow \pi$, $M \rightarrow \pi$ or $\pi \rightarrow \pi$ type; the symbol $n_{=S}$ refers to transitions involving free electron pairs on the sulfur atoms which are in the plane of the molecule but are at 120° to the M—S—C bond angles. In the monoanions, the partial filling of the $3b_{2g}$ orbital was not expected to cause serious divergences from the basic spectra of the neutral species, as was observed, but the $2b_{1u} \rightarrow 3b_{2g}$ transition did move to lower frequencies. Because the $2b_{1u}$ orbital, like the $3b_{2g}$, derived most of its character from ligand π molecular orbitals, its absolute energy was dependent on a substituent effect. Thus, a hypsochromic shift of the $2b_{1u} \rightarrow 3b_{2g}$ transition was expected and observed. The total occupancy of the $3b_{2g}$ level in the nickel group dianions caused all intense low-energy transitions to vanish, the first allowed interligand transition occurring at much higher energies. The absence of the low-energy charge-transfer bands permitted the observation of the weak ligand-field absorptions from which the "spectrochemical" series were devised. Two bands of medium intensity observed in the spectra of the dianions between 17,000 and 25,000 cm^{-1} were assigned to $n_{=S} \rightarrow M$ transitions.

Several other points in favor of this molecular orbital treatment of the dithiolenes were advanced. These included the observation of a proton resonance in the NMR spectrum of $[NiS_4C_4H_4]^0$ in the "aromatic" region as distinct from the "olefinic" region which would be expected if the complex was a true dithiolate of structure **22b**, and chemical evidence that $[NiS_4C_4H_4]^0$ could be alkylated in a classical Friedel-Crafts alkylation reaction. The latter was viewed as evidence for the "quasi-aromatic" nature of the neutral dithiolenes, that is, complexes having structure **22c**.

Molecular orbital calculations on the iron and cobalt dimeric dithiolenes have not been published, but the complexes have been described (96) in valence-bond resonance forms. It would seem that because it is possible to do this, these complexes should have delocalized electronic structures of a type similar to those proposed for $[MS_4C_4R_4]^{0, -1}$, almost by inference. Thus, the five limiting structures shown in **24** were regarded as the principal

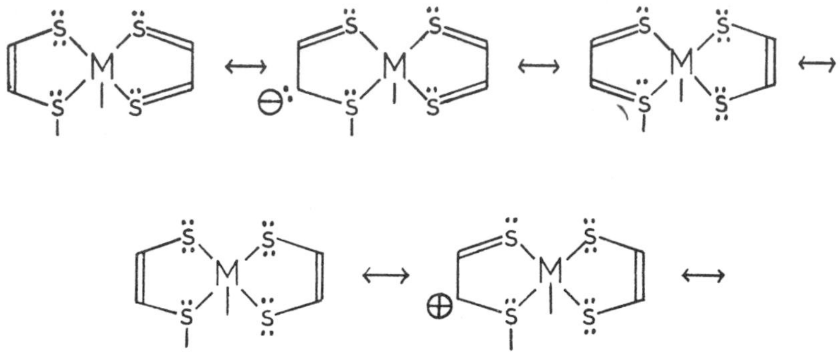

(24)

structural contributions to the ground state of these molecules. However, no account was taken of M=S bonding which could be important. Here, as in the nickel group complexes, the metal ion was regarded as in oxidation state +2. Assuming that these five structures contributed equally to the overall structure of the molecules, the bond distances calculated from the estimated double-bond characters were in fair agreement with the observed bond lengths.

In the dimeric sulfur-bridged complex, $[Fe_2S_2(S_2C_2Ph_2)_2]^0$, which underwent reversible electron transfer reactions, the main contributing structures, as represented by valence-bond resonance forms, were described as shown in **25**. It was proposed that the complex contained trivalent iron and that delocalization could take place across the sulfide bridges. Assuming that the molecule had D_{2h} symmetry, the $3p_\pi$ orbitals of the bridging S atoms transformed as B_{1u} and B_{2g}, and the B_{1u} combination was considered to overlap with the two $4p_z$ iron orbitals, thus connecting the B_{1u} π molecular orbitals of the chelating ligands.

(25)

Using the Wolfsberg-Helmholtz method, Schrauzer and Mayweg showed that several changes in the orbital sequence in $[MS_4C_4R_4]^0$ occurred but that the basic conclusion derived from the Hückel molecular orbital treatment was unaltered.

Gray and colleagues based their early proposals (123) for the electronic configurations of $[MS_4C_4R_4]^{-1,-2}$ on calculations of the modified Wolfsberg-Helmholtz type (129) on $[NiS_4C_4(CN)_4]^{-2}$. They calculated that the ground state of this molecule was

$$\cdots(3b_{1u})^2(2b_{1g})^2(4b_{2g})^2(4a_g)^2(3b_{1g})^0$$

and that the symmetries of the three upper levels shown were $(d_{xz,yz})$, $(d_{x^2-y^2})$, and (d_{xy}), respectively; it was estimated that the $4a_g$ and $3b_{1g}$ levels were 26.0 and 55.7% metal in character respectively. The most important features of the molecular orbital scheme are shown in Figure 10. Thus, in the nickel monoanions, the $4a_g$ level would be singly occupied. The support for this scheme and electronic configuration was drawn mainly from the electronic spectra of the di- and monoanionic dicyano–dithiolene complexes.

It was stated that the $4b_{2g}$ and $4a_g$ levels were "definitely more 'ligand' than 'metal'" while the $3b_{2g}$ and $2a_g$ levels, which were bonding, were localized on the metal. The $3b_{3g}$, $3a_g$, and $3b_{2g}$ orbitals were also largely metal in character, and the block of orbitals labeled d in Figure 10 were regarded as a filled d^8 shell in all dithiolene complexes of the cobalt, nickel, and copper groups. It was observed that in complexes in which the $4a_g$ and $4b_{2g}$ levels were not completely filled the metal ion could not be described correctly by classical oxidation-state formalism. An example of this would be $[Pd(S_2C_6H_3Me)_2]^{-1}$, which had a spin–doublet ground state and therefore would have the configuration

$$\cdots(4b_{2g})^2(4a_g)^1$$

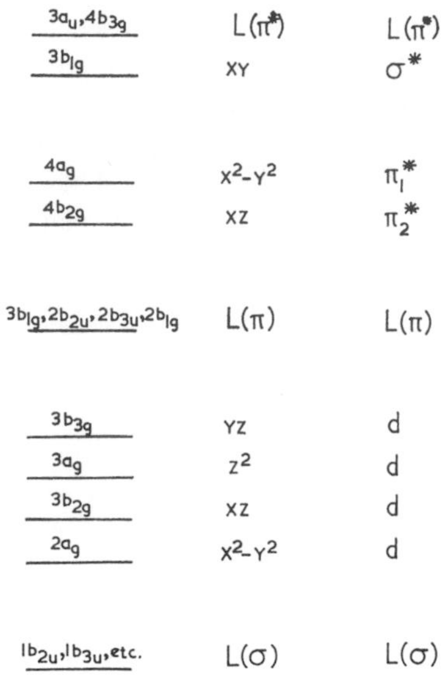

$3a_u, 4b_{3g}$	$L(\pi^*)$	$L(\pi^*)$
$3b_{1g}$	XY	σ^*
$4a_g$	x^2-y^2	π_1^*
$4b_{2g}$	XZ	π_2^*
$3b_{1g}, 2b_{2u}, 2b_{3u}, 2b_{1g}$	$L(\pi)$	$L(\pi)$
$3b_{3g}$	YZ	d
$3a_g$	z^2	d
$3b_{2g}$	XZ	d
$2a_g$	x^2-y^2	d
$1b_{2u}, 1b_{3u}, etc.$	$L(\sigma)$	$L(\sigma)$

Fig. 10. The MO scheme for $[NiS_4C_4(CN)_4]^{-1,-2}$ proposed by Gray et al. (129).

and it was suggested that this complex would *not* contain Pd(III) but was better represented as a derivative of Pd(II) (d^8) complexed by one dianionic dithiolato ligand and one spin-free radical monoanionic ligand. An alternative description of the ligands in this situation was to regard them collectively as a partially oxidized trianion, $[S_4C_4(CN)_4]^{-3}$. Those complexes in which the $4b_{2g}$ and $4a_g$ levels were filled, however, could be correctly represented as containing dianionic dithiolato ligands and the metal ion in the classical oxidation state appropriate to the overall charge on the complex. Thus, $[CuS_4C_4(CN)_4]^{-1}$, which had the predicted ground state

$$\cdots(4b_{2g})^2(4a_g)^2(3b_{1g})^0$$

was believed to be a complex of Cu(III).

The hyperfine splitting observed in the ESR spectra of $[^{61}NiS_4C_4Ph_4]^{-1}$ was considered to be very small and to be strongly indicative of the low metal character of the $4a_g$ orbital. It was known that the calculated normalized hyperfine splitting (98) in $[^{61}NiS_4C_4Ph_4]^{-1}$ was much smaller than that in (the formally isoelectronic) $[CuS_4C_4(CN)_4]^{-2}$ where the half-filled $3b_{1g}$ (σ^*) orbital was calculated to have nearly 56% metal character. It must be stated at this point that it is hard to reconcile the threefold anisotropy of the g tensor with the proposed ^2A_g ground state of the nickel group monoanions.

The molecular orbital scheme as proposed by Gray et al. for [Ni—S_4]$^{-1, -2}$ was extended to include other planar dithiolene complexes, and was used extensively in the interpretation of the electronic spectra of many for these complexes. In the generalized M.O. scheme, a simplified notation was employed and is shown in Figure 10. The separation between the π_1^* and σ^* was referred to as Δ_1 and that between the π_2^* and π_1^* levels as Δ_2.

The two weak bands at low energies in the spectra of complexes having the predicted ground states

$$\cdots(\pi_2^*)^2(\pi_1^*)^2$$

i.e., the nickel dianions and copper monoanions and their heavy metal analogs, were assigned as $\pi_1^* \rightarrow \sigma^*$ and $\pi_2^* \rightarrow \sigma^*$ transitions. The values of Δ_1 were used to place the dithiolate dianions in "spectrochemical" series relative to each other and to other more conventional ligands, as described in Sec. III-F. Δ_1 for [Cu(S$_2$C$_6$Cl$_4$)$_2$]$^{-1}$ was 500 cm^{-1} greater than that for [Cu(S$_2$C$_6$Me$_4$)$_2$]$^{-1}$ and the value of Δ_2 was some 1200 cm^{-1} greater. It was felt that this increase in Δ_2 in going from a complex containing a ligand with electron-releasing substituents (Me) to one containing electron-withdrawing substituents (Cl) was powerful evidence for the d_{xz} symmetry of the π_2^* level (and, by implication therefore, for the $d_{x^2 - y^2}$ symmetry of the π_1^* level). It was expected that the π_2^* level was predominantly of ligand π molecular orbital character and would therefore be susceptible to ligand substituent effects in the way previously discussed above in connection with $E_{1/2}$ values. Thus, the π_2^* level would become stabilized with respect to the π_1^* level when electron-withdrawing substituents were attached to the benzene ring. The same argument could be advanced for the observed dependence of the oxidation potentials in the reactions

$$[M—S_2]^{-2} \rightleftharpoons [M—S_4]^{-1} + e^-$$

except that here, the π_1^* level was involved which, in view of its proposed symmetry, would be altogether less sensitive to subtle changes in the energy of the ligand π molecular orbitals.

In those complexes having

$$\cdots(\pi_2^*)^2(\pi_1^*)^1$$

ground states, the electronic spectra were much more complicated, as described in Sec. III-F. The intense low-energy transition was assigned as $L(\pi) \rightarrow \pi_1^*$, and the very weak transitions observed at slightly lower frequencies in some complexes were described as the parity forbidden $\pi_2^* \to \pi_1^*$ transition. The $L(\pi) \to \pi_1^*$ transition was thought to obscure the $\pi_1^* \to \sigma^*$ band, but weak absorptions near 20,000 cm^{-1} may have been the $\pi_2^* \to \sigma^*$ transition. The cobalt complexes, $[\text{Co}\text{—}\text{S}_4]^{-1}$, whose ground states were predicted to be

$$\cdots(\pi_2^*)^2(\pi_1^*)^1$$

had complicated spectra which were not analyzed, and no ground-state predictions were made for the dimeric dianionic and neutral cobalt and iron species.

The calculations carried out by Maki, Edelstein, Davison, and Holm (106), on the basis of detailed ESR studies of a number of dicyano–dithiolene complexes having spin–doublet ground states, led to slightly different conclusions from those discussed above. The theoretical expressions for the g and A (^{61}Ni) tensors were calculated using perturbation theory and a basis set of pure metal $3d$ functions, and the probable ground states were deduced from the best fit of the measured data with these expressions. Covalence in the metal–sulfur bonds was accounted for by reducing the spin–orbit coupling parameter and the mean inverse cube electron–nuclear distance, which were deduced from the spectra of the free metal ions; the procedure was similar to that used by Griffith (130) for d^n strong-field complexes except that a basis set of real d orbitals was used. Thus, for $[\text{NiS}_4\text{C}_4(\text{CN})_4]^{-1}$, a ground state of

$$\cdots(b_{3g})^1(b_{1g})^0$$

was reached (note that the coordinate axes used in these calculations are identical to those in Fig. 6 and the reverse of those in Fig. 9). This half-filled orbital was calculated to be approximately 50% metal in character, and the complex was regarded as containing a d^7 Ni(III) metal ion. These results were justified by the view that it was better practice to assume an appropriate d-orbital configuration for the metal based on the interpretation of the ESR results and *then* to allow for mixing of the metal and ligand orbitals of the appropriate symmetry, the mixing coefficients being obtained from the measured g and A values, rather than to calculate these coefficients from preconstructed molecular orbitals of the complex produced by semiempirical procedures.

Other complexes which were investigated included $[\text{CoS}_4\text{C}_4(\text{CN})_4]^{-2}$, its rhodium analog, and $[\text{CuS}_4\text{C}_4(\text{CN})_4]^{-2}$. It was calculated that the unpaired electron in the copper complex was in an orbital of predominantly metal character having d_{xy} symmetry, but that there was a not inconsiderable degree of covalence in the Cu—S bonds. It should be noted that these remarks are in agreement with the independent proposals of Schrauzer and

Gray. Essentially similar conclusions were drawn about the $[Cu—O_4]^{-2}$ species, although the degree of covalence in the Cu—O bonds was somewhat less than that in the Cu—S bonds. The calculated ground state for $[CoS_4C_4(CN)_4]^{-2}$ was the same as that for the (isoelectronic) nickel monoanion, although the half-filled orbital was thought to have even greater metal character. The analogous rhodium complex, although regarded as formally containing Rh(II), had a slightly different ground state,

$$····(b_{3g})^1(a_g)^0$$

It was observed that the b_{3g} orbital could interact extensively with the ligand π-molecular orbitals.

While both Schrauzer and Mayweg, and Maki and his colleagues, were agreed that the monoanionic nickel complexes had $^2B_{3g}[····(b_{3g})^1(b_{1g})^0]$ ground states, they disagreed about the extent to which the b_{3g} level had predominantly ligand or metal character. Both of these groups, however, criticized the $^2A_g[····(a_g)^1(b_{1g})^0]$ ground-state formulation proposed by Gray and co-workers, principally on the grounds that it was incompatible with voltammetric (60) and ESR (98) data. However, Gray and his colleagues have conceded (65) that the ordering of the π_2^* and π_1^* levels in their scheme may be reversed.

Thus, there were apparently two different formulations for the $[Ni—S_4]^{-1}$ species and their analogs—one in which the complexes were regarded as containing a d^7 Ni(III) metal ion, and the other in which they were thought to consist of Ni(II) with the unpaired electron strongly delocalized over the ligands (23). The latter is tantamount to a description of the monoanions as cation-stabilized free radicals [csfr in the parlance of Holm, Davison, Maki, and their colleagues (44)]. However, in view of the method by which these formulations were reached, it was felt that they represented extreme descriptions and that perhaps each was appropriate to a particular type of ligand system. In other words, the degree of participation of ligand or metal orbitals in the molecular orbital of the unpaired electron could vary significantly with the type of ligand. This was well illustrated by a study of a number of nickel, copper, and zinc complexes of cyclic tetradentate ligands having the coordination unit $[cis$-$[M—N_2S_2]^z$ (44).

The complexes containing the coordination units $trans$-$[M—N_2S_2]^z$ and cis-$[M—N_2S_2]^z$ were examined voltammetrically (Sec. III-C) and by ESR techniques (Sec. III-E). The neutral zinc and cadmium complexes contained both metal and ligand closed-shell electronic configurations, and, on reduction of these to the monoanions, the electron was placed in a molecular orbital of nearly exclusive ligand character. In Table XVI it is seen that the g factors approach those of aromatic free radicals very closely, and the extent to which spin–orbit coupling with the sulfur atoms effected the g values was thought to be slight. Accordingly, it was proposed that complexes correctly described as cation-stabilized free radicals should exhibit isotropic g values less than ca. 2.01, and slight or unresolvable g-tensor anisotropies. Using these, and the previously described criteria, the paramagnetic complexes, $[Ni-S_4]^{-1}$ and its analogs, $[Ni-N_2S_2]^{-1}$, $[Cu-N_2S_2]^0$, and $[M-N_2S_2]^{-1}$ (M = Zn or Cd) were classified according to the ground-state descriptions shown in Table XXIV.

One of the most interesting features which emerged from this classification was that those complexes in which there was conjugation of the whole ligand system through an azomethine bridge (as, for example, in complexes of gma) exhibited markedly more isotropic g values than those of their nonconjugated analogs, e.g., $[Ni(H_2gma)]^{-1}$ or $trans$-$[Ni-N_2S_2]^{-1}$. This would seem to indicate an increase in the ligand character of the half-filled molecular orbital in the conjugated complexes.

By using the ESR results, qualitative molecular orbital models for the bonding scheme in these cis-$[M-N_2S_2]^z$ species were developed. Because these complexes possessed a common N_2S_2 cyclic chelate ring having idealized C_{2v} symmetry, it was possible to show that the ligand π molecular orbitals transformed as a_2 and b_2 and the σ molecular orbitals as a_1 and b_1. The partial qualitative M.O. diagram for the cis-$[M-N_2S_2]^z$ complexes, excluding $[Ni(H_2gma)]^z$, is shown in Figure 11.

TABLE XXIV

Proposed Ground-State Descriptions of Paramagnetic Nickel, Copper, and Zinc Group Dithiolenes, and Their Analogs

$^2B_{3g}$ or $^2B_g{}^a$ with appreciable metal–orbital admixture in orbital of unpaired electron, [Ni(III)]	Cation-stabilized free radical, Ni(II), d^8	Cu(II), d^9	Cation-stabilized free radical, M(II), d^{10}
$[Ni-S_4]^{-1}$	cis-$[Ni-N_2S_2)^{-1}$	cis-$[Cu-N_2S_2]^0$	cis-$[Zn-N_2S_2]^{-1}$
$[Ni-N_4]^{-1}$	$[Ni(gma)]^{-1}$	cis-$[Cu-N_2O_2]^0$	cis-$[Cd-N_2S_2]^{-1}$
$[Ni-N_2S_2]^{-1}{}^b$	cis-$[Ni-N_2O_2]^{-1}$	$[Cu-S_4]^{-2}$	
$[Ni-O_2S_2]^{-1}$		$[Cu-O_4]^{-2}$	
$[Ni(H_2gma)]^{-1}$			

[a] This ground state refers to $trans$-$[Ni-N_2S_2]^{-1}$.

[b] Includes complexes only of presumed $trans$-$[Ni-N_2S_2]^{-1}$, e.g., of ligands $C_6H_4(SH)(NH_2)$ and $HSC(Ph)N \cdot NH_2$.

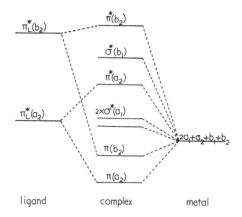

Fig. 11. The qualitative M.O. diagram for *cis*-[M—N$_2$S$_2$]^z complexes and [Ni(gma)]^z.

For the sake of simplicity in the ground-state descriptions of these molecules, it was assumed that the most important ligand orbitals capable of π mixing with the metal d orbitals were the highest filled and lowest empty orbitals in the dinegative ligands, *cis*-[N$_2$S$_2$]$^{-2}$, which were weakly antibonding. The form of these orbitals, designated $\pi_L^*(a_2)$ and $\pi_L^*(b_2)$,

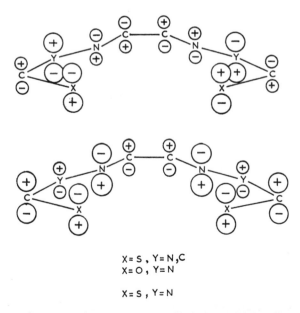

X = S , Y = N , C
X = O , Y = N

X = S , Y = N

Fig. 12. The $\pi_L^*(a_2)$ and $\pi_L^*(b_2)$ molecular orbitals in *cis*-[N$_2$S$_2$]^z ligand systems.

is shown in Figure 12. In terms of symmetry (and on the basis of its eigen-vector), $\pi_L^*(a_2)$ was expected to mix significantly with the metal d_{yz} orbital whereas the mixing of $\pi_L^*(b_2)$ with the metal d_{xz} was thought to be minimal. Thus, the $\pi_L^*(b_2)$ level, as shown in Figure 11 was predominantly ligand in character. It should be observed that the degree of mixing of the ligand π^* orbitals with the metal d orbitals depended not only on overlap factors, but also on the relative energies of the orbitals concerned.

In Table XXV, the complexes cis-$[M—N_2S_2]^z$ are classified into iso-electronic groups in order of decreasing number of valence electrons. All the molecular orbitals in the group A complexes are filled up to, and in-cluding $\pi^*(a_2)$, although the relative ordering of the filled molecular or-bitals could not be determined with certainty (note that the ligand orbital $\pi_L^*(a_2)$ is doubly occupied). It seems evident that they are of lower energy than the $\sigma^*(b_1)$ and $\pi^*(b_2)$ levels. It is clear from magnetic and other spectral data that the ground states of cis-$[Ni—N_2S_2]^{-1}$ and $[Ni(gma)]^{-1}$ are different from those of the isoelectronic cis-$[Cu—N_2S_2]^0$. In the latter, the ground-state configuration was quite definitely

$$\cdots[\sigma^*(b_1)]^1$$

The csfr formulation for the nickel monoanions, however, was perhaps best represented by a reversal of the order of $\sigma^*(b_1)$ and $\pi^*(b_2)$, so that the ground-state configuration of these became

$$\cdots[\pi^*(b_2)]^1$$

A small mixing of the metal d_{xz} orbital with the $\pi^*(b_2)$ molecular orbital

TABLE XXV

Isoelectronic Complexes Having Coordination Unit cis-$[M—N_2S_2]^z$

Groups	A	B	C	D	E
Nickel	$[Ni]^0$ [a]	$[Ni]^{-1}$	$[Ni]^{-2}$ [b]		
Copper	$[Cu]^{+1}$ [c]	$[Cu]^0$	$[Cu]^{-1}$ [b]		
Zinc[d]			$[Zn]^0$	$[Zn]^{-1}$	$[Zn]^{-2}$ [c]

[a] Diamagnetic species.

[b] Presumed diamagnetic, although spin–triplet ground state not impossible; actual ground state unknown because of lack of data.

[c] Presumed diamagnetic, although actual ground state unknown because of lack of data.

[d] Includes cadmium.

could account for the barely observable g-tensor anisotropy. The configuration for the group C complexes of zinc and cadmium

$$\cdots[\sigma^*(b_1)]^2$$

was reasonable but the possibility that the copper mono- and nickel dianions might have spin–triplet ground states, i.e., the configurations

$$\cdots[*(b_1)]^1[\pi^*(b_2)]^1$$

were recognized. The group D and E complexes were suggested to have the configurations,

$$\cdots[\pi^*(b_2)]^1 \quad \text{and} \quad \cdots[\pi^*(b_2)]^2$$

respectively.

It has already been pointed out that the g tensors of $[Ni(gma)]^{-1}$ and $[Ni(H_2gma)]^{-1}$ were markedly different and that those of the latter were similar to those of the (nonconjugated) trans-$[Ni—N_2S_2]^{-1}$ species. This lack of similarity in the g tensors was related to the lack of conjugation throughout the ligand system in $[Ni(H_2gma)]^{-1}$. Assuming that the latter had idealized C_{2v} symmetry (possible structures for this monoanion are

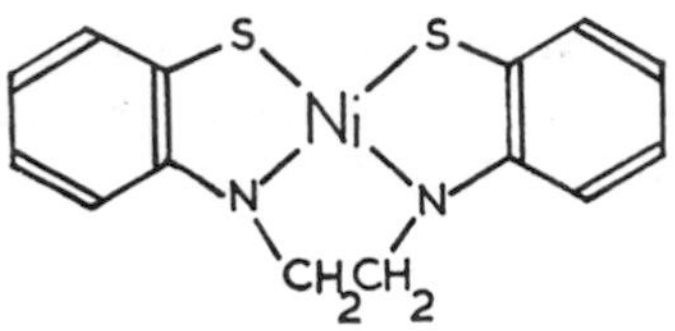

(26a)

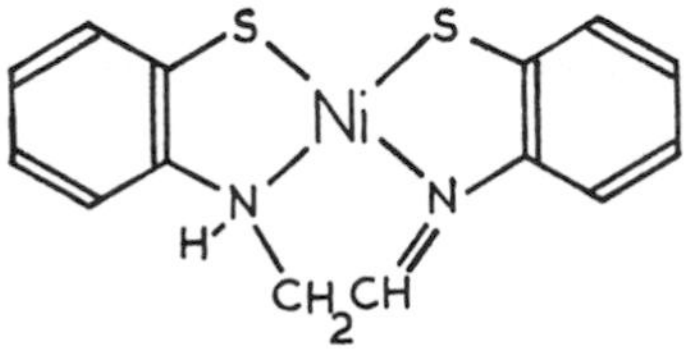

(26 b)

shown in **26**), the basic form of the $\pi_L^*(a_2)$ and $\pi_L^*(b_2)$ orbitals were very similar except that there were no contributions from the bridging carbon atoms. The energies of these two orbitals were taken to be nearly identical and the mixing of the $\pi_L^*(a_2)$ with the d_{yz} metal orbital would be much greater than that of the $\pi_L^*(b_2)$ with d_{xz}. The qualitative M.O. diagram is shown in

Figure 13. Unlike the situation in $[Ni(gma)]^{-1}$ the differential mixing of the $\pi_L^*(a_2)$ and $\pi_L^*(b_2)$ orbitals with the metal orbitals was thought to depend nearly completely on overlap, rather than on a combination of overlap and relative energy considerations, and this was believed to result in a reversal of the relative order of these ligand π^* orbitals, as shown in Figure 13. If the ordering of the molecular orbitals of the complex was as shown, the ground-state configuration of $[Ni(H_2gma)]^{-1}$ would be

$$\cdots[\sigma^*(b_1)]^1$$

which was definitely not consistent with the marked anisotropy of the g tensor in this species. Thus, the configuration must be

$$\cdots[\pi^*(a_2)]^1$$

that is, a reversal of the $\sigma^*(b_1)$ and $\pi^*(a_2)$ levels, with the order of those orbitals more stable than the $\pi^*(a_2)$ unknown. Since it was proposed that appreciable mixing of the metal d_{yz} orbital could be expected with the $\pi_L^*(a_2)$ level, the predicted ground-state would be consistent with the observed g tensor anisotropy and, therefore, the situation would be very similar to that in $[Ni-S_4]^{-1}$, $[Ni-N_4]^{-1}$, $[Ni-O_2S_2]^{-1}$, and trans-$[Ni-N_2S_2]^{-1}$, where the ground states were thought to be $^2B_{3g}$ or ^2B_g.

None of the foregoing discussion permits the controversy over the percentage ligand character of the half-filled orbital in these paramagnetic complexes to be resolved. However, it was possible to discern a relative trend of increasing ligand (and decreasing metal) orbital involvement in the

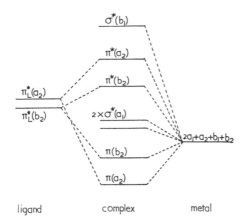

ligand   complex   metal

Fig. 13. The qualitative M.O. diagram for $[Ni(H_2gma)]^z$.

wave function of the unpaired electron. This was arrived at from a scrutiny of the g- and A-tensor anisotropies, and the following order was proposed:

$$[Co—S_4]^{-2} < [Ni—S_4]^{-1} \sim [Ni—N_4]^{-1} \sim [Ni—O_2S_2]^{-1} \sim \textit{trans-}[Ni—N_2S_2]^{-0}$$
$$< [Ni—gma)]^{-1} \sim \textit{cis-}[Ni—N_2S_2]^{-1} \sim \textit{cis-}[Ni—N_2O_2]^{-1}$$
$$< [Zn(gma)]^{-1} \sim \textit{cis-}[Zn—N_2S_2]^{-1}$$

The radical nature of the complexes increases from left to right while the metal character of the half-filled orbital decreases. Thus, $[Co—S_4]^{-2}$ is regarded as containing Co(II). However, the view that the monoanionic nickel complexes, $[Ni—S_4]^{-1}$, $[Ni—N_4]^{-1}$, and $[Ni—O_2S_2]^{-1}$, contain Ni(III) was extreme (although pragmatically useful for handling ESR data) particularly when it is remembered that it can logically be extended to such species as $[Ni—S_4]^0$, $[Ni—N_4]^{+1}$, etc., all of which have been detected and characterized. The use of formal oxidation states in such systems must be misleading.

IV. TRIS-1,2-DITHIOLENE AND RELATED METAL COMPLEXES

A. Syntheses and General Characteristics

The six-coordinate dithiolene complexes were prepared by methods generally similar to those outlined in Section III-A. It appears that of the known dithiolene complexes containing only the sulfur ligands (Table XXVI), tris-substituted complexes are formed exclusively by titanium, vanadium, and the group VI metals but that bis-substituted species, $[M—S_4]_n^z$ ($n = 1$ or 2), and tris complexes are formed by cobalt, iron, ruthenium, osmium, manganese, and rhenium.

1. Complexes Derived From Dianionic Dithiolates

The dianionic tris-dicyano–dithiolene complexes of titanium (24), vanadium (18), molybdenum (73,131), tungsten (73,131), manganese (73,131), rhenium (132), and iron (73,131) were obtained by addition of $Na_2S_2C_2(CN)_2$ to solutions containing the appropriate metal halides, acetylacetonates or aquo salts in polar solvents. It was important in the preparation of the iron complex $[FeS_6C_6(CN)_6]^{-2}$ (73,131) to ensure that the molar ratio of iron salt to dithiolate dianion was $1:3$; otherwise, the dimeric dianion, $[FeS_4C_4(CN)_4]_2^{-2}$, was formed in substantial yields; if this reaction was carried out under strictly anaerobic conditions, the trianion $[FeS_6C_6(CN)_6]^{-3}$ was formed, which was readily oxidized by air in solution to the dianion. Addition of two moles of $Na_2S_2C_2(CN)_2$ to $[FeS_4C_4(CN)_4]_2^{-2}$ or its cobalt analog caused fission of the dimeric units and formation (73,133) of the tris-trianions, $[M—S_6]^{-3}$; the cobalt tri-anion was also obtained (73) by treatment of Co(III) ammines with $Na_2S_2C_2(CN)_2$.

It may be recalled from Section III-A that the most stable complexes of copper, nickel, palladium, and cobalt containing the $S_2C_2H_2$ ligand were monoanionic or neutral species, whereas the most stable complexes containing the dicyano–ethylene–dithiolate group were dianionic and oc-casionally monoanionic. Similar behavior was encountered in the tris-complex systems where treatment of VO(acac)$_2$, Na_2MO_4 (M = Mo or W), Re_2O_7, or $ReCl_5$ with $Na_2S_2C_2H_2$ in HCl solution afforded (95) the neutral complexes $[MS_6C_6H_6]^0$. When Ti(acac)$_3$ was treated with Na_2S_2-C_2H_2 under similar conditions, transient blue colors attributed to $[TiS_6$-$C_6H_6]^{-2}$ were observed but no complexes were isolated.

Replacement of two moles of CO from $[Mo(CO)_2S_4C_4Ph_4]^0$ using $Na_2S_2C_2H_2$ pr $Na_2S_2C_2(CN)_2$ in methanol afforded (135) the neutral complexes $[MoS_6C_6Ph_4H_2]^0$ and $[MoS_6C_6Ph_4(CN)_2]^0$, respectively.

2. Complexes Derived from Dithietenes

Reaction of the group VI metal hexacarbonyls with $(CF_3)_2C_2S_2$ in high-boiling paraffins afforded the volatile, neutral complexes $[MS_6C_6$-$(CF_3)_6]^0$ (14,18). However, when V(CO)$_6$ was treated with the dithietene (136) the monoanion $[VS_6C_6(CF_3)_6]^{-1}$ was produced. The latter was also obtained when $[Ph_4As][V(CO)_6]$ was treated with $(CF_3)_2C_2S_2$, but if the sodium diglyme salt $[Na(diglyme)][V(CO)_6]$ was used, the dianion, $[V—S_6]^{-2}$, was formed. One of the many products of the reaction between $[\pi$-$C_5H_5Ru(CO)_2]_2$ and the dithietene (46) appeared to be $[(\pi$-$C_5H_5)_2Ru]_2$-$[RuS_6C_6(CF_3)_6]$ which may have contained the tris-dithiolene $[Ru—S_6]^{-2}$ stabilized by the ruthenocenium cation.

3. Complexes Derived from Arene Dithiols and Their Analogs

The neutral complexes $[M(S_2C_6H_4)_3]^0$ and $[M(S_2C_6H_3Me)_3]^0$, where M = Mo or W, and $[Re(S_2C_6H_3Me)_3]^0$ were obtained (137) by reaction of the appropriate metal penta- or hexachloride with the arene dithiol in CCl$_4$. The molybdenum complex $[Mo(S_2C_6H_3Me)_3]^0$ was also formed (11) in the reaction between molybdates and toluene-3,4-dithiol in acidic media. If aqueous acidic solutions containing Mo(V) salts were treated with this dithiol (11), mixtures of the green $[Mo(S_2C_6H_3Me)_3]^0$, and a red complex originally thought to contain Mo(IV), were obtained. However, a recent reinvestigation (138) of this reaction established that the red

TABLE XXVI

Tris-Substituted 1,2-Dithiolene Complexes and Related Species

Complex	z	Cation	Color	mp,°C	μ_{eff},BM	S^b	Refs.
$TiS_6C_6(CN)_6$	-2	Me_4N^+	Deep red		Dia.	1	24
$VS_6C_6(CN)_6$	-2	Ph_4As^+	Dark green		1.82	1	18,140
	-1	Ph_4As^+	Purple		Dia.	1,5	18
$VS_6C_6(CF_3)_6$	-2	Ph_4As^+	Purple	205–206	1.83	2	136
	-1	Ph_4As^+	Purple	175–177	Dia.	2	136
$V(S_2C_6H_4)_3$	-2	Bu_4N^+	Purple	201–204		3	136
$V(S_2C_6H_3Me)_3$	-2	Bu_4N^+	Red–purple	188–190	1.73	3	136
$VS_6C_6H_6$	-1	Ph_4As^+		245 d		1	134
	0			ca. 05 d		1	95,136
$VS_6C_6Ph_6$	-2	Et_4N^+		235 d	Dia.	4,5	95,136
	-1	Et_4N^+	Dark blue	300		4	95,136
	0		Purple	250–252	1.80	4	95,136
$VS_6C_6(p\text{-}MeC_6H_4)_6$	-1	Et_4N^+		240 d		4	95
$VS_6C_6(p\text{-}MeOC_6H_4)_6$	-1	Et_4N^+		203 d		4	95
$CrS_6C_6(CN)_6$	-3	Ph_4As^+	Brown	239–244	3.90	1	18
	-3	Ph_4P^+	Brown		3.91	1	73
	-2	Ph_4As^+	Yellow–brown	225–230	2.89	1,5	18
	-2	Ph_4P^+	Yellow–brown	213–216	2.83	1,5	73
$CrS_6C_6(CF_3)_6$	-2	Ph_4As^+	Olive–green	193–194	2.95	2,5	18
	-1	Ph_4As^+	Dark green	128–129	1.89	2,5	18
	0		Red–purple	210–215	Dia.	2	18
$CrS_6C_6Ph_6$	0		Red–purple	291 (241)	Dia.	4	
$MoS_6C_6(CN)_6$	-2	Ph_4P^+	Deep green	280–285	Dia.	1	73
$MoS_6C_6(CF_3)_6$	-2	Ph_4As^+	Dark blue	250	Dia.	2,5	18
	-1	Ph_4As^+	Dark blue	189–193	1.79	2,5	18

Complex	z	Cation	Color	M.p. (°C)	μ		Ref.
$MoS_6C_6(CF_3)_6$	0		Purple	245	Dia.	2	18,14
$Mo(S_2C_6H_4)_3$	0		Green	250		3	137
$Mo(S_2C_6H_3Me)_3$	0		Green	201–204 d		3	137
$MoS_6C_6H_6$	0		Green	ca. 76 d		1	95
$MoS_6C_6Ph_6$	−2	Et_4N^+	Green	235 d		4,5	95,137
$MoS_6C_6(p\text{-}MeC_6H_4)_6$	0		Green	360		4	95
$MoS_6C_6Me_6$	0		Green	306 d		4	95
$WS_6C_6(CN)_6$	−2	Ph_4P^+	Red-purple	360	Dia.	1	73
$WS_6C_6(CF_3)_6$	−2	Ph_4As^+	Red-purple	269–275	Dia.	2,5	18
	−1	Ph_4As^+	Dark blue	250	1.77	2,5	18
	0		Red-purple	178–181	Dia.	2	18
$W(S_2C_6H_4)_3$	0		Blue	—		3	137
$W(S_2C_6H_3Me)_3$	0		Blue	250	Dia.	3	137
$WS_6C_6H_6$	0			250		1	95
$WS_6C_6Ph_6$	0		Blue-green	ca. 103 d		4	95,137
$WS_6C_6(p\text{-}MeC_6H_4)_6$	0			360		4	95
$WS_6C_6(p\text{-}MeOC_6H_4)_6$	0			312 d		4	95
$WS_6C_6Me_6$	0		Blue-green	315 d		4	137,95
$MnS_6C_6(CN)_6$	−2	Ph_4P^+	Deep green	250	4.01	1	131
	−2	Ph_4As^+	Green	172–175	3.85	1	24
$Re(S_2C_6H_4)_3$	0		Green	232		3	137
$Re(S_2C_6H_3Me)_3$	0		Green	188–192 d	1.55	3	137
$ReS_6C_6H_6$	−1	Ph_4As^+		151 d		1,5	95
	0			ca. 105 d		1	95
$ReS_6C_6(p\text{-}MeC_6H_4)_6$	−1	Et_4N^+		262		4	95
	−3	Ph_4P^+	Red-brown	162–165	2.50	1,5	73
	−2	Ph_4P^+	Deep green	174–178	2.99	1	73
	−2	Ph_4As^+			3.00	1	24
$FeS_3C_6(CN)_6$	−2	Et_4N^+	Deep green	ca. 135		1	73

(continued)

TABLE XXVI (continued)

Complex	z	Cation	Color	mp,°C	μ_{eff},BM	S^b	Refs
$Fe(O_2C_6Cl_4)_3$	-3	Pr_4N^+	Violet	179–181	5.93	3	40
$RuS_6C_6Ph_6$	0		Blue–green	225		4	19,20
$OsS_6C_6Ph_6$	0		Red	190 d		4	19,20
$CoS_6C_6(CN)_6$	-3	Pr_4N^+	Green	224–227	Dia.	1	133
	-3	Ph_4P^+	Green		Dia.	1	73
$MoS_6C_6Ph_4(CN)_2$	0		Green	183 d		(1)	135
$MoS_6C_6Ph_4H_2$	0		Green	85 d		(1)	135
$Mo_2S_2(S_4C_4Ph_4)_4$	0		Deep red	275 d		6	135
$Mo_2S_2(S_4C_4(p\text{-}MeC_6H_4)_4)_4$	0		Deep red	187 d		6	135
$W_2S_2(S_4C_4Ph_4)_4$	0			229 d		6	135
$[MoS_{3.5}(S_2C_2Ph_2)]_x$	0		Violet	278 d		6	135

ᵃ At room temperature in the solid state.   ᵇ Syntheses as in Section IV-A.

species was $[Mo_2(S_2C_6H_3Me)_3]^0$ which could be prepared in good yield if the mole ratios of Mo(V) to dithiol in $0.4M$ hydrochloric acid solution were $2:5$.

The dianionic vanadium complexes $[V(S_2C_6H_4)_3]^{-2}$ and $[V(S_2C_6H_3Me)_3]^{-2}$, were obtained (136) from the reaction between VCl_3 and the disodium salts of the ligands in ethanol/THF mixtures. These reactions were initially carried out under nitrogen when deep green solutions were formed. Exposure of these solutions to air caused the color to turn first to brown and finally to reddish-purple; the dianionic tris complexes, themselves, were red–purple in solution and it is possible that the green or brown colors were due to initial formation of the air-unstable trianions $[V-S_6]^{-3}$.*

Reaction between Fe^{2+} salts and tetrachlorocatechol and n-Pr_4NOH in aqueous solution afforded (40) the red–brown $[Pr_4N][Fe(O_2C_6Cl_4)_3]$, the only tris-dithiolene analog to be reported. Although no authentic specimens of neutral species of the type $[Fe-O_6]^0$ have been prepared, complexes closely approximating this formulation were obtained when $Fe(CO)_5$ was reacted with o-chloranil. There are several examples, however, of complexes ostensibly having the formulation $[Fe-O_6]^{+3}$. These include compounds containing phenanthrene-9,10-diones and chrysene-quinone such as $[Fe(phenan)_3]X_3$, where $X = Cl$ or Br, and $[Fe(chrysen)_3]Br_3$. Attempts to prepare $[Fe(O_2C_6Cl_4)_3]Cl_3$ were unsuccessful.

4. Complexes Derived from Disubstituted Ethylene Dithiophosphoric Esters

Reaction of V(III) or V(IV) compounds with the dithiophosphoric esters derived from benzoin, toluoin, and anisoin afforded (95,134,136, 137) the monoanionic species $[V-S_6]^{-1}$ although, under acidic conditions, the corresponding neutral complexes were sometimes obtained.

Treatment of $Cr(CO)_6$ and its molybdenum and tungsten analogs with diphenylacetylene and sulfur in benzene or toluene under pressure (Cr,W) or at atmospheric pressure (Mo) afforded $[MS_6C_6Ph_6]^0$ (17,20).†

---

* *Note added to proof:* A series of mono- and dianionic tristetrachlorobenzene dithiolene complexes, $[M(S_2C_6Cl_4)_3]^z$, have been prepared where $M = V$, Cr, Fe, and Co ($z = -2$) and $M = V$, $z = 1$; $[W(S_2C_6H_3Me)_3]^{-1}$ has also been isolated as the salt of heavy organic cations.

†These complexes were initially described as bis-planar species, $[MS_4C_4Ph_4]^0$ (17) but were subsequently reformulated (19,20) as tris complexes. This confusion may have arisen because of the apparent coexistence of bis- and tris-substituted complexes of ruthenium and osmium (135) and because of the formation of "sulfur-rich" species (Sec. IV-A-6).

These complexes could also be prepared by treatment of the appropriate metal halides or oxyanions with the dithioester solutions, and in this way phenyl-substituted complexes and $[MS_6C_6Me_6]^0$ were obtained (95,137).

The rhenium complex $[ReS_6C_6Ph_6]^0$ was prepared (137,148) by treating $ReCl_5$ in ethanol with $[Ph_2C_2S_2P_2S_2]_2$ in xylene. Other tris-diphenyl–dithiolene complexes which have been mentioned briefly include $[RuS_6-C_6Ph_6]^0$ and $[OsS_6C_6Ph_6]^0$ (22,135).

5. Reduction and Oxidation of Preformed Dithiolene Complexes

According to voltammetric data (Sec. IV-C) the one-electron reduction of the dianionic dicyano–dithiolene complexes of vanadium, manganese, and iron should be relatively easy to achieve. However, treatment (73) of $[VS_6C_6(CN)_6]^{-2}$ with hydrazine hydrate in DMSO did not afford the expected $[VS_6C_6(CN)_6]^{-3}$, although it may have been formed transiently; the major product of this reaction was $[VOS_4C_4(CN)_4]^{-2}$ (140), the oxygen presumably being extracted from the solvent since air was rigorously excluded at all stages during the reaction. The manganese dianion was unaffected by mild reducing agents but was decomposed under more powerful reducing conditions. $[FeS_6C_6(CN)_6]^{-3}$ could easily be obtained from the corresponding dianion by reduction with aqueous sulfite ion; oxidation of the trianion to the dianion was speedily accomplished by air in solution. The trianionic chromium complex, $[CrS_6C_6(CN)_6]^{-3}$ was conveniently oxidized (73) by iodine, giving the corresponding dianion which was alternatively prepared (18) by oxidation of the trianion with $[MoS_6C_6(CF_3)_6]^0$, the molar ratios of the reactant being 2:1. The reaction between one mole of the trianion and two moles of $[CrS_6C_6(CF_3)_6]^0$ apparently produced the monoanion, $[CrS_6C_6-(CN)_6]^{-1}$, but this complex was not isolated.

When the neutral tris-bis-perfluoromethyl–dithiolene complexes of the Group VI metals were dissolved in dry, weakly basic solvents such as acetone or THF, their reduction to the corresponding monoanions took place. However, these complexes were, in general, impure and were better synthesized by prior reduction of the neutral species with hydrazine (or DMF) which afforded the dianions, $[M—S_6]^{-2}$, and treatment of the latter with one mole of the neutral species.

Oxidation of the monoanionic tris-diphenyl–dithiolene complexes of vanadium to the neutral species was accomplished using iodine; reduction of the latter with ethanolic hydrazine regenerated the monoanion. Borohydride reduction of the tris-diphenyl–dithiolene complexes of the group VI metals in diglyme afforded the air-unstable monoanions.

6. Sulfur-Rich Dithiolene Complexes

When the *cis*-dicarbonyl bis-diphenyl–dithiolene complexes of molybdenum and tungsten (135) were treated with sulfide ion, deep red disulfide-bridged complexes, $[M_2S_2(S_4C_4Ph_4)_2]^0$, were formed. The molybdenum complex was also prepared by a modified "dithioester" reaction, but by-products of this reaction, corresponding to the formulation $[Mo(S_{3.5})S_2C_2Ph_2]_x$, were also isolated.

7. General Characteristics

It seems probable that most of the neutral tris-substituted dithiolene complexes adopt trigonal prismatic structures (see Sec. IV-B). Little structural information is available on the charged species. All of the complexes which have so far been isolated undergo electron transfer reactions. The spin ground states of these complexes are summarized in Table XXVII.

The Mössbauer spectra of $[Ph_4P]_2[FeS_6C_6(CN)_6]$ and $[Ph_3P]_3[FeS_6-C_6(CN)_6]$ have been measured (83) (Table III). The increase in chemical shift on reduction of the dianion to the trianion was suggested to be consistent with a decrease in the formal oxidation state of the metal ion.

TABLE XXVII
Spin Ground States of Tris-Substituted
Dithiolene Complexes

| $[M-S_6]^z$ | z | | | |
	-3	-2	-1	0
Ti—S$_6$		0		
V—S$_6$		½	0	½
Cr—S$_6$	$\frac{3}{2}$ [a]	1	½	0
Mo—S$_6$		0	½	0
W—S$_6$		0	½	0
Mn—S$_6$		$\frac{3}{2}$		
Re—S$_6$		½ ?	0	½
Fe—S$_6$	½	1		
Fe—O$_6$ [b,c]	$\frac{5}{2}$			
Co—S$_6$	0			

[a] Only observed in $[CrS_6C_6(CN)_6]^{-3}$.
[b] Observed in $[Fe(O_2C_6Cl_4)_3]^{-3}$.
[c] $[Fe-O_6]^{+3}$, $S = \frac{5}{2}$.

The mass spectra of $[CrS_6C_6(CF_3)_6]^0$ and its molybdenum analog revealed (82) certain basic similarities. The principal fragmentation pattern is summarized as

$$[MC_{12}F_{18}S_6]^+ \longrightarrow [MC_8F_{12}S_4]^+ \longrightarrow [MC_4F_6S_2]^+$$

and

$$[MC_{12}F_{18}S_6]^+ \longrightarrow [MC_{12}F_{17}S_6]^+ \longrightarrow [MC_8F_{11}S_4]^+$$

B. Structural Data

The structures of four tris-substituted dithiolene complexes have been determined by x-ray techniques—$[ReS_6C_6Ph_6]^0$ (27,141), $[MoS_6C_6H_6]^0$ (22), $[VS_6C_6Ph_6]^0$ (23), and $[Me_4N]_2[VS_6C_6(CN)_6]$ (24). The first three adopted essentially trigonal prismatic structures whereas the last was best described as a very severely distorted octahedron.

In the neutral complexes, the metal atoms are surrounded by six sulfur donor atoms in a trigonal prismatic arrangement giving the $[M—S_6]$ coordination D_{3h} symmetry. The metal atoms are all well separated from each other, the closest approach in the rhenium complex being 9 Å and in in the vanadium complex 10 Å. The closest intermolecular contact in $[ReS_6C_6Ph_6]^0$ was with hydrogen atoms on the phenyl rings of an adjacent molecule, at ca. 3.2 Å.

The rhenium complex formed the most perfect trigonal prism, the sides of the prism being 3.043 Å. The metal–ligand planes radiated outwards from the metal in a "paddle-wheel" arrangement and the phenyl rings on these ligands were twisted out of the ligand planes and were skewed with respect to each other. Only one carbon atom in one of the ligands was slightly distorted out of the ligand plane. The intraligand and interligand S—S distances were 3.032 (Table XXVIII) and 3.080 Å, respectively. The overall symmetry of the molecule was approximately C_3. The idealized trigonal prismatic structure is shown in Figure 14.

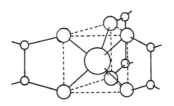

Fig. 14. The idealized trigonal prismatic structure adopted by $[MS_6C_6R_6]^z$.

TABLE XXVIII

Structural Data Obtained from Tris-Substituted Dithiolene Complexes

Complex	M—S[a]	S—M—S[b]	C—C[a,c]	C—S[a]	S—S[a,d]	S—S[a,e]
$[ReS_6C_6Ph_6]^0$	2.325 ±0.004	81.4	1.34 ±0.03	1.62 − 1.75 ±0.03	3.032 ±0.01	3.050 ±0.008
$MoS_6C_6H_6]^0$	2.33±0.02	82.5	1.34±0.04	1.70±0.03	3.10	3.11
$[VS_6C_6Ph_6]^0$	2.338 ±0.004	81.7±0.1	1.38 ±0.01[f] 1.46 ±0.02[g]	1.69±0.01	3.058	2.927, 3.088, 3.178 ±0.00
$[VS_6C_6(CN)_6]^{-2}$[j]	2.36±0.01		1.37±0.02[f] 1.29±0.03[g]	1.71±0.02	3.18 ±0.01[h] 3.12 ±0.01[i]	

[a] In Å.
[b] In degrees.
[c] "Ethylenic" bond length.
[d] Distance within each ligand (intra).
[e] Distance between S atoms related by threefold axis (inter).
[f] Two distances related by twofold axis.
[g] Other distance.
[h] On twofold axis.
[i] Other two ligands.
[j] Me_4N^+ salt.

The vanadium complex adopted a similar but more distorted trigonal prismatic structure. A twofold axis bisects one of the five-membered metal-chelate rings and the prism was slightly distorted about the threefold axis giving three different interligand S—S distances (Table XXVIII). The metal–chelate ring planes were approximately planar and appeared to coincide with the dihedral mirror planes about the threefold axis. The "ethylenic" bond lengths in the unique and other two ligands were 1.38 and 1.46 Å, respectively, suggesting, possibly, that the ligand on the two-fold axis had more "dithiolate" character than the other two.

The molybdenum compound had an essentially trigonal prismatic arrangement of both S and C atoms about the Mo atom. However, the carbon atoms were not exactly coplanar with the MoS_2 units but deviated by 18° from these planes, and the overall symmetry of the molecule was C_{3h}. The Mo—S bond distances were comparable with those in MoS_2 which, like WS_2, had a trigonal prismatic arrangement (143) of S atoms about the central metal atom. The C—S bond distances in this complex, and in the other neutral species, may be compared with the C—S bond length in thiophene, 1.72 Å (142).

The structure of $[Me_4N]_2[VS_6C_6(CN)_6]$ was much more complicated than the other three complexes which had been investigated. The vanadium atom lay on a twofold axis which bisected (Fig. 15) one of the three chelating sulfur ligands, but out of the planes determined by the ligands alone. The six sulfur donor atoms, at an average distance of 2.36 Å from the metal, were arranged in what has been described (24) as a very distorted octahedral arrangement. The intraligand S—S distances are larger than those in the other three neutral complexes (Table XXVIII). The average S—M—S bond angle involving pairs of donor atoms which were farthest apart (and in adjacent ligands) in the trigonal prisms was 136 ± 1°. Taking into account the constraint imposed on an idealized octahedron (where this angle would be 180°) by the chelating nature of the bidentate ligand, this calculated average angle would be ca. 173° which, indeed, was close to that in $[Cr(O_2C_2O_2)_3]^{-3}$ (144), where O—Cr—O is 172 °. The observed angle in the dianionic vanadium dicyano–dithiolene was 158.6° which was closer to the trigonally distorted octahedral structure than to the trigonal prism. The "ethylenic" bond length of the ligand bisected by the twofold axis was some 0.08 Å shorter than the average lengths in the other two ligands, suggesting, as in $[VS_6C_6Ph_6]^0$, that the unique ligand had more "dithiolate" character than the other two. It seeemed that if the geometrical parameters were determined solely by the ligands, then the structure of the dianion would be described as a trigonal prism, but if they were

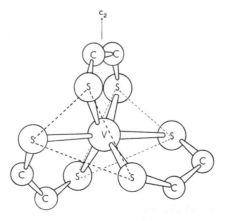

Fig. 15. The molecular structure of the anion in $[Me_4N]_2[VS_6C_6(CN)_6]$.

determined by the MS_6 framework, then an octahedral description would be more appropriate. It was suggested (24) that the resultant structure was a compromise between these two forms.

[ReS$_6$C$_6$Ph$_6$]^0 was isomorphous with its tungsten analog but not with the molybdenum, chromium, and vanadium complexes which were themselves isomorphous with each other and were presumed to be isostructural. The ruthenium and osmium complexes, [MS$_6$C$_6$Ph$_6$]^0 (22), were neither isomorphous with each other nor with any other known tris-diphenyl–thiolene. [MoS$_6$C$_6$H$_6$]^0 was isomorphous with [WS$_6$C$_6$H$_6$]^0 but not with its rhenium analog. It has been reported that the complexes [Ph$_4$As]$_2$-[MS$_6$C$_6$(CN)$_6$], where M = Ti, V, and Cr (24), are isomorphous and presumably isostructural. However, it has been shown (73) that the related complexes [Ph$_4$P]$_2$[MS$_6$C$_6$(CN)$_6$], where M = V, Cr, Mo, W, and Fe, were isomorphous with each other, but not with [Ph$_4$P]$_2$[MnS$_6$C$_6$(CN)$_6$].

C. Polarographic and Voltammetric Studies

The relationships between the half-wave oxidation or reduction potentials and the syntheses and chemical stabilities of the tris-dithiolene complexes are the same as those discussed in Section III-C. In the tris-complexes, it was possible to envisage a four-membered electron transfer series corresponding to species having the charge $z = -3, -2, -1$, or 0 and in certain cases, species having $z = +1$ were detected:

$$[M—S_6]^{-3} \longleftrightarrow [M—S_6]^{-2} \longleftrightarrow [M—S_6]^{-1} \longleftrightarrow [M—S_6]^0(\longleftrightarrow [M—S_6]^{+1})$$

In general, the potential data (Table XXIX) reflected the enhanced stabilization of the more highly charged species ($Z = -2$ or -3) by electron-withdrawing ligand substituents such as CN or CF$_3$. With the exception of the two rhenium complexes so far investigated, there was little evidence for trianionic species in the tris-arene- and tris-diphenyl–dithiolene complexes. Several vanadium, molybdenum, and tungsten complexes were investigated polarographically by Olson, Schrauzer, and Mayweg (60) who found that, for the electrode processes

$$[M—S_6]^0 + e^- \rightleftharpoons [M—S_6]^{-1} \quad \text{and} \quad [M—S_6]^{-1} + e^- \rightleftharpoons [M—S_6]^{-2}$$

the half-wave potentials were strongly dependent upon a ligand substituent effect. Indeed, a linear correlation of the $E_{1/2}$ values with Taft's σ^* inductive substituent constant was demonstrated. These results were very similar to those obtained from polarographic studies of the nickel group dithiolenes (Sec. III-C).

TABLE XXIX

Voltammetric Data Obtained from Tris-Substituted Dithiolene Complexes and Their Analogs

Complex	Solvent	Ref. cell[a]	Electrode process[b]			
			$-3 \rightarrow -2$	$-2 \rightarrow -1$	$-1 \rightarrow 0$	$0 \rightarrow +1$
$[CoS_6C_6(CN)_6$	CH_2Cl_2	*a*	+0.03			
	CH_3CN	*a*	+0.12			
	DMF	*d*	*+0.15*			
$[FeS_6C_6(CN)_6]$	CH_2Cl_2	*a*	−0.38[c]	+0.53[d]		
	CH_3CN	*a*	−0.29[c]	+0.70		
	DMF	*d*	*−0.26*	*+0.73*		
$[Fe(O_2C_6Cl_4)_3]$	CH_2Cl_2	*b*	+0.37[e]	+0.76	+1.10	
$[MnS_6C_6(CN)_6]$	CH_2Cl_2	*a*	−0.35[f]			
	CH_3CN	*a*	−0.36[f]			
	DMF	*d*	*−0.33*			
$[Re(S_2C_6H_3Me)_3]$	CH_3CN	*a*	*−2.06*	*−1.28*	*+0.19*	*+0.69*
	DMF	*c*	−2.591	— −1.812	−0.340	+0.163
	DMF	*d*	*−2.03*	*−1.25*	*+0.22*	*+0.72*
$[ReS_6C_6Ph_6]$	CH_3CN	*a*	*−1.85*	*−1.05*	*+0.47*	*+0.92*
	DMF	*c*	−2.375	−1.577	−0.065	+0.387
	DMF	*d*	*−1.82*	*−1.02*	*+0.50*	*+0.95*
$[CrS_6C_6(CN)_6]$	CH_2Cl_2	*a*	+0.05	+0.69		
	CH_3CN	*a*	+0.16	+0.76		
	DMF	*d*	*+0.08*	*+0.72*		
$[CrS_6C_6(CF_3)_6]$	CH_3CN	*a*			+1.14	
	DMF	*d*			+1.17	
$[MoS_6C_6(CN)_6]$	CH_2Cl_2	*a*	−1.15	+0.49		
	CH_3CN	*a*	−1.12[d]	+0.66		
	DMF	*d*	*−1.09*	*+0.69*		
$[MoS_6C_6(CF_3)_6]$	CH_3CN	*a*	(*−1.5* ?)[f]	+0.36	+0.95	
	DMF	*c*	(*−2.0* ?)[g]	*−0.17*	*+0.42*	
	DMF	*d*		*+0.39*	*+0.98*	
$[Mo(S_2C_6H_3Me)_3]$	CH_3CN	*a*	*−2.09*	*−0 57*	*+0.04*	
	DMF	*c*	−2.62	−1.095	−0.489	
	DMF	*d*	*−2.06*	*−0.54*	*+0.07*	
$[MoS_6C_6Ph_6]$	CH_3CN	*a*	−1.39[h]	*−0.65*[i]	*+0.04*	
	DMF	*c*	−2.92	−1.095	−0.489	
	DMF	*d*	−0.617	−0.617	+0.009	
$[MoS_6C_6H_6]$	CH_3CN	*a*		*−0.78*	*−0.12*	
	DMF	*d*		−0.745	−0.09	
$[MoS_6C_6Me_6]$	CH_3CN	*a*		*−0.97*	*−0.34*	
	DMF	*d*		−0.936	−0.307	
$[WS_6C_6(CN)_6]$	CH_2Cl_2	*a*	−1.52	+0.43		
	CH_3CN	*a*	—	+0.76		
	DMF	*d*		*+0.79*		
$[WS_6C_6(CF_3)_6]$	CH_3CN	*a*		+0.32	+0.88	
	DMF	*d*		*+0.35*	*+0.91*	
$[W(S_2C_6H_3Me)_3]$	CH_3CN	*a*		*−0.55*	*+0.28*	

(*continued*)

TABLE XXIX (*continued*)

Complex	Solvent	Ref. cell[a]	$-3 \rightarrow -2$	$-2 \rightarrow -1$	$-1 \rightarrow 0$	$0 \rightarrow +1$
					Electrode process[b]	
[W(S$_2$C$_6$H$_3$Me)$_3$]	DMF	c		−1.075	−0.247	
	DMF	d		*−0.52*	*+0.31*	
[WS$_6$C$_6$Ph$_6$]	CH$_3$CN	a		−0.71[i]	−0.07[i]	
	DMF	c		−1.135	−0.542	
	DMF	d		−0.684	−0.041	
[WS$_6$C$_6$H$_6$]	CH$_3$CN	a		−0.88	−0.16	
	DMF	d		−0.845	−0.133	
[WS$_6$C$_6$(p-MeC$_6$H$_4$)$_6$]	CH$_3$CN	a		−0.71	−0.12	
	DMF	d		−0.681	−0.091	
[WS$_6$C$_6$(p-MeOC$_6$H$_4$)$_6$]	CH$_3$CN	a		−0.78	−0.17	
	DMF	d		−0.751	−0.138	
[WS$_6$C$_6$Me$_6$]	CH$_3$CN	a		−1.02[i]	−0.36[i]	
	DMF	c		−1.405	−0.839	
	DMF	d		−0.994	−0.333	
[VS$_6$C$_6$(CN)$_6$]	CH$_2$Cl$_2$	a	−0.61	+0.48		
	CH$_3$CN	a	−0.49	+0.69		
	DMF	d	*−0.46*	*+0.72*		
[VS$_6$C$_6$(CF$_3$)$_6$]	CH$_3$CN	a		+0.47		
	DMF	d		*+0.50*		
[V(S$_2$C$_6$H$_4$)$_3$]	CH$_3$CN	a	−0.12			
	DMF	d	*−0.09*			
[V(S$_2$C$_6$H$_3$Me)$_3$]	CH$_3$CN	a	−0.18			
	DMF	d	*−0.15*			
[VS$_6$C$_6$Ph$_6$]	CH$_3$CN	a	−0.71	−0.71	+0.30	
	DMF	d		−0.68	+0.33	
[VS$_6$C$_6$H$_6$]	CH$_3$CN	a		−0.75	+0.22	
	DMF	d		−0.722	+0.25[d]	
[VS$_6$C$_6$(p-MeC$_6$H$_4$)$_6$]	CH$_3$CN	a		−0.78	+0.29	
	DMF	d		−0.745	+0.323	
[VS$_6$C$_6$(p-MeOC$_6$H$_4$)$_6$]	CH$_3$CN	a		−0.81	+0.23	
	DMF	d		−0.783	+0.26	
[Mo$_2$S$_2$(S$_4$C$_4$Ph$_4$)$_2$]	CH$_3$CN	a		−0.83[j]	−0.41	
	DMF	d		−0.80[j]	−0.38	

[a] Reference cells: *a*. SCE; *b*. Ag/AgI (40); *c*. Ag/AgClO$_4$; *d*. Ag/AgCl.

[b] In volts; estimated values in *italics*.

[c] Further reduction wave (with diffusion current halved) at −0.72 (CH$_2$Cl$_2$) or −0.97 (CH$_3$CN).

[d] Irreversible waves.

[e] Further reduction wave (−3 → −4) at −1.66 (CH$_3$CN, *a*).

[f] Further reduction wave (with diffusion current halved) at −1.19 (CH$_2$Cl$_2$) or −1.25 (CH$_3$CN).

[g] Estimated.

[h] Calculated on results obtained in DMF, *c*; conversion factor 0.53.

[i] Calculated on results obtained in DMF, *d*; conversion factor 0.03.

[j] Two-electron reduction.

As indicated above, only the rhenium complexes $[ReS_6C_6Ph_6]^0$ and $[Re(S_2C_6H_3Me)_3]^0$ exhibited (137,148) oxidation waves corresponding to the formation of monocations, $[Re—S_6]^{+1}$. Neither of these species has been isolated yet, although it may be anticipated that one-electron oxidation might be achieved using either $[MoS_6C_6(CF_3)_6]^0$ or its tungsten analog.

As in the bis-substituted complex series, the bis-perfluormethyl–dithiolene complexes were found to be the most powerful one-electron oxidizing agents. $[MoS_6C_6(CF_3)_6]^0$ was an effective oxidizing agent for generating $[VS_6C_6(CN)_6]^{-1}$ (18) from the corresponding dianion, and treatment of $[CrS_6C_6(CN)_6]^{-2}$ with one mole of $[CrS_6C_6(CF_3)_6]^0$ was believed (18) to produce the chromium dicyano–dithiolene monoanion. These bis-perfluoromethyl–dithiolene complexes were themselves unstable to reduction in weakly basic solvents, as their half-wave potential data indicated.

Attempts to effect the one-electron reduction of the dianionic tris-dicyano–dithiolene complexes $[MS_6C_6(CN)_6]^{-2}$, where M = V, Mn, and Fe, met with conspicuous lack of success (73), as has been described, except in the case of the iron complex where sulfite or hydrazine reduction afforded the trianionic species, $[Fe—S_6]^{-3}$. The voltammograms of the manganese and iron tris-dianions in dichloromethane solution displayed second cathodic waves having diffusion currents approximately one-half of those due to the reduction waves for the process $z = -2 \rightarrow -3$; the nature of these further reduced species is unknown. Although the voltammetric data relating to the one-electron oxidation of $[CoS_6C_6(CN)_6]^{-3}$ indicated that this reaction might be achieved using iodine, all attempts to form the dianion in this way resulted in the dissociation of the complex and formation of the dimeric dianion, $[CoS_4C_4(CN)_4]_2^{-2}$.

The only dithiolene analog system which has been investigated (40) was the complex formed by Fe(III) salts and tetrachlorocathechol. It was possible to envisage an eight-membered series related by one-electron transfer reactions in which the two most reduced species were formulated as containing Fe(II) and Fe(III) with three dianionic tetrachlorocatecholate ligands. The series would then be

$$[Fe—O_6]^{-4} \longleftrightarrow [Fe—O_6]^{-3} \longleftrightarrow [Fe—O_6]^{-2} \longleftrightarrow [Fe—O_6]^{-1} \longleftrightarrow$$
$$[Fe—O_6]^0 \longleftrightarrow [Fe—O_6]^{+1} \longleftrightarrow [Fe—O_6]^{+2} \longleftrightarrow [Fe—O_6]^{+3}$$

The most oxidized species would be represented by complexes of Fe(III) containing o-quinones. Voltammetric studies of $[Fe(O_2C_6Cl_4)_3]^{-3}$ in dichloromethane demonstrated the existence of three one-electron oxidation waves corresponding, presumably, to the generation of the species

$z = -2$, -1, and 0, whereas, in acetonitrile, a one-electron reduction wave was detected which may have been due to the formation of the species $z = -4$.

Reversible polarographic reduction of $[Mo_2S_2(S_4C_4Ph_4)_2]^0$ in acetonitrile indicated the existence of three anionic species having $z = -1$ -3, and -4.

D. Infrared and NMR Spectral Studies

The infrared data obtained from several different types of tris-dithiolene complexes (18,95) are presented in Table XXX. The assignments of these frequencies are the same as those described in Section III-D, namely, that ν_1 corresponds to the perturbed "$C{=}C$" stretching frequency, ν_2 to the perturbed "$C{=}S$" stretching mode, ν_3 to vibrational modes associated

with the $R{-}C\underset{C}{\overset{S}{\diagdown}}$ group, and ν_4 and ν_5 to the "M—S" stretching frequencies."

Changes in the "$C{=}C$" and "$C{=}S$" bond orders should be reflected in relative shifts of the frequencies ν_1, ν_2, and ν_3. Thus, as the "dithioketonic" nature of the sulfur ligands decreased, and the "dithiolate" nature increased, it might reasonably be expected that ν_1 would increase and ν_2 and ν_3 decrease. An investigation of the available infrared data revealed that ν_1 increased as the charge on the basic coordination unit $[M{-}S_6]$, z, increased, that it increased in any one complex series as the central metal atom became heavier, and that it decreased when the ligand substituent was varied, the order being $CF_3 > Me > Ph > H$; it was not clear why the bis-perfluoromethyl–dithiolene complexes appeared in this particular order. Less regular trends could be observed in ν_2, although it did increase as the central metal atom became progressively heavier; ν_3 varied in the predicted way.

From the accumulation of data described above, it would seem that as z increases, the sulfur ligands become more "dithiolate" in character. This could be explained in terms of d-orbital expansion in the charged species, relative to the neutral complexes, leading to increased back-donation from the metal to the ligand antibonding π-molecular orbitals. A similar explanation could be advanced for the relative trends in ν_1 as the central metal atom became heavier, that is, that metal–ligand overlap would become more favorable with a third-row metal in comparison with a first-row metal.

TABLE XXX

Infrared Date from Tris-Dithiolene Complexes

Complex	$\nu_1{}^b$	ν_2	ν_3	ν_4	ν_5	Medium	Cation
$[VS_6C_6H_6]^0$	1347	1113	894(401)[a]	385	361	KBr disk	
$[VS_6C_6H_6]^{-1}$	1416	1118	849(431)	392	363	KBr disk	Ph_4As^+ salt
$[VS_6C_6H_6]^{-2}$	1494	952	799	367	350	KBr disk	Ph_4As^+ salt
$[VS_6C_6Ph_6]^0$	1372	1150	892	406	346	KBr disk	
$[VS_6C_6Ph_6]^{-1}$	1428	1165	869(418)[a]	398	349	KBr disk	Et_4N^+ salt
$[CrS_6C_6(CF_3)_6]^0$	1446					Fluorolube mull	
$[CrS_6C_6(CF_3)_6]^{-1}$	1484					Fluorlube mull	Ph_4As^+ salt
$[CrS_6C_6(CF_3)_6]^{-2}$	1511					Fluorolube mull	Ph_4As^+ salt
$[CrS_6C_6Ph_6]^0$	1398	1160	891	421	356	KBr disk	
$[MoS_6C_6(CF_3)_6]^0$	1455					Fluorolube mull	
$[MoS_6C_6(CF_3)_6]^{-1}$	1508					Fluorolube mull	Ph_4As^+ salt
$[MoS_6C_6(CF_3)_6]^{-2}$	1538					Fluorolube mull	Ph_4As^+ salt
$[MoS_6C_6H_6]^0$	1402	1121	866	380	354	KBr disk	
$[MoS_6C_6Ph_4(CN)_2]^0$	1376 1400	1157 1177				KBr disk	
$[MoS_6C_6Ph_6]^0$	1400	1165	878	403	356	KBr disk	
$[MoS_6(S_2C_2Ph_2)]_2{}^0$	1406	1176	885 855	(387)[a]		KBr disk	
$[MoS_6C_6Me_6]^0$	1428	934	565	402	306	KBr disk	

Compound						Medium	Salt
$[WS_6C_6(CF_3)_6]^0$	1474					Fluorolube mull	
$[WS_6C_6(CF_3)_6]^{-1}$	1520					Fluorolube mull	Ph_4As^+ salt
$[WS_6C_6(CF_3)_6]^{-2}$	1541					Fluorolube mull	Ph_4As^+ salt
$[WS_6C_6H_6]^0$	1408	1118	854(430)[a]	369	329	KBr disk	
$[WS_6C_6Ph_6]^0$	1422	1165	872	403	359	KBr disk	
$[WS_6C_6Me_6]^0$	1470	930	562	398	300	KBr disk	
$[ReS_6C_6H_6]^0$	1418	1106	856(422)[a]	338	333	KBr disk	
$[ReS_6C_6H_6]^{-1}$	1450	1099	824(422)[a]	361(351)[a]	333	KBr disk	
$[ReS_6C_6Ph_6]^0$	1430	1172	879(398)[a]	373	359	KBr disk	Ph_4As^+ salt
$[ReS_6C_6Ph_6]^{-1}$	1481	1030	?(374)[a]	361	350	KBr disk	Et_4N^+ salt

[a] Low-intensity bands of uncertain origin, or shoulders.
[b] All frequencies in cm^{-1}.

Infrared studies of the mixed ligand complex $[MoS_6C_6Ph_4(CN)_2]^0$ revealed (135) that ν_1 and ν_2 were split into two components. This result was interpreted as evidence for a particular valence-bond structure (27) which was one of the principal contributing forms to the structure of this

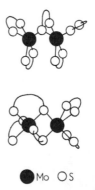

(27)

complex. It was indicated that such a structure would be consistent with the superior electron-withdrawing ability of the dicyano ligand in relation to the diphenyl ligands. In the analogous $[MoS_6C_6Ph_4H_2]^0$, no splitting of ν_1 or ν_2 was observed.

The proton resonance spectra of three dithiolene complexes $[MS_6C_6H_6]^z$, where M = V and Re and $z = -1$, and M = Mo and $z = -2$, were recorded (95). In the vanadium complex, the resonance occurred at $\tau = 0.8$, similar to that in $[NiS_4C_4H_4]^0$ (57). In the rhenium and molybdenum compounds, the resonances occurred at higher fields, at $\tau = 2.6$

● Mo   ○ S

Fig. 16. Two possible structures for $[Mo_2(S_2C_6H_3Me)_5]$.

and 2.39, respectively. These resonances, however, are well below that in $Na_2S_2C_2H_2$ suggesting that ring currents of the type experienced in aromatic systems may be operating to cause the resonances to shift to lower field.

The proton resonance spectrum of $[Mo_2(S_2C_6H_3Me)_3]^0$ in CS_2 (138) showed four distinct signals due to methyl protons. Two possible structures for this dinuclear complex are shown, Figure 16, and it would seem that the latter is favored by this NMR data.

E. ESR Spectral and Magnetic Properties

The magnetic data obtained from the tris-dithiolene complexes are relatively sparse (Table XXXI). However, such that is available revealed that the first-row transition metal complexes tended to exhibit high-spin behavior, whereas the second- and third-row complexes were low spin. While this remark, on its own, is hardly surprising within the context of general transition metal chemistry, it is of significance within the context of these dithiolene complexes and their proposed electronic configuration.

Only the tris-dicyano–dithiolenes have been studied magnetically over a temperature range (90–300°K)(73). It was found that all of the paramagnetic species, with the exception of $[CrS_6C_6(CN)_6]^{-2}$ and $[FeS_6C_6(CN)_6]^{-3}$, had moments close to the spin-only values expected for ground states with no orbital contributions to their susceptibilities, and were also virtually independent of temperature. These results were believed to be consistent with orbital singlet ground states (unsplit, to a first approximation, by spin–orbital coupling effects). The moment of $[CrS_6C_6(CN)_6]^{-2}$ experienced a small, but definite, temperature dependence and this was believed to be indicative of an orbital doublet ground state for this complex. It should be noted that $[CrS_6C_6(CF_3)_6]^{-2}$ was also high spin ($S = 1$) whereas all the molybdenum and tungsten complex dianions were diamagnetic. The moment of $[FeS_6C_6(CN)_6]^{-3}$, which could be compared with that of 2.28 BM for $[Fe(S_2C_2O_2)_3]^{-3}$ (145), in addition to being markedly temperature dependent, was also unusually high. While these facts might be consistent with an orbital doublet ground state for this trianion, other factors, such as spin-state equilibria, might be important and lead to a different ground state formulation. Interesting magnetic behavior also has been observed in the formally isoelectronic dithiocarbamates, $Fe(S_2CNR_2)_3$ (146) where the molecules existed in equilibria between spin–doublet and sextet states.

ESR signals were observed in solutions and in glasses from all species having $S = \frac{1}{2}$ (Table XXXI), but those complexes having more than one unpaired electron apparently did not give resonances. A fairly comprehensive analysis of the ESR data obtained from vanadium and group VI metal dithiolenes (particularly the bis-perfluoromethyl–dithiolenes) has been made (18,136). The ESR parameters were compared with those expected from [Ti(acac)$_3$] (147). It was concluded that the vanadium dianions and the monoanionic group VI metal dithiolenes could not be described

TABLE XXXI

Solution ESR Data for and Solid-State Magnetic Moments of Tris-Substituted
1,2-Dithiolene Complexes

Complexes	Cation	Solvent	$\langle g \rangle$	$\langle A \rangle$,[a]G	μ,BM[b]
[CoS$_6$C$_6$(CN)$_6$]$^{-3}$	Ph$_4$P$^+$				Dia.
[FeS$_6$C$_6$(CN)$_6$]$^{-3}$	Ph$_4$P$^+$				2.44[c]
[FeS$_6$C$_6$(CN)$_6$]$^{-2}$	Ph$_4$P$^+$				2.89[c]
	Ph$_4$As$^+$				3.00
[Fe(O$_2$C$_6$Cl$_4$)$_3$]$^{-3}$	Pr$_4$N$^+$				5.93
[MnS$_6$C$_6$(CN)$_6$]$^{-2}$	Ph$_4$P$^+$				3.90[c]
	Ph$_4$As$^+$				3.85
[Re(S$_2$C$_6$H$_3$Me)$_3$]^0	—	CHCl$_3$	2.010 ± 0.003		1.55
[ReS$_6$C$_6$H$_6$]^0	—	Solid	2.010		
[ReS$_6$C$_6$Ph$_6$]^0	—	CHCl$_3$	2.015 ± 0.003		1.79
		THF	2.015		
[CrS$_6$C$_6$(CN)$_6$]$^{-3}$	Ph$_4$P$^+$				3.90[c]
	Ph$_4$As$^+$				3.90
[CrS$_6$C$_6$(CN)$_6$]$^{-2}$	Ph$_4$P$^+$				2.85[c]
	Ph$_4$As$^+$				2.89
[CrS$_6$C$_6$(CF$_3$)$_6$]$^{-2}$	Ph$_4$As$^+$				2.95
[CrS$_6$C$_6$(CF$_3$)$_6$]$^{-1}$	Ph$_4$As$^+$	CH$_2$Cl$_2$	1.9941 ± 0.0003	16.3 ± 0.5	1.89
[CrS$_6$C$_6$Ph$_6$]$^{-1}$	—[d]	Diglyme	1.996	19.0 ± 0.5	
[MoS$_6$C$_6$(CN)$_6$]$^{-2}$	Ph$_4$P$^+$				Dia.
[MoS$_6$C$_6$(CF$_3$)$_6$]$^{-2}$	Ph$_4$As$^+$				Dia.
[MoS$_6$C$_6$(CF$_3$)$_6$]$^{-1}$	Ph$_4$As$^+$	CHCl$_3$	2.0097 ± 0.0003	12.2 ± 0.5	1.79
[MoS$_6$C$_6$(CF$_3$)$_6$]^0	—				Dia.
[Mo(S$_2$C$_6$H$_3$Me)$_3$]$^{-1}$	—[d]	Acetone	2.003	29.1 ± 1	
[MoS$_6$C$_6$Ph$_6$]$^{-1}$	—[d]	Diglyme	2.011	11.4 ± 0.4	
[WS$_6$C$_6$(CN)$_6$]$^{-2}$	Ph$_4$P$^+$				Dia.
[WS$_6$C$_6$(CF$_3$)$_6$]$^{-2}$	Ph$_4$As$^+$				Dia.
[WS$_6$C$_6$(CF$_3$)$_6$]$^{-1}$	Ph$_4$As$^+$	CHCl$_3$	1.9910 ± 0.0005		1.77
[WS$_6$C$_6$(CF$_3$)$_6$]^0	—				Dia.
[WS$_6$C$_6$Ph$_6$]$^{-1}$	—[d]	Diglyme	1,992		
[W(S$_2$C$_6$H$_3$Me)$_3$]$^{-1}$	—[d]	THF	1.974 ± 0.005	50 ± 10	

(continued)

TABLE XXXI (*continued*)

Complexes	Cation	Solvent	$\langle g \rangle$	$\langle A \rangle$,[a]G	μ,BM[b]
$[VS_6C_6(CN)_6]^{-2}$	Ph_4P^+				1.78[c]
	Ph_4As^+	$CHCl_3/$ DMF	1.980 ± 0.001	63.3 ± 0.5	1.82
$[VS_6C_6(CN)_6]^{-1}$	Ph_4As^+				Dia.
$[VS_6C_6(CF_3)_6]^{-2}$	Ph_4As^+	CH_2Cl_2	1.9829 ± 0.0005	62.4 ± 0.2	1.83
$[VS_6C_6(CF_3)_6]^{-1}$	Ph_4As^+	CH_3CN	1.9822 ± 0.0005	62.1 ± 0.2	
$[VS_6C_6(CF_3)_6]^{-1}$	Ph_4As^+				Dia.
$[V(S_2C_6H_3Me)_3]^{-2}$	Bu_4N^+	CH_2Cl_2	1.9782 ± 0.0005	65.8 ± 0.3	1.73
$[V(S_2C_6H_4)_3]^{-2}$	Bu_4N^+	CH_2Cl_2	1.9799 ± 0.0005	65.2 ± 0.3	
$[VS_6C_6H_6]^{-2}$	—[d]	$CHCl_3$	1.981	67.8 ± 0.2	
$[VS_6C_6H_6]^0$	—	$CHCl_3$	1.991	62.6 ± 0.2	
$[VS_6C_6Ph_6]^{-2}$	—[e]	CH_3CN	1.9811 ± 0.0005	63.9 ± 0.2	
$[VS_6C_6Ph_6]^{-1}$	Et_4N^+				Dia.
$[VS_6C_6Ph_6]^0$	—	CH_2Cl_2	1.9960 ± 0.0005	61.5 ± 0.2	1.80

[a] Metal hyperfine splittings: ^{53}Cr, $I = \frac{3}{2}$; $^{95,97}Mo$, $I = \frac{5}{2}$; ^{183}W, $I = \frac{1}{2}$; ^{51}V, $I = \frac{7}{2}$.

[b] Measured at room temperature unless otherwise stated.

[c] Measured over a temperature range $90°-300°K$ (73).

[d] Salt generated chemically in solution.

[e] Salt generated electrochemically in solution.

correctly as containing d^1 metal ion systems. This was deduced from the several inconsistencies in the ESR data when compared with theory, namely, that there was a near lack of anistotropy in the g tensors and that these, in solution, were close to the free electron value, particularly in $[MoS_6C_6(CF_3)_6]^{-1}$, that $g_\perp \gtrsim g_\parallel$ rather than the opposite, that in $[VS_6C_6(CN)_6]^{-2}$, $|A_\parallel| > |A_\perp|$ rather than the reverse according to theory, and that the observed anisotropic metal-hyperfine interactions in the vanadium and molybdenum complexes were much smaller than those calculated by theory.

In the neutral diphenyl–dithiolene complex, $[VS_6C_6Ph_6]^0$, the g value was 1.9900 and the isotropic metal-hyperfine constant was very similar to those observed in the dianionic species, $[V—S_6]^{-2}$ (136). It was suggested that the neutral complex had a ground state very similar to those of the dianionic compounds. The normalized hyperfine splittings of 42 G/nm for $[VS_6C_6Ph_6]^0$ and 60 G/nm for $[CrS_6C_6Ph_6]^{-1}$ (which contains two more electrons) indicated (19) that the partially filled molecular orbital in the latter had somewhat more metallic character than that in the former.

In the neutral rhenium complexes (137,148), the resonance signals were broad in the solid state and in solution, and the g tensors were close to the free electron value; no metal hyperfine interactions were observed.

The magnetic moment of $[Mo_2(S_2C_6H_3Me)_3]^0$ (138) was 0.2 BM at 17°, much below the spin-only value for Mo(V) but the complex displayed a weak ESR signal in dichloromethane at 1.999, and metal-hyperfine splitting, $\langle A \rangle = 32$ G, was observed.

F. The Electronic Spectra of Tris-Substituted 1,2-Dithiolene Complexes

The electronic spectra of the highly oxidized complexes $[M\text{---}S_6]^z$, where $z = 0, -1$, or -2, were characterized by intense transitions in the visible region (Table XXXII). The spectra of the trianionic species

TABLE XXXII

Electronic Spectral Data from Tris-Substituted Dithiolene and Other Compounds

Complexes	Solvent	Maxima (intensity), cm^{-1}	Assignments[a] (1)	(2)
$[VS_6C_6H_6]^0$	CH_2Cl_2	11,620 (525)	$\pi \rightarrow \pi$	
		14,990 (7,998)	$\pi \rightarrow \pi$	
		16,130 (6,457 sh)	$\pi \rightarrow \pi$	
		19,660 (11,400)	$M \rightarrow \pi$	
		20,700 (5,888 sh)	$M \rightarrow \pi$	
		24,750 (2,963)	$\pi \rightarrow \pi$	
		28,410 (4,677)	$\pi \rightarrow \pi$	
		32,680 (3,890)	$\pi \rightarrow \pi$	
		39,220 (15,140 sh)	$\pi \rightarrow \pi$	
$[VS_6C_6Ph_6]^0$	$CHCl_3$	9,850 (1,500)		
		13,310 (27,000)		
		17,940 (14,600)		
		23,500 (sh)		
		39,200 (65,300)		
	Solid	9,300		
		12,200		
		17,500		
		23,000		
$[CrS_6C_6Ph_6]^0$	$CHCl_3$	15,000 (11,200)	$\pi \rightarrow \pi$	
		17,200 (24,100)	$\pi \rightarrow \pi$	
		23,250 (3,981 sh)	$\pi \rightarrow \pi$	
		35,200 (63,000)	$\pi \rightarrow \pi$	
		40,700 (63,830)	$\pi \rightarrow \pi$	

(continued)

TABLE XXXII (*continued*)

Complexes	Solvent	Maxima (intensity), cm^{-1}	Assignments[a]	
			(1)	(2)
	Solid	13,900		
		16,800		
[MoS$_6$C$_6$H$_6$]^0	CHCl$_3$	16,580 (9,550)	$\pi \rightarrow \pi$	
		20,410 (3,890)	$n_{=S} \rightarrow \pi$	
		24,210 (9,210)	M $\rightarrow \pi$	
		27,000 (3,162 sh)	$\pi \rightarrow \pi$	
		33,000 (4,786)	$\pi \rightarrow \pi$	
		38,800 (6,166)	$\pi \rightarrow \pi$	
[MoS$_6$C$_6$Ph$_6$]^0	CHCl$_3$	14,490 (20,890)	$\pi \rightarrow \pi$	$\pi \rightarrow \pi$
		15,150 (2,582)	$\pi \rightarrow \pi$	
		22,270 (12,220)	M $\rightarrow \pi$	$\pi \rightarrow$ M
		31,440 (11,610 sh)	$\pi \rightarrow \pi$	
		34,960 (24,830 sh)	$\pi \rightarrow \pi$	
		40,000 (61,600)	$\pi \rightarrow \pi$	
	Solid	13,800		
		21,600		
[Mo(S$_2$C$_6$H$_3$Me)$_3$]^0	CHCl$_3$	11,500 (1,000 sh)		
		14,630 (20,900)		$\pi \rightarrow \pi$
		22,990 (17,400)		$\pi \rightarrow$ M
[WS$_6$C$_6$H$_6$]^0	CHCl$_3$	17,300 (6,699)	$\pi \rightarrow \pi$	
		20,000 (2,818 sh)	$n_{=S} \rightarrow \pi$	
		24,750 (4,130)	M $\rightarrow \pi$	
		27,170 (3,981)	$\pi \rightarrow \pi$	
		31,250 (2,188 sh)	$n_{=S} \rightarrow \pi$	
		35,970 (2,630 sh)	$\pi \rightarrow \pi$	
		39,200 (4,786 sh)	$\pi \rightarrow \pi$	
[WS$_6$C$_6$Ph$_6$]^0	CHCl$_3$	15,150 (27,900)	$\pi \rightarrow \pi$	$\pi \rightarrow \pi$
		18,520 (3,020)	$\pi \rightarrow \pi$	$\pi \rightarrow \pi$
		23,980 (11,700)	M $\rightarrow \pi$	$\pi \rightarrow$ M
		33,330 (17,780)	$\pi \rightarrow \pi$	
		40,000 (85,110)	$\pi \rightarrow \pi$	
	Solid	14,800		
		24,400		
[WS$_6$C$_6$Me$_6$]^0	CHCl$_3$	16,450 (20,200)		$\pi \rightarrow \pi$
		19,800 (1,800 sh)		$\pi \rightarrow \pi$
		24,750 (9,600)		$\pi \rightarrow$ M
[W(S$_2$C$_6$H$_3$Me)$_3$]^0	CHCl$_3$	12,400 (200 sh)		
		15,670 (23,400)		$\pi \rightarrow \pi$
		23,000 (900 sh)		$\pi \rightarrow \pi$
		25,890 (15,700)		$\pi \rightarrow$ M
[ReS$_6$C$_6$Ph$_6$]^0	CHCl$_3$	8,230 (1,090)		M $\rightarrow \pi$

(*continued*)

TABLE XXXII (*continued*)

Complexes	Solvent	Maxima (intensity), cm^{-1}	Assignments[a]	
			(1)	(2)
		14,050 (24,000)		$\pi \rightarrow \pi$
		20,800 (ca. 1,000 sh)		$\pi \rightarrow \pi$
		23,450 (12,300)		$\pi \rightarrow M$
	solid	8,000		
		13,500		
		23,000		
$[Re(S_2C_6H_3Me)_3]^0$	$CHCl_3$	7,920 (200)		
		9,220 (500)		
		14,450 (16,000)		
		24,930 (11,000)		
Monoanions				
$[VS_6C_6H_6]^{-1}$	$CHCl_3$	15,500 (891)	$\pi \rightarrow \pi$	
		17,240 (1,462)	$\pi \rightarrow \pi$	
		18,680 (1,462)	$M \rightarrow \pi$	
		27,500 (2,754)	$\pi \rightarrow \pi$	
		33,900 (3,388)	$\pi \rightarrow \pi$	
		40,820 (4,083)	$\pi \rightarrow \pi$	
$[VS_6C_6Ph_6]^{-1}$	$CHCl_3$	14,490 (12,020)	$\pi \rightarrow \pi$	
		16,750 (14,790)	$\pi \rightarrow \pi$	
		18,380 (10,470)	$M \rightarrow \pi$	
		30,760 (sh)	$\pi \rightarrow \pi$	
		33,500 (31,620 sh)	$\pi \rightarrow \pi$	
		42,010 (63,970)	$\pi \rightarrow \pi$	
Dianions				
$[VS_6C_6(CN)_6]^{-2}$	CH_2Cl_2	10,300 (2,800)		
		15,600 (2,100 sh)		
		17,200 (4,700)		
		18,900 (3,700 sh)		
		23,400 (6,600)		
		32,400 (21,600)		
$[CrS_6C_6(CN)_6]^{-2}$	CH_2Cl_2	4,800 (1,200)		
		9,900 (200)		
		14,400 (2,900)		
		17,400 (2,400 sh)		
		20,300 (4,800)		
		28,700 (12,700 sh)		
$[MoS_6C_6(CN)_6]^{-2}$	CH_2Cl_2	15,000 (5,600)		
		20,000 (2,700)		
		21,500 (2,600 sh)		
		25,600 (10,000)		

(*continued*)

TABLE XXXII (*continued*)

Complexes	Solvent	Maxima (intensity) cm^{-1}	Assignments[a] (1)	(2)
$[WS_6C_6(CN)_6]^{-2}$	CH_2Cl_2	17,500 (4,000)		
		20,200 (6,300)		
		23,400 (3,500)		
		26,300 (7,400)		
$[MnS_6C_6(CN)_6]^{-2}$	CH_2Cl_2	12,100 (2,400)		
		15,800 (2,800)		
		16,700 (2,400 sh)		
		31,900 (29,800)		
$[FeS_6C_6(CN)_6]^{-2}$	CH_2Cl_2	12,400 (3,100)		
		15,500 (1,000 sh)		
		24,400 (7,100)		

Trianions[b]

Complexes	Solvent	Maxima (intensity) cm^{-1}	Assignments[a] (1)	(2)
$[CrS_6C_6(CN)_6]^{-3}$	CH_2Cl_2	14,100 (420)		
		17,700 (1,700)		
		21,100 (6,000)		
		23,500 (8,500)		
$[Cr(S_2C=C(CN)_2)_3]^{-3}$	CH_3CN	15,300 (36)		
		20,250 (50 sh)		
		22,700 (227 sh)		
		26,800 (7,200)		
$[Cr(S_2CNEt_2)_3]$	CHCl	15,500 (380)		
		20,400 (360)		
		31,600		
$[Cr(S_2C_2O_2)_3]^{-3}$	Water	17,000 (300)		
		21,900 (590)		
$[FeS_6C_6(CN)_6]^{-3}$	CH_2Cl_2	12,500 (2,200)		
		24,400 (8,900 sh)		
		27,900 (16,300 sh)		
$[Fe(S_2C=C(CN)_2)_3]^{-3}$	CH_3CN	15,600 (8,900)		
		19,000 (5,800)		
		26,300 (27,000 sh)		
$[Fe(S_2CNEt_2)_3]$	$CHCl_3$	17,100 (1,200)		
		19,700 (1,700)		
		25,700 (7,000)		
$[Fe(S_2C_2O_2)_3]^{-3}$	Water	13,000 (124 sh)		
		18,300 (2,605)		
		20,900 (2,455)		
$[CoS_6C_6(CN)_6]^{-3}$	CH_2Cl_2	14,900 (1,100)		
		21,500 (6,800)		
$[Co(S_2C=C(CN)_2)_3]^{-3}$	CH_3CN	15,000 (517)		
		24,900 (36,000)		
		29,600 (62,000)		

(*continued*)

TABLE XXXII (*contiuued*)

Complexes	Solvent	Maxima (intensity) cm^{-1}	Assignments[a]	
			(1)	(2)
[Co(S$_2$CNEt$_2$)$_3$]	CHCl$_3$	15,400 (500)		
		21,000 (760 sh)		
		25,500 (10,000 sh)		
		27,800 (14,000)		
[Co(S$_2$C$_2$O$_2$)$_3$]$^{-3}$	Water	17,700 (860)		
		22,000 (3,020)		

[a] Assignments by (1) G. N. Schrauzer and V. P. Mayweg (95) and (2) E. I. Stiefel, R. Eisenberg, R. C. Rosenberg, and H. B. Gray (137).

[b] A comparison is made with isoelectronic metal dithio-oxalates, 1,1-dicyanoethylene-2,2-dithiolates, and dithiocarbonates.

[MS$_6$C$_6$(CN)$_6$]$^{-3}$, M = Cr, Fe, or Co (73), exhibited less intense absorptions in this region. The gross similarity between the spectra of the neutral dithiolenes [MS$_6$C$_6$R$_6$]^0, R = H or Ph, was in accord with the probability that all these complexes adopt trigonal prismatic structures.

The spectra of the neutral vanadium complexes resembled (23) those of the corresponding anions and it was suggested that these species had similar prismatic geometries. This was substantiated by the observation (23) that the spectra of [VS$_6$C$_6$Ph$_6$]$^{-1}$ was very similar to that of [CrS$_6$C$_6$Ph$_6$]^0, which had been shown to be isomorphous, and therefore probably isostructural, with [VS$_6$C$_6$Ph$_6$]^0.

According to the assignments of Schrauzer and Mayweg (95), five types of transitions would be expected in the electronic spectra of these tris-dithiolenes—$\pi \rightarrow \pi$, $M \rightarrow \pi$, $n_{=S} \rightarrow \pi$, $\pi \rightarrow M$, and $d \rightarrow d$—but the first three of these would obscure the last two which were expected to be weak. The first intense band in the vanadium and group VI dithiolene complexes was described as $\pi \rightarrow \pi$ transition analogous to the $2b_{1u} \rightarrow 3b_{2g}$ transition in the oxidized nickel dithiolene complexes (57). It was observed that bands of the type $M \rightarrow \pi$ should be metal dependent and would be expected to move to higher frequencies as the ionization potentials of the metal increased. Accordingly, an intense band at 18,680 cm^{-1} in [VS$_6$C$_6$H$_6$]$^{-1}$ was observed to shift to 24,120 and 24,750 cm^{-1} in the neutral molybdenum and tungsten analogs, and was therefore assigned to these metal–ligand charge-transfer transitions.

The spectra of the neutral rhenium complexes (137,148) were broadly similar to those of their molybdenum and tungsten analogs but had additional bands of low intensity at low frequency (Table XXXIII). It was observed that the transitions in the neutral tris-toluene–dithiolenes occurred at higher energy than in the corresponding diphenyl–dithiolenes.

The spectra of the dianionic dicyano–dithiolenes, $[MS_6C_6(CN)_6]^{-2}$, were less well resolved (73) than those of the monoanionic and neutral complexes discussed above. They exhibited transitions of medium to strong intensity in the visible region whereas the trianionic species, $[M—S_6]^{-3}$, did not have strong absorptions in this region. A comparison of the spectra of the formally isoelectronic $[MnS_6C_6(CN)_6]^{-2}$ and $[CrS_6C_6(CN)_6]^{-3}$ revealed few similarities, suggesting that the two did not have similar electronic structures.

A comparison (73) of the spectra of the chromium, iron, and cobalt complex trianions, $[MS_6C_6(CN)_6]^{-3}$, with analogous compounds containing more "conventional" ligands revealed that there were few similarities between the two groups. Jorgensen had reported (149) that the first spin-allowed band in low-spin square-planar Ni(II) complexes occurred at about the same energy as the first spin-allowed band in the corresponding Co(III) complexes. This correlation worked well in the dithiocarbamates (150), 1,1-dicyanoethylene-2,2-dithiolates (151), and dithiooxalates (145, 152). It did not work, however, in the dicyano–dithiolene systems where the first spin-allowed band in $[NiS_4C_4(CN)_4]^{-2}$ occurred at 11,690 cm^{-1} and that in $[CoS_6C_6(CN)_6]^{-3}$ at 14,930 cm^{-1}. Fackler and Coucouvanis (153) used this same low-energy transition in Ni(II) complexes to predict the position of bands in the spectra of six-coordinate Cr(III) complexes, and a satisfactory application of their methods to dithiooxalates was made (73). However, the predictions made for $[CrS_6C_6(CN)_6]^{-3}$ were quite unsatisfactory since the absorptions were calculated to appear at 12,000 and 16,800 cm^{-1} and were actually observed at 14,100 and 17,700 cm^{-1}. This could indicate distortion of the idealized D_3 (trigonally distorted octahedral) symmetry of the trianion in a way similar to that in $[Me_4N]_2$ $[VS_6C_6(CN)_6]$ (24).

G. Electronic Structures of the Tris-Substituted 1,2-Dithiolene Complexes

Schrauzer and Gray, and co-workers, have devised (95,137) molecular orbital schemes for the trigonal prismatic complexes $[M—S_6]^0$ and have used these schemes to interpret the physical and chemical properties of these and the reduced complexes.

Schrauzer and Mayweg (95) based their calculations on the tris-1,2-dithiolenes $[MS_6C_6H_6]^z$, and assumed D_{3h} symmetry throughout. Although the overall symmetry of $[MoS_6C_6H_6]^0$ was actually C_{3h} (22) (the

$MoS_2C_2H_2$ chelate rings deviating some 18° from planarity), it was felt that such a small deviation from ideality was relatively unimportant; the relative ordering of the energy levels in C_{3h} symmetry differed only slightly from that in D_{3h}, and the overall description of the molecules remained essentially the same. It was noted that there was a distinct possibility of delocalization involving the π- and σ-electron systems of the ligands, thereby accounting for the "aromaticity" of some of the prismatic complexes. The calculations were carried out for both first- and third-row metal complexes and, although the results indicated certain reversals in the order of the lower filled bonding (and higher unfilled antibonding orbitals), no essential difference in the overall electronic descriptions of the tris complexes of the group VI metals were found.

The important energy levels are shown in Figure 17. The $4e'$ level was

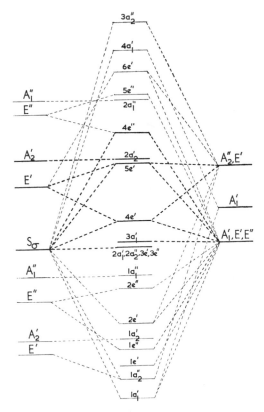

Fig. 17. The molecular orbital scheme for $[MS_6C_6H_6]^z$ according to Schrauzer et al. (95). The important levels are underlined in heavy type.

described as consisting of 41% ligand π molecular orbital, 17% lone pair (sp^2) sulfur orbital, and 25 and 17% metal d and p orbital chatacter, respectively. In other words, this molecular orbital was heavily ligand in character. The $5e'$ level was similar but had a higher degree of metal d-orbital participation. The $3a_1'$ orbital was predominantly metal (d_{z^2} symmetry) in character, whereas the $2a_2'$ was exclusively of ligand π molecular orbital character. The sequence of $5e'$ and $2a_2'$ was found to be highly dependent on the input parameters in the calculations, but, on the basis of ESR and limited magnetic data, it was proposed that the ordering was $5e' < 2a_2'$.

The configuration of the group VI neutral dithiolenes, $[M—S_6]^0$, and of the isoelectronic $[V—S_6]^{-1}$, was therefore

$$\cdots(3a_1')^2(4e')^4$$

whereas that for the paramagnetic complexes $[M—S_6]^{-1}$ (M = Cr, etc.) and $[V—S_6]^{-2}$ was

$$\cdots(3a_1')^2(4e')^4(5e')^1$$

The dianionic group VI complexes would be described as

$$\cdots(3a_1')^2(4e')^4(5e')^1$$

and the most oxidized vanadium complexes, $[V—S_6]^0$, as

$$\cdots(3a_1')^2(4e')^3$$

These proposed configurations, and some extensions to other known tris-dithiolene complexes, are summarized in Table XXXIII.

At this stage, some objections must be raised to these M.O. proposals. Firstly, while the $4e'$ and $5e'$ levels were defined as consisting of both metal and ligand character, it is difficult to see why metal-hyperfine splitting was observed in the ESR spectra of $[V—S_6]^{0,-2}$ or $[M—S_6]^{-1}$ (M = Cr or Mo) (Sec. IV-E); in symmetry terms, partial occupancy of an orbital of a_1' symmetry would have been expected. Secondly, according to the proposed scheme, the dianionic group VI metal complexes would be paramagnetic ($S = 1$); indeed, the chromium species are, but their molybdenum and tungsten analogs are diamagnetic. In fairness to the authors of this scheme, it must be said that they did suggest that there were many mechanisms available for the relief of the degeneracies of these levels of reputed e' symmetry. Clearly, the loss of trigonal prismatic symmetry in these charged species is of no little significance in this respect, and it may be recalled that $[Ph_4P]_2[MoS_6C_6(CN)_6]$ and its tungsten analog were reported (73) to be isomorphous with $[Ph_4P]_2[VS_6C_6(CN)_6]$, and, therefore, possibly isostructural.

TABLE XXXIII

Proposed Electronic Configurations of Tris-1,2-Dithiolene Complexes

Schrauzer[a]	$....(4e')^3$	$....(4e')^4$	$....(4e')^4(5e')^1$ $....(4e')^4(2a_2')^1$	$....(4e')^4(5e')^2$ $....(4e')^4(2a_2')^2$
Gray[b]	$....(2a_2')^1$	$....(2a_2')^2$	$....(2a_8')^2(3a_1')^1$	$....(2a_8')^2(3a_2')^2$
Isoelectronic complexes	$[V—S_6]^0$	$[V—S_6]^{-1}$ $[Cr—S_6]^0$ $[M—S_6]^{0\ c}$ $[Re—S_6]^{+1}$	$[V—S_6]^{-2}$ $[Cr—S_6]^{-1}$ $[M—S_6]^{-1}$ $[Re—S_6]^0$	$[V—S_6]^{-3}$ $[Cr—S_6]^{-2\ d}$ $[M—S_6]^{-2}$ $[Re—S_6]^{-1}$

Schrauzer[a]	$....(4e')^4(5e')^3$ $....(4e')^4(2a_2')^2(5e')^1$	$....(4e')^4(5e')^4$ $....(4e')^4(2a_2')^2(5e')^2$
Gray[b]	$....(2a_2')^2(3a_1')^2(5e')^1$	$....(2a_2')^2(3a_1')^2(5e')^2$
Isoelectronic complexes	$[Cr—S_6]^{-3\ e}$ $[M—S_6]^{-3}$ $[Mn—S_6]^{-2\ e}$ $[Re—S_6]^{-2\ f}$ $[Fe—S_6]^{-1}$	$[Mn—S_6]^{-3}$ $[Re—S_6]^{-3}$ $[Fe—S_6]^{-2\ g}$

Schrauzer[a]	$....(4e')^4(5e')^4(2a_2')^1$ $....(4e')^4(2a_2')^2(5e')^3$	$....(4e')^4(5e')^4(2a_2')^2$ $....(4e')^4(2a_2')^2(5e')^4$
Gray[b]	$....(2a_2')^2(3a_1')^2(5e')^3$	$....(2a_2')(^23a_1')^2(5e')^4$
Isoelectronic complexes	$[Fe—S_6]^{-3\ h}$ $[Co—S_6]^{-2}$	$[Co—S_6]^{-3}$

[a] Ref. (95).
[b] Ref. (137).
[c] M = Mo and W.
[d] $S = 1$.
[e] $S = \frac{3}{2}$.
[f] $S = \frac{1}{2}$.
[g] $S = 1$.
[h] $S = \frac{3}{2}$ with orbital contributions.

Some consideration was given to the possible oxidation states of the metal ions in these highly-oxidized dithiolene complexes. On complexation, the "ethylenic" bond in the sulfur ligands was observed to increase in length (see Sec. IV-B), and the "C=S" bond length to decrease, thus strongly suggesting that the ligands had considerable "dithioketonic" character, especially in the neutral species. This implied that a formal $+6$ oxidation state for the metal was incorrect, and, in $[V—S_6]^0$, this would be patently absurd. Assuming that the $4e'$ level was predominantly ligand in character as suggested, then the complexes would contain d^2 metal ions (the $3a_1'$ level always being filled), and an oxidation state of $+4$ for the group VI metals, and $+3$ for the vanadium, would be appropriate. The overall structures of the complexes could then be represented by the valence bond forms shown in **28** and **29**. These representations were believed to be

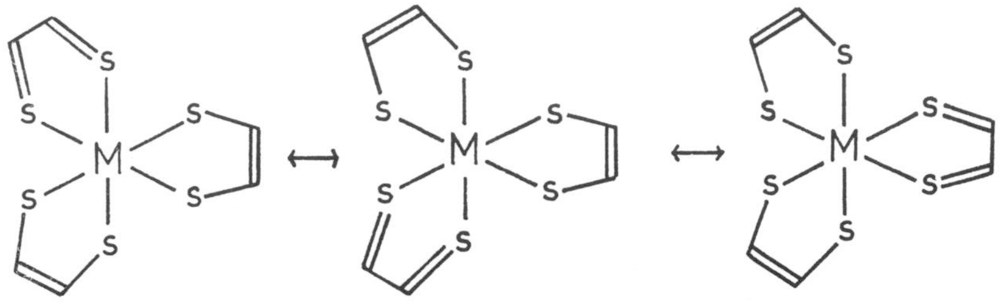

(28)

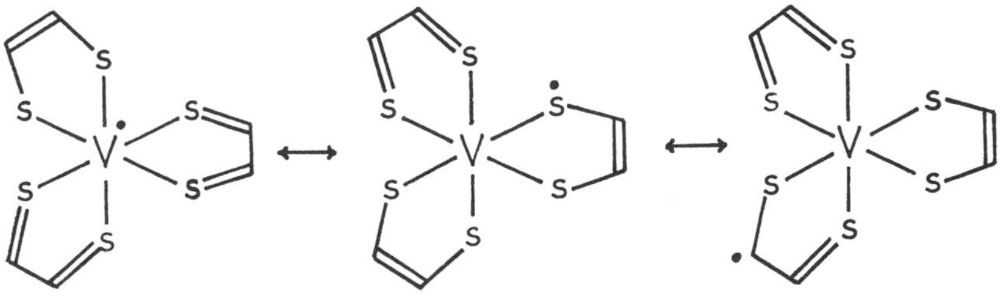

(29)

consistent with the M.O. descriptions of these complexes, and some support for this belief was drawn from the observation of the distortion of the molecular structure of $[VS_6C_6Ph_6]^0$ from ideal D_{3h} symmetry.

Calculations performed (95) using a model with D_3 symmetry demonstrated that the same set of orbital sequences could be produced. Thus, the reason why these highly oxidized systems adopted trigonal prismatic geometries was not immediately apparent. Since similar structures had already been observed in MoS_2 and WS_2 (143), it was suggested that some form of collective interaction by the sulfur donor atoms along the triangular faces of the prisms was responsible for the structure.

Gray and his colleagues (137) performed very similar calculations but reached a different conclusion about the relative ordering of the most important levels (Fig. 18). The three levels $2a_2''$, $3e'$, and $3e''$ consisted mainly

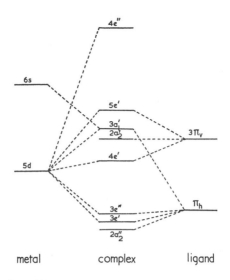

Fig. 18. The molecular orbital scheme for $[MS_6C_6R_6]^z$ according to Gray et al. (137) (only the important levels are shown).

of the sulfur sp_2 hybrid orbitals which were at $120°$ to the σ orbitals pointing at the metal atom. The $2a_2'$ orbital was nonbonding ($3\pi_v$) and the $3a_1'$ level was described as principally metal in character, having symmetry d_{z^2}. The $4e'$ and $5e'$ orbitals were both metal (symmetry $d_{x^2-y^2}$, d_{xy}) and ligand ($3\pi_v$) in character, and it was suggested that both were strongly delocalized over both metal and sulfur ligands. The predicted ground state of $[Re-S_6]^0$ was therefore

$$\cdots(4e')^4(2a_2')^2(3a_1')^1$$

which was thought to be consistent with ESR and spectral properties. The group VI complexes, $[M-S_6]^0$, and $[V-S_6]^{-1}$ had the configurations

$$\cdots(4e')^4(2a_2')^2$$

and the corresponding dianions

$$\cdots(4e')^4(2a_2')^2(3a_1')^2$$

The neutral and dianionic vanadium complexes, $[V-S_6]^{0,-2}$, had predicted configurations

$$\cdots(4e')^3 \quad \text{and} \quad \cdots(4e')^4(2a_2)^1$$

respectively. Other predicted configurations are summarized in Table XXXIII.

In terms of formal oxidation states for the central metal ions, two extreme descriptions could be visualized. If the $4e'$ level was predominantly metal in character, then the rhenium complex would be described as a d^5 metal system, that is, containing Re(II). The d-orbital splitting would then be $xz,yz > z^2 > x^2-y^2,xy$. The occupancy by two electrons of the $2a_2'$ level (which was mainly ligand $3\pi_V$) would then correspond to a collectively charged ligand set $(-S_6)^{-2}$ and the complex could be represented by $[Re(II)(-S_6)]^0$. The other extreme was that in which the $4e'$ level was predominantly ligand in character implying that, of the five electrons occupying the $4e'$ and $3a_1'$ levels, only one had metal d character. Thus the complex would be represented as $[Re(VI)(-S_6)]^0$. The calculations favored neither of these two extremes, demonstrating the inappropriateness of the use of formal oxidation states in these dithiolene complexes.

While these schemes may be valid for the description of some of the neutral complexes, it is clear that some modification must be made to account for the loss of trigonal prismatic symmetry (or distortion from octahedral symmetry) on reduction of the neutral species to dianions or trianions. Such a loss of symmetry would undoubtedly lead to a relief of degeneracy of the e'-type orbitals thereby providing ground-state electronic descriptions more consistent with the observed ESR spectral and magnetic data. However, even the ground-state descriptions of some of the neutral complexes, especially $[V-S_6]^0$, must be questioned since they would not be expected to give rise to substantial metal-hyperfine splittings in their ESR spectra.

Gray and co-workers (137) devoted some attention to the reason why the trigonal prism was the preferred coordination geometry for the neutral tris-dithiolene complexes. From the several x-ray structural determinations of various dithiolene complexes (Sec. III-B and IV-B), it was apparent that the interligand S—S distance was always close to 3.05 Å. This observation was taken as an indication of some form of interligand bonding interaction which was considerably stronger in the neutral dithiolenes than it was in conventional octahedral, planar, or tetrahedral complexes. Further contributing factors to the stability of the trigonal prisms may have been the interaction between the sulfur π_h orbitals and the metal d_{z^2} orbital which led to the stable bonding orbital $2a_1'$, which was filled in all of the neutral species, and also the important interaction between the metal d_{xy}, $d_{x^2-y^2}$ orbitals with the $3\pi_v$ ligand orbitals which were strongly delocalized over the ligand framework.

In an extrapolation of the remarks about the apparent constancy of the interligand S—S distance in the dithiolene complexes, it was suggested (137) that eight-coordinate cubic (or approximately cubic) complexes might exist. In this situation, each edge of the S_8 cube would be 3.0–3.1 Å, thus requiring an M—S distance in $[M—S_8]^2$ of about 2.6 Å, a strong possibility in the complexes of actinides.

Davison, Edelstein, Holm, and Maki (18), in rationalizing their results obtained from detailed ESR studies of the bis-perfluoromethyl–dithiolenes, $[MS_6C_6(CF_3)_6]^{-1}$, and various vanadium complexes $[V—S_6]^{0,-1}$, assumed D_3 symmetry (trigonal prismatic geometry for these complexes had not been discovered at that time) and concluded that, according to their calculations, the unpaired electron was localized in a molecular orbital predominantly of ligand character. Thus, for the group VI monoanionic dithiolenes and for $[V—S_6]^{-2}$, the configuration $(a_2^*)^2$ $(a_2^*)^1$ was proposed, where the a_1^* level had metal ($d_z{}^2$) symmetry and the a_2^* level was a ligand π-molecular orbital only capable of interacting with the metal $4p_z$ orbital. For the neutral vanadium complex, the configuration $(a_2^*)^1$ was proposed since the ESR spectral parameters were very similar to those of the dianionic species and would not be consistent with the $(a_1^*)^1$ configuration. Other proposals, such as $(a_1^*)^2(a_2^*)^2(e^*)^{1,3}$ for the neutral and dianionic vanadium species, were considered but were discounted in view of the magnetic properties of the dianions ($S = \frac{1}{2}$) and also because of the probability that Jahn-Teller distortions, or even solvent interactions, would lift the degeneracies of the orbitals of e^* symmetry.

V. DITHIOLENE COMPLEXES CONTAINING TWO, THREE, OR FOUR ELECTRON DONOR LIGANDS

A. Carbonyl Complexes

UV or visible light irradiation (135) of mixtures of the group VI metal carbonyls and nickel bis-dialkyl- or diaryl-dithiolenes afforded the carbonyl complexes $[M(CO)_4S_2C_2R_2]^0$ and $[M(CO)_2S_4C_4R_4]^0$. The chromium complexes were too unstable to isolate, and no tetracarbonyl derivatives of molybdenum could be detected. However, $[W(CO)_4S_2C_2Me_2]^0$ had an infrared spectrum (Table XXXIV) consistent with the expected C_{2v} symmetry, Figure 19. The "C=C" stretching frequency occurred at 1515 cm^{-1}, much higher than that in $[NiS_4C_4Me_4]^0$ (1333 cm^{-1}) or $[WS_6C_6Me_6]^0$ (1470 cm^{-1}), indicating that there was a considerable degree of double bond character in this carbon–carbon bond and that the metal had a formal oxidation state of $+2$.

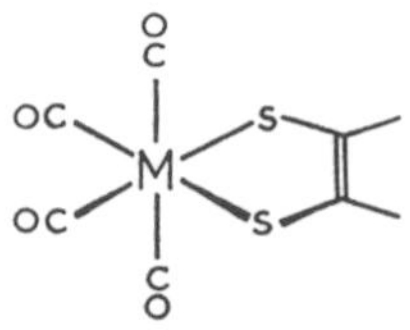

Fig. 19. The probable structure of $[M(CO)_4S_4C_4R_4]^z$.

The dicarbonyls, $[M(CO)_2S_4C_4R_4]^0$, clearly contained *cis* carbonyl groups, and treatment of these complexes with mono- and diphosphines resulted in the replacement of the CO by the phosphines giving such species as $[M(CO)(PR_3')S_4C_4R_4]^0$ and $[M(R_2'PCH_2CH_2PR_2)S_4C_4R_4']^0$ (see Sec. V-C). Pyridine and CN^- also effected the removal of CO, but the mixed ligand complexes so formed were too unstable to be isolated. The mixed-ligand dithiolenes, $[MS_6C_6R_4R_2']^0$, where $R' = H$ or CN, were readily formed by treating the dicarbonyls with $[S_2C_2R_2']^{-2}$. With sulfide ion, the novel sulfur-bridged complexes $[M_2S_2(S_4C_4Ph_4)_2]^0$, believed to have the structure shown in Figure 20, was formed. Polarographic investigations of $[W(CO)_4S_2C_2Me_2]^0$ and the dicarbonyls provided no evidence of redox behavior, but $[Mo(CO)(PPh_3)S_4C_4Ph_4]^0$ could be reduced in three stages, the first wave being a reversible one-electron process which may have corresponded to the generation of the monoanion $[M(CO)(PPh_3)—S_4]^{-1}$.

On the basis of a simple molecular orbital calculation using parameters similar to those previously employed in calculations on the trigonal prismatic dithiolenes (95), a trigonal prismatic structure was proposed for these dicarbonyl complexes (Fig. 21). It was estimated that the ligands had "approximately one third S=C—C=S and two thirds S—C=C—S character." The "C=C" stretching frequencies were of the same order as those in the tris-substituted complexes, $[MS_6C_6R_6]^0$, and quite different from those in $[W(CO)_4S_2C_2Me_2]^0$ and $[NiS_4C_4R_4]^{-2}$.

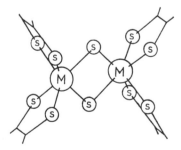

Fig. 20. The proposed structure for the sulfide-bridged complexes $[M_2S_2(S_4C_4R_4)_2]^0$.

TABLE XXXIV
Metal Carbonyl Dithiolate and Dithiolene Complexes

Compound	Color	mp	ν_1	ν_2	ν_3	ν_4	ν_{CO}
$[W(CO)_4S_2C_2Me_2]^0$	Yellow	92 d	1515	943	546	392	2070
						300	2024
							1988
							1969
$[Mo(CO)_2S_4C_4Me_4]^0$	Brown–black	151 d	1449	943	565	400	2008
					555	295	1960
$[Mo(CO)_2S_4C_4Ph_4]^0$	Violet	150 d	1426	1168	875	404	2032
						358	1992
$[W(CO)_2S_4C_4Me_4]^0$	Dark green	162 d	1470	934	562	398	2040
						300	1960
$[W(CO)_2S_4C_4Ph_4]^0$	Red–brown	165 d	1428	1156	824	407	2050
					874	374	1990
$[Mo(CO)_2S_4C_4(p\text{-MeOC}_6H_4)_4]^0$	Violet–black	135 d					
$[W(CO)_2S_4C_4(p\text{-MeOC}_6H_4)_4]^0$	Dark blue	140 d					
$[Mo(CO)(PPh_3)S_4C_4Ph_4]^0$	Red	155d	1443	1173	868	403	1980
						359	
$[Mo(CO)(PBu_3)S_4C_4Ph_4]^0$	Red	191 d					
$[W(CO)_2(P\text{—}P)S_2C_2Ph_2]^{0\ b}$	Orange–yellow	229 d					1919
							1845
$[Fe_2(CO)_6S_2C_2Ph_2]$	Red	126					2090
							2045
							1990
$[Fe_2(CO)_6S_2C_2Me_2]$	Red	76–77					2090
							2040
							2010
$[Fe_2(CO)_6S_2C_2H_2]$	Red	54–55					2070
							2030
							1995
$[Fe_2(CO)_6S_2C_2(p\text{-MeC}_6H_4)_2]$	Red	128					2090
							2040
							2010
$[Fe_2(CO)_6S_2C_2(p\text{-MeOC}_6H_4)_2]$	Red	129					2080
							2040
							1995
$[Fe_2(CO)_6S_2C_6H_3Me]$	Red	123–124					2080
							2045
							2030
							2000
							1990

(*continued*)

TABLE XXXIV (*continued*)

Compound	Color	mp	Infrared data[a]				
			ν_1	ν_2	ν_3	ν_4	ν_{CO}
$[Mn(CO)_4S_2C_2(CN)_2]^{-1}$ [c]	Red						2075
							2010
							1982
							1933
$[Mn(CO)_4S(NH_2)C_6H_4]$	Yellow						2021
							1928
							1902
							1889
$[Mn(CO)_3(HS)_2C_6H_3Me]$	Yellow						2024
							1953
							1929
$[Mn(CO)_3S_2C_6H_3Me]_2$	Dark brown	130					2033
							2016
							1973
							1935
$[Co(CO)S_2C_2(CF_3)_2]_3$	Black–violet						2110
							2080

[a] In cm^{-1}.
[b] P—P = $Ph_2PCH_2CH_2PPh_2$.
[c] As Ph_3MeP^+ salt.

The simple monoanionic complexes $[Mn(CO)_4$—$S_2]^{-1}$ were obtained from the reaction of $Mn(CO)_5Cl$ (or the bromide) with either $Na_2S_2C_2(CN)_2$ (154) or $Na_2S_2C_2H_2$ (139), and also by treatment of the carbonyl halide with the dimethyltin salts of these dianions, $Me_2SnS_2C_2R_2$ (139); the *o*-amino-thiophenol analog, $[Mn(CO)_4(S)(NH_2)C_6H_4]$, was obtained by treatment of the pentacarbonyl bromide with the ligand in methanol (155). All complexes

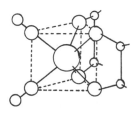

Fig. 21. The proposed structure for $[M(CO)_2S_4C_4R_4]^0$.

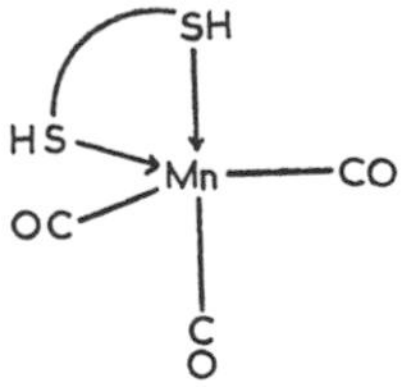

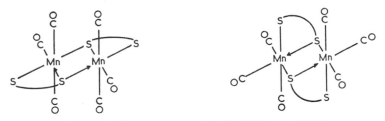

Fig. 22. The proposed structure of [MeC$_6$H$_3$(SH)$_2$Mn(CO)$_3$].

of this type must have either C_{2v} symmetry (where the ligand is a dithio-late) or C_s symmetry (where the ligand is o-aminothiophenolate) and, consequently, four infrared-active CO stretching frequencies (Table XXXIV) were observed.

In a comparison of the CO stretching frequencies of [Mn(CO)$_4$S$_2$C$_2$-(CN)$_2$]$^{-1}$ with those of the isoelectronic [Mn(CO)$_4$S$_2$CNEt$_2$] (154), it was suggested that the dithiolate ligand functioned as a π acceptor when com-plexed to a metal in a formally low oxidation state. This postulate rested on the fact that the CO stretching frequencies in the anion were not de-pressed relative to those in the neutral complex as much as might have been expected.

Fig. 23. Idealized structures for [Mn(CO)$_3$S$_2$C$_6$H$_3$Me]$_2$.

Reaction of Mn(CO)$_5$Br with toluene-3,4-dithiol (155) provided two interesting complexes—the paramagnetic monomer [MeC$_6$H$_3$(SH)$_2$Mn-(CO)$_3$], and the diamagnetic dimer [Mn(CO)$_3$S$_2$C$_6$H$_3$Me]$_2$. The course of the reaction, which was carried out using the ligand as solvent, was thought to be that shown below.

$$2MeC_6H_3(SH)_2 + 2Mn(CO)_5Br \longrightarrow 2MeC_6H_3(SH)_2Mn(CO)_3Br \text{ (I)} + 2CO$$

$$\downarrow -2HBr$$

$$2MeC_6H_3S(SH)Mn(CO)_3$$

$$\downarrow$$

$$2MeC_6H_3(SH)_2Mn(CO)_3 \xleftarrow{+2(I)} [MeC_6H_3S(SH)Mn(CO)_3]_2$$
$$+$$
$$[Mn(CO)_3S_2C_6H_3Me]_2$$

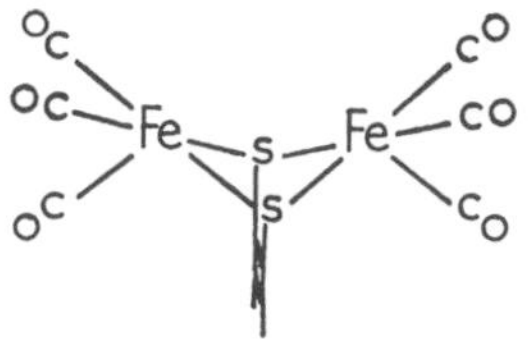

Fig. 24. The structure of $[Fe_2(CO)_6S_2C_2Ph_2]$.

The monomer, which had a magnetic moment of 1.5 BM, had an infrared spectrum consistent with the structure shown in Figure 22. The dimer could have two structures (Fig. 23) where one has C_{2h} and the other C_i symmetry; unfortunately, both point groups give rise to three active carbonyl stretching frequencies, and since four were actually observed, one particular structure could not be proposed.

The dinuclear μ-mercaptide complexes, $[Fe(CO)_3]_2S_2C_2R_2$, were obtained (96) by treating $[FeS_4C_4Ph_4]_2^0$ or $[NiS_4C_4R_4]^0$ with $Fe(CO)_5$, but the toluene-3,4-dithiol analog was prepared directly from the ligand and either $Fe(CO)_5$ or $Fe_3(CO)_{12}$ in refluxing methylcyclohexane (156). The structures of these species are undoubtedly similar to that of $[Fe(CO)_6$-$S_2C_2Ph_2]$, Figure 24 (157). Treatment of $[Fe_2(CO)_6S_2C_2Ph_2]$ with sulfur resulted in the formation of $[Fe_2S_2(S_2C_2Ph_2)_2]^0$, but thermal degradation afforded only FeS, CO, and diphenylacetylene.

When $Co_2(CO)_8$ was treated with $S_2C_2(CF_3)_2$ in refluxing methylcyclohexane (45), black crystals of empirical formula $[Co(CO)S_2C_2(CF_3)_2]_n$ were formed. It will be recalled that under different conditions (99), all CO was expelled and the dimer $[CoS_4C_4(CF_3)_4]_2^0$ was produced. The carbonyl was initially described (45) as a dimer ($n = 2$), but was later reformulated on the basis of mass spectrometric evidence (158) as a paramagnetic trimer, $n = 3$, $\mu = 1.84$ BM, with the proposed structure shown in Figure 25. However, such a symmetrical structure would be expected to exhibit only one CO stretching frequency, and two were actually observed.

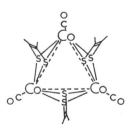

Fig. 25. The proposed structure of $[Co(CO)S_2C_2(CF_3)_2]_3$.

B. Nitrosyl Complexes

McDonald, Phillips, and Mower (159) have described the ESR spectra of species formed in aqueous solution when salts of Fe(III) and Fe(II) were treated with NO and either $S_2C_2(CN)_2^{-2}$, $S_2C_6H_4^{-2}$, or $S_2C=C(CN)_2^{-2}$. By use of isotopic substitution, i.e., treatment of these solutions with ^{14}NO or ^{15}NO, or by using ^{57}Fe, they established that the composition of these complexes which they had formed was probably represented by $[Fe(NO)-(S_2C_2(CN)_2)_2]^z$, $[Fe(NO)(S_2C_6H_4)_2]^z$, and $[Fe(NO)(S_2C=C(CN)_2)_2]^z$. Although the exact charge, z, on the complexes was not specified (the complexes were not isolated), from comparisons made with the very similar bis-dithiocarbamates, $[Fe(NO)(S_2CNR_2)_2]$, (160) it was clear that $z = -2$.

An examination of the ESR parameters obtained from the 1,2-dithiol and 1,1-dithiol complexes (Table XXXV) revealed that the g tensors of the two classes of complexes were markedly different and that both the ^{14}N and ^{57}Fe hyperfine coupling constants were larger in the 1,2-dithiol system than those in the 1,1-dithiol system, the former by about 2.6 G and the latter by about 1 G. Thus, the chelate ring size appeared to be an important factor in the determination of the g factors and hyperfine splittings. It was suggested that the proportionality between $\langle A_N \rangle$ and $\langle A_{Fe} \rangle$ could reflect the distribution of unpaired electron spin density between the dithiol ligands and the Fe(NO) group. It was also postulated that the ESR data was consistent with a d^7 configuration for the metal ion, i.e., that the complexes contained Fe(I).

A subsequent reexamination of the iron nitrosyl dithiolate system by McCleverty and co-workers (49,50) revealed that many other five-coordinate species, $[Fe(NO)-S_4]^z$, existed, and could be isolated (Table XXXV). The most useful intermediates in the synthesis of these mono-nitrosyls proved to be the dimeric dianions, $[Fe-S_4]_2^{-2}$, and treatment of these dimers with NO in nonpolar or weakly coordinating solvents resulted in the formation of the monoanionic mononitrosyls, $[Fe(NO)-S_4]^{-1}$. These species were stable to aerial oxidation but were often attacked by two-electron donors such as DMF, pyridine, or PPh$_3$ resulting in the expulsion of NO and formation of the appropriate adducts $[Fe(L)-S_4]^{-1}$ (see Sec. V-C). Reduction of the monoanions to the paramagnetic dianions, $[Fe(NO)-S_4]^{-2}$, could be effected using a wide variety of reducing agents, the specific one required to carry out the reduction being predicted from the half-wave potentials for the reaction,

$$[Fe(NO)-S_4]^{-1} + e^- \rightleftharpoons [Fe(NO)-S_4]^{-2}$$

The dianions, notably those of the dicyano and toluene–dithiolene ligands, displayed ESR spectra very similar to those observed by McDonald, Phillips, and Mower (159) for the same (or similar) systems. From the half-wave potentials for the oxidation process,

$$[Fe(NO)—S_4]^{-1} \rightleftharpoons [Fe(NO)—S_4]^0 + e^-$$

it was clear that many of the monoanions could be oxidized to neutral species under the appropriate chemical conditions. However, only one complex, $[Fe(NO)S_4C_4Ph_4]^0$, was obtained by oxidation of the corresponding monoanion with iodine, although an unidentified green nitrosylated species was formed when $[Fe(NO)(S_2C_6H_3Me)_2]^{-1}$ was treated with iodine.

Cobalt mononitrosyls, $[Co(NO)—S_4]^{-1}$, were obtained by treatment of the dimeric dianions, $[Co—S_4]_2^{-2}$, or $[Co(S_2C_6H_3Me)_2]^{-1}$, with NO in nonpolar solvents. These species were considerably less stable than the analogous iron complexes, and appeared to exist in an equilibrium,

$$[Co—S_4]^{-1} + NO \rightleftharpoons [Co(NO)—S_4]^{-1} \text{ or}$$
$$[Co—S_4]_2^{-2} + 2NO \rightleftharpoons 2[Co(NO)—S_4]^{-1}$$

The equilibrium could be pushed to the right hand side by maintaining NO gas in, and above, solutions containing the dithiolene. The only complex to be successfully isolated was $[PyrMe][Co(NO)S_4C_4(CN)_4]$. Voltammetric studies of this revealed that a dianion could be prepared and, indeed, $[Co(NO)S_4C_4(CN)_4]^{-2}$ was synthesized in a variety of ways; there was also some evidence for $[Co(NO)S_4C_4Ph_4]^{-2}$.

It was established by voltammetric studies in dichloromethane that all the isolable mononitrosyls underwent one-electron transfer reactions. Thus, the existence of neutral, monoanionic, and dianionic species was detected or confirmed. However, a second cathodic wave in the voltammograms (Table XXXVI) of $[Fe(NO)S_4C_4(CN)_4]^{-1}$ and $[Fe(NO)S_4C_4(CF_3)_4]^{-1}$ and a cathodic wave in that of $[Co(NO)S_4C_4(CN)_4]^{-2}$ and the corresponding iron analog, was attributed to the formation of trianionic species, $[M(NO)—S_4]^{-3}$. An anodic wave in the voltammogram of $[Fe(NO)S_4C_4Ph_4]^0$ was observed and may have been due to the generation of the monocation, $[Fe(NO)—S_4]^{+1}$. Neither the trianions nor the monocation was isolated.

The study of the half-wave potentials in the light of the generalizations made in Section III-C was very useful in systematizing the chemical behavior of these nitrosyl species. The correlation of the half-wave potentials for the electron transfer reactions in planar and trigonal prismatic dithiolene complexes has already been discussed in Sections III-C and

TABLE XXXV. Nitrosyl Dithiolene and Dithiolate Complexes

Complex	Cation	mp,°C	Color	ν_{NO}, cm^{-1}	μ_{eff},[a] BM	$\langle g\rangle$[b]	$\langle A\rangle$,[b]G
$[Fe(NO)S_4C_4(CN)_4]^{-2}$	Et_4N^+	109–115	Brown[c]	1645			
			Green–brown[d]	1650		2.028	15.4
	Ph_4P^+			1633[c]	1.84		
				1647[d]			
$Fe(NO)S_4C_4(CF_3)_4]^{-2}$	—	168 d	Green–brown[c]	1624	1.77	2.027	15.5[c]
	Ph_4As^+		Yellow–green	1615		2.027	15.0
$[Fe(NO)(S_2C_6Cl_4)_2]^{-2}$	Bu_4N^+	198 d	Green[c]	1624	—	2.027	15.4
			Yellow–green				
$[Fe(NO)(S_2C_6H_4)_2]^{-2}$	—		Brown[c]	1620	—	2.028	15.1[f]
$[Fe(NO)(S_2C_6H_3Me)_2]^{-2}$	Ph_4P^+	185 d	Yellow–green[d]	1600	(2.62)	2.028	15.4[g]
$[Fe(NO)S_4C_4Ph_4]^{-2}$	—						
$[Co(NO)S_4C_4(CN)_4]^{-2}$	Ph_4P^+	225–228	Red–Brown[c]	1620[h]	Dia.	2.026	15.4
				(1589 sh)			
			Orange–red[d]	1617			
				1607			
				(1690 sh)			
$[Co(NO)S_4C_4Ph_4]^{-2}$	—			1600,	Dia.		
				1630[c,h]			
$[Fe(NO)S_4C_4(CN)_4]^{-1}$	Et_4N^+	200–202	Black[c]	1867[c]			
			Red–brown[d]	1840[d]			
	Ph_4P^+	211–213	Red[c]	1814[c]			
			Red–brown[d]	1857[d]			

Complex	Cation	m.p.	Color	IR	μ	g	
[Fe(NO)S$_4$C$_4$(CF$_3$)$_4$]$^{-1}$	Et$_4$N$^+$	150 d	Black[c] / Yellow–brown	1829[c] 1840[d]	Dia.		
	Ph$_4$P$^+$			1814,[c] 1857[d]			
[Fe(NO)(S$_2$C$_6$Cl$_4$)$_2$]$^{-1}$	Bu$_4$N$^+$	262	Black[c] / Red[d]	1824[c] 1825[d]	Dia.		
[Fe(NO)(S$_2$C$_6$H$_3$Me)$_2$]$^{-1}$	Bu$_4$N$^+$	152 d	Deep red[c] / Deep red[d]	1765[c] 1802[d]	Dia.		
[Fe(NO)S$_4$C$_4$Ph$_4$]$^{-1}$	Ph$_4$P$^+$ / Et$_4$N$^+$	116–117	Brown[c] / Orange-red[d]	1790,[c]1796[d] 1770[c] 1777[d]	Dia.		
[Co(NO)S$_4$C$_4$(CN)$_4$]$^{-1}$	PyrMe$^+$		Black[c]	1672, 1700[c]	1.55		
[Co(NO)S$_4$C$_4$(CF$_3$)$_4$]$^{-1}$	—		Green[d]	1730[d]		2.063	31.9[i]
[Co(NO)(S$_2$C$_6$H$_3$Me)$_2$]$^{-1}$	—		Green[c,h]	1712[c]		2.059	32.9[i]
[Co(NO)S$_4$C$_4$Ph$_4$]$^{-1}$	—		Green[c,h]	1655[c,h]		n.o.[j]	29.4[i]
[Fe(NO)S$_4$C$_4$Ph$_4$]^{0}	—	135–140	Green[c,h] Brown[c] / Yellow–brown[d]	1697[c] 1802[d] 1805[d]	1.94	2.050 2.009	–
[Fe(NO)(S$_2$CNMe$_2$)$_2$]^{0}	—		Green[c]	1690,[c] 1714[d]	1.74	2.041	12.8
[Fe(NO)(S$_2$CC(CN)$_2$)$_2$]$^{-2}$	—					2.041	12.8[h,j]
[Co(NO)(S$_2$CNMe$_2$)$_2$]^{0}	—		Green[c]	1624,[c] 1650[d]	Dia.		
[Mn(NO)(S$_4$C$_4$CF$_3$)$_4$]$^{-2}$	Et$_4$N$^+$	165–171	Green[c]	1660[d]	1.30		
[Mn(NO)S$_4$C$_4$(CN)$_4$]$^{-2}$	Bu$_4$N$^+$		Red				
[Mn(NO)S$_4$C$_4$(CN)$_4$]$^{-2}$	Et$_4$N$^+$		Green[c]	1690	Para.		
[Mn(NO)(S$_2$C$_6$Cl$_4$)$_2$]$^{-2}$	Et$_4$N$^+$		Green[c]	1670	Para.		

(continued)

TABLE XXXV (*continued*)

Complex	Cation	mp, °C	Color	ν_{NO}, cm⁻¹	μ_{eff},[a] BM	$\langle g \rangle$[b]	$\langle A \rangle$,[b] G
$[Mo(NO)_2S_4C_4(CN)_4]^{-2}$	Ph_4P^+		Green[d]	1742, 1633[d]	Dia.		
$[W(NO)_2S_4C_4(CN)_4]^{-2}$	Ph_4As^+		Brown–red[d]	1710,[d]	Dia.		
$Mo(NO)_2(S_2CNMe_2)_2]^0$	—			1770,[d] 1670	Dia.		
$[W(NO)_2(S_2CNMe_2)_2]^0$	—			1720,[d] 1605	Dia.		
$[Fe(NO)(S_2C_2(CF_3)_2]_3{}^k$	—		Black	1777,[c] 1799, 1815, 1832, 1882			
$[Fe(NO)_2S_6C_6Ph_6]^0$	—	156–160	Purple[c,d]	1820,[l] 1817, 1817,[m] 1773	Dia.		
$[Fe_2(NO)_2S_6C_6Ph_6]^{-1}$	Et_4N^+	135–137	Green[c,d]	1766,[l] 1735, 1718, 1765,[m,n] 1740, 1714	2.00/ Fe	2.020[n] 2.021[m]	11.2[e,n] 10.4[e,m]
$[Fe_2(NO)_3S_6C_6Ph_6]^0$	—		Grey– purple[d]	1713[m]		2.008	n.o.[j]
$[Fe(NO)S_4C_4Ph_2(CN)_2]^{-1}$	Et_4N^+		Red[c,d]	1760[d]	Dia.		

[a] In the solid state at room temperature. [b] In dichloromethane solution. [c] In the solid state. [d] In solution. [e] ^{14}N hyperfine splitting. [f] $\langle A_{Fe} \rangle = 9.3$ G, ^{57}Fe, $I = \frac{1}{2}$ (164). [g] In DMSO solution. [h] Not isolated from solution. [i] ^{59}Co hyperfine splittings, ^{59}Co, $I = \frac{7}{2}$. [j] $\langle A_{Fe} \rangle = 8.4$ G [k] $\langle A_{Fe} \rangle = 9.4$ G. [l] $CHCl_3$ solution. [m] CH_3CN solution. [n] Controlled potential electrolysis.

TABLE XXXVI

Voltammetric Studies of Nitrosyl Dithiolene Complexes

Process: $[M(NO)—S_4]^z + e^-[M \rightleftharpoons (NO)—S_4]^{z-1}$

Complex	Electrode process[a] $(z \rightleftharpoons z-1)$			
	$+1 \rightarrow 0$	$0 \rightarrow -1$	$-1 \rightarrow -2$	$-2 \rightarrow -3$
$[Fe(NO)S_4C_4(CN)_4]$			$+0.03$	-1.34
$[Fe(NO)S_4C_4(CF_3)_4]$		$+0.84$	-0.07	-0.36
$[Fe(NO)(S_2C_6Cl_4)_2]$		$+0.74$	-0.24	
$[Fe(NO)(S_2C_6H_3Me)_2]$		$+0.27$	-0.64	
$[Fe(NO)S_4C_4Ph_4]$	$+0.71$	-0.02	-0.83	
$[Co(NO)S_4C_4(CN)_4]$			$+0.16$	-1.32
$[Fe_2(NO)_2S_6C_6Ph_6]$	—[b]	-0.21	-0.84	

[a] Measured in dichloromethane using a rotating Pt electrode vs. standard calomel electrode; half-wave potentials in volts.

[b] Additional two-electron oxidation wave at ~ 0.9 V.

IV-C. Similar effects were observed in the nitrosyl dithiolenes, the order of stability of the reduced species to oxidation in the couples

$$[Fe(NO)—S_4]^0 + e^- \rightleftharpoons [Fe(NO)—S_4]^{-1}$$

and

$$[Fe(NO)—S_4]^{-1} + e^- \rightleftharpoons [Fe(NO)—S_4]^{-2}$$

being

$$CN > CF_3 > S_2C_6Cl_4 > S_2C_6H_3Me > Ph$$

It was predicted from this order and by comparison with the data in Section III-C that the mono- and dianions of the $[Fe(NO)S_4C_4Me_4]^z$ system would be less stable to oxidation than the diphenyl–dithiolenes. Similarly, the mono- and dianions of the $[Fe(NO)S_4C_4H_4]^z$ system would be slightly less stable than the corresponding diphenyl–dithiolenes but more stable than the dimethyl–dithiolenes. It was thought that isolation of $[Fe(NO)S_4C_4R_4]^{+1}$ would be feasible when R = Me.

The NO stretching frequencies of the iron and cobalt nitrosyls are listed in Table XXXV. It can be seen that ν_{NO} depends mainly upon two factors—the overall charge, z, on the complexes and the nature of the ligand, particularly the ligand substituent. The type of cation and the medium in which the spectra were recorded clearly also had some effect on ν_{NO} but these were of secondary importance.

The NO stretching frequencies fell within the range generally associated with metal-coordinated "NO$^+$" (164). Those of the dianionic species occurred between 1650 and 1620 cm^{-1}, those of the monoanions between 1870 and 1770 cm^{-1}, and that of [Fe(NO)S$_4$C$_4$Ph$_4$]^0 at ca. 1804 cm^{-1}. An increase in ν_{NO} as negative charge was removed from the [M(NO)—S$_4$]^z coordination unit would be anticipated purely on the grounds that back-donation to the NO group in the monoanions would be less than in the dianions. The difference in ν_{NO} between the dianionic and monoanionic iron complexes was ca. 170 cm^{-1} and between the corresponding cobalt complexes ca. 120 cm^{-1}; the difference in ν_{NO} between [Fe(NO)S$_4$C$_4$Ph$_4$]$^{-1}$ and [Fe(NO)S$_4$C$_4$Ph$_4$]^0 was only 25 cm^{-1}. The relative values of these differences was surprising and suggested that there was some significant difference in the electronic structure of the dianions on the one hand and the monoanions and neutral species on the other.

The M—N—O bond angle in nitrosyl bis-dithiocarbamates of iron and cobalt (which are isoelectronic with the dithiolene dianions) seems to vary between 135 and 170° (161,162) and it has been suggested (163,164) that NO stretching frequencies within the approximate range 1600–1700 cm^{-1} were characteristic of such nonlinear M—N—O groups. It was proposed that the M—N—O bond angle in the dianionic dithiolenes was also bent and this was confirmed by a single-crystal x-ray analysis (165) of [Et$_4$N]$_2$ [Fe(NO)S$_4$C$_4$(CN)$_4$] where the angle was found to be ca. 168° (Fig. 26) (see Table XXXVII for other structural parameters.)

The type of ligand and the electronic nature of the substituent on the ligand ring system had a pronounced effect on ν_{NO} in both the mono- and dianionic species. As the relative acceptor ability of the dithiolate ligand decreased, so the extent of back-donation to the NO group increased, and ν_{NO} decreased. The order of decrease of ν_{NO} was therefore

$$CN > CF_3 > S_2C_6Cl_4 > S_2C_6H_3Me > Ph$$

precisely the same as that observed for the $E_{1/2}$ values discussed above. Similar electronic effects on the position of ν_{NO} had been observed in the spectra of the N,N'-bis(salicylidene)ethylenediamine cobalt mononitrosyl (166) and its ligand-substituted analogs.

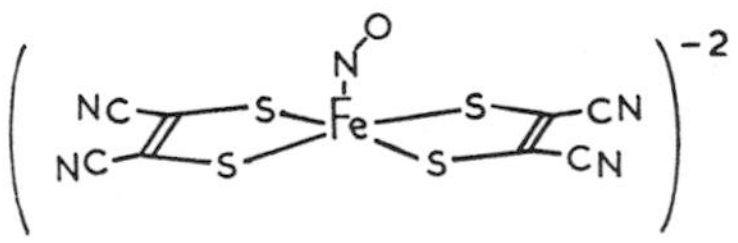

Fig. 26. The idealized molecular structure of [Fe(NO)S$_4$C$_4$(CN)$_4$]$^{-2}$.

The electronic spectra of both the neutral diphenyl–dithiolene iron and monoanionic dithiolene iron nitrosyls of complexes were characterized by the presence of intense absorptions in the visible region, a feature strongly reminiscent of the oxidized nickel dithiolenes, $[NiS_4X_4R_4]^{0,-1}$ (Sec. III-G), and the five-coordinate dithiolene phosphine complexes of iron and cobalt (96,101). These intense bands were generally absent from the spectra of the iron dianions and of all of the cobalt nitrosyl species. $[Fe(NO)S_4C_4Ph_4]^0$ and $[Co(PPh_3)S_4C_4Ph_4]^0$ are isoelectronic, and spectral comparisons of the two revealed a probable similarity in electronic structure. Similar comparisons between the iron monoanions and $[Co(PPh_3)\text{-}S_4C_4(CN)_4]^{-1}$ and its analogs and $[Co(pyr)S_4C_4(CN)_4]^{-1}$ did not disclose such a close resemblance in spectra, although the electronic configurations may have been similar. There was no likeness between the spectra of $[Fe(NO)S_4C_4Ph_4]^0$ and $[Fe(pyr)S_4C_4(CN)_4]^{-1}$ which was hardly surprising since the spin ground states of the two were different. The electronic spectra of the series of monoanionic complexes $[Fe(NO)\text{—}S_4]^{-1}$ did not vary greatly with changes in the ligands indicating that, within the series, the electronic structures of the complexes were similar. This is also true for the series of complexes $[Fe(NO)\text{—}S_4]^{-2}$ and $[Co(NO)\text{—}S_4]^{-1}$. However, there were obvious spectral differences between the isoelectronic $[Fe(NO)\text{—}S_4]^{-2}$ and $[Co(NO)\text{—}S_4]^{-1}$, suggesting that the electronic structures of these were not alike. Also, the spectra of the iron dianions were comparable with those of the corresponding nitrosyl bis-dithiocarbamates whereas those of the analogous cobalt compounds were not.

The complexes $[Fe(NO)S_4C_4Ph_4]^0$, $[Fe(NO)\text{—}S_4]^{-2}$, and $[Co(NO)\text{—}S_4]^{-1}$ were paramagnetic and the magnetic moments, where measured, were generally consistent with the expected spin–doublet ground state (Table XXXV). The iron dianions in solution at room temperature exhibited characteristic ESR spectra which consisted of a triplet due to ^{14}N $(I = 1)$ hyperfine splittings were ca. 15 G (Table XXXVII). These results were in general agreement with those obtained by McDonald et al (159). The cobalt series, $[Co(NO)\text{—}S_4]^{-1}$, also displayed ESR signals in solution at room temperature. These consisted of eight lines due to hyperfine interaction with ^{59}Co $(I = 7/2)$; g factors of ca. 2.058 were observed and the metal hyperfine splittings averaged 30 G; no ^{14}N hyperfine splittings were resolved, again suggesting a difference in electronic structure between the cobalt monoanions and iron dianions. The ESR spectrum of $[Fe(NO)\text{-}S_4C_4Ph_4]^0$ consisted of a single line with $\langle g \rangle = 2.009$, and ^{14}N splittings were not resolved. [This situation was reminiscent of that in $[Co(PPh_3)\text{-}S_4C_4Ph_4]^0$ and its tributyl phosphine analog (96,101) where ^{59}Co, but not

^{31}P, couplings were detected.] In DMF/CHCl$_3$ glasses, a threefold anisotropy of the g tensor was observed and $g_1 = 2.000$, $g_2 = 2.013$, and $g_3 = 2.021$.

A qualitative molecular orbital scheme was presented and used to explain the following features:

1. The observation of ^{14}N hyperfine splittings in [Fe(NO)—S$_4$]$^{-2}$. but their apparent absence in [Fe(NO)—S$_4$]^0 and [Co(NO)—S$_4$]$^{-1}$.

2. The shift of the g-factor from ca. 2.027 in [Fe(NO)—S$_4$]$^{-2}$ to 2.009 in [Fe(NO)—S$_4$]^0, i.e. the trend towards the free electron value on oxidation of the nitrosyls.

3. The marked shift of ν_{NO} on moving from [Fe(NO)—S$_4$]$^{-2}$ to [Fe(NO)—S$_4$]$^{-1}$ when compared with the smaller shift on moving from [Fe(NO)—S$_4$]$^{-1}$ to [Fe(NO)—S$_4$]^0.

4. The dependence of $E_{1/2}$ and ν_{NO} on the nature of the ligands.

5. The occurrence of intense low-energy transitions in the electronic spectra of [Fe(NO)—S$_4$]^0 and [Fe(NO)—S$_4$]$^{-1}$, and their virtual absence in all other nitrosyl dithiolene complexes which have been prepared.

The orbitals of importance in the proposed molecular orbital scheme are presented in Figure 27. (The labels refer to C_{2v} symmetry.) The b_1', b_2', and a_1' levels, which were completely filled in all of the complexes, were predominantly metal in character (having the symmetries d_{xz}, d_{yz}, and $d_{x^2-y^2}$, respectively.) Next in energy above these were placed, somewhat arbitrarily, a number of orbitals, L, which were primarily sulfur ligand in character. The exact number and form of these orbitals were not specified but it was stressed that the orbital immediately below the b_1'' and b_2'' levels,

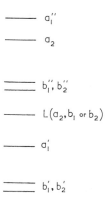

Fig. 27. The proposed molecular orbital scheme for [Fe(NO)—S$_4$]^z.

which were derived predominantly from the π^*(NO) orbitals, was of this type; orbitals of ligand character frequently had energies comparable with those of the metal d orbitals (Sec. III-G and IV-G). The b_1'' and b_2'' levels were the highest orbitals to be occupied in any of the complexes, and the remaining levels, a_2 and a_1'', which were always empty, were predominantly metal d_{xy} and d_{z^2} in character, respectively. With the exception of the inclusion of the L orbitals, the model was similar to that proposed by Gray and Manoharan (167) for $[Fe(NO)(CN)_5]^{-2,-3}$ and related compounds.

According to the scheme, the configurations of the iron complexes were

$$\cdots(L)^2(b_1'',b_2'')^1$$

for the dianions,

$$\cdots(L)^2$$

for the monoanions, and

$$\cdots(L)^1$$

for the neutral species. Although the broad features of the scheme were expected to be applicable to all three different charge types, $z = 0, -1$. and -2, it was observed that in the dianionic species, the near orbital degeneracy of the ground state arising from the configuration

$$\cdots(L)^2(b_1'',b_2'')^1$$

would be relieved by a suitable distortion, corresponding to a bent Fe—N—O arrangement, which had been found (165). The monoanionic and neutral species in the scheme had orbitally nondegenerate ground states and linear Fe—N—O bond arrangements were anticipated. Thus, the correlation of the properties of the neutral and single-charged species within the framework of the M.O. scheme as proposed was expected, the difference between the two species being merely the addition (or removal) of a single electron. However, the dianions could not be regarded as arising from the monoanions by placing an extra electron in the lowest unoccupied orbital of the scheme appropriate for the neutral and monoanionic molecules, and this distinction was of great importance in the interpretation of certain properties of the complexes.

The ^{14}N hyperfine splitting in the dianions was assumed to arise by spin polarization of electrons in the s orbitals of the nitrogen by spin density in the π^* orbitals of NO, the situation being analogous to that in $[Fe(NO)(CN)_5]^{-3}$ (167). In the neutral complex, much more spin density

TABLE XXXVII

Dithiolene Complexes Containing Phosphines, Pyridines, and Related Ligands

Complex	M	R	z	Color	S^a	μ_{eff},^{b} BM	$\langle g \rangle$	$\langle A \rangle$,^{c} G	$\nu_{C=C}$, cm^{-1}
$(PPh_3)_2MS_2C_2R_2$	Ni	CN	0	Brown	0	Dia.			
	Pd	CF_3	0	Olive–green	0	Dia.			1553
	Pd	CN	0	Pink	0	Dia.			
	Pd	CF_3	0	Pink	0	Dia.			1548
		CH_3	0	Yellow	0	Dia.			
		Ph	0	Yellow	0	Dia.			
	Pt	CN	0	Yellow	0	Dia.			
		CF_3	0	Lemon–yellow	0	Dia.			1541
		Ph	0	Yellow	0	Dia.			
$(PBu_3)_2MS_2C_2R_2$	Pd	Ph	0	Red	0	Dia.			
$(PBu_3)_2M_2(S_2C_6H_3Cl)_2$	Pd		0	Red	0	Dia.			
$(PBu_3)_2M_2(S_2C_2R_2)_2$	Pd	Ph	0	Red	0	Dia.			
$(o\text{-phen})M(S_2C_6H_3Me)_2$	Zn		0	Cream	0	Dia.			
	Ni		0	Red	0	Dia.			
$(dipyr)MS_2C_2R_2$	Cu	CN	−1	Red	0	Dia.			
	Ni	CN	0	Red	0	Dia.			
$(pyr)MS_4C_4R_4^d$	Co	CN	−1	Green	0	Dia.			
		CF_3	−1	Green	0	Dia.			
	Fe	CN	−1	Brown–orange	$3/2$	3.90			
		CF_3	−1	Red	$3/2$				
$(pyr)M(S_2C_6H_3Me)_2$	Fe	CN	−1	Red–mauve	$3/2$	3.84			
$(pyr)M(S_2C_6Cl_4)_2$	Fe	CN	−1	Red	$3/2$	—			
$(PEt_3)MS_4C_4R_4$	Co	CN	−1	Red–brown	0	Dia.			
		CF_3	−1	Green	0	Dia.			
	Fe	CF_3	−1	Blue–violet	$1/2$	Dia.	2.044	27.9^e	

Complex	M	R	z	Color	S	μ	g		IR
(PEt₃)M(S₂C₆H₃Me)₂	Co	Ph	−1	Orange–brown	0	Dia.	2.020	23.4^f	1415
(PBu₃)MS₄C₄R₄	Co	H	0	Red	½	—			1371
	Fe	Ph	0	Green	0	Dia.			1392
(PPh₃)MS₄C₄R₄	Co	CN	0	Green	0	Dia.			
		CN	−1	Yellow–brown	½	—	2.046	28.1^e	
		CF₃	−1	Red–brown	0	Dia.			
		CF₃	−1	Yellow–brown	0	Dia.			
	Fe	Ph	0	Brown	½	1.75	2.0192	28.83^f	1403
		Ph	0	Red	½	—	2.013	26.4^f	1380
		CF₃	0	Green	0	Dia.			
		CF₃	0	Green	0	Dia.			
(PPh₃)M(S₂C₆Cl₄)₂	Co	CF₃	−1	Red–brown	3/2 ?	3.97			
[P(OPh)₃]MS₄C₄R₄	Co	CF₃	−1	Red	0	Dia.			
		CN	−1	Orange–brown	0	Dia.			
		CF₃	−1	Lime–green	0	Dia.			
		CF₃	0	Green	{ ½	1.70^j	2.0165	26.4^{f,i}	
					0	Dia.^{k}			
	Fe	Ph	0	Red	½	—	2.014	24.1^g	
		CN	−1	Green	½	—			
		CF₃	−1	Blue–green	½	—			
		CF₃	0	Blue	0	Dia.			
		Ph	0	Green	0	Dia.			
[P(OEt)₃]MS₄C₄R₄	Co	CN	−1	Orange–brown	½	1.73	2.0197	31.15	
(AsPh₃)MS₄C₄R₄	Co	CF₃	0	Brown	0	Dia.			
	Fe	CF₃	0	Black	0	Dia.			
(Pr₃As)S₄C₄R₄	Co	CN	−1	Red	½	1.78	2.044	28.0^e	
	Fe	CN	−1	Green	0	Dia.			
(SbPh₃)MS₄C₄R₄	Co	CN	−1	Red–brown	0	Dia.			
		CF₃	−1	Green	½	1.79			
		CF₃	0	Brown	0	Dia.	2.0302	—^{l}	

(continued)

TABLE XXXVII (continued)

Complex	M	R	z	Color	S[a]	μ_{eff},[b] BM	$\langle g \rangle$	$\langle A \rangle$,[c] G	$\nu_{C=C}$, cm^{-1}
$(CN)MS_4C_4R_4$	Fe	CN	-2	Yellow–brown	$\frac{1}{2}$	1.90	2.054[h]		
	Fe	CF_3	-2	Lilac	$\frac{1}{2}$	—			
	Co	CN	-2	Red	$\frac{3}{2}$	—			
$(N_3)MS_4C_4R_4$	Fe	CN	-1	Lime–green	0	Dia.			
$(dipyr)MS_4C_4R_4$	Co	CN	-1	Green	0	Dia.			
$(o\text{-}phen)MS_4C_4R_4$	Fe	CN	-1	Orange–brown	$\frac{1}{2}$	2.12	2.085[h]		
	Co	CN	-1	Green	0	Dia.			
	Co	CF_3	-1	Brown–green	0	Dia.			
	Fe	CN	-1	Orange–brown	$\frac{1}{2}$	1.92	2.083[h]		
	Fe	CF_3	-1	Blue–black	$\frac{1}{2}$	—	2.083[h]		
$(o\text{-}phen)M(S_2C_6Cl_4)_2$	Co		-1	Green–brown	0	Dia.			
$(o\text{-}phen)M(S_2C_6Me_4)_2$	Co		-1	Gray–green	0	Dia.			
$(en)MS_4C_4R_4$	Co	CN	-1	Gray–green	0	Dia.			
	Co	CF_3	-1	Blue–green	0	Dia.			
$(dmg)MS_4C_4R_4$	Co	CN	-1	Green	0	Dia.			
$(PPh_3)MS_4C_4R_4$	Ni	Ph	0		0	Dia.			
$(PBu_3)_2MS_4C_4R_4$	Ni	Ph	0		0	Dia.			
	Pt	Ph	0	Blue	0	Dia.			
$(Ph_2PCH_2)_2MS_4C_4R_4$	Mo	Ph	0	Red	0	Dia.			
	W	Ph	0	Brown	0	Dia.			1436

[a] Spin ground state.
[b] In the solid state.
[c] ^{59}Co or ^{31}P hyperfine splitting.
[d] Other pyridine-like ligands include picolines, quinolines, 4-amino, and 4-vinyl-pyridines.
[e] ^{31}P hyperfine splitting.
[f] ^{59}Co hyperfine splitting.
[g] Small ^{31}P hyperfine splitting (ca. 8.5 G) observed.
[h] Broad.
[i] ^{31}P hyperfine splitting 9.3 G.
[j] In CH_2Cl_2 solution.
[k] Complex in solid state probably dimeric.
[l] Complex spectra not analyzed.

on the sulfur ligands would be expected, and if the uppermost, singly occupied L orbital did not have a_1 symmetry, it seemed reasonable that the ^{14}N hyperfine splitting should be very much smaller or even zero.

In the dianions, spin–orbit coupling would allow admixture of the excited configurations

$$(a_1')^1 \cdots (b_1'', b_2'')^2, (b_1'', b_2'')^0 \cdots (a_1'')^1 \quad \text{and} \quad (b_1'', b_2'')^0 \cdots (a_2)^1$$

of which the first was thought to be the most important. The net result would be a positive shift of the g factor, as observed. In the neutral complex, however, a situation similar to that used to explain the value of $g_\perp$ in vanadocene (168) would operate, namely the near cancellation of positive and negative contributions to the g factor.

The marked changes in ν_{NO} could also be simply explained in terms of the M.O. scheme. In the dianions, the high population of the $\pi^*(NO)$ orbitals compared with the singly-charged and neutral species would lead directly to a weakening of the NO bond, with concomitant decrease in the force constant and ν_{NO}. The change in population of the $\pi^*(NO)$ orbitals on moving from the monoanions to the neutral species would be very small and consequently there should be no great change in ν_{NO}.

It might be expected that the half-wave oxidation potentials for the monoanions would show a linear correlation with the orbital energies of the highest occupied L level, and to this extent, their dependence on the nature of the sulfur ligand could be understood. However, the dependence on sulfur ligands of the half-wave oxidation potentials of the dianions was not so simply rationalized. It was observed that a direct correlation with the orbital energy of (b_1'', b_2'') would not be expected, but that a relationship with the energy difference between the ground states of the mono- and dianions would be reasonable, even though the dianions probably required a modified M.O. scheme by virtue of the nonlinear Fe—N—O bond angle. Similar remarks were applied to the interpretation of the dependence of ν_{NO} on the nature of the ligands, and to the interpretation of the electronic spectra. It was felt inadvisable to go farther than to state that the neutral and monoanionic iron complexes might be expected to have some transition in common [particularly in comparison with [Ni—$S_4]^{0,-1}$ (Sec. III-F)], but that the dianions would behave differently because of differences in the finer details of the M.O. scheme as applied to these complexes.

While the molecular orbital scheme appeared to be moderately satisfactory in explaining the majority of the properties of the iron nitrosyl dithiolenes, it was less satisfactory when applied to the corresponding cobalt complexes. It was suggested that in the cobalt systems, the relative ordering of the uppermost filled energy levels was different and certainly the chemical evidence cited previously seems to indicate electronic structures significantly different from those of their iron analogs.

An assignment of formal oxidation states to the iron complexes on the basis of the proposed M.O. scheme proved to be an irresistible temptation. As already stated, the b_1', b_2', and a_1' levels in all of the complexes would be filled, and since these were defined as being predominantly metal–d-orbital in character, i.e., the complexes were believed to contain d^6 iron, the formal oxidation state of the iron in all species was $+2$. Thus, the dianions were formally described as $[Fe^{2+}(NO^{\cdot})(-S_4)^{-4}]$, where the two sulfur ligands were considered collectively as $(-S_4)^z$. The monoanions were represented as $[Fe^{2+}(NO^+)(-S_4)^{-4}]$ and the neutral species as $[Fe^{2+}(NO^+)(-S_4)^{-3}]$. It was known that the dicyanoethylene-1,2-dithiolate dianion could be oxidized (54) to *cis* and *trans*-$[S_4C_4(CN)_4]^{-2}$ via a radical anion intermediate, and so the existence of the neutral species, and a monocationic species, formally represented as $[Fe^{2+}(NO^+)(-S_4)^{-2}]$, did not seem unreasonable (the latter was detected voltammetrically). It was also possible to envisage the acceptance by NO of an electron giving NO^-, and the predicted $[Fe^{2+}(NO^-)(-S_4)^{-4}]$, or $[Fe(NO)-S_4]^{-3}$, was also identified voltammetrically. However, it was emphasized that the assumption of pure metal, sulfur ligand, or $\pi^*(NO)$ character for these various energy levels was an oversimplification and the assignment of formal oxidation states to the nitrosyl systems, although attractive and intriguing, was viewed with caution.

Other five-coordinate nitrosyl compounds containing dithiolate ligands have recently been identified. Ethanolic hydrazine reduction (46) of $[\pi\text{-}C_5H_5Mn(NO)S_2C_2(CF_3)_2]^0$ (Sec. VI-A) afforded the paramagnetic $[Mn(NO)S_4C_4(CF_3)_4]^{-2}$ (Table XXXVII).

The six-coordinate dinitrosyls, $[M(NO)_2S_4C_4(CN)_4]^{-2}$, M = Mo or W, were obtained (154) in the same way as their bis-dialkyldithiocarbamate analogs, $[M(NO)_2(S_2CNR_2)_2]^0$ (169), namely by treating the polymeric nitrosyl halides with the sodium salt of the ligands. The NO stretching frequencies of these complexes occurred in the region normally associated with "metal-coordinated NO^+" (Table XXXV). The occurrence of two frequencies of approximately equal intensity indicated a *cis* configuration for the NO groups in either octahedral or trigonal prismatic stereochemistries. A significant feature emerged from a comparison of the values of ν_{NO} in the dithiolate complexes and in the dithiocarbamate species, namely that the average decrease in the NO stretching modes on going from the neutral $[Mo(NO)S_4]^0$ (dithiocarbamate) coordination unit to the dinegatively charged $[Mo(NO)S]_4^{-2}$ (dithiolate) coordination unit was only

30 cm^{-1}, and the comparable average decrease in the tungsten analogs was 0 cm^{-1}. Normally, a decrease of between 50 and 100 cm^{-1} might have been anticipated (164) and these results were interpreted as evidence for the dithiolate ligand's considerable π-acceptor ability when complexed to a metal ion in a "low oxidation state."

By refluxing in benzene a mixture of $Hg[Fe(CO)_3(NO)]_2$ and S_2C_2-$(CF_3)_2$ (170), a black nitrosyl complex, initially formulated as $[Fe(NO)S_2$-$C_2(CF_3)_2]_2$, was obtained. A later mass spectrometric and magnetic examination (158) of this complex revealed that it was a mixture composed principally of a paramagnetic trimer, $[Fe(NO)S_2C_2(CF_3)_2]_3$, with traces of a more complex species, $\{Fe_3(NO)_4S[S_2C_2(CF_3)_2]_2\}$. A broad ESR signal was detected in solutions reputed to contain the paramagnetic trimer (171).

Treatment of a suspension of $[FeS_4C_4Ph_4]_2{}^0$ in chloroform with NO resulted in the formation of the brown, paramagnetic five-coordinate monomer, $[Fe(NO)S_4C_4Ph_4]^0$ (49,50,96) and a purple diamagnetic complex initially formulated as $[Fe(NO)S_2C_2Ph_2]_2{}^0$ (49). As a consequence of the reformulation of the "dimeric" bis-perfluoromethyl–dithiolene complex as a trimer, the purple compound was reinvestigated (171) and found to be a dinuclear species containing one mole of chloroform of crystallization, $[Fe_2(NO)_2S_6C_6Ph_6] \cdot CHCl_3$. This compound could also be produced by allowing $[Fe(NO)S_4C_4Ph_4]^0$ to stand in chloroform solutions exposed to air, or by passing NO into chloroform solution of $[Fe_2S_2(S_2C_2Ph_2)_2]^0$. (This latter reaction afforded other unidentified nitrosyl dithiolenes.) Voltammetric studies of the purple compound revealed the existence of two negatively charged species, $[Fe_2(NO)_2S_6C_6Ph_6]^{-1,-2}$ (Table XXXVI) and there was also some evidence for a dicationic species of the same formulation. Borohydride reduction of the purple compound in acetone afforded a green monoanion, $[Fe_2(NO)_2S_6C_6PH_6]^{-1}$, isolated as the salts of heavy organic cations. This complex had one voltammetric oxidation (and one reduction) wave at potentials comparable with those of the parent purple complex.

The infrared spectra of the purple dinuclear species (Table XXXV) consisted of two bands in the NO stretching frequency region, except in pyridine where four frequencies were detected, suggesting, possibly, that reaction with the solvent had taken place. The spectra of the green monoanion consisted of three absorptions, two strong and one of medium intensity, which were displaced to lower frequencies than those of the parent purple complex. The green monoanion was paramagnetic and had an ESR spectrum in solution at room temperature which consisted of a three-line multiplet, $\langle A_N \rangle = 11.2$ G (in CH_2Cl_2) (Table XXXV). This result may be compared with those obtained from $[Fe(NO)S_4C_4Ph_4]^0$ and $[Fe(NO)S_4C_4Ph_4]^{-2}$. The Mössbauer spectra (83) indicated the presence of two iron sites in the purple species and one in the green species.

If NO was passed into acetone solutions containing the purple complex, gray–purple solutions were formed. Although no new solid product could be isolated from the solutions ($[Fe_2(NO)_2S_6C_6Ph_6]^0$ was recovered quantitatively), it was believed that they contained $[Fe_2(NO)_3-S_6C_6Ph_6]^0$. The latter would be expected to be paramagnetic, and indeed, broad ESR signals were detected from the gray–purple solutions.

A compound tentatively described (96) as $[Fe_3(NO)_2S_6C_6Ph_6]\cdot C_6H_6\cdot C_6H_{12}$ (decomposing at 100°) was obtained in the reaction of $[FeS_4C_4Ph_4]_2^0$ with NO in benzene. It was isolated from the reaction mixture by chromatography on silica gel using 1:1 benzene-cyclohexane mixtures as eluant. This work was repeated (171) and it was shown that the compound was identical to $[Fe_2(NO)_2S_6C_6Ph_6]^0$, although it clearly contained solvent of crystallization. It is obvious that many other interesting nitrosyl species are present in this system.

The structure of $[Fe_2(NO)_2S_6C_6Ph_6]^0$ and its related monoanion is difficult to envisage in the absence of definitive structural data. It is clear that the neutral species contains two different iron sites, thereby ruling out a symmetrical structure as suggested by Figure 28. This would be attractive for the monoanion also if it were not for the observation of a triplet signal in the compound's ESR spectrum; a quintet would have been anticipated because of coupling of the electron with the two nitrogen atoms of the NO groups. It seems futile to speculate on other possible structures until a single crystal x-ray structural determination of at least one of these complexes has been carried out.

If $[Fe_2(NO)_2S_6C_6Ph_6]^0$ was treated with $Na_2S_2C_2(CN)_2$, itself a mild reducing agent, the green monoanion, $[Fe(NO)S_4C_4Ph_4]^{-1}$ and a mixed ligand complex, $[Fe(NO)S_4C_4Ph_2(CN)_2]^{-1}$, were formed (171). Similar complexes to that last-mentioned were obtained (172) when solutions containing two different monoanions were allowed to remain in contact over

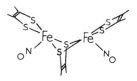

Fig. 28. A possible, but unlikely, structure for $[Fe_2(NO)_2S_6C_6Ph_6]^z$.

long periods or were refluxed. The reaction scheme, which was investigated using ESR methods, appeared to be

$$[Fe(NO)S_4C_4R_4]^{-1} + [Fe(NO)S_4C_4R'_4]^{-1} \longrightarrow [Fe(NO)S_4C_4R_4]^{-2}$$
$$+ [Fe(NO)S_4C_4R'_4]^{0} \longrightarrow 2[Fe(NO)S_4C_4R_2R'_2]^{-1}$$

C. Phosphine, Pyridine, and Related Complexes

1. Four-Coordinate Species

Phosphine, dipyridyl, and o-phenanthroline complexes of zinc (173), copper (62), and the nickel group metals (79,99) (Table XXXVII) were prepared by treating the appropriate metal halide complex, e.g., $(PPh_3)_2$-$PtCl_2$, $[Cu(dipyr)PPh_3I]$, or $(o$-Phenan$)ZnBr_2$, with the dithiol or dithiolate, by displacement of CO or PPh_3 from zerovalent nickel group complexes or by direct reaction of $[MS_4C_4R_4]^{0}$ with the phosphine or heterocyclic ligand.

The infrared and electronic spectra of these complexes indicated that they contained "dithiolate" ligands. None of them has been investigated polarographically yet.

2. Five-Coordinate Species

The anomalous magnetic properties of the supposed "monoanionic" dithiolenes complexes of cobalt and iron were discussed in Section III-E. There, it was suggested that the low-spin (Fe) or diamagnetic (Co) dimeric dianions were dissociated in strongly polar solvents, many of which (the solvents) could function as two-electron donor molecules. The effect of dissociation was to cause the iron complexes to become high-spin ($S = \frac{3}{2}$), and adducts containing the "solvent," such as pyridine, were isolated (101,110) and shown to have magnetic moments consistent with a spin–quartet ground state. The dissociated cobalt complexes, with the exception of $[CoS_4C_4(CN)_4]^{-1}$ in DMSO ($\mu_{eff} = 2.81$ BM), were diamagnetic, and there was strong evidence for the formation of five- and even six-coordinate solvent adducts, $[Co(L)_{1,2}$—$S_4]^{-1}$. The cobalt arene-1,2-dithiolenes, which were monoanionic in the solid and in solution, were high-spin ($S = 1$) and there was little evidence for axial perturbation by donor solvents.

In an extensive study of the reactions of charged iron and cobalt dithiolenes with Lewis bases (101), it was found that treatment of the dimeric dianions with a wide variety of monodentate ligands such as pyridines, phosphines, phosphites, stibines, isonitriles, CN^-, N^{3-}, or NCO^- caused dissociation of the complexes and formation of five-coordinate adducts, $[M(L)$—$S_4]^{-1}$, $[M(CN)$—$S_4]^{-2}$, etc. (Table XXXVII). Some of these, notably the adducts of $[Fe(S_2C_6H_3Me)_2]^{-1}$ and its tetrachlorobenzene–dithiolene analog, were labile in solution. Those five-coordinate iron adducts containing "weak-field" ligands were high-spin ($S = \frac{3}{2}$) and had magnetic moments in the solid state consistent with three unpaired electrons. Those iron adducts containing "strong-field" ligands were low-spin ($S = \frac{1}{2}$) and had magnetic moments entirely consistent with spin–doublet ground states. All of the cobalt species were diamagnetic. Gray and his colleagues (133,175) had previously observed color changes when solutions of " $[CoS_4C_4(CN)_4]^{-1}$ " were treated with pyridine, triphenylphosphine, or arsine and had noted that some of the adducts formed were labile. In a kinetic study, it was established that the order of ligand affinity was $PPh_3 > pyr > AsPh_3$, corresponding to the Ahrland, Chatt, and Davies "Type B" behavior for the metal ion. This was thought to be unusual for a complex containing (formally) Co(III) and it was suggested that the kinetic data was evidence for the incorrectness of the formal oxidation state description of this type of complex.

A series of neutral complexes, $[M(L)S_4C_4R_4]^0$, where M = Co or Fe, R = Ph or CF_3, and L is a phosphine, phosphite, stibine, or arsine, were isolated (96,186) from the reactions between the donor ligands and the neutral dimeric dithiolenes in refluxing benzene (Ph) or pentane (CF_3). Some bis-toluoyl-dithiolene adducts and $[Fe(PBu_3)S_4C_4H_4]^0$ were also obtained (96). The iron complexes were diamagnetic whereas the cobalt species were paramagnetic (one unpaired electron) with the exception of $\{Co[P(OPh)_3]S_4C_4(CF_3)_4\}^0$ which was diamagnetic and may have been dimeric.

All those complexes having spin–doublet ground states displayed ESR signals in solution at room temperature (Table XXXVIII). Thus the monoanionic phosphine complexes, $[Fe(PR_3)—S_4]^{-1}$, exhibited doublet signals due to ^{31}P nuclear hyperfine splittings, and had g values and hyperfine splittings averaging 2.045 and 28.0 G, respectively. These data were in direct contrast to those obtained from the formally isoelectronic cobalt complexes, $[Co(PR_3)—S_4]^0$, where ^{59}Co, but not ^{31}P hyperfine splittings were detected. (An exception to this occurred in $[Co(P(OPh)_3S_4C_4Ph_4]^0$ where $\langle A_{Co}\rangle = 24.1$ G and $\langle A_P\rangle = 8.5$ G.) The cyanides, $[Fe(CN)—S_4]^{-2}$, displayed broad resonance signals but neither ^{13}C nor ^{14}N hyperfine splittings were resolved; the isonitriles behaved similarly. The width and resolution of some of the resonance signals were found to be dependent on temperature and concentration of the fifth ligand.

Polarographic and voltammetric studies of the stable monoanions (101) and neutral species (96,101,186) established that they all underwent electron transfer reactions, many of the electrode processes being reversible.

Nearly all of the complexes revealed voltammetric waves (Table XXXVIII) in dichloromethane (101,186) corresponding to the generation of mono-cationic, neutral, mono-, and possibly dianionic species; the existence of the latter is uncertain since many of the free ligands undergo one-electron reduction at, or close to, the potentials at which the formation of the dianions might occur. From the half-wave potential data, it was clear that one-electron oxidation of certain species, e.g. $[M(PEt_3)S_4C_4(CF_3)_4]^{-1}$, should be possible (see Sec. III-C) and, indeed, the corresponding neutral species were produced by treatment of the monoanions with iodine in acetone. From the polarographic studies of the neutral phosphine diphenyl-dithiolene adducts in DMF (96), it was thought that the first, reversible reduction wave corresponded to the generation of a monoanion, $[M(PR_3)-S_4C_4Ph_4]^{-1}$, but that on addition of the second electron, in an irreversible electrode process, removal of the phosphine ligand occurred. The half-wave oxidation and reduction potentials were dependent on four factors—the nature of the sulfur ligand, the metal atom, the fifth ligand donor atom, and the groups attached to the donor atom. The first two exerted the greatest effect on the potentials, the third exerted a smaller effect, and the fourth was only just detectable.

No detailed study of the electronic structures of these five-coordinate dithiolene complexes has been described. However, a study of the ESR spectrum of $\{Co[P(OPh)_3]S_4C_4Ph_4\}^0$ in solutions and in glasses (187) has been carried out, and it was suggested that the unpaired electron was in an orbital of ligand antibonding character having a $3d_{xy}$ metal component. It was observed that the calculations performed using these ESR param-eters were not inconsistent with the view that the complex could be described as containing Co(II) $(3d^7)$, two monovalent bidentate sulfur ligands, $[Ph_2C_2S_2]^-$, and $P(OPh)_3$. However, it has been observed that there were certain spectral similarities between the neutral and mono-anionic nickel bis-dithiolenes, $[NiS_4C_4R_4]^{0,-1}$ and $[Fe(PR_3)S_4C_4R_4]^0$ and $[Co(PR_3)S_4C_4R_4]^0$, respectively, suggesting a similarity in their electronic "make-up." The phosphine adducts were represented by the limiting structures shown in **30** and **31**. However, these suggestions implied the

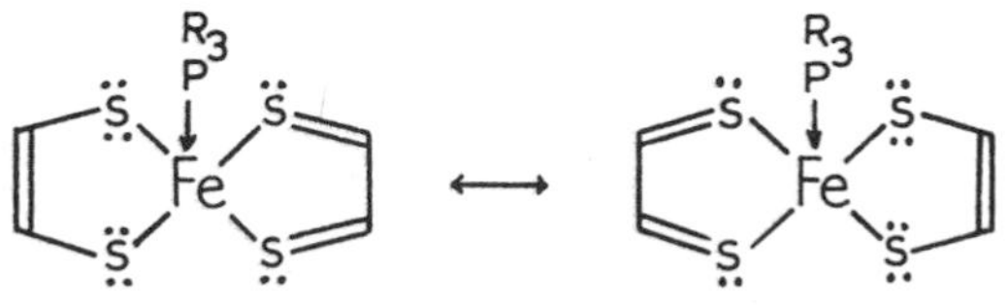

(30)

TABLE XXXVIII

Voltammetric Data Obtained from Some Five- and Six-Coordinate Lewis Base Adducts of Bis-dithiolene Complexes

Couple: $[ML\!-\!S_4]^z + e^- \rightleftharpoons [ML\!-\!S_4]^{z-1}$

| Complex | Ligand | Couple[a] | | | Ref. cell |
		$-2 \rightleftharpoons -1$	$-1 \rightleftharpoons 0^b$	$0 \rightleftharpoons +1^b$	
$[CoS_4C_4(CN)_4]^{-1}$	PEt$_3$	-0.84^d	$+0.46$	$+1.53$	A
	PEt$_2$Ph	-0.78^d	$+0.44$	$+1.46$	A
	PEtPh$_2$	-0.59^d	$+0.45$	$+1.51$	A
	PPh$_3$	—	$+0.77$	—	A
	P(CH$_2$CH$_2$CN)$_3$	-0.43^d	$+0.70$	—	A
	P(i-C$_3$H$_7$)$_3$	-0.65^d	$+0.53$	$+1.50$	A
	P(OEt)$_3$	-0.68^d	$+0.51$	—	A
	P(OPh)$_3$	-0.51^d	$+0.55$	—	A
	As(n-C$_3$H$_7$)$_3$	-0.65^d	$+0.43$	$+1.50$	A
	AsPh$_3$	—	$+0.75$	—	A
	SbPh$_3$	—	$+0.73$	—	A
$[CoS_4C_4(CF_3)_4]^{-1}$	PEt$_3$	-1.22^d	$+0.15$	$+1.45$	A
$CoS_4C_4(CF_3)_4]^0$	PPh$_3$	—	$+0.41$	$+1.50$	A
	P(OPh)$_3$	—	$+0.52$	$+1.56$	A
	AsPh$_3$	—	$+0.46$	$+1.50$	A
	SbPh$_3$	—	$+0.40$	$+1.50$	A
$[CoS_4C_4Ph_4]^0$	PBu$_3$	—	-0.58	$+0.28$	A
		-1.73^d	-0.43	—	B
	PPh$_3$	-1.04^d	-0.42	$+0.30$	A
	P(OPh)$_3$	-1.04^d	-0.29	—	B
$[CoS_4C_4(p\text{-}MeC_6H_4)_4]^0$	PBu$_3$	-1.28^d	-0.43	—	B
	PPh$_3$	-1.06^d	-0.27	—	B
$[FeS_4C_4(CN)_4]^{-1}$	Pyridine	-0.51^d	—	—	A
	PEt$_3$	-0.82^d	$+0.37$	—	A
	PBu$_3$	-0.80^d	$+0.38$	—	A
	PEt$_2$Ph	-0.83^d	$+0.38$	—	A
	PEtPh$_2$	-0.59^d	$+0.38$	—	A
$[FeS_4C_4(CN)_4]^{-1}$	PPh$_3$	—	$+0.69$	—	A
	P(CH$_2$CH$_2$CN)$_3$	—^{e}	$+0.62$	—	A
	As(n-C$_3$H$_7$)$_3$	—^{e}	$+0.38$	—	A
	AsPh$_3$	—	$+0.77$	—	A
	SbPh$_3$	—	$+0.70$	—	A
$[FeS_4C_4(CF_3)_4]^{-1}$	PEt$_3$	-1.22^d	$+0.02$	—	A
$[FeS_4C_4(CF_3)_4]^0$	PPh$_3$	—	$+0.30$	—	A
	P(OPh)$_3$	—	$+0.37$	—	A
	AsPh$_3$	—	$+0.47$	—	A
	SbPh$_3$	—	$+0.41$	—	A

(continued)

TABLE XXXVIII (*continued*)

Complex	Ligand	Couple[a]			Ref. cell[c]
		$-2 \rightleftharpoons -1$	$-1 \rightleftharpoons 0$[b]	$0 \rightleftharpoons +1$[b]	
FeS₄C₄Ph₄]⁰	PBu₃	—	−0.72	+0.55	A
		−1.09[d]	−0.50	—	B
	P(OPh)₃	−1.09[d]	−0.37	—	B
FeS₄C₄H₄]⁰	PBu₃	−1.05[d]	−0.56	—	B
CoS₄C₄(CN)₄]⁻¹	o-Phen	−0.91[d]	+0.60	+1.26	A
	Dipyr	−0.94[d]	+0.59	+1.21	A
	En[f]	−0.92[d]	+0.55	—	A
	Dmg[g]	−0.90[d]	—[h]	—[h]	A
CoS₄C₄(CF₃)₄]⁻¹	o-Phen	−1.39[d]	+0.25	+1.21	A
FeS₄C₄(CN)₄]⁻¹	o-Phen	−0.68[d]	+0.40	—	A
	Dipyr	−0.64[d]	+0.41	—	A
FeS₄C₄(CF₃)₄]⁻¹	o-Phen	−1.06[d]	+0.03	—	A
MoS₄C₄Ph₄]⁰¹	PPh₃, CO	−1.40[d,j]	−0.71	—	B
	(Ph₂PCH₂)₂	−1.96	−1.15	—	B

[a] In volts.
[b] Reversible waves.
[c] Reference cell and solvent: A. SCE using CH_2Cl_2 and RPE; B. Ag/AgCl using DMF and DME.
[d] Irreversible waves, possibly corresponding to reduction of free ligand.
[e] Electrode coating caused wave to disappear.
[f] En = ethylenediamine.
[g] Dmg – dimethylglyoxime.
[h] Multiple oxidation waves.
[i] cis-[Mo(PPh₃)(CO)S₄C₄Ph₄]⁰.
[j] Irreversible reduction wave at −2.38 V (B).

localization of the odd electron in the cobalt species in a ligand-based orbital with a $3d_{xz}$ component, which was shown to be unlikely from the ESR calculations (187) on the phosphite adduct. The addition of the donor ligand to the dimeric diphenyl-dithiolenes caused a slight increase in $\nu_{C=C}$ relative to that in the parent dimers (Table XXXVII) and the proton resonance spectrum of $[Fe(PBu_3)S_4C_4H_4]^0$ exhibited resonances due to the

(31)

phosphine alkyl groups and to the protons on the chelate rings, the latter at $\tau = 0.1$, i.e., in a region similar to that in $[NiS_4C_4H_4]^0$.

There were clearly no electronic similarities between $[Fe(PR_3)—S_4]^{-1}$ (101) and the formally isoelectronic $[Fe(NO)—S_4]^0$ and $[Co(PR_3)—S_4]^0$ and the same was true for $[Co(PR_3)—S_4]^{-1}$ and $[Fe(NO)—S_4]^{-1}$.

3. Six-Coordinate Species

The formation of four-coordinate phosphine complexes of the nickel group, $(PR_3)_2MS_2C_2R'_2$, has already been discussed. However, if the reaction conditions required to produce these compounds were altered slightly (174), six-coordinate complexes, $(PR_3)_2MS_4C_4R'_4$ and $(R_2RCH_2-CH_2PR_2)MS_4C_4R'_4$, were formed. The blue platinum species $(PBu_3)_2-PtS_4C_4Ph_4$ and $(Ph_2PCH_2CH_2PPh_2)PtS_4C_4Ph_4$ decomposed into their components, the former on heating at 162° (the melting point of the complex) and the latter on UV irradiation. Thermal instability was also observed when the nickel species, $(PR_3)_2NiS_4C_4Ph_4$, were melted, the phosphine and original bis-diphenyl–dithiolene being recovered quantitatively. It was interesting that the green color of $[NiS_4C_4Ph_4]^0$ persisted in mixtures of phosphines and the complex even up to the fusion point, but that adduct formation took place *only* in the cold.

The infrared and electronic spectra of these adducts revealed that the ligands had not lost their "dithiolate" character and that the complexes could be regarded formally as containing the metal in the +4 oxidation state. These species were not investigated voltammetrically.

Gray and co-workers had observed (133) that the "monoanionic" cobalt bis-dicyano–dithiolene formed a crystalline adduct with o-phenanthroline, $[Co(o-phen)S_4C_4(CN)_4]^{-1}$. An extensive study of the reactions of dimeric cobalt and iron dithiolenes with bidentate four-electron donors (101) showed that six-coordinate complexes were very readily formed. The most stable adducts were those containing α,α' dipyridyl and o-phenanthroline whereas those of ethylenediamine and dimethylglyoxime, which complexed only with the cobalt dithiolenes, were labile in solution.

The six-coordinate iron adducts were paramagnetic, having magnetic moments consistent with spin–doublet ground states (Table XXXVII), and exhibited broad ESR signals whose resolution and width were dependent on temperature; ^{14}N hyperfine interactions were not resolved. The cobalt adducts were diamagnetic.

Voltammetric investigations (101) of these stable dipyridyl and phenanthroline complexes disclosed the existence of monocationic, neutral, and possibly dianionic species. Treatment of some complexes, e.g.,

$[Fe(o\text{-}phen)\text{-}S_4C_4(CF_3)_4]^{-1}$, with iodine in acetone resulted in the formation of the corresponding neutral complexes.

Treatment of the carbonyls $[M(CO)_2S_4C_4Ph_4]^0$ (135) with the chelating diphosphine $Ph_2PCH_2CH_2PPh_2$ caused the removal of the CO and formation of the diphosphine complex $[M(Ph_2PCH_2CH_2PPh_2)S_4C_4Ph_4]^0$, which was also obtained by the thermal decomposition of $[MS_6C_6Ph_6]^0$ in the presence of the phosphine. Polarographic reduction of this compound revealed the existence of two reversible one-electron waves presumably corresponding to the formation of mono- and dianions. A trigonal prismatic structure for this phosphine complex was suggested. Other labile adducts, $[M(CN)_2S_4C_4Ph_4]^{-2}$ and $(M(pyr)_2S_4C_4Ph_4)^0$, were also reported.

D. Vanadyl, Chromyl, Molybdenyl, and Related Dithiolene Complexes

The five-coordinate vanadyl complex, $[VOS_4C_4(CN)_4]^{-2}$, was obtained (140) either by treatment of aqueous vanadyl salts with $Na_2S_2C_2$-$(CN)_2$, or by prior reduction of $[VS_6C_6(CN)_6]^{-2}$ with hydrazine in DMSO followed by exposure of the reduced species to air. The light-brown complex (which was red–brown in solution) was paramagnetic (one unpaired electron) and displayed an eight-line multiplet ESR spectrum (Table

TABLE XXXIX

Vanadyl, Molybdenyl, and Chromyl Bis-dithiolene Complexes

Complex	Color	$\nu_{M=O}$, cm^{-1}	$E_{1/2}$, V	g	A^a	μ_{eff},[b] BM
$[VOS_4C_4(CN)_4]^{-2}$	Red–brown[c]	978[d]	+0.49[f,g]	1.986	77	1.83
		960[e]	+0.40[h]			
$[MoOS_4C_4(CN)_4]^{-2}$	Pale green[c]	936[e]	+0.48[f]			Dia.
			+0.35[h]			
$[CrO(S_2C_6H_3Me)_2]^{-1}$	Purple[i]			1.994	14.9	
$[VO(S_2C_6Cl_4)_2]^{-1}$	Red	950				17.3

[a] Metal hyperfine splitting: ^{51}V, $I = \frac{7}{2}$; ^{53}Cr, $I = \frac{1}{2}$.
[b] In the solid state.
[c] Ph_4P^+ salt.
[d] In mulls.
[e] In solution.
[f] In CH_3CN vs. SCE.
[g] Irreversible wave.
[h] In CH_2Cl_2 vs. SCE.
[i] Bu_4N^+ salt.

XXXIX) (^{51}V, $I = \frac{7}{2}$) whose parameters were similar to those of $[VS_6C_6-(CN)_6]^{-2}$. Treatment of this oxy-species with excess $Na_2S_2C_2(CN)_2$ resulted in the removal of the oxygen and formation of $[VS_6C_6(CN)_6]^{-2}$. Attempts to prepare analogous bis-diphenyl– and toluene–dithiolene complexes were unsuccessful. However, the tetrachlorobenzene–dithiolene, $[VO(S_2C_6Cl_4)_2]^{-2}$ has been obtained (172).

The molybdenyl compound, $[MoOS_4C_4(CN)_4]^{-2}$, was isolated (176) from the reaction between molybdates and $Na_2S_2C_2(CN)_2$ in hot water. The light-brown complex, which formed pale green solutions in donor solvents, was diamagnetic, and was not isomorphous with its vanadium analog.

Both of these complexes were studied voltammetrically and mono-anionic species $[MOS_4C_4(CN)_4]^{-1}$, were detected in dichloromethane (Table XXXIX). There appeared to be no evidence for other charged complexes.

In the reaction between hydrated titanium(III) sulfate and $Na_2S_2C_2-(CN)_2$ in aqueous acetone, a deep red species, which appeared to be the five-coordinate $[TiOS_4C_4(CN)_4]^{-2}$, was formed. The complex was very unstable to air and a Ti=O stretching frequency was not detected in the infrared. A similar rhenium complex, $[ReOS_4C_4(CN)_4]^{-2}$, has also been prepared (132) but details are not available.

Treatment of ethanolic solutions of chromic chloride with the toluene-3,4-dithiolate dianion and oxygen resulted (137) in the formation of the chromyl complex $[CrO(S_2C_6H_3Me)_2]^{-1}$, which was paramagnetic (one unpaired electron) and displayed a sharp ESR signal at $\langle g \rangle = 1.994$ with ^{53}Cr hyperfine splitting (Table XXXIX).

VI. ORGANOMETALLIC DITHIOLENE COMPLEXES

A. π-Cyclopentadienyl Complexes

The π-cyclopentadienyl complexes are of two main types—uncharged species and anionic species. The majority of complexes which have been prepared contain $S_2C_2(CF_3)_2$ and $S_2C_2(CN)_2$ ligands, those species with the former being predominantly uncharged and those with the latter mainly anionic (Table XL).

1. Simple π-Cyclopentadienyl Dithiolenes

Reaction of $(\pi\text{-}C_5H_5)_2TiCl_2$ with one mole of $Na_2S_2C_2(CN)_2$ (53) or $C_6H_3Me(SH)_2$ (177) afforded the monomeric complexes $(\pi\text{-}C_5H_5)_2Ti(—S_2)$, whose structure, Figure 29, is probably similar to that of $(\pi\text{-}C_5H_5)_2MoH_2$

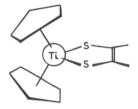

Fig. 29. The probable structure of $(\pi\text{-}C_5H_5)_2M(\text{---}S_2)$.

(178). Similar tungsten and molybdenum complexes are also known (179). Treatment of the titanium dicyanoethylene–dithiolate with a further mole of the sulfur ligand resulted in the cleavage of one C_5H_5 ring and the formation of the apparently seven-coordinate monoanion, $[\pi\text{-}C_5H_5TiS_4C_4(CN)_4]^{-1}$. The structure of this complex is probably similar to that of $\pi\text{-}C_6H_5Nb(CO)_4$, where the cyclopentadienyl ring is assumed to occupy three coordination positions (180).

Treatment of the π-cyclopentadienyl metal carbonyls $\pi\text{-}C_5H_5V(CO)_4$, $[\pi\text{-}C_5H_5Cr(CO)_3]_2Hg$, and $[\pi\text{-}C_5H_5Mo(CO)_3]_2$ with $S_2C_2(CF_3)_2$ afforded (45) the dimeric complexes of general formula $[\pi\text{-}C_5H_5MS_2C_2(CF_3)_2]_2$. The vanadium complex was paramagnetic, having a moment of 0.6 BM per vanadium atom, comparable with that of the analogous μ-mercaptide, $[\pi\text{-}C_5H_5V(SMe)_2]_2$ [0.9 BM (181)]. The reduction of these moments from the expected 1.7 BM per metal atoms suggested that the bond order of the V—V bond was greater than one, and that the degree of multiple bond character was greater in the dithiolene. The chromium and molybdenum complexes were diamagnetic but their structures, shown in Figures 30 and 31, revealed distinct differences. The unsymmetrical sulfur bridges in the chromium complex (46), at first sight, made the stereochemistry of the Cr

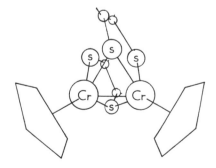

Fig. 30. The structure of $[\pi\text{-}C_5H_5CrS_2C_2(CF_3)_2]_2$.

TABLE XL

π-Cyclopentadienyl Dithiolenes

Complex	Color	mp,°C	μ_{eff},BM	τ_{op}	δ_{19F}	$\nu_{C=C}$, cm^{-1}	$\nu_{CO/NO}$, cm^{-1}
Cp$_2$TiS$_2$C$_2$(CN)$_2$	Green	260		3.57[a]		1442[h]	
Cp$_2$TiS$_2$C$_6$H$_3$Me	Green						
[CpVS$_2$C$_2$(CF$_3$)$_2$]$_2$	Green		0.6/V				
[CpCrS$_2$C$_2$(CF$_3$)$_2$]$_2$	Purple–brown				−420--−460[b]	1590[l],1430	
[CpMoS$_2$C$_2$(CF$_3$)$_2$]$_2$	Purple–brown				−385[c]	1530[l],1435; 1650[l],1630, 1420	
[CpFeS$_2$C$_2$(CF$_3$)$_2$]	Black	150					
[CpCoS$_2$C$_2$(CF$_3$)$_2$]	Violet	147	Dia.		−710[c]	1480[l],1417	
[CpCoS$_2$C$_2$(CN)$_2$]	Purple	282 d		4.34[d]		1347[h]	
[CpRhS$_2$C$_2$(CF$_3$)$_2$]	Red	179–180				1512[l]	
[CpIrS$_2$C$_2$(CF$_3$)$_2$]	Red–brown	187–188				1512[l]	
[CpNiS$_2$C$_2$(CF$_3$)$_2$]	Green–black		1.67			1520[l],1440, 1405	
[CpTiS$_4$C$_4$(CN)$_4$]$^{-1}$	Magenta	204–205[j]		3.69[a]			
[CpMoS$_4$C$_4$(CN)$_4$]$^{-1}$	Blue–green	282–285[k]		4.87[e]			
[CpWS$_4$C$_4$(CN)$_4$]$^{-1}$	Magenta	269–272[k]		4.87[d]			
[CpWS$_4$C$_4$(CF$_3$)$_4$]$^{-1}$	Red–violet d	231–233 d		5.20[g]	+54.3[f]	1540[l]	
[CpWS$_4$C$_4$(CF$_3$)$_4$]^{0}	Green	224–225	1.68	4.11[d]		1517[l],1496	
[CpMo(NO)S$_2$C$_2$(CF$_3$)$_2$]$_2$	Brown						1661[l]
[CpMn(NO)S$_2$C$_2$(CF$_3$)$_2$]$_2$^{0}	Brown	142–144	1.73			1535[l]	1797[l]
[CpMn(NO)S$_2$C$_2$(CN)$_2$]$^{-1}$	Red[l]	300	Dia.			1486[l]	1705
[CpMn(NO)(S$_2$C$_6$H$_3$Me)]	Purple	300	1.74				1774
[CpW(CO)S$_2$C$_2$(CF$_3$)$_2$]$_2$	Brown					1540[l],1530	2048[l],2011, 1983

	Color	M.p.				
[CpFe(CO)$_2$]$_2$S$_2$C$_2$(CN)$_2$	Red–brown	136–138 d		4.64[a]	1440[h]	2050,2010, 1960
Cp$_2$MoS$_2$C$_2$(CN)$_2$	Red–orange	ca. 180	Dia.			
Cp$_2$WS$_2$C$_2$(CN)$_2$	Yellow–orange	ca. 220	Dia.			
Cp$_2$MoS$_2$C$_6$H$_3$Me	Purple	229	Dia.			
Cp$_2$WS$_2$C$_6$H$_3$Me	Orange	230	Dia.			

[a] Acetone.
[b] THF; reference (CFCl)$_2$.
[c] CH$_2$Cl$_2$; reference (CFCl)$_2$.
[d] CHCl$_3$.
[e] DMSO.
[f] Acetone; reference CFCl$_3$.
[g] (CD$_3$)$_2$CO.
[h] KBr disk.
[i] Halocarbon mull.
[j] Et$_4$N$^+$ salt.
[k] Ph$_4$P$^+$ salt.

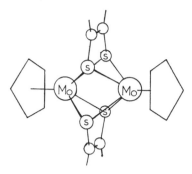

Fig. 31. The structure of $[\pi\text{-}C_5H_5MoS_2C_2(CF_3)_2]_2$.

atoms octahedral but the existence of a direct Cr–Cr interaction (2.98 Å) effectively rendered each metal atom seven-coordinate. The symmetrical bridging of the sulfur ligands in the molybdenum compound (46) made each Mo atom clearly seven-coordinate.

The monomeric $[\pi\text{-}C_5H_5WS_4C_4(CF_3)_4]^0$ was obtained (46) by treatment of $\pi\text{-}C_5H_5W(NO)(CO)_2$ with the dithietene, and no evidence for dimeric species analogous to the chromium and tungsten complexes was discovered. The complex was paramagnetic with a solid-state magnetic moment of 1.68 BM. It was reduced with ethanolic hydrazine affording the diamagnetic monoanion, $[\pi\text{-}C_5H_5WS_4C_4(CF_3)_4]^{-1}$, which was similar to the dicyano–dithiolenes, $[\pi\text{-}C_5H_5MS_4C_4(CN)_4]^{-1}$, where M = Mo or W (53), prepared by treating the metal cyclopentadienyl carbonyl halides with $Na_2S_2C_2(CN)_2$. While these complexes may be described formally as containing the metals in oxidation states $+5$ ($z = 0$) or $+4$ ($z = -1$), it is probably more realistic to regard them as a resonance hybrid of the two forms shown in Figure 32, i.e., as represented by Figure 33.

There was some mass spectrometric evidence for $\pi\text{-}C_5H_5FeS_2C_2$-$(CF_3)_2$ among the products of the reaction (46) between $[\pi\text{-}C_5H_5Fe(CO)_2]_2$

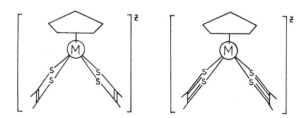

Fig. 32. Resonance hybrid forms for $[\pi\text{-}C_5H_5MS_4C_4R_4]^z$.

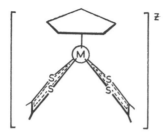

Fig. 33. A simple representation of the π-cyclopentadienyl dithiolenes,
$[\pi\text{-}C_5H_5MS_4C_4R_4]^z$.

and $S_2C_2(CF_3)_2$. The other products of this reaction, and of a similar one involving $[\pi\text{-}C_5H_5Ru(CO)_2]_2$, were complex and are discussed later. Reaction of $\pi\text{-}C_5H_5M(CO)_2$, M = Co or Rh, $\pi\text{-}C_5H_5IrC_8H_{12}$, and $[\pi\text{-}C_5H_5Ni(CO)]_2$ with $S_2C_2(CF_3)_2$ afforded (45,46) the monomeric species, $[\pi\text{-}C_5H_5MS_2C_2(CF_3)_2]^0$, and the analogous dicyano–dithiolene, $[\pi\text{-}C_5H_5\text{-}CoS_2C_2(CN)_2]^0$, was obtained by treatment of $\pi\text{-}C_5H_5Co(CO)I_2$ with $Na_2S_2C_2(CN)_2$ (53). The cobalt group complexes were diamagnetic whereas $[\pi\text{-}C_5H_5NiS_2C_2(CF_3)_2]^0$ had a magnetic moment of 1.67 BM and exhibited an ESR spectrum in solution at room temperature which consisted of a single line (Table XLI). On the basis of infrared studies, it was initially suggested (45,53) that the $CoS_2C_2R_2$ group in the cobalt complexes was bent whereas in the nickel complex it was linear. This bending of the metal–ligand group would have facilitated interaction between the carbon–carbon double bond and the metal atom thereby permitting the effective atomic number of the cobalt atom to be 36, assuming the ligands to be dithiolates and the metal atom oxidation state to be $+3$. However, it was observed that if the sulfur ligand had considerable dithioketonic character, then the metal–ligand group would probably be planar. (The E. A. N. of the cobalt atom would still be 36 if its oxidation state was $+1$.) The problem was solved by the x-ray structural determination of $[\pi\text{-}C_5H_5CoS_2C_2(CF_3)_2]^0$ (182) where it was found that the metal–ligand group was planar. The Co—S bond lengths were 2.08 Å, which may be compared with the calculated single-bond distance of 2.20 Å and the distance in $[CoS_4C_4(CF_3)_4]_2^0$, viz., 2.16 Å. The estimated bond order of the Co—S bond was 1.22. The C—S bond distances averaged 1.74 Å and the "C=C" distance was 1.48 Å. The latter was 0.07 Å longer than the "ethylenic" bonds in any other reported dithiolene complexes, and may be compared with the C—C single bond distance of 1.479 Å. Thus, it is clear

TABLE XLI

Voltammetric and ESR Data Obtained from π-Cyclopentadienyl Dithiolenes

Complex	$E_{\frac{1}{2}}^1$,V[a]	n^1 [b]	$E_{\frac{1}{2}}^2$,V[a]	g_{av}[c]	A_{av}[c]	ΔH[c]	Remarks
$[Cp_2TiS_2C_2(CN)_2]$	-0.73[f]	2	—				
$[Cp_2TiS_2C_6H_3Me]$	-1.7	2	-3.4	1.9857	9.58	2.0	$^{47}Ti(I = \frac{5}{2})$; $^{49}Ti(I = \frac{7}{2})$
$[CpVS_2C_2(CF_3)_2]_2$	-1.4	2	—[d]			578	$^{51}V(I = \frac{7}{2})$; 64 lines
$[CpMoS_2C_2(CF_3)_2]_2$	-1.7	2	-2.3				
$[CpTiS_4C_4(CN)_4]^{-1}$	-0.73[f]	1	-0.98[f,h]				
$[CpMoS_4C_4(CN)_4]^{-1}$	-1.42[f]	1	$+0.77$[f,h]				
$[CpWS_4C_4(CN)_4]^{-1}$	-1.70[f]		$+0.77$[f,h]				
$[CpFeS_2C_2(CF_3)_2]$	-2.0	2	—				
$[CpCoS_2C_2(CF_3)_2]$	-1.1	1	-2.9	2.454	41.0	288.1	$^{59}Co(I = \frac{5}{2})$; 8 lines
$[CpCoS_2C_2(CN)_2]$	-0.45[f]	1	—				
$[CpRhS_2C_2(CF_3)_2]$	$+1.4$	1	—				
$[CpIrS_2C_2(CF_3)_2]$	-1.7	1	—				
$[CpNiS_2C_2(CF_3)_2]$	-0.96	1	-2.4	2.0479[e]			
$[CpMo(NO)S_2C_2(CF_3)_2]$	-1.3	2	-1.9				
$[CpMn(NO)S_2C_2(CN)_2]^{-1}$	-0.50[f]	1	$+0.20$[f,g]				
$[CpMn(NO)S_2C_6H_3Me]$	-0.58[f]	1	$+0.70$[f,g]				

[a] Data obtained in dimethoxyethane with Bu_4NClO_4 as supporting electrolyte. Reference cell Ag/AgClO$_4$ suspended in Bu_4NClO_4 in dimethoxyethane.

[b] Number of electrons in first reduction step.

[c] In gauss.

[d] Complex wave.

[e] Neutral species paramagnetic.

[f] Data obtained in dichloromethane using Et_4NClO_4 as supporting electrolyte, reference cell SCE, with rotating Pt electrode.

[g] Possibly two-electron oxidation wave; irreversible.

[h] One-electron oxidation wave.

that there is multiple bonding between the cobalt and sulfur atoms, and that the ligand may be described as having considerable dithioketonic character.*

* *Note added in proof.* Recently, complexes of ethylene–dithiol, $(\pi\text{-}C_5H_5)_2\text{-}MS_2C_2H_2$ (M = Ti or Zr), $[Ph_4As][\pi\text{-}C_5H_5WS_4C_4H_4]$, $[\pi\text{-}C_5H_5Fe(CO)_2]_2S_2C_2H_2$, and $\pi\text{-}C_5H_5MS_2C_2H_2$ (M = Co or Rh), have been prepared (188). These complexes are very similar to the dicyanodithiolenes (53). The proton resonance spectra of $(\pi\text{-}C_5H_5)_2MS_2C_2H_2$ indicated that the sulfur ligand is truly represented as a dithiolate, the metals having the formal oxidation states $+4$, whereas the spectra of $\pi\text{-}C_5H_5MS_2\text{-}C_2H_2$ suggested that the sulfur ligand had considerable dithioketonic character.

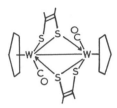

Fig. 34. A possible structure for $[\pi\text{-}C_5H_5W(CO)S_2C_2(CF_3)_2]_2$.

2. π-Cyclopentadienyl Carbonyl and Sulfur Dithiolenes

When $\pi\text{-}C_5H_5W(CO)_3H$ was treated with $S_2C_2(CF_3)_2$ (46), a dimeric carbonyl complex, $[\pi\text{-}C_5H_5W(CO)S_2C_2(CF_3)_2]_2$, Figure 34, was formed together with a small amount of $[\pi\text{-}C_5H_5WS_4C_4(CF_3)_4]^0$. The mass spectrum of this carbonyl was consistent with its formulation but also provided evidence for trace impurities of sulfide, $[\pi\text{-}C_5H_5WS(S_2C_2(CF_3)_2]_2$. Other carbonyl- and sulfur-containing bis-perfluoromethyl–dithiolene complexes containing Fe or Ru were detected by mass spectrometric analysis (46) of the products obtained from the reactions of the sulfur ligand with $[\pi\text{-}C_5H_5M(CO)_2]_2$. Among the species observed were ions generated from $[\pi\text{-}C_5H_5M(S)_xS_2C_2(CF_3)_2]_2, (\pi\text{-}C_5H_5Ru)_2(CO)S_2C_2(CF_3)_2$, and $[\pi\text{-}C_5H_5Ru)_3(S)_2S_2C_2(CF_3)_2$.

The dimeric dicyano–ethylene–dithiolate, $[\pi\text{-}C_5H_5Fe(CO)_2]_2S_2C_2(CN)_2$, was formed (53) in the reaction between $\pi\text{-}C_5H_5Fe(CO)_2Cl$ and $Na_2S_2C_2(CN)_2$. On the basis of infrared evidence it was suggested that the structure of this complex was similar to that of $\pi\text{-}C_5H_5Fe(CO)_2SMe$ where each S atom in the ligand was monodentate and that the dithiolate ligand had the *cis* configuration, Figure 35.

There was mass spectrometric evidence for a μ-bridged dithiolate, $[\pi\text{-}C_5H_5Fe(CO)]_2S_2C_2(CN)_2$, which was presumably formed on thermal degradation of the dicarbonyl.

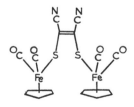

Fig. 35. The probable structure of $[\pi\text{-}C_5H_5Fe(CO)_2]_2S_2C_2(CN)_2$.

3. π-Cyclopentadienyl Nitrosyl Complexes

The dimeric nitrosyl complex, $[\pi\text{-}C_5H_5Mo(NO)S_2C_2(CF_3)_2]_2$, was obtained (46) by the treatment of $\pi\text{-}C_5H_5Mo(NO)(CO)_2$ with $S_2C_2(CF_3)_2$. The structure of this complex is probably similar to that of the tungsten carbonyl, Figure 34, although a metal–metal bond is unlikely. Reaction of $[\pi\text{-}C_5H_5Mn(NO)(CO)]_2$ with the dithietene afforded the monomeric mononitrosyl $[\pi\text{-}C_5H_5Mn(NO)S_2C_2(CF_3)_2]^0$, Figure 36, which was paramagnetic, having a magnetic moment of 1.73 BM. Similar dithiolene species, $[\pi\text{-}C_5H_5Mn(NO)\text{—}S_2]^z$, $z = 0, -1$, where the ligands were dicyano– or tetrachlorobenzene–dithiolene, were obtained by treatment (51) of $[\pi\text{-}C_5H_5Mn(NO)(CO)_2]PF_6$ with the sodium, potassium, or triethylammonium salts of the ligands. The neutral species are paramagnetic (one unpaired electron), exhibit six-line ESR spectra (^{55}Mn, $I = \frac{5}{2}$) in solution, and undergo one-electron oxidations and reductions, presumably generating species having $z = \pm 1$. The monoanions could be readily oxidized to the neutral species. The structure of these complexes is probably represented by Figure 36. Attempted formation of charged bis-perfluoro–methyl–dithiolene complexes by reduction of the neutral compound with ethanolic hydrazine resulted in the cleavage of the cyclopentadienyl ring and formation of $[Mn(NO)S_4C_4(CF_3)_4]^{-1}$.

The mass spectrum of the dimeric molybdenum nitrosyl described above contained the expected parent ion together with a fragmentation pattern consistent with the molecule's formulation and, in addition, traces of another species tentatively described as $(\pi\text{-}C_5H_5Mo)_2[S_2C_2(CF_3)_2]_3$.

4. Cleavage of the Cyclopentadienyl Rings

In several cases, treatment of the π-cyclopentadienyl metal carbonyls and their derivatives with the appropriate dithiolate ligand resulted in the cleavage of the C_5H_5 ring. This was particularly noticeable in the reactions with $Na_2S_2C_2(CN)_2$ when in excess and thus $[MoS_6C_6(CN)_6]^{-2}$, $[FeS_4C_4(CN)_4]_2^{-2}$, $[CoS_4C_4(CN)_4]^{-2}$, and $[NiS_4C_4(CN)_4]^{-2}$ were obtained.

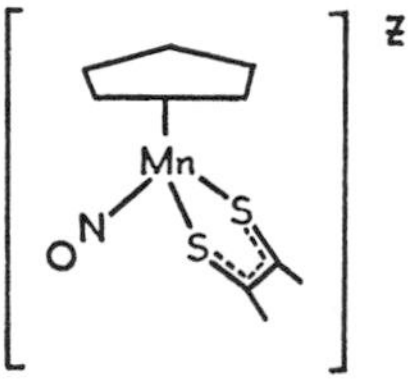

Fig. 36. The proposed structure of $[\pi\text{-}C_5H_5Mn(NO)(\text{—}S_2)]^z$.

Similar ring cleavage was observed when some of the bis-perfluoromethyl–dithiolenes were treated with ethanolic hydrazine or potassium ethoxide.

5. Electrochemical Studies

Several π-cyclopentadienyl dithiolene complexes have been investigated voltammetrically (47,48,50). In most cases examined, reduction was observed and many of the species produced electrochemically were also studied by ESR methods. The electrochemical and ESR data are summarized in Table XLI. Apparent one-electron oxidation and reduction were detected voltammetrically in $[\pi\text{-}C_5H_5MS_4C_4(CN)_4]^-$ (Table XLI) and it was presumed that dianionic and neutral species were being formed.

The proposed structure of the radical anion $[\pi\text{-}C_5H_5TiS_2C_6H_3Me]^{\cdot-}$ is shown in Figure 37. It was suggested that if the sulfur chelate ring had remained intact and bonded to the metal by both S atoms, proton hyperfine coupling with the odd electron spin would have been observed, but, in view of the lack of proton hyperfine interactions in simple toluene–dithiolene complexes, this does not seem strong evidence for the proposed structure. The two-electron reductions of the titanium species were reversible and the ESR data were similar to those obtained from $[(\pi\text{-}C_5H_5)_2TiCl_2]^{\cdot-}$, suggesting, perhaps, similar electronic structures.

The dimeric species $[\pi\text{-}C_5H_5MS_2C_2(CF_3)_2]_2$, where M = V and Mo, and $[\pi\text{-}C_5H_5Mo(NO)S_2C_2(CF_3)_2]_2$, all reduced in two discrete two-electron steps. The reduced vanadium complex gave an ESR spectrum in solution at room temperature consisting of 64 lines suggesting that there may have been proton and/or fluorine, hyperfine splittings in addition to the expected metal hyperfine splittings.

The monomeric complexes, $[\pi\text{-}C_5H_5MS_2C_2(CF_3)_2]^0$, where M = Fe, Co, Rh, Ir, Ni, and $\pi\text{-}C_5H_5CoS_2C_2(CN)_2$ and $\pi\text{-}C_5H_5Co(S_2C_6H_3Me)$ all exhibited reversible one-electron reductions with the exception of the iron complex which apparently was reduced in a two-electron process. That these radical anions were formed was conclusively demonstrated by the

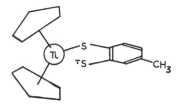

Fig. 37. The proposed structure of $[\pi\text{-}C_5H_5TiS_2C_6H_3Me]^{\cdot-}$.

appearance of ESR signals in solution containing the reduced cobalt species, and by the gradual disappearance of the ESR signal from electrolyzed solutions containing $[\pi\text{-}C_5H_5NiS_2C_2(CF_3)_2]^0$. It is noteworthy that the reduction of the complexes became more cathodic in the order Fe < Co < Ni and Co > Rh > Ir. There was no observable correlation between these orders and those in the simple dithiolene series.

6. Mass Spectrometric Studies

In all of the cyclopentadienyl bisperfluoromethyl–dithiolenes, there was a tendency for the parent ion to lose a fluorine atom forming an ion of mass difference 19. This behavior was similar to that displayed by the simple dithiolenes (Sec. III-A). Furthermore, degradation of the ion having m/e corresponding to $[C_5H_5MS_2C_2(CF_3)_2]^+$ to $[C_5H_5MS_2]^+$, by loss of hexafluorobut-2-yne, was always observed.

B. Olefin Complexes

The bis-diphenyl–dithiolene complexes of the nickel group absorbed certain unsaturated hydrocarbons forming 1:1 adducts (79). The hydrocarbons included substituted but-1,3-dienes, cyclic 1,3-dienes, 1,3,5-cyclooctatriene, cyclooctatetraene, and norbornene. The adducts and their characteristics are listed in Table XLII. These compounds were generally insoluble in organic solvents, were frequently thermally unstable (the order of stability with respect to the metal atom being Pd > Ni > Pt), and often decomposed on exposure to light.

The electronic spectra of these adducts showed that the intense absorptions in the visible region associated with the oxidized nickel group dithiolenes (Sec. III-F) were absent. Furthermore, the infrared frequencies ν_1, ν_2, and ν_3, also associated with the oxidized dithiolenes, were absent. Thus the complexes were regarded as containing "dithiolate" ligands. However, the spectral data and the observed lability of these complexes indicated that they were not simple transition metal π-complexes. In the unlikely event of the formation of new (and presumably robust) C—C bonds, it was postulated that an interaction with the sulfur atoms of one dithiolene ligand had occurred, thus forming labile C—S bonds. An examination of the diene adducts in the "olefinic" region of the infrared spectrum revealed that the characteristic frequency associated with free C=C groups were absent, but that a new frequency at ca. 1540 cm^{-1} had appeared. This new band was assigned to the stretching mode of a C=C group interacting with the metal atoms as illustrated in Figure 38.

TABLE XLII

Olefin Adducts of Nickel Group Bis-diphenyl–dithiolenes

Complex	Olefin	mp,°C	$\nu_{C=C}$,cm^{-1}
NiS$_4$C$_4$Ph$_4$	2,3-Dimethyl-1,3-butadiene	100	
	Norbornadiene	165	1538
PdS$_4$C$_4$Ph$_4$	Butadiene[a]	165	
	Isoprene	159	
	2,3-Dimethyl-1,3-butadiene	200	
	1,3-Cyclohexadiene	110	
	Norbornadiene	252	1515
	Norbornene	236	—[b]
PtS$_4$C$_4$Ph$_4$	2,3-Dimethyl-1,3-butadiene	100	

[a] Other Pd olefin adducts formed but not characterized included 2-chloroprene, 1,3-cyclooctadiene, 1,3,5-cyclooctatriene, cycle-octatetraene, and piperylene. Olefins which did not form adducts included methylsorbate, muconic ester, acrylonitrile, and 2,5-dimethyl-2,4-hexadiene.

[b] The 1515 cm^{-1} absorption present in the norbornadiene is absent in this adduct.

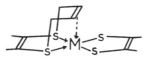

Fig. 38. The proposed structure of the adduct of MS$_4$C$_4$R$_4$ and a buta-1,3-diene.

The infrared spectra of the norbornadiene adducts of [NiS$_4$C$_4$Ph$_4$]^0 and its palladium analog exhibited bands at 1538 and 1515 cm^{-1}, respectively, which were assigned to the perturbed C=C stretching frequency; the characteristic nortricyclene absorptions at 800 cm^{-1} were absent. The correctness of these assignments was justified by the absence of bands at ca. 1540 cm^{-1} in the norbornene palladium bis-diphenyl–dithiolene

Fig. 39. The proposed structure of norbornadiene adducts of MS$_4$C$_4$R$_4$.

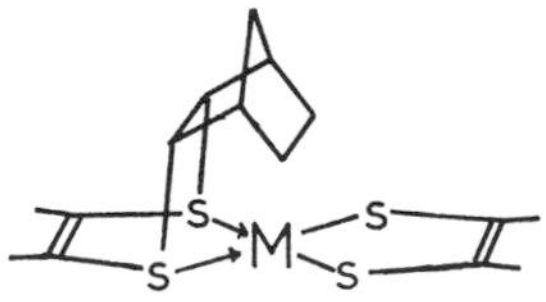

Fig. 40. The proposed structure of norbornene adducts of $MS_4C_4R_4$.

adduct. The proposed structure for the norbornadiene and norbornene derivatives are shown in Figures 39 and 40.

The isolation of the norbornene complex indicated that conjugation or homoconjugation was not a necessary prerequisite for adduct formation. The adducts were too insoluble to be investigated by NMR techniques.

No olefin adducts were obtained with the bis-dimethyl–dithiolene complexes.

C. Reactions with Alkynes

The nickel group bis-dimethyl– and bis-diphenyl–dithiolenes reacted with alkynes (79) at, or near, 160° forming substituted thiophenes. The yields using $[NiS_4C_4Ph_4]^0$ were close to 80% but those with the platinum and palladium analogs were much lower. It was found that the reactivity of the alkynes toward the dithiolenes was dependent on the alkyne substituents, diphenylacetylene, and the acetylene dicarboxylic esters being

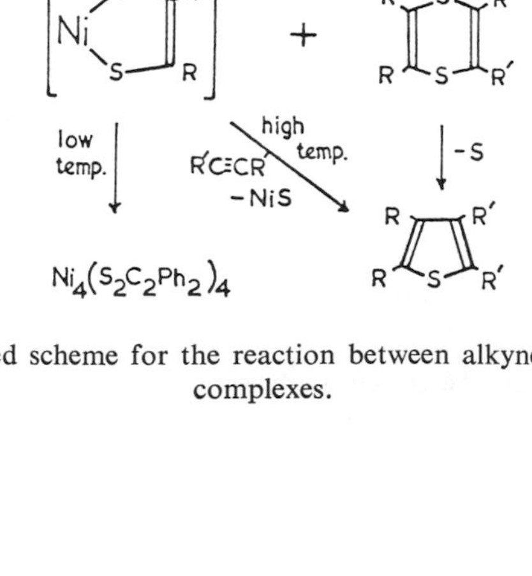

Fig. 41. The proposed scheme for the reaction between alkynes and bis-dithiolene complexes.

among the most reactive. Under carefully controlled reaction conditions at lower temperatures (60°), dithiadienes could be isolated and under the very mildest conditions which would still sustain the reaction, the yields of dithiadiene dropped to less than 50%, suggesting that only one sulfur ligand per molecule underwent reaction with the alkyne. The metal-containing residue of these reactions was not NiS as might have been expected, but was the tetramer $Ni_4(S_2C_2Ph_2)_4$. Thus, a reaction scheme, Figure 41, was proposed.

VII. RECENT DEVELOPMENTS

A. Potential Dithiolene Systems

The redox properties of 1,2-dithiolenes, that is, metal–sulfur complexes containing five-membered (unsaturated) metal-chelated rings, have been thoroughly investigated. The possibility that compounds containing four- and six-membered metal-chelated ring systems could undergo electron transfer reactions cannot be ignored, and several workers have investigated the potential redox properties of the four-membered ring system. Thus Fackler and co-workers, in extensive studies (183) of complexes the ligand of which were derived from CS_2, e.g., 32 and 33, estab-

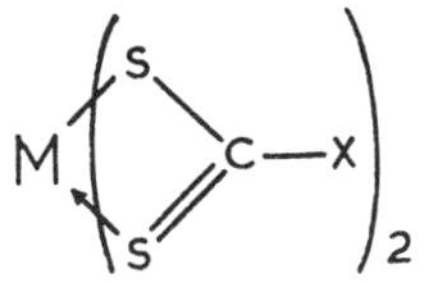

(32) 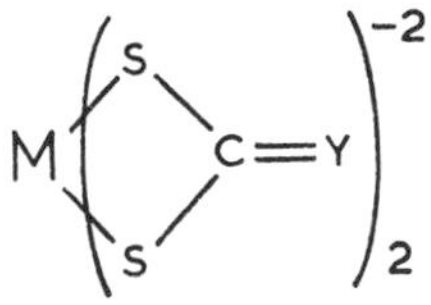   (33)

lished that these systems did not take part in any well-defined electron transfer reactions. However, they did find that chemical oxidation of many of the complexes, especially of type 33, resulted in a novel sulfur insertion reaction and formation of compounds containing five-membered chelate rings (34). These compounds did not possess redox properties. It has been

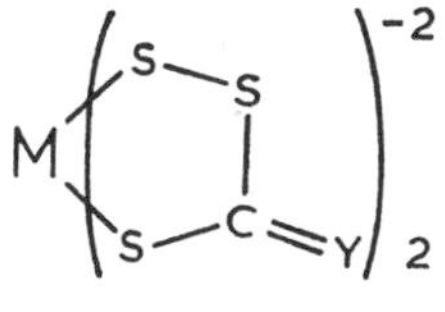

(34)

suggested (184) that the inability of these four-membered chelate ring systems to engage in electron transfer reactions could be related to the impossibility of representing them in Kekule resonance forms.

Compounds containing dithioacetylacetone (35) can be written in "suitable" resonance forms, so it is possible that these may be oxidized or reduced in much the same way as the 1,2-dithiolenes.

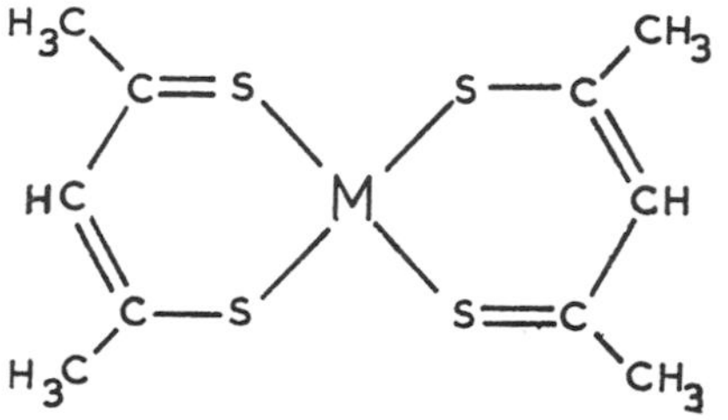

(35)

B. Metal Diselenolenes

Davison and Shawl have recently reported the preparation of bis-perfluoromethyldiselenetene, $Se_2C_2(CF_3)_2$ (185). They used this ligand to prepared nickel and vanadium complexes, $[NiSe_4C_4(CF_3)_4]^{0,-1,-2}$ and $[VSe_6C_6(CF_3)_6]^{-1,-2}$, and investigated their polarographic and magnetic properties. The half-wave potentials for the reversible one-electron transfer

TABLE XLIII

Voltammetric and ESR Data Obtained from Nickel and Vanadium
Bis-perfluoromethyl–diselenolenes

Complex	$[ML_n]^z + e^- \rightleftharpoons [ML_n]^{z-1}$ [a]			ESR data[b]	
	$0 \leftrightarrow -1$	$-1 \leftrightarrow -2$	$-2 \leftrightarrow -3$	$\langle g \rangle$	$\langle A_{Se} \rangle$[c]
$[NiSe_4C_4(CF_3)_4]^z$	$+0.89$	-0.17		2.008 (g_1)	69 ± 2 (A_1)
				2.116 (g_2)	
				2.193 (g_3)	
$[VSe_6C_6(CF_3)_6]^z$		$+0.07$	-1.05		

[a] In volts vs. SCE; solvent, CH_2Cl_2.
[b] In $DMF/CHCl_3$ glass at $-170°$.
[c] ^{77}Se hyperfine splitting; $I = \frac{1}{2}$, abundance 7.5%.

reactions which these species underwent were very similar to those of their sulfur analogs. The ESR data clearly showed $^{77}Se(I = \frac{1}{2})$ hyperfine interactions which will be invaluable in calculations of spin densities on these donor atoms.

Zinc and cadmium glyoxal bis(2-selenoanil) (36) (76) also underwent reversible one-electron transfer reactions corresponding to the formation of mono- and dianionic species.

(36)

Acknowledgments

I would like to thank my friends and colleagues, particularly Professors Alan Davison, Dick Holm, Harry Gray, Bruce King, and G. N. Schrauzer for communicating to me many of their results prior to publication and thereby making my task as reviewer much easier. I wish to express my particular gratitude to Dr. Sid Kettle for his valued comments on this manuscript. To my students John Locke, Eric Wharton, Brian Ratcliff, Neil Connelly, Tom James, and Chris Jones, I owe my continuing interest in this field of chemistry.

References

1(a). S. E. Livingstone, *Quart. Rev.*, *19*, 386 (1965); (b) C. M. Harris and S. E. Livingstone, "Bidentate Chelates," in *Chelating Agents and Metal Chelates*, F. P. Dwyer and D. P. Mellor, Eds., Academic Press, New York, 1964, p. 95.

2. C. K. Jorgensen, *Inorganic Complexes*, Academic Press, London, 1963.

3. H. B. Gray, "Electronic Structures of Square Planar Metal Complexes," in *Transition Metal Chemistry*, Vol. 1, R. L. Carlin, Ed., Edward Arnold, London, 1965, p. 240.

4. W. H. Mills and R. E. D. Clark, *J. Chem. Soc.*, *1936*, 175.

5. R. E. D. Clark, *Analyst*, *60*, 242 (1936); *Ibid.*, *62*, 661 (1937): *Tech. Publ. Intern. Tin Res. Dev. Council*, A, *41*, 1936.

6. J. Hamence, *Analyst*, *65*, 152 (1940); C. C. Miller and A. J. Lowe, *J. Chem. Soc.*, *1940*, 1258; J. E. Wells and R. Pemberton, *Analyst*, *72*, 185 (1947).

7. H. G. Short, *Analyst*, *76*, 710 (1951).

8. G. Bähr and H. Schleitzer, *Chem. Ber.*, *90*, 438 (1957).

9. D. B. Stevancevic and V. C. Drazic, *Bull. Inst. Nucl. Sci. "Boris Kidrich,"* *9*, 69 (1959).

10. G. H. Ayres and H. F. Janota, *Anal. Chem.*, *31*, 1985 (1959).
11. T. W. Gilbert, Jr., and E. B. Sandell, *J. Am. Chem. Soc.*, *82*, 1087 (1960).
12. G. N. Schrauzer and V. P. Mayweg, *J. Am. Chem. Soc.*, *84*, 3221 (1962).
13. H. B. Gray, R. Williams, I. Bernal, and E. Billig, *J. Am. Chem. Soc.*, *84*, 3596 (1962).
14. R. B. King, *Inorg. Chem.*, *2*, 641 (1963).
15. A. Davison, N. Edelstein, R. H. Holm, and A. H. Maki, *J. Am. Chem. Soc.*, *85*, 2029 (1963).
16. H. B. Gray and E. Billig, *J. Am. Chem. Soc.*, *85*, 2019 (1963).
17. G. N. Schrauzer and V. P. Mayweg, *Z. Naturforsch.*, *19b*, 192 (1964); G. N. Schrauzer, V. P. Mayweg, H. W. Finck, U. Müller-Westerhoff, and W. Heinrich, *Angew. Chem. Intern. Ed. Engl.*, *3*, 381 (1964).
18. A. Davison, N. Edelstein, R. H. Holm, and A. H. Maki, *J. Am. Chem. Soc.*, *86*, 2799 (1964).
19. J. H. Waters, R. Williams, H. B. Gray, G. N. Schrauzer, and H. W. Finck, *J. Am. Chem. Soc.*, *86*, 4198 (1964).
20. G. N. Schrauzer, H. W. Finck, and V. P. Mayweg, *Z. Naturforsch.*, *19b*, 1080 (1964).
21. R. Eisenberg and J. A. Ibers, *J. Am. Chem. Soc.*, *87*, 3776 (1965).
22. A. E. Smith, G. N. Schrauzer, V. P. Mayweg, and W. Heinrich, *J. Am. Chem. Soc.*, *87*, 5798 (1965).
23. R. Eisenberg, E. I. Stiefel, R. C. Rosenberg, and H. B. Gray, *J. Am. Chem. Soc.*, *88*, 2874 (1966).
24. E. I. Stiefel, Z. Dori, and H. B. Gray, *J. Am. Chem. Soc.*, *89*, 3353 (1967).
25. A. L. Balch and R. H. Holm, *Chem. Commun.*, *1966*, 552.
26. J. H. Enemark and W. N. Lipscomb, *Inorg. Chem.*, *4*, 1729 (1965).
27. A. Davison, D. V. Howe, and E. T. Shawl, *Inorg. Chem.*, *6*, 458 (1967).
28. A. L. Balch, F. Röhrscheid, and R. H. Holm, *J. Am. Chem. Soc.*, *87*, 2301 (1965).
29. R. Willstäter and A. Pfannenstiel, *Chem. Ber.*, *38*, 2348 (1905).
30. F. Feigl and M. Fürth, *Monatsh.*, *48*, 445 (1927).
31. Z. Bardodej, *Coll. Czech. Chem. Commun.*, *20*, 176 (1955).
32. R. S. Nyholm, *Chem. Rev.*, *53*, 263 (1953).
33. A. L. Balch and R. H. Holm, *J. Am. Chem. Soc.*, *88*, 5201 (1966).
34. E. I. Stiefel, J. H. Waters, E. Billig, and H. B. Gray, *J. Am. Chem. Soc.*, *87*, 3016 (1965).
35. R. Weinland, A. Döttinger, *Z. Anorg. Allgem. Chem.*, *102*, 223 (1918); R. Weinland and A. Döttinger, *Ibid.*, *111*, 167 (1920); H. Reihlen, *Ibid.*, *123*, 173 (1922); E. Spacu and M. Kuras, *J. Prakt. Chem.*, *141*, 201 (1934); R. Weinland and E. Walter, *Z. Anorg. Allgem. Chem.*, *126*, 141 (1923); P. Pfeiffer, H. Simons, and E. Schmidt, *Ibid.*, *256*, 318 (1948); E. Selles, *Anales. Real Soc. Espan. Fis. Quim. (Madrid)*, *37*, 114 (1941).
36. D. R. Eaton, *Inorg. Chem.*, *3*, 1268 (1964).
37. J. D. Bu'lock and J. Harley-Mason, *J. Chem. Soc.*, *1951*, 2248.
38. R. Criegee and K. Klonk, *Ann.*, *564*, 1 (1949).
39. K. H. Meyer, *Chem. Ber.*, *41*, 2567 (1908); F. R. Japp and A. E. Turner, *J. Chem. Soc.*, *1890*, 4; J. Knox and H. R. Innes, *Ibid.*, *1914*, 1451; P. J. Crowley and H. M. Haendler, *Inorg. Chem.*, *1*, 904 (1962).
40. F. Röhrscheid, A. L. Balch, and R. H. Holm, *Inorg. Chem.*, *5*, 1542 (1966).
41. W. Hieber and R. Brück, *Z. Anorg. Allgem. Chem.*, *269*, 13 (1952).

42. L. F. Larkworthy, J. M. Murphy, and D. J. Phillips, *J. Am. Chem. Soc.*, *88*, 1570 (1966).
43. K. A. Jensen and J. F. Miquel, *Acta Chem. Scand.*, *6*, 189 (1952).
44. R. H. Holm, A. L. Balch, A. Davison, A. H. Maki, and T. E. Berry, *J. Am. Chem. Soc.*, *89*, 2866 (1967).
45. R. B. King, *J. Am. Chem. Soc.*, *85*, 1587 (1963).
46. R. B. King and M. B. Bisnette, *Inorg. Chem.*, *6*, 469 (1967).
47. R. E. Dessy, F. E. Stary, R. B. King, and M. Waldrop, *J. Am. Chem. Soc.*, *88*, 471 (1966).
48. R. E. Dessy, R. B. King, and M. Waldrop, *J. Am. Chem. Soc.*, *88*, 5112 (1966).
49. J. Locke, J. A. McCleverty, E. J. Wharton, and C. J. Winscom, *Chem. Commun.*, *1966*, 677.
50. J. A. McCleverty, N. M. Atherton, J. Locke, E. J. Wharton, and C. J. Winscom, *J. Am. Chem. Soc.*, *89*, 6082 (1967).
51. J. A. McCleverty, T. A. James, J. Locke, and C. J. Winscom, to be published.
52. G. Bähr, and G. Schleitzer, *Chem. Ber.*, *88*, 1771 (1955).
53. J. Locke and J. A. McCleverty, *Inorg. Chem.*, *5*, 1157 (1966).
54. H. E. Simmons, D. C. Blomstrom, and R. D. Vest, *J. Am. Chem. Soc.*, *84*, 4756 (1962).
55. R. J. H. Clark and W. Errington, *Inorg. Chem.*, *5*, 650 (1966).
56. H. E. Simmons, D. C. Blomstrom, and R. D. Vest, *J. Am. Chem. Soc.*, *84*, 4782, (1962).
57. G. N. Schrauzer and V. P. Mayweg, *J. Am. Chem. Soc.*, *87*, 3585 (1965).
58. W. Schroth and J. Peschel, *Chimia*, *18*, 171 (1964).
59. E. Hoyer, W. Dietsch, and W. Schroth, *Proceedings 9th I.C.C.C.*, Switzerland, 1966, p. 316.
60. D. C. Olson, V. P. Mayweg, and G. N. Schrauzer, *J. Am. Chem. Soc.*, *88*, 4876 (1966).
61. C. G. Krespan, B. C. McKusick, and T. L. Cairns, *J. Am. Chem. Soc.*, *82*, 1515 (1960).
62. A. Davison, unpublished work; private communication.
63. G. N. Schrauzer, V. P. Mayweg, and W. Heinrich, *Inorg. Chem.*, *4*, 1615 (1965).
64. R. Adams and A. Ferretti, *J. Am. Chem. Soc.*, *81*, 4927 (1959); *Org. Synth.*, *42*, 22, 54 (1962).
65. M. J. Baker-Hawkes, E. Billig, and H. B. Gray, *J. Am. Chem. Soc.*, *88*, 4870 (1966).
66. R. Leuckart, *J. Prakt. Chem.*, *41*, 192 (1890); L. Haitinger, *Monatsh.*, *4*, 166 (1883); C. le Fevre and C. Degrez, *Compt. Rend.*, *198*, 1432 (1934).
67. C. L. Jackson and R. D. MacLaurin, *J. Am. Chem. Soc.*, *38*, 127 (1907).
68. H. Jadamus, Q. Fernando, and H. Freiser, *J. Am. Chem. Soc.*, *86*, 3056 (1964).
69. G. Bähr and G. Schleitzer, *Z. Anorg. Allgem. Chem.*, *280*, 161 (1955).
70. H. Von Pechmann and W. Bauer, *Chem. Ber.*, *42*, 659 (1909).
71. B. Holmberg, *Arkiv. Kemi*, *4*, 33 (1951).
72. J. A. McCleverty and J. Locke, unpublished work.
73. J. A. McCleverty, J. Locke, E. J. Wharton, and M. Gerloch, *J. Chem. Soc. A*, 1968, in press.
74. J. A. McCleverty and N. M. Atherton, unpublished work.
75. I. Bernal, unpublished work; private communication.
76. J. A. McCleverty and B. Ratcliff, to be published.

77. A. H. Maki, T. E. Berry, A. Davison, R. H. Holm, and A. L. Balch, *J. Am. Chem. Soc.*, **88**, 1080 (1966).

78. F. Lalor, M. F. Hawthorne, A. H. Maki, K. Darlington, A. Davison. H. B. Gray, Z. Dori, and E. I. Stiefel, *J. Am. Chem. Soc.*, **89**, 2278 (1967).

79. G. N. Schrauzer and V. P. Mayweg, *J. Am. Chem. Soc.*, **87**, 1483 (1965).

80. A. Davison, J. A. McCleverty, E. T. Shawl, and E. J. Wharton, *J. Am. Chem. Soc.*, **89**, 830 (1967).

81. S. M. Bloom and G. O. Dudek, *Inorg. Nucl. Chem. Letters*, **2**, 183 (1966).

82. J. A. McCleverty, B. Ratcliff, M. G. H. Wallbridge, and E. J. Wharton, to be published.

83. T. Birchall, N. N. Greenwood, and J. A. McCleverty, *Nature*, **215**, 625 (1967), and work to be published.

84. F. A. Cotton, C. Oldham, and R. A. Walton, *Inorg. Chem.*, **6**, 214 (1967).

85. G. N. Schrauzer and V. P. Mayweg, unpublished work.

86. R. Eisenberg, J. A. Ibers, R. J. H. Clark, and H. B. Gray, *J. Am. Chem. Soc.*, **86**, 113 (1964).

87. J. D. Forrester, A. Zalkin, and D. H. Templeton, *Inorg. Chem.*, **3**, 1500 (1964).

88. H. B. Gray, unpublished work; private communications.

89. D. Sartain and N. R. Truter, *Chem. Commun.*, **1966**, 382; *J. Chem. Soc. A*, 1264, 1967.

90. M. R. Taylor, E. J. Gabe, J. P. Glusker, J. A. Minkin, and A. L. Patterson, *J. Am. Chem. Soc.*, **88**, 1845 (1966).

91. J. D. Forrester, A. Zalkin, and D. H. Templeton, *Inorg. Chem.*, **3**, 1507 (1964).

92. C. J. Fritchie, *Acta Cryst.*, **20**, 107 (1966).

93. J. F. Weiher, L. R. Melby, and R. E. Benson, *J. Am. Chem. Soc.*, **86**, 4329 (1964).

94. J. G. Hamilton and R. Spratley, *Abstracts of the 7th Intern. Congress and Symp. on Crystal Growth*, 1966, *A150*, 9.26; W. C. Hamilton and I. Bernal, *Inorg. Chem.*, **6**, 2003 (1967).

95. G. N. Schrauzer and V. P. Mayweg. *J. Am. Chem. Soc.*, **88**, 3235 (1966).

96. G. N. Schrauzer, V. P. Mayweg, H. W. Finck, and W. Heinrich, *J. Am. Chem. Soc.*, **88**, 4604 (1966).

97. E. Billig, R. Williams, I. Bernal, J. H. Waters, and H. B. Gray, *Inorg. Chem.*, **3**, 663 (1964).

98. A. Davison, N. Edelstein, A. H. Maki, and R. H. Holm, *Inorg. Chem.*, **2**, 1227 (1963).

99. A. Davison, N. Edelstein, R. H. Holm, and A. H. Maki, *Inorg. Chem.*, **3**, 814 (1964).

100. A. Davison and R. H. Holm, *Inorg. Synth.*, **10**, 8 (1967).

101. N. G. Connelly, J. A. McCleverty, and C. J. Winscom, *Nature*, **216**, 999 (1967); J. A. McCleverty, N. M. Atherton, N. G. Connelly, and C. J. Winscom, to be published.

102. L. R. Melby, R. J. Harder, W. R. Hertler, W. Mahler, R. E. Benson, and W. E. Mochel, *J. Am. Chem. Soc.*, **84**, 3374 (1962); R. E. Merrifield and W. D. Phillips, *Ibid.*, **80**, 2778 (1958).

103. A. A. Vlček, in *Progress in Inorganic Chemistry*, Vol. 5, F. A. Cotton, Ed., Interscience, New York, 1963, p. 211; *Z. Anorg. Allgem. Chem.*, **304**, 109 (1960).

104. G. J. Hoijtink, *Rec. Trav. Chim.*, **77**, 555 (1958).

105. K. Nakamoto, *Infrared Spectra of Inorganic and Coordination Compounds*, Wiley, New York, 1963, p. 214.

106. A. H. Maki, N. Edelstein, A. Davison, and R. H. Holm, *J. Am. Chem. Soc.*, *86*, 4580 (1964).
107. A. H. Maki and B. R. McGarvey, *J. Chem. Phys.*, *29*, 31, 35 (1958).
108. R. Petterson and T. Vänngård, *Arkiv. Kemi*, *17*, 249 (1961).
109. H. R. Gersmann and J. D. Swalen, *J. Chem. Phys.*, *36*, 3221 (1962).
110. R. Williams, E. Billig, J. H. Waters, and H. B. Gray, *J. Am. Chem. Soc.*, *88*, 43 (1966).
111. J. H. Waters and H. B. Gray, *J. Am. Chem. Soc.*, *87*, 3534 (1965).
112. T. Vänngård and S. Åkerström, *Nature*, *184*, 183 (1959).
113. R. L. Rich and H. Taube, *J. Phys. Chem.*, *58*, 6 (1954); A. MacCragh and W. S. Koski, *J. Am. Chem. Soc.*, *87*, 2496 (1965).
114. M. Bonamico,, G. Dessy, A. Mugnoli, A. Vaciago, and L. Zambonelli, *Acta Cryst.*, *19*, 886 (1965).
115. F. A. Cotton and G. Wilkinson, *Advanced Inorganic Chemistry*, Interscience, New York, 1966.
116. W. G. Hodgson, S. A. Buckler, and G. Peters, *J. Am. Chem. Soc.*, *85*, 543 (1963); J. J. Windle, A. K. Wiersma, and A. L. Tappel, *J. Chem. Phys.*, *41*, 1996 (1964); A. Zweig and W. G. Hodgson, *Proc. Chem. Soc.*, *1964*, 417, and reference therein; A Zweig and A. K. Hoffman, *J. Org. Chem.*, *30*, 3997 (1965); W. Rundel and K. Scheffler, *Angew. Chem.*, *77*, 220 (1965); K. Akasaka, *J. Chem. Phys.*, *45*, 90 (1966), and references therein.
117. G. N. Schrauzer and H. Thyret, *J. Am. Chem. Soc.*, *82*, 6420 (1960); *Idem.*, *Z. Naturforsch.*, *16b*, 352 (1961); *Ibid.*, *17b*, 72 (1962); *Chem. Ber.*, *96*, 1755 (1963); *Angew. Chem.*, *74*, 488 (1962).
118. A. Davison, N. Edelstein, R. H. Holm, and A. H. Maki, *J. Am. Chem. Soc.*, *85* 3049 (1963).
119. B. N. Figgis and R. S. Nyholm, *J. Chem. Soc.*, *1954*, 12; R. Havemann, W. Haberditzl, and K.-H. Mader, *Z. Physik. Chem. (Leipzig)*, *218*, 71 (1961).
120. F. A. Cotton and R. H. Holm, *J. Am. Chem. Soc.*, *82*, 2979 (1960).
121. F. A. Cotton and R. C. Elder, *Inorg. Chem.*, *4*, 1145 (1965); G. J. Bullen, *Acta Cryst.*, *12*, 703 (1959).
122. E. Billig, S. I. Shupack, J. H. Waters, R. Williams, and H. B. Gray, *J. Am. Chem. Soc.*, *86*, 926 (1964).
123. S. I. Shupack, E. Billig, R. J. H. Clark, R. Williams, and H. B. Gray, *J. Am. Chem. Soc.*, *86*, 4594 (1964).
124. H. B. Gray and C. J. Ballhausen, *J. Am. Chem. Soc.*, *85*, 260 (1963).
125. J. P. Fackler, Jr., F. A. Cotton, and D. W. Barnum, *Inorg. Chem.*, *2*, 97, (1963); J. P. Fackler, Jr., and F. A. Cotton, *Ibid.*, *2*, 102 (1963).
126. F. A. Cotton and J. P. Fackler, Jr., *J. Am. Chem. Soc.*, *83*, 2818 (1961); J. P. Fackler, Jr., and F. A. Cotton, *Ibid.*, *83*, 3775 (1961); L. Wolf and E. Butter, *Z. Anorg. Allgem. Chem.*, *339*, 191 (1965).
127. F. A. Cotton and R. H. Soderberg, *Inorg. Chem.*, *3*, 1 (1964).
128. R. W. Taft, Jr., *Steric Effects in Organic Chemistry*, M. S. Newman, Ed., Wiley, New York, 1956, Chap. 13.
129. A. Viste and H. B. Gray, *Inorg. Chem.*, *3*, 1113 (1964); C. J. Ballhausen and H. B. Gray, *Ibid.*, *1*, 111 (1962).
130. J. S. Griffith, *The Theory of Transition Metal Ions*, Cambridge University Press, London, 1961.
131. M. Gerloch, S. F. A. Kettle, J. Locke, and J. A. McCleverty, *Chem. Commun.*, *1966*, 29.

132. H. B. Gray, private communication.
133. C. H. Langford, E. Billig, S. I. Shupack, and H. B. Gray, *J. Am. Chem. Soc.*, *86*, 2958 (1964).
134. G. N. Schrauzer, V. P. Mayweg, and W. Heinrich, *Chem. Ind.* (*London*), *1965*, 1464.
135. G. N. Schrauzer, V. P. Mayweg, and W. Heinrich, *J. Am. Chem. Soc.*, *88*, 5174 (1966).
136. A. Davison, N. Edelstein, R. H. Holm, and A. H. Maki, *Inorg. Chem.*, *4*, 55 (1965).
137. E. I. Stiefel, R. Eisenberg, R. C. Rosenberg, and H. B. Gray, *J. Am. Chem. Soc.*, *88*, 2956 (1966).
138. A. Butcher and P. C. H. Mitchell, *Chem. Commun.*, *1967*, 176.
139. E. W. Abel, unpublished work; private communication.
140. N. M. Atherton, J. Locke, and J. A. McCleverty, *Chem. Ind.* (*London*), *1965*, 1300.
141. R. Eisenberg and J. A. Ibers, *Inorg. Chem.*, *5*, 411 (1966).
142. R. Bak, D. Christensen, J. Rastrup-Andersen, and E. Tannenbaum, *J. Chem. Phys.*, *25*, 892 (1956).
143. R. Dickinson and L. Pauling, *J. Am. Chem. Soc.*, *45*, 1466 (1923).
144. J. N. van Niekirk and F. R. L. Schoenig, *Acta Cryst.*, *6*, 499 (1952).
145. R. L. Carlin and F. Canziani, *J. Chem. Phys.*, *40*, 371 (1964).
146. L. Cambi and L. Szego, *Chem. Ber.*, *64*, 2591 (1931); A. H. Ewald, R. L. Martin, I. G. Ross, and A. H. White, *Proc. Roy. Soc.* (*London*), *A280*, 235 (1964).
147. B. R. McGarvey, *J. Chem. Phys.*, *38*, 388 (1963).
148. E. I. Stiefel and H. B. Gray, *J. Am. Chem. Soc.*, *87*, 4012 (1965).
149. C. K. Jorgensen, *Absorption Spectra and Chemical Bonding in Complexes*, Pergamon-Addison-Wesley, Reading, Mass., 1962, p. 123.
150. C. K. Jorgensen, *J. Inorg. Nucl. Chem.*, *24*, 1571 (1962).
151. B. G. Werden, E. Billig, and H. B. Gray, *Inorg. Chem.*, *5*, 78 (1966).
152. A. R. Latham, V. C. Hascall, and H. B. Gray, *Inorg. Chem.*, *4*, 788 (1965).
153. J. P. Fackler, Jr., and D. Coucouvanis, *J. Am. Chem. Soc.*, *88*, 3913 (1966).
154. J. Locke and J. A. McCleverty, *Chem. Commun.*, *1965*, 102.
155. W. Hieber and M. Gscheidmeier, *Chem. Ber.*, *99*, 2312 (1966).
156. R. B. King, *J. Am. Chem. Soc.*, *85*, 1584 (1963).
157. R. F. Bryan and H. P. Weber, *Chem. Commun.*, *1966*, 329.
158. R. B. King and T. F. Korenowski, *Chem. Commun.*, *1966*, 771.
159. C. C. McDonald, W. D. Phillips, and H. F. Mower, *J. Am. Chem. Soc.*, *87*, 3319 (1965).
160. L. Cambi and A. Cagnasso, *Atti Accad. Naz. Lincei*, *13*, 254, 809 (1931).
161. P. R. Alderman, P. G. Owston, and J. M. Rowe, *J. Chem. Soc.*, *1962*, 668.
162. G. R. Davies, R. H. B. Mais, and P. G. Owston, *Chem. Commun.*, *1968*, 81; M. Colapietro, A. Domencano, L. Scaranuzza, A. Vaciago, and L. Zambonelli, *Chem. Commun.*, *1967*, 583; L. F. Dahl, private communication.
163. P. Gans, *Chem. Commun.*, *1965*, 144
164. B. F. G. Johnson and J. A. McCleverty in *Progress in Inorganic Chemistry*, Vol. 7 F. A. Cotton, Ed., Interscience, New York, 1966.
165. A. I. M. Rae, *Chem. Commun.*, *1967*, 1245.
166. A. Earnshaw, P. C. Hewlitt, and L. F. Larkworthy, *J. Chem. Soc.*, *1966*, 4718.
167. P. T. Manoharan and H. B. Gray, *J. Am. Chem. Soc.*, *87*, 3340 (1965).

168. R. Prins, P. Biloen, and J. D. W. van Voorst, *J. Chem. Phys.*, *46*, 1216 (1967).
169. B. F. G. Johnson and J. A. McCleverty, to be published.
170. R. B. King, *Inorg. Chem.*, *2*, 1275 (1963).
171. J. Locke, J. A. McCleverty, E. J. Wharton, and C. J. Winscom, *Chem. Commun.*, *1967*, 1289; and work to be published.
172. E. J. Wharton and J. A. McCleverty, to be published.
173. H. Krebs, E. F. Weber, and H. Fassbender, *Z. Anorg. Allgem. Chem.*, *276*, 128 (1954).
174. V. P. Mayweg and G. N. Schrauzer, *Chem. Commun.*, *1966*, 640.
175. E. Billig, H. B. Gray, S. I. Shupack, J. H. Waters, and R. Williams, *Proc. Chem. Soc.*, *1964*, 110.
176. J. A. McCleverty, J. Locke, E. J. Wharton, B. Ratcliff, and C. J. Winscom, to be published.
177. H. Kopf and M. Schmidt, *J. Organometal. Chem.*, *4*, 426 (1965).
178. M. Gerloch and R. Mason, *J. Chem. Soc.*, *1964*, 296.
179. M. L. H. Green and W. E. Lindsell, *J. Chem. Soc. A*, *1967*, 1455.
180. L. F. Dahl, private communication.
181. R. H. Holm, R. B. King, and F. G. A. Stone, *Inorg. Chem.*, *2*, 219 (1963).
182. H. W. Baird and B. M. White, *J. Am. Chem. Soc.*, *88*, 4744 (1966).
183. J. P. Fackler, Jr. and D. Coucouvanis, *Chem. Commun.*, *1965*, 556; D. Coucouvanis and J. P. Fackler, Jr., *J. Am. Chem. Soc.*, *89*, 1346 (1967).
184. G. N. Schrauzer, private communication.
185. A. Davison and E. T. Shawl, *Chem. Commun.*, *1967*, 670.
186. A. L. Balch, *Inorg. Chem.*, *6*, 2158 (1967).
187. E. E. Crenser, *Inorg. Chem.*, *7*, 13 (1968).
188. R. B. King and C. A. Eggers, *Inorg. Chem.*, *7*, 340 (1968).

Complexes of Simple Carboxylic Acids

By C. Oldham

Chemistry Department, University of Lancaster, England

INTRODUCTION

Most of the chemical elements are known to form monocarboxylate, or at least acetate, derivatives. Many of these have been known almost as long as inorganic chemistry itself and many may be readily purchased commercially. Despite their formidable history one can choose almost any current inorganic chemical journal and reasonably expect to find papers devoted to such compounds. Thus this review does not claim in any way to be exhaustive but rather to bring together most of the salient chemistry of such compounds in an effort to bring a little system to a subject which surprisingly has not been reviewed in its own right previously.

As there is such a variety of monocarboxylic acids this review will use acetic acid as a basis to discuss carboxylate compounds. Significant differences from other carboxylates, where they exist, will be indicated. Generally, formate and acetate ions form many stable coordination complexes. Propionates and derivatives of higher acids show the same chemical characteristics as acetates except that as the carbon chain of the acid grows, the tendency to coordinate apparently decreases. The physical properties of acetic acid are suitable to enable its use as a convenient nonaqueous solvent medium. For details of this use the reader may consult the standard books on the subject of nonaqueous solvents. Mention should be made of the useful, recently published, monograph by Holliday and Massey (1).

One important feature of the organic chemistry of acetic acid which may well assume a greater importance in inorganic chemistry is the ability to form dimeric molecules via hydrogen bonds.

An increasing number of complexes e.g., Cu(II), Cr(II), Re(III), etc. have been shown to have a structure which, superficially at least, may be regarded as a replacement of those enolic hydrogens (such as occurs with β-diketones) giving a polymeric acetate bridge structure. To what extent this behavior is governed by the metal and what part the ligand plays is at present poorly understood.

I. STRUCTURE

Compounds of carboxylic acid derivatives can be structurally divided into five groups, as in Table I.

The ionic nature of the formate grouping in sodium formate is readily shown by the x-ray analysis (2). The equivalence of the two CO bonds (1.27 Å) contrasts markedly with the situation in lithium acetate dihydrate (4). Here the acetate ion functions as a unidentate ligand, the bond length in the —C—O—M unit is 1.33 Å while that of the free carbonyl portion is 1.22 Å.

The intricate chemistry of copper(II) formate complexes provides examples of carboxylate systems in which metal–metal interaction is made untenable by the *syn–anti* or *anti–anti* forms of the ligand. For example, in the royal blue anhydrous formate the *anti–syn* configuration (6) of the ligand results in a closest approach of copper atoms of 3.44 Å. The tetragonal–pyramidal coordination around each copper atom requires a "sharing" of one of the oxygen atoms and this is reflected in the bond lengths (1). Similarly, in the tetrahydrated formate, the *anti–anti* configu-

TABLE I

Type	Description	Example	Ref.
R—C with O (double/dotted) M	Ionic	Na(HCOO)	2
R—C with O—M and O	Monomeric, where each RCOO group uses only one of the two available atoms to form a bond to the metal	Co(O$_2$CCH$_3$)$_2$4H$_2$O Li(O$_2$CCH$_3$)2H$_2$O	3 4
M—O R—C—O M	i.e., *syn–anti* or *anti–anti* forms; no possibility of M—M bond formation	Cu(O$_2$CH)$_2$4H$_2$O (*anti–anti*) Cu(O$_2$CH)$_2$ (*anti–syn*)	5 6
R—C with O—M and O—M	i.e., *syn–syn* form possibility of M—M bond formation	[Cu(O$_2$CCH$_3$)$_2$H$_2$O]$_2$ [Cr(O$_2$CCH$_3$)$_2$H$_2$O]$_2$	7 8
R—C with O / O forming ring to M	Bidentate ligand forming four-membered ring	Zn(O$_2$CCH$_3$)$_2$2H$_2$O Na[UO$_2$(O$_2$CCH$_3$)$_3$]	9 10

ration (5,279) results in the closest approach of the metal atoms of 5.80 Å.

There is an increasing number of complexes, particularly within the transition series, which have bridging acetate cage structures. These arise

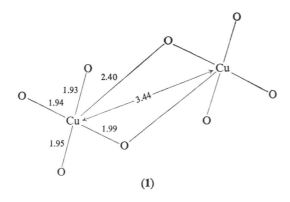

(1)

from the *syn–syn* configuration of the ligand. It has been observed that such structures appear to be favored by metals which have, in the octahedral stereochemistry, one unpaired electron in the *eg* orbitals (11). This has led to the postulate of metal–metal interaction or even metal–metal bonding in some cases (12). The reason that one needs this distinction between direct bonding on the one hand and loose coupling of spins on the other, is that acetic acid exists as a hydrogen bonded dimer. Thus, these dimeric complexes may be regarded as simple metal replacement of these hydrogen atoms, and in many instances, e.g., copper(II) acetate, the fact that the metal atoms are "adjacent" is simply a condition demanded by the bridging acetate ligand. When, however, the metal–metal distance is short enough to cause a distortion from the hydrogen-bonded dimer dimensions, then obviously metal–metal bonding has to be considered.

Lowry and French (13) were the first to postulate that the acetate group functioned as a bidentate chelate giving a four-membered ring. There are, however, few confirmed examples of the acetate ion chelating to the metal atom. In the case of the zinc (9) and the uranyl (10) complexes this has been confirmed by x-ray crystallography. Presumably of all the possible modes of attachment to a metal this would be the least favored. As of necessity a four membered ring must be formed and although there are examples of these scattered around inorganic chemistry their formation is more the exception than the rule.

II. SPECTRAL PROPERTIES

It has been established by crystallographic studies that the carboxylate ligand may show five distinct patterns of behavior. Thus, a prime requirement of infrared spectroscopy is as a means of differentiating between these different structural patterns. We shall examine how far this technique meets this requirement. The situation to 1963 has been reviewed (295,296).

A major difficulty when one considers infrared spectroscopy and carboxylate chemistry arises from the already low symmetry of the free ion (C_{2v}). The consequence of this is that in coordination to a metal, the usual lowering of symmetry together with splitting of degenerate modes does not occur. [Contrast with the high symmetry of the sulfate free ion (T_d).] Nevertheless, one can fairly readily identify type (II) i.e., unidentate coordination. It has been observed by many workers that complexes known to have unidentate carboxylate groupings show an increase in the antisymmetric COO stretching frequency together with a decrease in the symmetric COO stretching frequency. The reason for this is the break-down in equality of the carbonyl groups. Such dissimilarity of the carbonyl groups is unique to the unidentate form of coordination. Nevertheless, even this rule does not have universal validity. Donaldson and co-workers (297) have shown for a series of acetate and formate complexes of metals possessing a lone pair of electrons, e.g., Tl(I), that the carbonyl group and the lone pair interact with a resultant lowering of stretching frequency in the infrared spectrum. A similar effect may be observed in the presence of intramolecular hydrogen bonding (298) such as occurs in nickel acetate tetrahydrate.

The presence of ionic carboxylates is possibly best detected by other methods, but many such salts have a very intense antisymmetric COO stretching frequency within the range 1560–1600 cm^{-1}.

The technique of infrared spectroscopy has been of little use in dif-ferentiating between chelating and bridged forms of carboxylate ligands. knowledge of the structure, from other techniques, has enabled some workers to observe trends in peak position. For example, in some binuclear chloroacetate complexes, the absorptions due to the COO grouping were observed to shift together (299), on charging the metal. Nakamoto and co-workers (300) have observed a similar feature in some binuclear acetate complexes. However, use of infrared spectroscopy to differentiate between complexes of this type seldom yields meaningful results, although inspection based on acid type has sometimes been useful (301,302).

Recently, (303) low-frequency infrared spectroscopy has been used to examine metal–oxygen vibrations in a series of metal formates and acetates of known structure. Some correlations between spectra and structure were found but again differences were such as to make structural deductions of unknown acetate complexes hazardous (230).

The electronic spectra of metal carboxylates have so far been limited, often being confined to an empirical one-metal study. In some cases empirical correlations between color and terminal groups in the binuclear series of acetates have been found.

III. GROUP I

In addition to forming normal salts with monocarboxylic acids, many of the group I elements form acid salts $M \cdot H(O_2CCH_3)_2$ (2). This, however, has been shown to be a solid state phenomenon only (14,15), as in solution they yield the free acid and ions of the normal salt (16,17).

In the solid state, these acid salts appear to be of two types; in one X^- and HX are crystallographically equivalent with a short, symmetrical hydrogen bond between the two (18). In the second group of compounds it has been shown that the acid radicals are crystallographically non-equivalent. In addition, this second class of compounds, in contrast to those above, has an infrared spectrum (19) which may be reproduced by superimposing the spectra of the free acid and the neutral salt [e.g., MH(ortho nitro benzoate)$_2$ where M = K or Rb].

It has been recently suggested (20) on the basis of comparative infrared spectroscopy that all alkali metal formates, with the exception of lithium, have a similar ionic structure.

IV. GROUP II

The formation of so-called basic acetates of the type $Be_4O(O_2CCH_3)_6$ and anhydrous salts of the type $Be(O_2CCH_3)_2$ is characteristic of beryllium. The oxycarboxylates are obtained by the reaction (21,22):

$$4Be(OH)_2 + 6CH_3COOH \rightarrow B_4O(O_2CCH_3)_6 + 7H_2O$$

from which the normal salt, $Be(O_2CR)_2$, may be obtained by treatment with acetyl chloride in the appropriate acid.

An x-ray analysis (23) of the oxy complex shows it to be an example of the class of compounds where the carboxyl groups form a bridge between two metal atoms. In the case of the beryllium complex there are four tetrahedrally arranged beryllium atoms [a feature suggested (24) as long ago as 1923] around a central oxygen atom. The tetrahedron around each beryllium is completed by three oxygen atoms from bridging carboxylate groups. One feature of such structures which has been investigated in the case of beryllium but neglected with other cage species is the ability of cage structures derived from aryl carboxylic acids to form clathrate compounds with aromatic hydrocarbons (25). The smaller acetate complex has been reported (26) to form clathrate compounds with the smaller molecules SO_2, H_2S, CH_4. Explanations (27) based on molecular dimensions have also been used to explain the increase in the heats of sublimation as one moves from aliphatic to arylic carboxylate complexes of beryllium.

Amine derivatives of beryllium acetate approximating to $Be(O_2CCH_3)_2$ (amine)$_2$ have been isolated and various formulas suggested (305) but none was conclusively proved to be correct. NMR studies may well be of use here. A calcium complex $2Ca(O_2CCH_3)_2(NH_3)_2$ is also mentioned briefly.

In contrast to beryllium, the literature on the other alkaline earth carboxylates is not extensive. Magnesium formate $Mg(O_2CH)_2 2H_2O$ is reported to be isomorphous with the zinc and cadmium formates (28). Calcium, strontium, and barium formates have been reported by Nitta and co-workers (29). It has been suggested (30,31) that the high insolubility of magnesium acetate may be explained by the formation of acetate bridged polymers but the evidence for this is not conclusive. The unusual coordination number of twelve for barium and one of six for calcium is reported to occur in calcium barium propionate (32).

V. GROUP III

No systematic study of carboxylic acid derivatives of boron have been made. Boron triacetate is the product of the reaction between acetic anhydride and boric acid if the temperature is below 60°; above this temperature pyroboron acetate $[B(O_2CCH_3)_2]_2O$ is formed (34,76). Pyroboron acetate is also reported to be the main product of the reaction of acetic anhydride and boron trichloride (35). Boron triacetate in acetic acid reacts with hydroxyquinones yielding quinone diacetate derivatives of boron which rapidly lose acetic anhydride (36). $M^I[B(O_2CCH_3)_4]$ complexes, which infrared evidence indicates contain unidentate attachment of at least some of the acetate groups, may be prepared (37) by heating basic boron acetate with $M^I(O_2CCH_3)$.

The heat of formation of solid aluminum triacetate, which may be prepared from the chloride in acetic anhydride has been calculated (38) to be -451.8 kcal/mole. Acetic anhydride is also reported (39) to react with the trichloride or bromide yielding the complexes $2AlX_3(CH_3CO)_2O$. Aluminum triacetate is readily hydrolyzed to give the basic salt $[Al_3(OH)_2-(O_2CCH_3)_6]^+$ which presumably may be similar to basic chromium acetate, i.e., with bridging acetate groups. Recently, Russian workers (40) have isolated other basic acetate complexes of aluminum and have suggested similar acetate bridged species. Although these may well be correct, one would feel more satisfied if criteria other than infrared evidence had been used.

Although tris acetate compounds of gallium and indium are reported, there has been no successful attempt to prepare basic acetates analogous to those of aluminum. In the case of gallium compounds, derivatives with one, two, and finally three acetate groups have been reported (41).

In accord with the general chemistry of this group, thallium is the only element with a stable unipositive acetate derivative. This finds use as a standard in NMR studies (42). Evidence for the species $[Tl(O_2CCH_3)_2]^-$

has also been presented but is not as yet confirmed (43). Bridging carboxylate groups have been postulated (44) in some thallium(III) carboxylates.

VI. GROUP IV

Carboxylic acid derivatives in this group are, for the most part, confined to the later members of the group, although a tetraacetate of silicon is reported (45). This is monomeric (46) and nonconducting in acetic acid solution which raises the interesting question as to the stereochemistry around the silicon. In mixed silicon β-diketone–carboxylate complexes evidence has been found for *cis* and *trans* isomerism (47). Germanium dicarboxylates are mentioned in a patent publication (48) but their existence does not seem to have been confirmed.

There are excellent recent reviews of tin(II) acetates (49,100). Three coordination appears to persist throughout much of their chemistry (50), the carboxylate behaving as a unidentate ligand (51). Tin(IV) carboxylates have recently been investigated (52); the tetracarboxylate may be prepared by heating tetravinyl tin with the anhydrous acid (53). Although these are monomeric in solution, a polymeric structure containing both unidentate and bidentate bridging carboxylate groups, and a coordination number of six for the tin atom has been postulated in the solid state. Various basic acetates of tin have also been reported (54). Pentachlorobenzoate complexes of tin have recently been reported (77) to have different structures in the solid state and in solution.

Lead tetracarboxylates are among the most accessible lead(IV) compounds and find frequent use in synthetic organic chemistry. Although the complexes are nonelectrolytes in acetic acid, exchange with radioactive lead occurs in this solvent (55). No plausible explanation to reconcile these two has been put forward. The diacetate is also readily formed but most of the data on these compounds is old and some of the many modern physical techniques may well give an insight into these structures which are at present uncertain. See, for example, reference 56.

As has been noted with other main group elements, there is also a series of basic acetates of lead; again, structural data are sparse but a recent review (57) on these compounds has appeared together with one on organic lead carboxylates (58).

A nice gradation of behavior has been observed with group IV elements. Mehrotra and co-workers (59) investigated the reaction of the appropriate tetrachlorides with acetic acid and obtained, in the case of silicon, a monochlorotriacetate derivative (with titanium), a dichlorodiacetate, and (with zirconium and thorium), the tetraacetates.

VII. GROUPS V AND VI

The chemistry of the carboxylate derivatives of these metals is very limited. The carboxylates of antimony(III) and bismuth(III) may be prepared by refluxing the oxides with the appropriate carboxylic acid (60). The readily hydrolyzed antimony derivative may also be prepared by treating antimony metal with acetyl peroxide (61). The so-called Cadet reaction (62) involves the formation of high molecular weight polymers obtained when As_2O_3 is treated with acetic anhydride and potassium acetate. Few physical measurements on these compounds have been made (63). The carboxylates of bismuth and bismuthates have recently been reviewed (64). Tellurium benzoates have been briefly reported (65). For a review of compounds of selenium, tellurium, and polonium, see reference 66.

VIII. GROUP IIB

Zinc, like beryllium, forms two acetate complexes, $Zn_4O(O_2CCH_3)_6$ and $Zn(O_2CCH_3)_2$. The crystal structure of the basic acetate (67,68) has been shown to be similar to that of beryllium. The analogy between zinc and beryllium does not extend, except in chemical formula, to the normal acetates. A single-crystal examination (9) of the more stable $Zn(O_2CCH_3)_2 \cdot 2H_2O$ shows the acetate groups to be chelating and the distorted octahedral stereochemistry around the zinc atom to be completed by two water molecules. In addition to bridging and bidentate behavior, unidentate attachment of the acetate group has recently been observed in the chemistry of zinc (69). In bis(acetate)bis(thiourea)zinc(II), the approximate tetrahedral stereochemistry of the metal is formed using two thiourea sulfur atoms and an oxygen atom from each acetate ligand (Zn—O = 1.96 Å). The two remaining acetate oxygen atoms are nonbonded (Zn—O = 2.99 Å).

There are two recent reports (70) of species $Zn_2(O_2CCH_3)(C_5H_7O_2)_3$ which suggests that it may be ionic i.e., $[Zn(O_2CCH_3)][Zn(C_5H_6O_2)_3]$. A similar but insoluble compound is formed by the reaction of mercury(II) acetate with acetylacetone (71). An equally reasonable structure, particularly in view of the polymeric nature of $Zn(C_5H_7O_2)_2$, may well be a dimeric species with either bridging acetylacetone, or acetate, or both.

The chemistry of both cadmium and mercury carboxylates has been little investigated. Warming of the mercury(II) formate or acetate in aqueous solution results in precipitation of the mercury(I) complex, a reaction which curiously does not occur with the propionate. For a recent survey of the chemistry of mercury carboxylates see references 72 and 101. A compound $C_2H_4Hg(O_2CCH_3)_2$ has also recently been reported (73). An extension of the thermal decomposition of mercury(II) carboxylates has provided yet another route to perfluoro organomercury compounds (74). Dimeric mercury(I) derivatives of chloroacetic acids have long been known and an open-chain, unidentate carboxylate structure was suggested (75). Further studies on these compounds may prove rewarding.

A. Scandium, Yttrium, and the Lanthanides

Scandium(III) carboxylates may be readily prepared from the nitrate, chloride, or oxide by treatment with the acid anhydride (78). The properties of these white monomeric complexes have not been extensively investigated, but the infrared spectra (79) and thermal decomposition (80) have been reported.

Lanthanide carboxylates may be used in the electrochemical separation of these elements. Most of these are prepared (81) in a similar manner to scandium carboxylates although neither Sm, Tb, or Ho acetates have been prepared anhydrous. If instead of acetic anhydride, acetic acid is used, basic rare earth acetates were claimed (82) which converted to the triacetates on prolonged refluxing with acetic anhydride. It was suggested that these triacetates are dimeric and have an acetate bridged structure. Later workers (83) were unable to repeat the basic acetate preparation and also report that the triacetate, at least in benzene, is monomeric.

The stability constants of most of the acetate complexes of the lanthanides have been measured (84). As may be anticipated from the effect of the lanthanide contraction, there is a steady increase for the mono-acetate complexes from La^{3+} to Sm^{3+}; however there is then a decrease for Gd^{3+}, Dy^{3+} after which they remain effectively constant. A similar break at Gd^{3+} was also observed for the di- and triacetates. This behavior is not peculiar to acetate complexes but is a feature of many rare earth complexes and it has been suggested (85) that a configurational change also occurs in the region of Gd^{3+}. Perhaps significantly, the break also occurs where polynuclear acetate formation has been postulated. Further details of rare earth carboxylate complexes are given in reference 86 but much of their chemistry still remains to be investigated.

B. Thorium, Uranium, and Actinides

Russian workers (87–89,91,92) have been very active in preparing an impressive array of carboxylic acid complexes of thorium. Complexes both basic and nonbasic have been isolated with up to a maximum of eight carboxylic acids per thorium (90). Unfortunately, despite the large number of complexes, few physical measurements have been made and consequently the structures (which may well provide examples of eight or greater coordination for the metal atom as well as bridging bidentate or unidentate carboxylate coordination) remain a matter of speculation only.

Carboxylates of uranium(IV) may be prepared by zinc amalgam reduction of uranyl carboxylates (93) when the first product is $ZnU(O_2CCH_3)_6$ which may readily be converted to the green tetracetate. This ten-coordinate polymeric tetraacetate material, built up of square face-centered antiprisms sharing two edges, is held together by bridging acetate groups (94), as shown in structure 2. Uranyl acetate dihydrate $UO_2(O_2CCH_3)_2$

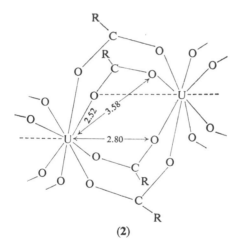

(2)

can be obtained by solution of UO_3 in acetic acid followed by evaporation of the solution. It has been suggested (95) that the two water molecules are coordinated to the uranium atom but these may be readily removed by warming (96) implying simply water of crystallization.

One of the few organic uranyl salts insensitive to light is sodium uranyl acetate $Na[UO_2(O_2CCH_3)_3]$ which may be readily prepared from uranyl nitrate and sodium acetate. For a long time the structural data remained incomplete (97) but a comparatively recent redetermination of the structure (10) has shown that the three carboxyl groups are bidentate (i.e., giving four-membered rings), equivalent, and have their six oxygen atoms lying approximately in a plane which also contains the uranium atom. The neptunium, americium, and plutonium complexes are isostructural (98).

Monomeric penta- and tetraacetate complexes of UO_2^{2+} are also known, as are several binuclear species (see for example ref. 99), but the limited data available on such complexes at the moment precludes a meaningful discussion of these.

C. Titanium, Zirconium, and Hafnium

Complexes between titanium and carboxylic acids have been extensively reviewed (102). The full range of complexes from $n = 1$ to $n = 4$ have now been prepared in the series $(RCOO)_n TiX_{4-n}$. The complex when $n = 4$ proved to be the most elusive (103–107) since the usual reaction between the anhydrous metal halide and a mixture of the corresponding acid and its anhydride, as was the case with boron, yielded only the basic triacetate derivatives. The complexes may, however, be prepared by a more careful control of temperature (108,109).

There is a report (110) that titanium tetrachloride reacts with esters of monobasic acids to form $TiCl_4$ (ester). It is suggested that Ti(IV) has the coordination number of 5 in these complexes. The reaction between the tetrachloride and unsaturated acids has also been briefly reported recently (113).

The reaction between alkali carboxylates and titanium(III) salts is reported (111) to yield complexes of composition $KTi_2(RCOO)_5(OH)_2 \cdot 4H_2O$ which have not been further studied and whose formulation may be questionable. More recently, a dark green d^1 titanium(III) formate, $Ti(O_2CH)_3$ has been reported (112) by the reaction of formic acid with basic titanium carbonate. A similar method (114) when applied to the attempted preparation of the acetate in the series resulted in the mixed titanium(III) complex $Ti(OH)(O_2CCH_3)_2$; possibly, as in the latter preparation, excess basic carbonate was used.

In view of the recent interest in binuclear carboxylates, careful preparative procedures may well result in the isolation of other d^1 titanium complexes of the type $Ti(O_2CR)_3$ which could well show interesting magnetic properties.

Other Russian workers have reported (115) organometallic acetates of titanium $(\pi\text{-}C_5H_5)Ti(O_2CCH_3)_2$ which, when reacted with iron(II) salts, readily yield ferrocene. Bidentate carboxylate behavior has recently been observed (126) in organometallic carboxylate complexes of titanium(III).

Several tetracarboxylate complexes of zirconium have been reported (116–118). Mainly through lack of work, few hafnium analogs are known although undoubtedly these can be made. One feature which has surely prompted the research is the current use of these in space exploration (119) as a base for adhesive cement in the manufacture of rocket nose cones. Physical measurements on the zirconium complexes are not extensive (120), but in view of the frequency of coordination number eight in zirconium chemistry, the tetraacetate complexes may well contain four chelating carboxylate ligands.

Zirconium and hafnium oxycarboxylates have also been reported (121–123). These are polymeric (formates are tetrameric while the acetates are dimeric) and from a study employing infrared spectroscopy (not a particularly useful technique in carboxylate chemistry), it was concluded that the monomers are held together via hydrogen bonding. It is surprising therefore that the authors did not specify whether the carboxylates are uni- or bidentate.

As was noted with titanium, carboxylic acids replace both the chlorine atoms and one of the cyclopentadiene rings in dicyclopentadienyl zirconium(IV) chloride (124).

D. Vanadium, Niobium, and Tantalum

In many respects the chemistry of vanadium carboxylates is representative of vanadium chemistry in general. Complexes have been isolated with the two predominant oxidation states V(III) and V(IV) while evidence for V(V) species is more nebulous.

By the reaction between acetic or benzoic acids and vanadium diborate at elevated temperatures, Greenwood and co-workers (125) obtained light-green, dimeric, nonconducting solids, $V_2(O_2CR)_6$ (where $R = CH_3$ or C_6H_5). A hydrated acetate was known as long ago as 1931 (127). For the anhydrous species the presence of two types of R groups in the proton magnetic resonance spectrum, together with a magnetic moment of 0.77 BM is consistent with the proposed copper acetate type structure, involving unidentate as well as bridging ligand groupings. As with copper acetate, one pyridine molecule may be added per metal atom; this was interpreted as replacement of the terminal carboxylate group by pyridine although no solid complex was isolated.

It was reported (128) that on refluxing vanadium(III) chloride in formic acid, a vanadium(IV) carboxylate, $V(O_2CH)_4$, was formed. However, Seifert (129) repeated this work and found the complex to have a magnetic moment of 2.7 BM; thus he formulated this as d^2 system (i.e., $V(O_2CH)_3HCOOH$). Heating *in vacuo* results in loss of formic acid leaving $V(O_2CH)_3$ still with a magnetic moment of 2.73 BM. Thus there appear to be two different series of vanadium(III) complexes, one dimeric and formed by acetic and higher acids, the other monomeric derived from

R—C, R—C, R—C, V, V, O, C, R, etc.

(3)

formic acid. Further studies on these systems may well prove rewarding particularly as there is also a brief reference to a monomeric vanadium(III) acetate (130). An alleged trimeric vanadium(III) complex $V_3(O_2CCH_2Cl)_7$-$(OH)_24H_2O$ has also been reported (131).

There is again a difference of opinion concerning the solvolysis of vanadyl(V) chloride in carboxylic acids. Seifert (132) isolated a brown complex reported to be $VO(O_2CCH_3)_2$ from the solvolysis of $VOCl_3$ in acetic acid, i.e., vanadium(IV) which he noted was insoluble in water. More recently, the reaction was reported (133) to yield the yellow, readily hydrolyzed vanadium(V) complex $VO_2(O_2CCH_3)$. The situation is no less complicated upon refluxing acetic anhydride containing $VOCl_3$. Siefert reports that this yields the brown–violet vanadium(IV) complex VCl_2-$(O_2CCH_3)_2$ while the other workers claim a gray complex containing no chlorine, formulated as $VO(O_2CCH_3)_2$, a complex which has also been mentioned in infrared studies (136). In the absence of magnetic data, which may be anticipated to be anomalous (134,135), it is difficult to judge the merits of these various claims.

Paul and co-workers (133) also report a complex of the form $VOCl_3$-3RCOOH for which they tentatively suggest the possibility of eight coordination for the vanadium. In view of the current interest in eight coordination, further studies on this complex would be illuminating.

With the exception of chelate complexes with α-hydroxy acids (137), no simple carboxylates of niobium and tantalum have been reported.

E. Chromium, Molybdenum, and Tungsten

1. Chromium Complexes

Several unusual chromium carboxylates have been reported. For example, Fischer (33), by treatment of the sodium salt of dibenzene chromium(0) with carbon dioxide and dimethylsulfate, was able to isolate a red complex $CrC_{12}H_{10}(CH_3COO)_2$ and Burger (138) prepared the yellow benzoic acid analog which on esterification gives the methyl benzoate biphenylchromium salt. The ketone carboxyl region of the infrared spectra has bands above 1700 cm^{-1} suggesting unsymmetrical coordination of the carboxylic acid.

The reaction of acetic acid with chromium(II) salts was first investigated by Peligot (139); more convenient preparations of the red diamagnetic $Cr_2(O_2CCH_3)_4 2H_2O$ have been reported (140). In order to explain the diamagnetism of the complex, it was originally suggested that the molecule was tetrahedral (141) but an x-ray structural investigation (8) showed the structure to be of the copper acetate type such that each chromium atom is in a distorted octahedral environment formed by the four acid oxygen atoms, a water molecule, and the second chromium atom. The close approach of the two chromium atoms, 2.64 Å, implies an interaction between the two metal atoms which presumably is evidently sufficient to pair the d electrons on each metal and hence explain the observed diamagnetism of the complex. The diamagnetic benzoate and higher homologs are presumed to have a similar binuclear structure (142).

Three chromium(II) formate complexes have been isolated (143). One is a blue form which, with a room temperature magnetic moment of 4.8 BM, does not appear to show direct metal–metal interaction. A red monohydrate is also reported which, with a room temperature moment of 0.6 BM decreasing to 0.3 BM at 90°K, appears to involve metal–metal interaction; there is also a red anhydrous form (144) (1.7 BM). Unfortunately, small traces of chromium(III) impurity are difficult to remove from the system and it may well be possible that a similar series of chromium formates exists as in the case with the often analogous copper(II).

As has been observed in the case of the copper(II) acetate molecule, it is possible to replace the terminal water molecules by various other end groups (145–147) giving complexes of the general formula $[Cr(O_2CCH_3)_2A]_2$ where A = CH_3OH, dioxane, and tetrahydrofuran.

Rupture of the dimeric molecule giving a black crystalline paramagnetic (μ_{eff} = 2.83 BM) product $Cr(dipy)_2(O_2CCH_3)_2$ is reported (148) to occur on treatment of the dimeric acetate in THF with 2,2'-dipyridyl. If this product is indeed still a chromium(II) complex, then the question of how its probable octahedral stereochemistry is achieved would be of interest. Two possibilities which appear feasible are: that there is unidentate acetate coordination or perhaps more likely, that these complexes involve one molecule of dipyridyl of solvation particularly as one of the two dipyridyl molecules may be replaced by other solvent molecules.

Chromium(III) carboxylic acid derivatives find industrial use as catalysts for the polymerization of α-olefins and also in the preparation of chrome tanning solutions (149). It was claimed (150) that chromium triacetates could be conveniently prepared by the reaction between chromium(III) nitrate and acetic anhydride. Later, extensive investigations by Weinland (151) and co-workers showed that the green compounds obtained by this method contained the cation $[Cr_3(CH_3COO)_6(OH)_2(H_2O)]^+$ and even by heating these in acetic anhydride, it was not possible to replace the hydroxy groups (152). As long ago as 1928, Welo (153) realized that such molecules were probably trimeric with bridging acetate groups and many structures for the above were postulated, all of which suffered from the disadvantage that they were unable to explain why only three ligands (e.g., amines, thiourea, etc.) could be added giving molecules of the form $[Cr_3(CH_3COO)_6(OH)(NH_3)_3]^+$. This ambiguity was recently resolved when x-ray crystallography (154) confirmed the postulated structure of Orgel (155). The structure was postulated using only the magnetic properties [recently reexamined (156)] and the assumption of an octahedral environment for each chromium atom and is shown in structure **4**. The

(4)

three carboxylate bridged chromium atoms lie at the apices of an equilateral triangle centered around a shared oxygen atom. The vertex of each octahedral *trans* to the shared oxygen is, in the case of the trihydrate, occupied by a water molecule which may be readily replaced by other donor ligands.

A series of acetate derivatives of chromium(III) phthalocyanines have

been reported (157). In these it was suggested that acetate bridging occurred but the presence in the infrared spectra of absorption in the 1700
cm^{-1} region merits further study.

Chromyl complexes of the type $CrO_2(O_2CCH_3)_2$ have also been
reported (158).

2. Molybdenum Complexes

With the intention of preparing arene molybdenum carbonyl sandwich structures, Wilkinson (159) reacted molybdenum hexacarbonyl with
benzoic acid and was able to isolate a yellow diamagnetic sublimable
compound $Mo(O_2CC_6H_5)_2$. The isolation of a similar acetate (160) and
higher carboxylate analogs (161) eliminated the possibility of a sandwich
structure for the benzoate and it was suggested that all these complexes
were dimeric with some, or all, of the acid units serving as bridging molecules. This was confirmed by an x-ray analysis (162) which showed that
the structure of the acetate closely resembles that of the copper and
chromium acetate dimers. Four acetate groups act as bridges between the
metal atoms but in contrast to the weak Cu—Cu and Cr—Cr interactions
in the cases of $[Cu(O_2CCH_3)_2]_2$ and its Cr analog, the short Mo—Mo
distance of 2.10 Å must be accounted for by a strong Mo—Mo interaction
implying a multiple bond. The multiplicity should await further similar
crystallographic studies before meaningful discussions as to the occurrence
of double, triple, or even quadruple bonds between metal atoms can be
undertaken. These molybdenum carboxylates undergo what is rapidly
becoming a characteristic reaction of this type of caged structure, i.e.,
adduct formation with many donor ligands e.g., Ph$_3$P, py, etc. Such ligands
coordinate along the axis of the metal–metal bond.

Two extremely interesting and potentially related molybdenum carboxylic acids have recently been isolated. Green and co-workers (163)
prepared complexes of the form $(\pi\text{-}C_5H_5)Mo(CO)_3CH_2COOH$. These
dimeric molecules contain a metal–carbon linkage (structure **5**).

2.67 Å

O·············O

$(\pi\text{-}C_5H_5)(CO)_3$—Mo—C—C ⟨ ⟩ C—C$\underline{2.41}$Mo(CO)$_3(\pi\text{-}C_5H_5)$

O·············O

(**5**)

Tricarboxyl π-cyclopentadienyl-trifluoroacetate molybdenum was isolated as a red solid by the decomposition of $(\pi\text{-}C_5H_5)Mo(CO)_3(O_2CCF_3)$
in trifluoroacetic acid (164). Infrared evidence suggests a unidentate

attachment of the trifluoroacetate group. The tungsten analog was also isolated.

3. Tungsten Complexes

Tungsten complexes appear to have been little investigated. There is a report of colored insoluble complexes $W(O_2CR)_6$ formed on warming tungsten hexachloride with the appropriate acid. The complexes were not further characterized (165).

F. Manganese, Technetium, and Rhenium

Little is known about the pink manganese(II) formate $Mn(HCOO)_2 \cdot 2H_2O$ except that it is isomorphous with the monoclinic formates of magnesium(II), copper(II), zinc(II), and cadmium(II) and that it may contain two kinds of metal atom in the crystal (166). Such a suggestion has been considered for $Mn(CH_3COO)_2 \cdot 4H_2O$ on the basis of magnetic studies over a temperature range in order to explain the weak ferromagnetic interaction observed (167–169). Thus, crystallographic data would be of interest. Measurements of the equilibrium constants for the dissociation of manganese(II) acetate have also been reported (170).

Although benzoic acid does not appear to react, dithiobenzoic acid is reported (171) to form a stable, red, monomeric complex with manganese carbonyl bromide. Spectral measurements indicate bidentate attachment of the thiobenzoate group.

With manganese(III), the brown $Mn(CH_3COO)_3 \cdot 2H_2O$ is made via the manganese(II) complex by oxidation in hot acetic acid. This complex is a ready source for the preparation of other manganese(III) complexes. It has been suggested (172) that there may be some magnetic exchange occurring in the system; further study of the magnetic structure may prove rewarding. This has the d^4 spin-free configuration and as nothing is known of the structure, the presence of Mn—Mn interaction cannot be excluded. Complex derivatives also include phenylhydrazine derivatives (173) $Mn(CH_3COO)_2(C_6H_5NHNH_2)_2$ and some curious polynuclear complexes (174) apparently containing four atoms of manganese which have been little investigated but in view of (a) their resemblance to the chromium complexes and (b) the ease with which acetate can serve as a bridging ligand, these complexes may well be acetate-bridged polynuclear complexes.

Unfortunately, no technetium derivatives of carboxylic acids have been described. Orange dimeric carboxylate complexes, $Re_2(RCOO)_4X_2$ where R = alkyl or aryl groupings, and X = Cl, Br, NCS, i.e., isoelectronic with $Mo_2(O_2CCH_3)_4$ may be prepared from either the trimeric (175) or dimeric (176) rhenium(III) halides. Chemical properties (177) and later x-ray crystallography (178) showed that these species have a structure similar to that of molybdenum acetate with terminal chlorine atoms along axis of the Re—Re bond. As was observed with the molybdenum complex a short Re—Re distance (2.24 Å) again implies strong metal–metal interaction explaining the observed diamagnetism; however, this may be more a feature of rhenium chemistry than of carboxylate bridging. In this connection, mention should be made of the paramagnetic complexes prepared by Wilkinson and co-workers (179) by refluxing $ReX_3O[P(R_6H_5)_3]_2$ (X = Cl or Br) in carboxylic acids with the exclusion of air. These have been formulated (179) as containing the unit in structure 6, and, in fact, the

$$
\begin{array}{c}
R \\
| \\
C \\
\diagup \quad \diagdown \\
O \quad\ \ Cl\ \ \quad O \\
|\quad \diagup\diagdown \quad | \\
Re \qquad Re
\end{array}
$$

(6)

entire proposed molecular structure, including this feature, has been confirmed by x-ray crystallography (180). Another complex which merits further investigation is the black, crystalline, diamagnetic complex $ReCl_3(O_2CCH_3)_2$ obtained by the reaction between $ReCl_5$ and acetic acid. The insolubility of the complex has limited a study of this, but acetate bridging and seven coordination has been postulated (175). Nuclear magnetic resonance measurements might well be possible if aromatic acids were used.

The visible spectra of many rhenium(III) carboxylates have been measured. A rough correlation is observed between the inductive effect of the R group, as measured by Taft's σ constant, and spectral position which strongly suggests that the orbitals of the carboxylate group contribute to these transitions (177).

Many of these tetracarboxylate complexes have orange colors but there also exists a closely analogous, but blue, series of rhenium carboxylates originally formulated (181) as complexes of rhenium(II) but now clearly established as rhenium(III) intermediates in the preparation of the tetracarboxylate complexes. It is believed at the present time that the blue color may be associated with either weakly coordinated or a complete absence of groups along the axis of the Re—Re bond (182). More experimental results perhaps with other complexes having this type of structure may further elucidate the presently ill-understood spectra of metal carboxylates.

A species, pentacarbonyltrifluoroacetaterhenium, analogous to the molybdenum and tungsten complexes mentioned earlier, forms during the reaction of pentacarbonylmethylrhenium with trifluoroacetic acid (182).

G. Iron, Ruthenium, and Osmium

The carboxylate chemistry of the iron group elements is not well understood. Formate complexes with iron(II) have been described (183) as green and of the general formula $Fe(O_2CH)_2 \cdot 2X$ where $X = H_2O$, py, etc. Magnetic measurements have been reported and the Mössbauer spectrum shows two pairs of lines indicating inequivalent ferrous ion sites (184). A solvated white formate complex is said to be precipitated on addition of formic acid to an acetic acid solution of iron(II) acetate. This may well be a basic formate of iron(III).

Boiling anhydrous acetic acid will, in an inert atmosphere, attack metallic iron giving a white iron(II) acetate, $Fe(O_2CCH_3)_2$. This can be oxidized in anhydrous acetic acid giving the red basic iron(III) complex, or in methanol giving the dimethoxy acetate. With pyridine, a green complex $Fe(O_2CCH_3)_2Py_4$ may be isolated which on heating loses two molecules of pyridine. An air-stable, green complex, $Fe_2(O_2CCH_3)_5$, results on solution of iron(II) acetate and the basic iron(III) acetate in acetic acid (185). If this formulation is correct, the possibility of metal–metal interaction is intriguing.

The iron(III) carboxylate system is, in general, similar to that of the chromium(III) system. The formate complex $Fe(O_2CH)_3$ has been reported (186) by the reaction of formic acid on moist ferric hydroxide but attempts to reproduce this have yielded only the basic formate (187). Other iron(III) carboxylates have been prepared and used as intermediates in the preparation of alkoxide derivatives (188,189) but have been little investigated in their own right.

Iron(III) forms analogous basic carboxylates (190–192) to the chromium complexes, i.e., $[Fe_3(O_2CR)_x(OH)_{8-x}]^+$. Although little x-ray data is available on these complexes, the fact that chromium can replace iron isomorphously in these salts strongly supports the suggestion of a structure similar to that of the basic chromium(III) complexes and not the linear, unidentate, hydroxy-bridged structure suggested recently (193). Two further pieces of evidence reinforcing this suggestion are provided by the Mössbauer spectrum which shows evidence for only one type of iron atom in the molecule (194) and magnetic data (195) which strongly suggests the persistence of the trinuclear species in solution.

Both ferrous and ferric trichloroacetates have the remarkable property of dissolving in *n*-heptane (196). Other interesting complexes of iron(III) were isolated by Rosenheim (197) and other workers (198) who, in addition to some polynuclear Fe_4 species, found that ferric chloride or bromide will recrystallize from anhydrous acetic acid as $Fe(O_2CCH_3)_2X$ (X = Cl, Br).

The reaction of acetic acid yielding ruthenium(III) complexes was first reported by Mond (199), who isolated apparently dinuclear black complexes. However, because of the possibility of some ruthenium(IV) oxide in his starting products, these claims may be questionable. By deliberately starting with the tetroxide, Martin (200) was able to reduce this to the ruthenium(III) species using acetaldehyde in acetic acid solution and isolate blue basic trinuclear complexes exactly analogous to those of iron(III). Although again no crystallographic data are available, one might reasonably anticipate these to have a triangular structure analogous to the basic iron(III) acetates.

The product obtained by refluxing ruthenium trichloride with a carboxylic acid/acid anhydride mixture is the reddish-brown $[Ru_2(O_2CR)_4Cl_2]$. This preparation is curiously aided by oxygen, yet no evidence for Ru=O or Ru—O groupings is found. Carboxylate bridging together with spin-free ruthenium (μ_{eff} = 2.7–3.4 BM/ruthenium) is suggested for these complexes (201).

No osmium complexes have been reported.

H. Cobalt, Rhodium, and Iridium

Cobalt(II) carboxylates are usually reddish-pink, often hydrated, materials which may be prepared in a number of ways, e.g., by refluxing the appropriate acid with either the metal carbonate, metal carbonyl or even the powdered metal itself (202). In the last two examples, oxygen is also passed through the reaction mixture. $Co(O_2CH)_2 \cdot 2H_2O$ is isostructural with the formates, $M(O_2CH)_2 2H_2O$ where M = Ni, Cu, in which the ligand behaves as a bridging group giving rise to polymeric species (203). There is also an unconfirmed report (204) of a red formate complex, $[Co(O_2CH)_6]^{4-}$.

An x-ray structural investigation of cobalt(II) acetate dihydrate has shown the presence of unidentate acetate groupings (205) strongly hydrogen bonded to a coordinated water molecule as shown in structure **7**.

This complex is isomorphous with the acetates of magnesium, manganese, nickel, and zinc. However, there is some evidence that the molecule may be dimeric in acetic acid solution (206). Magnetic data to show if the

(7)

complex is low-spin in this solution would be of interest as this then has the $t_{2g}^6 e_g^1$ configuration in which dimerization via metal–metal interaction might be anticipated. The magnetic susceptibility of $Co(O_2CCH_3)_2 \cdot 4H_2O$ down to $0.38°K$ obeys the Curie-Weiss Law (207,208), the anomaly previously observed at $0.6°K$ being erroneous (209). The treatment of this complex with ethylene thiourea (210) results in the purple $Co(etu)_2(O_2-CCH_3)_2$.

Anhydrous cobalt(II) acetate (211) has been much less studied but appears to have a strong tendency to form adducts of the general formula $Co(O_2CCH_3)_2 \cdot 2X$ where $X = py$, N_2H_4 (212), $(NH_4RCOO)_{1/4}$ (213), or $B_2O(O_2CCH_3)_4$ (214). Unidentate attachment of the trifluoroacetate ligand occurs (215) in the tetrahedral $[Co(O_2CCF_3)_4]^{2-}$.

The chemistry of cobalt(III) carboxylates is in a confused state. It has been claimed (216) that electrolytic oxidation of cobalt(II) acetate yields an apple-green compound with the analysis and molecular weight required for $Co(O_2CCH_3)_3$. In a subsequent publication (217), the molecular weight was said to correspond to $Co_3(O_2CCH_3)_9$. Later workers (218) were only able to prepare a sample of cobalt(III) acetate contaminated with cobalt(II). The close similarity of cobalt(III) acetate to potassium tetraoxalate-μ-dihydroxodicobalt(III) resulted in the postulate of a binuclear structure for the acetate. The magnetic data, although possibly inaccurate, do seem to indicate a large magnetic susceptibility for the complex which might not be anticipated if a bridging system were present. Cobalt(III) acetate is also reported (219) to be present in dilute acetic acid solutions of cobalt(II) acetate which have been oxidized with ozone. However, the absorption spectrum differs appreciably from that reported by other workers (220).

An interesting reaction of a mixed cobalt(III) acetate–picolinic acid complex, $Co(O_2CCH_3)(PH)Br$ where PH = picolinic acid-N-oxide, occurs on treatment with water, molecular bromine, and a cobalt complex are obtained (221). This was not extensively investigated but clathration of the bromine as occurs (222) in some glyoxime complexes was thought not to occur.

Passing mention must also be made of the numerous carboxylate-pentamine cobalt(III) complexes. The literature in this field has recently been reviewed (223,224). The more recent work has been mainly concerned with variation of the basicity and steric effects of the carboxylate ligand. It has been suggested (226,227) that steric effects are much less important than electronic effects in governing the properties of these complexes.

I. Rhodium

Ammonium chlororhodate reacts with formic and acetic acids yielding green diamagnetic crystals (228) of $Rh(O_2CR)_2H_2O$, where R = H, CH_3. Similar products are formed by using $Rh(OH)_3H_2O$ (229,230,247) or by reaction of rhodium(III) chloride with sodium acetate (231). X-ray diffraction studies of the acetate member of the series (232) and later the formate complex (233) have shown that these are dimeric rhodium(II) compounds containing a metal–metal bond (2.4 Å) with a H_2O—Rh—Rh—H_2O unit. As is frequently observed with systems of this type, the water molecules can be reversibly substituted by other groupings (234,236). With the single exception of the ethylenediamine complex (229), all the adducts may be reconverted to the anhydrous acetate upon heating. Ethylenediamine was the only bidentate ligand used and it was suggested that this may chelate with the metal atoms causing rupture of the acetate cage structure. Alternatively, the diamine may bridge dimeric molecules together in a similar way as in, for example, platinum(IV) chemistry.

An orange binuclear acetate complex of rhodium(I) is obtained on boiling $(C_8H_{12}RhCl)_2$ with potassium acetate in acetone solution (235). Infrared spectra suggest symmetrical bridging of the acetate. It has recently been observed (237) that $RhCl_3(Et_2PhP)_2$ is able to extract a CO grouping from acetic acid, via a carboxylate intermediate giving $RhCOCl(Et_2PhP)_2$. The mode of attachment of the carboxylate group in this intermediate is of interest and the presence of metal–carbon linkages cannot be discounted.

J. Iridium

One of the few reported complexes between carboxylic acids and iridium is that obtained by treating $IrH_3(PR_3)_3$ with acetic acid (238).

Hydrogen is evolved and $IrH_2(O_2CCH_3)(PR_3)_3$ is formed. Only kinetic studies (239) have been made of the complex $[Ir(NH_3)_5(O_2CCH_3)]^{2+}$ but as similar rhodium and cobalt complexes can be prepared, such complexes provide an accessible way to compare carboxylate complexes of the 1st, 2nd, and 3rd row transition metals.

Recently, diamagnetic iridium formates and acetates have been reported (304) $[IrCl(O_2CR_3)_2]_2$ where $R = CH_3, H$. By analogy with the similar rhodium complexes, the diamagnetism has been attributed to the presence of metal–metal interaction.

K. Nickel, Palladium, and Platinum

The different structural patterns between formate and acetate complexes so frequently observed previously are again exhibited in the chemistry of nickel(II) carboxylates. $Ni(O_2CH)_2 2H_2O$ is reported (240) to have two different kinds of octahedral nickel atoms; one nickel atom is surrounded by six formate oxygen atoms while the other octahedron is completed by four water oxygen atoms together with two *trans* formate oxygen atoms. Thus as is the case with copper(II), there are formate bridges between metal atoms allowing an infinite polymer to be built up.

Anhydrous nickel acetate is reported to be yellow but little is known of its properties or structure. In contrast, the green tetrahydrate is isostructural with the cobalt(II) analog and contains two *trans* unidentate acetate linkages (205). Magnetic measurements on both the acetate ($\mu_{eff} = 3.30$ BM) and the benzoate (241) ($\mu_{eff} = 3.2$ BM) together with temperature range studies are consistent with an octahedral structure for this complex (242). Acetate bridging has been postulated (243) in complexes of the type $[Ni_2(amine)_4OCR]_2^{2+}$ but this has not been unequivocally confirmed. It has also been observed (244) that solution of the tetrahydrate in anhydrous acetic acid yields various solvates some of which have magnetic moments in the range associated with a tetrahedral stereochemistry of the nickel atom. Further studies may prove interesting.

The reaction of π-allyl palladium chloride with sodium acetate yields the dimeric π-allyl palladium acetate (245). The Pd—Pd distance in the complex of 2.94 Å suggests some metal–metal interaction (246) probably of the same strength as in the copper(II) acetate case, i.e., ~ 1 kcal/mole but undoubtedly the main feature holding the molecule together is the presence of two acetate bridging groups. Carboxylate bridging is probably also present in the trimeric brown palladium(II) acetate obtained by the interaction of an acid solution of palladium(II) nitrate and glacial acetic acid (247). The order of stability of the trimeric molecule is apparently benzoate > acetate > perfluorocarboxylates. These are monomeric in benzene at 37°. With various nitrogen and phosphorus donor ligands, monomeric complexes $Pd(O_2CR)_2L_2$ are obtained; a zero dipole moment together with the presence of a band 1626 cm^{-1} in the infrared spectrum suggests a *trans* structure with unidentate carboxylate coordination. When L = phosphines, dimeric acetate bridged complexes $[Pd(O_2CCH_3)_2L]_2$ may be formed by using less phosphine (248). Subsequent reaction of these dimers with additional phosphine results in bridge breaking and the isolation of the monomeric complexes.

Interesting acetate-bridged binuclear alkoxypalladium(II) olefin adducts are the products (249) of the reaction between silver carboxylates and 1:5 cyclooctadienedichloropalladium(II). However, in acetic acid solution these are monomeric.

A much more stable platinum(II) acetate trimer is obtained by reduction of $Na_2Pt(OH)_6$ in acetic acid solution (247). Higher carboxylates have not yet been prepared. This acetate complex again forms solid adducts with various N and P donors; their stoichiometry is not clear, but they do not appear to be simple adducts (1:1). In view of these chemical differences, structural determinations of both of these trimers are needed.

L. Copper, Silver, and Gold

The first reported preparation (250) of "cuprous formate" was in 1904 when $Cu_2(O_2CH)_2 \cdot 4NH_3$ was described as blue, suggesting some copper(II) impurity. An authentic sample of white copper(I) formate was prepared two years later by solution of copper(I) oxide in formic acid (251). The acetate member of the series is formed as a sublimate by heating copper(II) acetate *in vacuo* (252) or by reduction of the copper(II) complex with hydroxylamine sulfate (253). Their chemistry has not been extensively investigated in contrast to that of copper(II) carboxylates which have been the subject of many investigations; more than three hundred papers have appeared on the subject, mainly prompted by the novel magnetic characteristics of the system. The task of this reviewer has been made less tedious by the recent appearance of two excellent reviews which survey the literature until 1964 (254,255), Consequently, only work after this date will be considered here except where it is necessary to draw correlations with other transition metal complexes.

The structure of copper(II) acetate dihydrate has been mentioned earlier. The question of the mechanism of the interaction between unpaired electrons on adjacent copper ions has been and still is the subject of many publications (256–258). The review in 1964 concerning complexes with subnormal magnetic moments (254) conveys the impression that the copper-to-copper magnetic interaction can mainly be accredited to a direct bond between the copper atoms (using either d_{z^2} orbitals and a σ bond, or $d_{x^2-y^2}$ orbitals and a δ bond (259–262). They do, however, comment that a superexchange mechanism via the ligand carboxylate is a possibility. At the time of their writing fewer dimeric transition metal carboxylates were known and so a Cu—Cu distance of 2.64 Å was reasonably interpreted as representing considerable metal–metal bonding. However, the recent determination of other metal–metal distances in dimeric carboxylates, e.g., Cr, Rh, Re, etc., has shown that shortening with respect to the antici-pated M—M distance may be substantial. Consequently, an internuclear distance of 2.64 Å in the case of Cu(II) need not indicate appreciable metal–metal interaction. Indeed, the strength of the interaction has been estimated (12) to be as low as 1 kcal/mole.

The physical properties of copper(II) formates have been extensively investigated (254,288). In many of these, e.g., $Cu(O_2CH)_2 \cdot 4H_2O$ (279), Cu—Cu = 6.35 Å, only ligand exchange can be reasonably postulated. The antiferromagnetic coupling in the tetrahydrate is explained by spin coupling between formate bridged copper(II) ions (291–293). A similar situation has recently been found in one form of copper(II) nitrate (294). The consequence of this is that magnetic moments may be lowered up to 50% of the "spin only" value. Where there is appreciable metal–metal interaction, for example in the case of chromium complexes, diamagnetic complexes are found. Consequently, it is obvious that there are at least two mechanisms for magnetic exchange in these systems, probably inter-related.

In conjunction with the magnetic data, electronic absorption spectra have also been extensively used as a probe to investigate the exact nature of the interaction of the molecule (254). The crystal spectrum of copper(II) acetate shows three main bands at $\nu = 1400$ cm^{-1}, $\nu = 27,000$ cm^{-1}, and $\nu = 40,000$ cm^{-1}. The most adequate account of position and polariza-tions of the bands is that of Ballhausen and co-workers (263) using a weak coupling scheme between the two parts of the dimer.

As has frequently been observed for cage molecules of this type, it is apparently possible to have an infinite variety of unidentate, sometimes charged, ligands coordinated along the axis of the two metal atoms (264–267,287). Unfortunately, little has been tried with polydentate ligands. A 4,4′-dipyridyl adduct shows magnetic exchange (268) but here the dipyridyl probably serves as a bridging group, bridging copper(II) benzoate molecules together in much the same way as has been suggested for the urea adducts (269). Some ill-characterized complexes with benzidine have been reported (286). With a tetradentate ligand, complete replacement of the acetate group occurs (270), and a similar reaction with 8-hydroxyquinoline is reported (271). Work on other normally chelating ligands may be of interest, particularly with regard to the work on chromium(II) derivatives. More studies using thioacetic acid as a ligand may also prove illuminating as the copper(II) derivative is diamagnetic (289).

The structure of aryl carboxylic acid complexes of copper(II) has been correlated with pK_a of the acid (257). A dimeric structure (shown by magnetic data) is favored by high pK_a values, i.e., $pK_a > 4.2$. X-ray analysis has shown (272) that for one form of copper benzoate trihydrate, the copper atoms are 3.15 Å apart (structure **8**) and there is only one carboxylate bridge between the metal atoms. Consequently, magnetic exchange

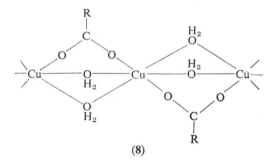

(8)

would be expected to be small as has been observed ($\mu_{eff} = 1.87$ BM). However, one modification (273) prepared by heating the "magnetically normal benzoate" appears to show magnetic exchange ($\mu_{eff} = 1.40$ BM, $\sigma = 15°$) as do many of the substituted benzoic acids and their adducts (256–258, 269). Thus one can safely say that more than simply electronic factors (as measured by pK_a) govern the structure which copper aryl carboxylates adopt, e.g., solvent effects (274).

The situation is little better in the alkyl series. Attempts to apply pK_a data to the structures adopted in this series of complexes has met with little success (275,290) which is hardly surprising as (a) electronically these are two very different systems and (b) steric factors may be expected to play a much greater role in the aliphatic series. Better correlations with the aryl series might be anticipated with propiolic acid and its derivatives as here we have more comparable systems. Preliminary experiments on copper(II) complexes of unsaturated carboxylic acids (276), e.g., vinylacetic acid, have recently been reported. Molecular weight and magnetic data suggest that these two are binuclear and in the anhydrous series it is suggested that the double bond in the acid coordinates along the axis of the Cu—Cu interaction, a suggestion that may readily be confirmed by NMR studies of a diamagnetic carboxylate.

The complex $Ca[Cu(O_2CCH_3)_4]6H_2O$ has been known for many years (277) but only very recently has this been shown (278) to be an example of the unusual eight-coordinate copper(II). The acetate ligands act not only as bridging ligands between copper and calcium atoms but also as chelating agents. It would be of interest to investigate the structure of $K_4[Cu(O_2CCH_3)_6]12H_2O$ and some of the amine complexes between copper(II) and acetic acid (225).

Silver acetate finds use as a catalyst in urethan-polymer manufacture and has been suggested as a convenient source of derivatives with which to characterize carboxylic acids (280). Silver(I) carboxylates are obtained by mixing equivalent amounts of silver oxide and an aqueous solution of the appropriate acid. The perfluorobutyrate member of the series has a dimeric structure with an eight-membered ring (9) (281).

$$CF_3{-}CF_2{-}CF_2{-}C \underset{\displaystyle O{-}Ag{-}O}{\overset{\displaystyle O{-}Ag{-}O}{\big\langle}} \underset{2.90}{\updownarrow} \overset{\displaystyle }{\big\rangle} C{-}CF_2{-}CF_2{-}CF_3$$

(9)

An Ag—Ag distance of 2.90 Å suggests some metal–metal interaction. A similar structure has been suggested for $IAg(O_2CR)_2$ but as yet is unconfirmed (282). Silver carboxylates are soluble in the parent acid and the solubility is found (283) to increase on addition of alkali acetates, which has led to the speculation that a double acetate $M[Ag(O_2CCH_3)_2]$ may exist in solution, but attempts to isolate this so far have been unsuccessful (284).

The only reported reaction between gold salts and carboxylic acids is by Weigand (285), who suggested the formation of a double acetate $Mg[Ag(O_2CCH_3)_4]_2$ but this has not been further investigated.

Acknowledgment

The author wishes to gratefully acknowledge the interest and encouragement of Professor F. A. Cotton. Part of this review was prepared while at Massachusetts Institute of Technology, Cambridge, to which the author is indebted.

References

1. A. K. Holliday and A. G. Massey, *Inorganic Chemistry in Non-Aqueous Solvents*, Pergamon, London, 1965, p. 70.

2. W. H. Zachariasen, *J. Am. Chem. Soc.*, *62*, 1011, (1940).
3. J. N. van Niekerk and F. R. L. Schoening, *Acta Cryst.*, *6*, 609 (1953).
4. V. Amirlhalingam and V. M. Padmanabhan, *Acta Cryst.*, *11*, 896 (1958).
5. R. Kiriyama, H. Ibamoto, and R. Matsuo, *Acta Cryst.*, *7*, 482 (1954).
6. G. A. Barclay and C. H. L. Kennard, *J. Chem. Soc.*, *1961*, 3289.
7. J. N. van Niekerk and F. R. L. Schoening, *Acta Cryst.*, *6*, 227 (1953).
8. J. N. van Niekerk, F. R. L. Schoening, and J. F. de Wet, *Acta Cryst.*, *6*, 501, (1953).
9. J. N. van Niekerk, F. R. L. Schoening, and J. H. Talbot, *Acta Cryst.*, *6*, 720 (1953).
10. W. H. Zachariasen and H. A. Plettinger, *Acta Cryst.*, *12*, 526 (1959).
11. J. Lewis and R. S. Nyholm, *Sci. Progr.*, *52*, 577 (1964).
12. F. A. Cotton, *Rev. Appl. Chem.*, *17*, 24 (1967).
13. J. Lowry and W. French, *Proc. Roy. Soc.* (*London*) A, *106*, 489 (1924).
14. J. C. Speakman, *J. Chem. Soc.*, *1949*, 3357.
15. J. C. Speakman, *J. Chem. Soc.*, *1951*, 185.
16. J. C. Speakman, J. M. Skinner, and G. M. D. Stewart, *J. Chem. Soc.*, *180*, 787 (1954).
17. J. C. Speakman and H. H. Mills, *J. Chem. Soc.*, *1961*, 1164.
18. R. Blinc, D. Hadzi, and A. Novak, *Z. Elektrochem.*, *64*, 567 (1960).
19. H. N. Shrivastava and J. C. Speakman, *J. Chem. Soc.*, *1961*, 1151.
20. J. D. Donaldson, J. F. Knifton, and S. D. Ross, *Spectrochim. Acta*, *20*, 847 (1964).
21. K. N. Semenko and G. M. Kurdyumov, *Zh. Neorgan. Khim.*, *7*, 1512 (1962).
22. J. Besson and H. D. Hardt, *Z. Anorg. Allgem. Chem.*, *277*, 188 (1954).
23. A. Tulinsky, C. R. Worthington, and E. Pignataro, *Acta Cryst.*, *12*, 623, 626, 634 (1959).
24. W. Bragg and G. T. Morgan, *Proc. Roy. Soc.* (*London*) A, *104*, 437 (1923).
25. K. N. Semenko and G. M. Kurdyumov, *Russ. J. Inorg. Chem.* (*Eng. Trans.*), *6*, 1298 (1961).
26. K. N. Semenko, *Russ. J. Inorg. Chem.* (*Eng. Trans.*), *5*, 1298 (1960).
27. K. N. Semenko, G. M. Kurdyumov, and I. V. Gordeev, *Zh. Neorgan. Khim.*, *6*, 2025 (1961).
28. K. Osaki, Y. Nakai, and T. Watanabe, *J. Phys. Soc.* (*Japan*), *18*, 919 (1963).
29. I. Nitta, K. Osaka, and Y. Saito, *X-rays* (*Osaka*), *5*, 37 (1948); *6*, 85 (1951).
30. E. A. Goode, N. S. Bayliss, A. Cherbury, and D. Rivett, *J. Chem. Soc.*, *1928*, 1950.
31. E. A. Goode, N. S. Bayliss, A. Cherbury, and D. Rivett, *J. Chem. Soc.*, *1926*, 1026.
32. L. F. Biefield and P. M. Harris, *J. Am. Chem. Soc.*, *57*, 396 (1935).
33. E. O. Fischer and M. Brunner, *Z. Naturforsch.*, *16B*, 406 (1961).
34. Y. Y. Zenlyanskii and V. V. Malyavkin, *Navchn. Tr. Mosk. Tekhnol. Inst. Legkoi Prom.*, *24*, 46 (1962).
35. H. Meerwein and H. Maier-Huser, *J. Prakt. Chem.*, *134*, 51 (1932).
36. O. Dimroth and T. Faust, *Ber.*, *54*, 3020 (1921).
37. U. Kibbel, *Z. Chem.*, *5*, 425 (1965).
38. C. T. Mortimer and P. W. Sellers, *J. Chem. Soc.*, *1963*, 1978.
39. J. Boeseken and F. Cluwen, *Rec. Trav. Chem.*, *31*, 367 (1912).

40. A. I. Grigor'ev and V. N. Maksimov, *Russ. J. Inorg. Chem. (Eng. Trans.)*, *9*, 559 (1964).
41. H. Funk and A. Paul, *Z. Anorg. Allgem. Chem.*, *330*, 70 (1964).
42. E. B. Baker and L. W. Burd, *Rev. Sci. Inst.*, *34*, 238 (1963).
43. M. C. Pai, C. M. Chang, H. J. Hu, and C. H. Chang, *Mua Hsuek Msueh Pao*, *27*, 163 (1961).
44. G. B. Deacon, *Aust. J. Chem.*, *20*, 459 (1967).
45. H. K. Mueller and K. Walter, *Z. Chem.*, *4*, 5 (1964).
46. G. Rudakoff and H. Dunken, *Z. Naturforsch. 17b*, 623 (1962).
47. C. E. Holloway, R. R. Luongo, and R. M. Pike, *J. Am. Chem. Soc.*, *88*, 206 (1966).
48. British Pat. 961730.
49. J. D. Donaldson, in *Progress in Inorganic Chemistry*, Vol. 8, F. A. Cotton, Ed., Interscience, New York, 1967.
50. J. D. Donaldson and J. F. Knifton, *J. Chem. Soc. A, 1966*, 332
51. J. D. Donaldson, J. F. Knifton, and S. D. Ross, *Spectrochim. Acta*, *21*, 1043 (1965).
52. R. C. Olberg and M. Stammler, *J. Inorg. Nucl. Chem.*, *26*, 565 (1964).
53. A. Henderson and A. K. Holliday, *J. Organomet. Chem.*, *4*, 377 (1965).
54. E. H. Baker and F. C. Tomkins, *J. Chem. Soc.*, *1952*, 4518.
55. G. Hevesy and L. Zechmeisher, *Z. Elektrochem.*, *26*, 151 (1920).
56. J. M. Rocard, M. Bloom, and L. B. Robinson, *Can. J. Phys.*, *37*, 522 (1959).
57. W. Kwestroo and C. M. Langereis, *J. Inorg. Nucl. Chem.*, *27*, 2533 (1965) and references therein.
58. L. C. Willemsens and C. M. Vander Kerk, *Investigations in the Field of Organolead Chemistry*, Intern. Lead and Zinc. Res. Org., New York, 1965.
59. R. N. Kapoor, K. C. Pande, and R. C. Mehrotra, *J. Indian Chem. Soc.*, *35*, 153 (1958).
60. J. D. Donaldson, J. F. Knifton, and S. D. Ross, *Spectrochim. Acta*, *20*, 847 (1964).
61. G. A. Razuvaev, B. N. Moryganov, E. P. Dlin, and Y. A. Oldekop, *Zh. Obshch. Khim.*, *24*, 262 (1954).
62. A. Titov and B. Blevin, *Sb. Statei Obshch. Khim.*, *2*, 1469 (1953).
63. H. Gutbier and H. Plust, *Chem. Ber.*, *88*, 1777 (1955).
64. *Gmelin's Handbuch der Anorganischen Chemie*, *19*, 831 (1964).
65. M. Campos, E. Suranyi, H. Andrade, and N. Petragnani, *Tetrahedron*, *20*, 2797 (1964).
66. K. Bagnall, *The Chemistry of Selenium, Tellurium, and Polonium*, Elsevier, Amsterdam, 1966, p. 145.
67. J. Wyart, *Bull. Soc. Franc. Mineral.*, *49*, 148 (1926).
68. H. Koyama and Y. Saito, *Bull. Chem. Soc. (Japan)*, *27*, 112 (1954).
69. L. Cavaka, G. F. Gasparri, G. D. Andreetti, and P. Domiano, *Acta Cryst.*, *22*, 90 (1967).
70. G. Rudolph and M. C. Henry, *Inorg. Chem.*, *3*, 1317 (1964); *4*, 1076 (1965).
71. J. Lewis and C. Oldham, unpublished results.
72. *Gmelin's Handbuch der Anorganischen Chemie*, *34b*, 277 (1965).
73. T. D. Smith, *Nature*, *209*, 907 (1966).
74. J. E. Connett, A. G. Davies, G. B. Deacon, and J. H. S. Green, *J. Chem. Soc. C*, *1966*, 106.

75. N. R. Davidson and L. E. Sutton, *J. Chem. Soc.*, *1942*, 565.
76. A. K. Rai and R. C. Mehrotra, *J. Indian Chem. Soc.*, *40*, 339 (1963).
77. G. B. Deacon and P. W. Felder, *Aust. J. Chem.*, *20*, 1587 (1967).
78. H. Funk and B. Koehler, *Z. Anorg. Allgem. Chem.*, *325*, 67 (1963).
79. A. I. Grigor'ev and V. N. Maksimov, *Zh. Neorgan. Khim.*, *9*, 1060 (1964).
80. E. L. Mead and C. E. Holley, *J. Inorg. Nucl. Chem.*, *26*, 525 (1964).
81. J. R. Witt and E. I. Onstott, *J. Inorg. Nucl. Chem.*, *24*, 637 (1962) and references therein.
82. J. S. Seaton, F. G. Sherif, and L. F. Audrieth, *J. Inorg. Nucl. Chem.*, *9*, 222 (1959).
83. S. N. Misra, T. N. Misra, and R. C. Mehrotra, *J. Inorg. Nucl. Chem.*, *25*, 201 (1963).
84. A. Sonesson, *Acta Chem. Scand.*, *12*, 165, (1958), 1937; *13*, 1437 (1959).
85. I. Grenthe and W. C. Fernelius, *J. Am. Chem. Soc.*, *82*, 6258 (1960).
86. S. P. Sinha, *Complexes of the Rare Earths*, Pergamon, London, 1966, p. 36.
87. V. A. Golovnya and O. M. Ivanova, *Russ. J. Inorg. Chem.* (*Eng. Trans.*), *8*, 1290 (1963).
88. I. A. Tserkovnitskaga and A. K. Charykov, *Izv. Vysshikh Uchebn. Zavedenii Khim. i Khim. Tekhnol.*, *7*, 544 (1963).
89. K. N. Kovatenko, D. V. Kazachenko, and O. S. Samsonova, *Russ. J. Inorg. Chem.* (*Eng. Trans.*), *8*, 1163 (1963).
90. J. Selbin, M. Schober, and J. Dortego, *J. Inorg. Nucl. Chem.*, *28*, 1385 (1966).
91. A. K. Molodkin and O. M. Ivanova, *Russ. J. Inorg. Chem.* (*Eng. Trans.*), *12*, 57 (1967).
92. G. A. Skotnikova, *Russ. J. Inorg. Chem.* (*Eng. Trans.*), *11*, 1056 (1967).
93. R. C. Paul, J. S. Ghotra, and M. S. Bains, *J. Inorg. Nucl. Chem.*, *27*, 265 (1965).
94. I. Jelenic, D. Grdenic, and A. Bezjak, *Acta Cryst.*, *17*, 758 (1964).
95. V. Golovnya and K. Shubochkin, *Russ. J. Inorg. Chem.* (*Eng. Trans.*), *8*, 579 (1963).
96. A. Colani, *Bull. Soc. Chem.*, *41*, 1291 (1912).
97. I. Fankuchen, *Z. Krist.*, *91*, 473 (1935).
98. J. K. Dawson, *J. Chem. Soc.*, *1952*, 2705.
99. B. Sahoo, *Indian J. Chem.*, *2*, 75 (1964).
100. Y. Maeda and R. Olcawara, *J. Organomet. Chem.*, *10*, 247 (1967).
101. W. Kitching and P. R. Wells, *Aust. J. Chem.*, *20*, 2029 (1967).
102. R. Feld and P. L. Cowe, *The Organic Chemistry of Titanium*, Butterworths, London, 1965, p. 81.
103. W. Gerrard and M. A. Wheelans, *Chem. Ind.* (*London*), *1954*, 758.
104. K. C. Pande and R. C. Mehrotra, *Z. Anorg. Allgem. Chem.*, *290*, 95 (1957).
105. K. C. Pande and R. C. Mehrotra, *J. Prakt. Chem.*, *5*, 101 (1957).
106. R. C. Mehrotra, *J. Indian Chem. Soc.*, *38*, 509 (1961).
107. K. C. Pande and R. C. Mehrotra, *Chem. Ind.* (*London*), *1957*, 114.
108. S. Prasad and R. C. Srivastava, *J. Indian Chem. Soc.*, *39*, 9 (1962).
109. K. Javra, H. Banga, and R. L. Kavshik, *J. Indian Chem. Soc.*, *39*, 531 (1962).
110. O. Osipov and Y. A. Lysenko, *Zh. Neorgan. Khim.*, *5*, 1840 (1960).
111. A. Stahler, *Ber.*, *1911*, 248, 251.
112. E. I. Semenova, *Dokl. Akad. Nauk SSSR*, *143*, 1568 (1962).
113. E. N. Kharlamova, E. N. Gur'yanova, and N. A. Slovakhotora, *Zh. Obshch. Khim.*, *37*, 303 (1967).

114. E. I. Semenova, *Russ. J. Inorg. Chem. (Eng. Trans.)*, *11*, 1156 (1966).
115. G. G. Dvoryantseva, N. A. Lazareva, and V. A. Dubovitskii, *Dokl. Akad. Nauk SSSR*, *161*, 603 (1965).
116. W. Bruger, U.S. Pat. 3, 076, 831, 1963.
117. K. Sono, K. Mori, and C. Kajisaki, *Nagoya Kogyo Gijutsu Shikensho Hokoku*, *5*, 142 (1956).
118. R. N. Kapoor, K. C. Pande, and R. C. Mehrotra, *J. Indian Chem. Soc.*, *35*, 157 (1958).
119. J. R. Dowell, *Adhesives Age*, *7*, 26 (1964).
120. R. C. Mehrotra and R. A. Misra, *J. Chem. Soc.*, *1965*, 43.
121. E. Every, U.S. Pat. 2,481,241, 1949.
122. L. N. Komissarova, M. V. Savel'eva, and V. E. Plyushchev, *Russ. J. Inorg. Chem. (Eng. Trans.)*, *8*, 27 (1963).
123. L. N. Komissarova, V. Spitzyn, and Z. Prozorovskaya, *Russ. J. Inorg. Chem. (Eng. Trans.)*, *11*, 1089 (1966).
124. E. M. Brainina and R. K. Freidlina, *Ivz. Akad. Nauk SSSR, Otd. Khim. Nauk*, *1963*, 756, 835.
125. N. N. Greenwood, P. V. Parish, and P. Thornton, *J. Chem. Soc. A*, *1966*, 320.
126. R. S. P. Coots and P. C. Wailes, *Aust. J. Chem.*, *20*, 1579 (1967).
127. A. Rosenheim, E. Hilzheimer, and I. Wolff, *Z. Anorg. Chem.*, *201*, 162 (1931).
128. P. H. Crayton and R. N. Vance, *J. Inorg. Nucl. Chem.*, *23*, 154 (1961).
129. H. J. Seifert, *J. Inorg. Nucl. Chem.*, *27*, 1269 (1965).
130. D. Nicholls, *Coord. Chem. Rev.*, *1*, 1966, 379.
131. J. Trzebiatowska and L. Pajdowski, *Roczniki Chem.*, *32*, 1061 (1958).
132. H. J. Seifert, *Z. Anorg. Chem.*, *317*, 123 (1962).
133. R. C. Paul and A. Kumer, *J. Inorg. Nucl. Chem.*, *27* 2537 (1965).
134. V. V. Zelentsov, *Russ. J. Inorg. Chem. (Eng. Trans.)*, *1962*, 670.
135. V. V. Zelentsov, V. T. Kalinnikov, and M. N. Volkov, *Zh. Strukt. Khim.*, *6*, 647 (1965).
136. V. T. Kalinnikov, V. V. Zelentsov, M. N. Volkov, and S. M. Shostakovskii, *Dokl. Akad. Nauk SSSR*, *159*, 882 (1964).
137. F. Fairbrother, D. Robinson, and J. B. Taylor, *J. Inorg. Nucl. Chem.*, *8*, 296 (1958).
138. T. F. Burger and M. Zeiss, *Chem. Ind. (London)*, *1962*, 183.
139. L. Peligot, *Ann. Chim. Phys.*, *12*, 528 (1844).
140. M. R. Hatfield, *Inorg. Syn.*, *3*, 149 (1950); *6*, 145 (1950).
141. W. R. King and C. S. Garner, *J. Chem. Phys.*, *18*, 689 (1950).
142. V. S. Herzog and W. Kalies, *Z. Anorg. Allgem. Chem.*, *329*, 83 (1964).
143. A. Earnshaw, L. F. Larkworthy, and K. S. Patel, *Proc. Chem. Soc.*, *1963*, 261.
144. A. Earnshaw, private communication.
145. D. N. Hume and H. W. Stone, *J. Am. Chem. Soc.*, *63*, 1200 (1941).
146. V. S. Herzog and H. Oberender, *Z. Chem.*, *3*, 67 (1963).
147. V. S. Herzog and W. Kalies, *Z. Anorg. Allgem. Chem.*, *351*, 237 (1967).
148. V. S. Herzog, H. Oberender, and S. Paul, *Z. Naturforsch.*, *18B*, 158 (1963).
149. A. Weiss and U. Hofmann, *J. Chem. Educ.*, *37*, 2 (1960).
150. E. Spath, *Monatsh.*, *33*, 242 (1912).
151. R. Weinland and H. Reihlen, *Z. Anorg. Chem.*, *82*, 426 (1913) and references therein.

152. K. Starke, *J. Inorg. Nucl. Chem.*, *13*, 254 (1960).
153. L. A. Welo, *Phil. Mag.*, *6*, 481 (1928).
154. B. N. Figgis and G. B. Robertson, *Nature*, *205*, 694 (1965).
155. L. E. Orgel, *Nature*, *187*, 504 (1960).
156. A. Earnshaw, B. N. Figgis, and J. Lewis, *J. Chem. Soc. A*, *1966*, 1656.
157. J. A. Elvidge and A. B. P. Lever, *J. Chem. Soc.*, *1961*, 1257.
158. H. L. Krauss, *Angew. Chem.*, *70*, 502 (1958).
159. E. W. Abel, A. Singh, and G. Wilkinson, *J. Chem. Soc.*, *1959*, 3097.
160. E. Bannister and G. Wilkinson, *Chem. Ind. (London)*, *1960*, 318.
161. T. A. Stephenson, E. Bannister, and G. Wilkinson, *J. Chem. Soc.*, *1964*, 2538.
162. R. Mason and D. Lawton, *J. Am. Chem. Soc.*, *87*, 921 (1965).
163. M. H. Green, J. K. P. Arlyaratne, A. M. Bjerrum, M. Ishaq, and C. K. Prout, *Chem. Commun.*, *1967*, 430.
164. A. Davidson, W. McFarlane, L. Pratt, and G. Wilkinson, *J. Chem. Soc.*, *1962*, 3653.
165. S. Prasad and K. S. R. Krishnaiah, *J. Indian Chem. Soc.*, *38*, 153 (1961).
166. P. Groth, *Chem. Krystallographie*, Vol. 3, Engielman, Leipzig, 1909.
167. K. Osaki, Y. Nakai, and T. Watanabe, *J. Phys. Soc. (Japan)*, *18*, 919 (1963).
168. R. B. Flippen and S. A. Friedberg, *Phys. Rev.*, *121*, 1591 (1961).
169. H. Abe and H. Morigaki, *Proc. Intern. Conf. Jerusalem*, *2*, 567 (1962).
170. S. K. Siddhauta and S. N. Banerjee, *J. Indian Chem. Soc.*, *35*, 339, 419 (1958).
171. I. A. Cohen and F. Basolo, *Inorg. Chem.*, *3*, 1641 (1964).
172. J. W. Haas and B. M. Schultz, *Physica*, *6*, 481 (1939).
173. M. J. Moitessier, *Compt. Rend.*, *125*, 611 (1897).
174. R. F. Weinland and S. Fischer, *Z. Anorg. Allgem. Chem.*, *120*, 161 (1921).
175. F. Taha and G. Wilkinson, *J. Chem. Soc.*, *1963*, 5406.
176. F. A. Cotton, N. F. Curtis, B. F. G. Johnson, and W. R. Robinson, *Inorg. Chem.*, *4*, 326 (1965).
177. F. A. Cotton, C. Oldham, and W. R. Robinson, *Inorg. Chem.*, *5*, 1798 (1966).
178. F. A. Cotton, W. R. Robinson, and W. K. Bratton, private communication.
179. G. Rouschias and G. Wilkinson, *J. Chem. Soc.*, *A*, *1966*, 465.
180. F. A. Cotton and B. M. Foxman, private communication.
181. A. S. Kotel'nikova and G. A. Vinogradova, *Dokl. Akad. Nauk SSSR*, *152*, 621 (1963).
181a. F. A. Cotton, C. Oldham, and R. A. Walton, *Inorg. Chem.*, *6*, 214 (1967).
182. A. Davidson, W. McFarlane, L. Pratt, and G. Wilkinson, *J. Chem. Soc.*, *1962*, 3653.
183. N. Yryu, *J. Chem. Phys.*, *42*, 235 (1965).
184. G. R. Hoy and F. de S. Barros, *Phys. Rev.*, *139A*, 929 (1965).
185. H. D. Hardt and W. Moeller, *Z. Anorg. Allgem. Chem.*, *313*, 57 (1961).
186. H. Lubwig, *Arch. Pharm.*, *107*, 1 (1861).
187. O. F. Towers, *J. Am. Chem. Soc.*, *32*, 953 (1910).
188. K. Starke, *J. Inorg. Nucl. Chem.*, *25*, 823 (1963).
189. A. Falecki, *Bull. Acad. Caracas*, *1913*, 573.
190. R. F. Weinland, *Einfuhrung in die Chemie der Complex Verbindunger*, Stuttgart, 1919, p. 345.
191. K. Starke, *J. Inorg. Nucl. Chem.*, *13*, 254 (1960).
192. L. Sommer and K. Pliska, *Coll. Czech. Chem. Commun.*, *26*, 2754 (1961).
193. W. Schneider and C. Wekle, *Arch. Pharm.*, *299*, 289 (1966).

194. J. F. Duncan, R. M. Golding, and K. F. Mok, *J. Inorg. Nucl. Chem.*, *28*, 1116 (1966).
195. A. Earnshaw, B. N. Figgis, and J. Lewis, *J. Chem. Soc. A*, *1966*, 1656.
196. D. Sharefield, *J. Inorg. Nucl. Chem.*, *24*, 1014 (1962).
197. A. Rosenheim and P. Muller, *Z. Anorg. Allgem. Chem.*, *39*, 175 (1904).
198. W. Traube, F. Kuhbier, and H. Marting, *Ber.*, *66*, 1545 (1933).
199. A. W. Mond, *J. Chem. Soc.*, *1930*, 1247.
200. F. S. Martin, *J. Chem. Soc.*, *1952*, 2682.
201. T. A. Stephenson and G. Wilkinson, *J. Inorg. Nucl. Chem.*, *28*, 2285 (1966).
202. E. de Bie and P. Doyen, *Cobalt*, *15*, 3 (1962).
203. A. S. Ancyskina, *Acta Cryst.*, *21*, A135 (1966).
204. W. Lossen and G. Voss, *Ann. Chem.*, *266*, 45 (1891).
205. J. N. van Niekerk and F. R. L. Schoening, *Acta Cryst.*, *6*, 609 (1963).
206. D. Benson, P. J. Proll, L. W. Sutcliffe, and J. Walkley, *Discussions Faraday Soc.*, *29*, 60 (1960).
207. S. A. Friedberg and J. T. Schriempf, *J. Appl. Phys.*, *35*, 1000 (1964).
208. A. Mookherjii and S. C. Mather, *Physica*, *31*, 1547 (1965).
209. R. B. Flippen and S. A. Friedberg, *Phys. Rev.*, *121*, 1591 (1961).
210. R. L. Carlin and S. L. Holt, *Inorg. Chem.*, *2*, 849 (1963).
211. E. de Bie and P. Doyen, *Cobalt*, *15*, 3 (1962).
212. T. L. Davis and A. V. Logan, *J. Am. Chem. Soc.*, *1940*, 1276.
213. W. P. Tappmeyer and A. W. Davidson, *Inorg. Chem.*, *2*, 823 (1963).
214. H. V. Kibbel, *Z. Chem.*, *4*, 104 (1964).
215. J. G. Bergman and F. A. Cotton, *Inorg. Chem.*, *5*, 1420 (1966).
216. C. Schall and H. Markgraff, *Trans. Am. Electrochem. Soc.*, *45*, 161 (1924).
217. C. Schall and C. Thiemewiedtmarcker, *Z. Elektrochem.*, *35*, 337 (1929).
218. J. A. Sharp and A. G. White, *J. Chem., Soc.*, *1952*, 110.
219. G. R. Hill, *J. Am. Chem. Soc.*, *71*, 2434 (1949).
220. G. Hargreaves and L. H. Sutcliffe, *Trans. Faraday Soc.*, *51*, 786 (1955).
221. A. P. B. Lever, J. Lewis, and R. Nyholm, *J. Chem. Soc.*, *1962*, 5262.
222. A. S. Foust and R. H. Soderberg, *J. Am. Chem. Soc.*, *89*, 5507 (1967).
223. *Gmelin's Handbuch der Anorganischen Chemie*, Cobalt, *1964*, 443.
224. K. Kuroda and P. S. Gentile, *J. Inorg. Nucl. Chem.*, *27*, 155 (1965).
225. G. Narain, *Can. J. Chem.*, *44*, 895 (1966).
226. K. Kuroda and P. S. Gentile, *Bull. Chem. Soc.* (*Japan*), *38*, 2159 (1965).
227. K. Kuroda and M. Goto, *Bull. Chem. Soc.* (*Japan*), *39*, 197 (1966).
228. I. I. Chernyaev, E. V. Shenderetskaya, L. A. Nazarova, and A. Antsyshkina, Proceedings of the 7th I.C.C.C., Stockholm, 1962.
229. S. A. Johnson, H. R. Hunt, and H. M. Neuman, *Inorg. Chem.*, *2*, 960 (1963).
230. V. Lorenzilli and F. Gesmundo, *Atti. Accad. Naz. Lincei Mem. Classe Fiz. Mat. Nat.*, *36*, 485 (1964).
231. V. G. Winkhaus and P. Zeigler, *Z. Anorg. Allgem. Chem.*, *350*, 51 (1967).
232. M. A. Porai-Koshits and A. S. Antsyshkina, *Dokl. Akad. Nauk SSSR*, *142*, 1102 (1962).
233. I. I. Chernyaev, E. V. Shenderetskaya, A. G. Maiorova, and A. A. Koryagina, *Russ. J. Inorg. Chem.*, *1965*, 290.
234. I. I. Chernyaev, E. V. Shenderetskaya, A. G. Maiorova, and A. A. Koryagina, *Russ. J. Inorg. Chem.*, *11*, 1383, 1387 (1966).
235. J. Chatt and L. M. Venanzi, *J. Chem. Soc.*, *1957*, 4735.

236. L. M. Dirkareva, *Acta Cryst.*, *21*, A140 (1966).
237. R. H. Prince and K. A. Raspin, *Chem. Commun.*, *1966*, 1556.
238. L. Malatesta, M. Angoletta, A. Araneo, and F. Canziani, *Angew. Chem.*, *73*, 273 (1961).
239. F. Manacelli, F. Basolo, and R. G. Pearson, *J. Inorg. Nucl. Chem.*, *24*, 1241 (1962).
240. K. Korgmann and R. Mattes, *Z. Krist.*, *118*, 291 (1963).
241. A. Mookcherji, *Indian J. Phys.*, *9*, 205 (1946).
242. S. Bhatnager, M. Khanna, and M. Nevgi, *Phil. Mag.*, *25*, 234 (1938).
243. C. J. Ballhausen and A. D. Liehr, *J. Am. Chem. Soc.*, *81*, 538 (1959).
244. H. D. Hardt and H. Pohlmann, *Z. Anorg. Allgem. Chem.*, *343*, 92 (1966).
245. M. Shupin, S. D. Robinson, and B. L. Shaw, Proceedings of the 8th I.C.C.C., Vienna, 1964, 223.
246. R. Mason and M. R. Churchill, *Nature*, *204*, 777 (1964).
247. T. A. Stephenson, S. M. Morehouse, A. R. Powell, J. P. Heffer, and G. Wilkinson, *J. Chem. Soc. A*, *1965*, 3632.
248. T. A. Stephenson and G. Wilkinson, *J. Inorg. Nucl. Chem.*, *29*, 2122 (1967).
249. C. B. Anderson and B. J. Burreson, *J. Organomet. Chem.*, *7*, 181 (1967).
250. P. Joannis, *Compt. Rend.*, *138*, 1498 (1904).
251. A. Angel, *J. Chem. Soc.*, *89*, 345 (1906).
252. A. Pechard, *Compt. Rend.*, *131*, 504 (1903).
253. A. Angel and A. V. Harcourt, *J. Chem. Soc.*, *81*, 1385 (1902).
254. M. Kato, H. B. Jonassen, and J. C. Fanning, *Chem. Rev.*, *64*, 99 (1964).
255. J. Lewis, *Pure Appl. Chem.*, *10*, 27 (1965).
256. W. E. Hatfield, C. S. Fountain, and R. Whyman, *Inorg. Chem.*, *5*, 1855 (1966).
257. J. Lewis, Y. C. Lin, L. K. Royston, and R. C. Thompson, *J. Chem. Soc.*, *1965*, 6464.
258. C. S. Fountain and W. E. Hatfield, *Inorg. Chem.*, *4*, 1368 (1965).
259. L. S. Foster and C. J. Ballhausen, *Acta Chem. Scand.*, *16*, 1385 (1962).
260. B. N. Figgis and R. L. Martin, *J. Chem. Soc.*, *1956*, 3837.
261. M. L. Tonnet, S. Yamada, and I. G. Ross, *Trans. Faraday Soc.*, *60*, 840 (1964).
262. D. J. Royer, *Inorg. Chem.*, *4*, 1830 (1965).
263. A. E. Hansen and C. J. Ballhausen, *Trans. Faraday Soc.*, *61*, 631 (1965).
264. D. M. L. Goodgame and D. F. Marsham, *J. Chem. Soc. A*, *1966*, 1167.
265. E. Kokot and R. L. Martin, *Inorg. Chem.*, *4*, 1306 (1964).
266. R. D. Gillard, D. M. Harris, and G. Wilkinson, *J. Chem. Soc.*, *1964*, 2838.
267. F. Hanic, D. Stempelova, and K. Harris, *Acta Cryst.*, *17*, 633 (1964).
268. J. Lewis and F. Mabbs, *J. Chem. Soc.*, *1965*, 3894.
269. M. Kishita, M. Inoue, and M. Kubo, *Nippon Kagaku Zasshi*, *84*, 758 (1963).
270. G. Donarvina, *J. Chem. Eng. Data*, *9*, 379 (1964).
271. J. C. Fanning and H. B. Jonassen, U.S. Dept. Commerce, Office of Technical Service, AD 266846.
272. H. Kolzumi, K. Osaki, and T. Watanabe, *J. Phys. Soc. (Japan)*, *18*, 117 (1963).
273. M. Inoue, M. Kishita, and M. Kubo, *Inorg. Chem.*, *3*, 239 (1964).
274. A. Yingst and D. M. McDaniel, *J. Inorg. Nucl. Chem.*, *28*, 2922 (1966).
275. W. E. Hatfield, H. M. McGuire, J. S. Paschal, and R. Whyman, *J. Chem. Soc. A*, *1966*, 1194.
276. B. J. Edmondson and A. P. B. Lever, *Inorg. Chem.*, *4*, 1608 (1965).
277. F. Rudorff, *Ber.*, *21*, 279 (1888).

278. D. A. Langs and C. R. Nare, private communication.
279. K. Okada, M. I. Kay, D. T. Cromer, and I. Almodovar, *J. Chem. Phys.*, *44*, 1648 (1966).
280. W. J. Bonner and J. DeGraw, *J. Chem. Educ.*, *1962*, 639.
281. A. E. Blakelese and J. L. Hoard, *J. Am. Chem. Soc.*, *78*, 3029 (1956).
282. I. R. Beatie and D. Bryce Smith, *Nature*, *179*, 577 (1957).
283. J. Knox and H. R. Will, *J. Chem. Soc.*, *115*, 853 (1919).
284. F. H. MacDougall and M. Allen, *J. Phys. Chem.*, *46*, 730, 736 (1942).
285. F. Weigand, *Z. Angew. Chem.*, *19*, 139 (1906).
286. G. Macarovici and G. Schimdt, *Rev. Roumaine Chem.*, *9*, 693 (1964).
287. M. Bukowska-Strzyzewska, *Roczniki Chem.*, *39*, 507 (1965).
288. L. S. Kravchuk, A. L. Poznyak, and B. Erofeev, *Zh. Strukt. Khim.*, *6*, 645 (1965).
289. V. Nortia, *Suomen Kemistileht.*, *33*, 120 (1960).
290. W. R. May and M. M. Jones, *J. Inorg. Nucl. Chem.*, *24*, 511 (1962).
291. H. Kobayashi and T. Haseda, *J. Phys. Soc. (Japan)*, *18*, 541 (1963).
292. R. B. Flippen and S. A. Friedberg, *J. Phys. Soc.*, *38*, 2652 (1963).
293. A. Mookherjii and S. C. Mather, *Bull. Am. Phys. Soc.*, *10*, 524 (1965).
294. G. Pellizer and G. De Alti, *J. Inorg. Nucl. Chem.*, *29*, 1565 (1967).
295. K. Nakamoto, *Infrared Spectra of Inorganic Coordination Compounds*, Wiley, New York, 1962.
296. L. L. Shevchenko, *Russ. Chem. Rev.*, *32*, 201 (1963).
297. J. D. Donaldson, J. F. Knifton, and S. D. Ross, *Spectrochim. Acta*, *21*, 275 (1965).
298. K. Nakamoto, J. Fujita, S. Tanaka, and M. Kobayashi, *J. Am. Chem. Soc.*, *79*, 4904 (1957).
299. A. V. R. Warrier and P. S. Narayanan, *Spectrochim. Acta*, *23A*, 1061 (1967).
300. K. Nakamoto, Y. Morimoto, and A. E. Martell, *J. Am. Chem. Soc.*, *83*, 4528 (1961).
301. S. Kirschner, *J. Am. Chem. Soc.*, *78*, 2372 (1956).
302. K. J. Eisentraut, *Dissertation Abstr.*, *26*, 53 (1965).
303. Y. Kuroda and M. Kubo, *Spectrochim. Acta*, *23A*, 2779 (1967).
304. I. I. Chernyaev and Z. M. Novozhenyuk, *Russ. J. Inorg. Chem.*, *11*, 1004 (1966).
305. A. I. Grigor'ev, E. G. Pogodilova, and A. V. Novoselova, *Russ. J. Inorg. Chem.*, *10*, 416 (1965).

Absorption Spectra of Crystals Containing Transition Metal Ions

By N. S. Hush and R. J. M. Hobbs

Department of Inorganic Chemistry, University of Bristol, Bristol, England

I. INTRODUCTION

The electronic spectra of transition metal ion complexes have until fairly recently most been studied in solution. [For reviews, see Jørgensen (191,192), Ballhausen (17), Griffith (159), and Dunn (102).] The transitions studied fall into three broad classes, viz:

1. Transitions among the components of the (predominantly) d^n configurations of the central metal ion;

2. Metal-to-ligand or ligand-to-metal charge transfer;

3. Internal excitations of the ligands.

While such a division can only be approximate, it reflects general qualitative differences in the nature of the spectra and in the theoretical methods used in interpreting them. Attention has naturally been most concentrated on the first type of transition, as the fundamentally simple ideas of crystal field theory suffice to interpret the broad features of the spectrum. In a cubic environment, the spectra are usually fitted using a four-parameter model. These parameters are the Condon-Shortley (F_2 and F_4) (73) or Racah (341) electron interaction integrals B and C (341), the one-electron spin–orbit coupling constant ζ, and the crystal field splitting parameter Dq. The first three parameters are sometimes close to those observed for the free ion, but more typically are smaller in magnitude. This is interpreted as evidence of covalent bonding, a notion which has quantitative significance in the interpretation of magnetic data (cf. e.g., ref. 8).

The interpretation of empirical crystal field or ligand field parameters has been discussed in particular by Jørgensen, who has also recently (193) proposed an alternative qualitative interpretation of the cubic splitting constant Dq. In fields of symmetry lower than cubic, additional crystal field or ligand field parameters have to be introduced. Transitions of types *2* and *3*, which imply extensive mixing of orbitals on different atomic centers, can only be discussed in terms of a many centered [usually molecular orbital (MO)] approach.

The information that can be obtained from solution spectra is, however, limited in at least two ways. Firstly, the spectra are necessarily obtained at fairly high temperatures. This means that the bands will be vibrationally broadened, thereby obscuring the fine structure of the transitions and also possibly masking weaker absorptions. To a small extent, this can be alleviated by working with rigid solutions at low temperatures. However, such spectra are complicated by the fact that in a glassy medium, there are normally a large number of different possible orientations of the complex ion, which will lead to environmental line broadening. Secondly, the detailed interpretation of fine structure depends on the possibility of selectively examining light absorption along defined molecular axes, and of studying the changes of band shapes and intensities down to very low temperatures in a well-defined environment.

The simplest way to overcome both these difficulties is to work with single crystals of known structure at temperatures preferably down to 4°K. Complications from resonance interaction between neighboring complexes can, in principle, be avoided by using the technique of homogeneous dilution in a host lattice. The principal kinds of new information that can be obtained in this way include:

1. A much more complete energy spectrum;

2. Detailed interpretation of the electronic structures of ground and excited states;

3. Positive identification of transitions by study of spin–orbit effects and band polarizations;

4. Understanding of the mechanisms of excitation;

5. Understanding of the coupling of electronic and nuclear motion from a study of the vibrational structure.

Interest in these results is not confined to spectroscopists; in particular, *1, 2,* and *5* are of general concern to inorganic chemists working in the transition metal field, as well as to solid state physicists concerned with the magnetic properties of inorganic crystals.

In this review, we present a brief survey of the work in this field, which it is hoped will provide a reasonably complete guide to the kinds of problems that are attacked and the degree of success so far obtained. The body of experimental material is now quite large, and we have space only for very brief comment on the relationship of one particular piece of work to another. Previous reviews of transition-ion crystal spectra have been given by Runciman (350), McClure (230,231), Ferguson (119), and Yamada (449); many aspects are discussed by Jørgensen (191), Ballhausen (17), and Dunn (102). The general theory of $d \rightarrow d$ transitions is outlined in references 17, 159, 191, and 411 and applications to crystal spectra are discussed in detail particularly by McClure. The review by Ferguson (119) is particularly interesting for its discussion of the role of magnetic dipole transitions.

The scope is indicated in Table I. The ions for which crystal absorption spectra data are available, in the indicated oxidation states, are shown in italic. The majority of these known so far occur in the first transition series. This merely reflects the natural trend of the work, rather than any fundamental differences; Jørgensen has emphasized the essential similarity of the spectra of ions of the first, second, and third transition series.

The fundamental problem in spectral analysis is assignment of the observed lines or bands. A necessary condition for this is that the ion site symmetry should be accurately known, and crystal structures are therefore given in detail. In a centrosymmetric environment, $d \to d$ electronic transitions can take place only by magnetic dipole or electric quadrupole mechanisms, both of which obey the parity conservation rule $g \leftrightarrow g$. Electric quadrupole transitions for localized $d \to d$ transitions are estimated to have very low oscillator strengths and have not yet been identified for transition ions. Magnetic dipole oscillator strengths, although several orders of magnitude below those for allowed electric dipole transitions, can be larger. The magnetic dipole oscillator strength is estimated (159) as 10^{-5} to 10^{-6} for the $^3A_2(^3F) \to {}^3T_2(^3F)$ transition of a Ni^{2+} ion in typical cubic environments. If, for this ion, an absorption line corresponding to excitation to one of the multiplet components of $^3T_2(^3F)$ were observed, it would represent a true electronic origin. A $d \to d$ transition in a centrosymmetric environment can, however, also gain intensity through excitation to or from a state containing one quantum of a nontotally symmetrical vibration, sometimes referred to as an "enabling" vibration.* This has the effect of mixing ungerade character into the excited state, and intensity is "stolen" from allowed transitions. In this case, the vibronic transition will have a finite electric dipole moment; this is considered to be the usual source of intensity where there is inversion symmetry at the ion site. In favorable cases, the mechanisms can be distinguished in a definite and simple manner (230,353). In a uniaxial crystal, an axial (a) spectrum is observed with the light directed along the optic axis c. In the orthoaxial spectra, the light may be polarized with the electric vector parallel (π, $\|c$) to the optic axis or perpendicular (σ, $\perp c$) to it. Correspondingly, the magnetic vector is perpendicular to c in the π spectrum and parallel to it in the σ spectrum. If the axial and π spectra coincide, the transition must be magnetic dipole. If the axial and σ spectra coincide, the transition is either electric dipole or electric quadrupole; these two possibilities can be distinguished by further examination of the polarization behavior where the

* For a brief account of the symmetry relationships involved, see ref. 74.

TABLE I. Diagram Showing the Known (75) Positive Oxidation States of the Transition Metal Ions. Those ions for which crystal spectra are known are shown in italic.

| Oxidation state | Transition series | Number of *d* electrons | | | | | | | | | | | Transition series |
		0	1	2	3	4	5	6	7	8	9	10	
I	1						Cr	Mn	Fe	Co	Ni	Cu	1
	2						Mo	Tc	Ru	Rh		Ag	2
	3						W	Re		Ir	Pt	Au	3
II	1	Ca	Sc	Ti	V	Cr	Mn	Fe	*Co*	*Ni*	*Cu*	Zn	1
	2	Sr		Zr	Nb	Mo	Tc	*Ru*	*Rh*	*Pd*	Ag	Cd	2
	3	Ba			Ta	W	Re	*Os*	*Ir*	*Pt*		Hg	3
III	1		*Ti*	*V*	*Cr*	*Mn*	*Fe*	*Co*	*Ni*	Cu			1
	2		Zr	Nb	Mo	Tc	*Ru*	Rh		Ag			2
	3		Hf	Ta	W	Re	*Os*	Ir		Au			3
IV	1	*Ti*	*V*	Cr	Mn	Fe	Co	Ni					1
	2	*Zr*	Nb	Mo	Tc	Ru	Rh	Pd					2
	3	*Hf*	Ta	W	Re	Os	Ir	Pt					3
V	1	V	Cr	Mn	Fe								1
	2	Nb	*Mo*	Tc	Ru								2
	3	Ta	W	Re	Os	Ir	Pt						3
VI	1	Cr	Mn	Fe									1
	2	Mo	Tc	Ru	Rh								2
	3	W	Re	Os	Ir	Pt							3
VII	1	*Mn*											1
	2	Tc	Ru										2
	3	Re	Os										3
VIII	1												1
	2	Ru											2
	3	Os											3

symmetry is high enough, as the components of the electric dipole and quadrupole operators have the transformation properties of the translations (x, y, z) and the polarizability components $(x^2, y^2, z^2, xy, xz, \text{ and } yz)$, respectively. An alternative method of distinguishing between electric dipole and magnetic dipole mechanisms is based on measurement of circular dichroism and the absorption spectrum of one stereoisomer of an optically active complex (227). In the vibronic mechanism, unlike these, a marked temperature dependence of the band intensity is expected. In the simplest theory, the vibronic intensity for a single ground-state enabling vibration of frequency ν is proportional to coth $(h\nu/2kT)$ (17). Values of $\omega (= 2\pi\nu)$ in the range 100–400 cm^{-1} are frequently found.

Where there is no inversion symmetry at the ion site, ungerade character can be acquired by the electronic wavefunctions by direct mixing, e.g., with p functions. This often results in relatively high intensities. It is frequently found that the site symmetry is almost centrosymmetric, but that there is a small component of the crystal potential (e.g., a trigonal component) which formally removes it. The appropriate selection rules are then often quite difficult to determine.

In general, the principal difficulties in interpretation of the crystal spectra of the essentially $d \rightarrow d$ type arise on the one hand from uncertainty about the relative order of importance of spin–orbit splitting and low-symmetry field components, and on the other from lack of knowledge of the *effective* ion site symmetry, without which polarization behavior cannot be predicted. This also complicates the assignment of transitions which are formally spin-forbidden.

It is usually assumed that the electronic states of a complex ion in a crystal are independent of those of neighboring ions—i.e., that the crystal is like an "oriented gas."

For ions substituted in simple host lattices like MgO, it is similarly usually assumed that the states of neighboring MO_6 complexes are independent. This is probably a reasonable assumption for the relatively weak $d \rightarrow d$ transitions, as the excited state interaction energy for nearest neighbors is roughly proportional to the square of the transition moment. However, for more intense transitions of the types *2* and *3*, p. 263 resonance interaction cannot be ignored, and should lead to appreciable Davydov splittings (25), analogous to those observed in organic molecular crystals [cf. Wolf (419)].

In considering transitions in which the electrons are strongly delocalized, a variety of molecular orbital approaches have been followed, references to which will be found throughout the text. Most of these have been of the one-electron Hückel or extended Hückel type, which broadly follow the method of Wolfsberg and Helmholz (420), sometimes with allowance for dependence of the parameters on the degree of orbital occupancy, as in the "self-consistent charge" (SCC) approach (153). It is now generally appreciated that the results of such calculations have only a qualitative significance. Many attempts are being made to deal explicitly with electron interaction, and to develop an approximate self-consistent field approach analogous to that of the Pariser-Parr-Pople method (311,331) for π systems. This work is gathering momentum, and promising results have already been obtained (83).*

In discussing crystal spectra, we have limited the field to *absorption* phenomena. Also, there is space only for incidental reference to solution or powder spectra of any of the systems discussed; in general, such reference is only made when the spectra supply evidence of bands not seen in the crystal spectrum, or when there is a marked difference from the crystal spectrum. Similarly, the results of circular dichroism and electron spin resonance (ESR) studies have been quoted only where these results complement those of the absorption spectra. For a detailed review of ESR studies on transition metal compounds, the reader is referred to McGarvey's review (240).

II. NOTATION AND UNITS

Observed and calculated frequencies are given in wave numbers (cm^{-1}). Extinction coefficients (ϵ) have units of liter mole-cm^{-1}; for some crystals the absorption coefficient (α) is quoted, usually in mm^{-1}. Where it is known, the oscillator strength (f) for an experimental band is given. For experimental data in the Tables, the measurements were made at room temperature unless a specific temperature is given. The symbols ϵ and α are also used for the Koide-Pryce covalence parameter (204) and the Trees angular momentum correction (394), respectively. However, no confusion is likely to arise from this in practice.

Many different terminological conventions are currently used in theoretical discussions of the energy levels. As far as possible, we have used the Mulliken (277) rather than the Bethe (39) notation in classifying electronic and vibronic states. However, where spin–orbit coupling is included, the Bethe notation is retained if it has been used in the original paper.

* It may be useful here to draw attention to the Quantum Chemistry Program Exchange project at Indiana University (supervisor: Dr. Richard W. Counts). This project lists and makes available programs covering a number of types of calculations and useful subroutines.

The electron interaction parameters have been expressed in the Racah notation, and Condon-Shortley parameters, F_2 and F_4, have been recalculated in terms of this ($B = F_2 - 5 F_4$; $C = 35 F_4$). The one-electron spin–orbit coupling constant ζ_d, more briefly ζ, is quoted. Values originally given in terms of λ for the ground state of a d^n ion have been recalculated as $\zeta = 2s \mid \lambda \mid$ where s is the electron spin quantum number for the particular configuration. The cubic field d-orbital splitting parameter is expressed as Dq. The notations for lower-symmetry parameters are somewhat variable, but as far as possible a consistent notation has been used, as is discussed in the text. All energy parameters are in cm^{-1}.

In the following sections we shall discuss the crystal spectra of ions belonging to each of the d^n configurations in each transition group in order of the number of d electrons they contain. Within each configuration, ions will be discussed in increasing order of valence and, as far as possible, complexes will be considered in decreasing order of symmetry at the central ion. For each complex, the crystal structure will be mentioned if it is known; this is followed by the observed absorption bands and the interpretation of spectra given by the author. When there is disagreement over the interpretation of results, these will be discussed in more detail. It should be noted that the crystal structures are nearly all determined at room temperature and it is assumed, unless indicated, that no alteration in the symmetry of the environment of the central ion occurs at lower temperatures. Such phase changes could lead to false interpretations of spectra recorded at low temperatures.

III. CONFIGURATION d^1

A. $3d^1$

In a cubic field, the only LS state ^2D in the free ion splits into the ground state $^2T_{2g}$ and an excited state ^2E_g.

1. $CsTi(SO_4)_2 12H_2O$

The crystal spectrum of $CsTi(SO_4)_2 12H_2O$ has been determined by Hartmann and Schläfer (165). No polarization data were recorded so the crystal structure of the alum will not be discussed in detail. Lipson and Beevers (216) first accurately determined the crystal structure of potassium alum, $KAl(SO_4)_2 12H_2O$. Later Lipson (215) studied the relation between the alum structures concluding that there were three different structures all based upon the space group $Pa3$, $T_h{}^6$, but dependent upon the mono-valent ion contained in the alum. The α structure is typical of medium sized monovalent ions such as $NH_4{}^+$, the β structure of larger ones like Cs^+, and the γ structure of the small Na^+ ion. More recently, Haüssuhl (176) has carried out a large number of crystallographic observations on a wide series of alums including $CsTi(SO_4)_2 12H_2O$, which has a lattice constant of 12.466 Å. The trivalent Ti^{3+} ion is surrounded by six octahedrally-coordinated H_2O molecules. The broad absorption band in the visible region at 20,300 cm^{-1}, $\epsilon = 4$, half width $(\Delta\nu_{1/2}) = 4600$ cm^{-1}, is interpreted as the $^2T_{2g} \rightarrow {}^2E_g$ transition (165). The absorption spectra of the hydrated ions of the first transition series, both in solution and in crystals, have been discussed by Holmes and McClure (182). Since the Ti^{3+} ion and its immediate environment of six water molecules have a center of symmetry, this must be removed by a perturbation before a transition can occur, assuming it is an electric dipole transition. This leads to the band being fairly strongly forbidden and hence to the resulting low extinction coefficient. The broadness of the band may be due to Jahn-Teller distortion of the excited state, but this has not been conclusively established.

2. α-TiCl$_3$, α-TiBr$_3$

The crystal spectra of α-TiCl$_3$ and α-TiBr$_3$ have been determined by Dijkgraaf and Rousseau (91). There is some confusion over the structure of TiCl$_3$, but a study (285) of the different forms of the compound has been carried out and in α-TiCl$_3$ the Ti^{3+} ions are in an octahedron of Cl$^-$ ions. The observed bands are tabulated in Table II.

Measurements with polarized light showed that the bands are isotropically polarized in the (001) plane of the crystals.

The relatively weak absorption bands at 14,300 cm^{-1} of α-TiCl$_3$ and at 12,500 cm^{-1} of α-TiBr$_3$ are assigned to the Laporte forbidden $^2T_{2g} \rightarrow {}^2E_g$ transition. The stronger bands at higher wavelengths are attributed to charge transfer transitions which are Laporte-allowed (91).

TABLE II
Observed Absorption Bands in Crystals of α-TiCl$_3$ and α-TiBr$_3$ (91)

α-TiCl$_3$		α-TiBr$_3$		
Band max., cm^{-1}	OD[a]	Band max., cm^{-1}	OD	Transition
14,300	0.23	12,500	0.41	$^2T_{2g} \rightarrow {}^2E_g$
18,900	1.55	15,500	1.01	
27,500	—	23,300	0.82	

[a] OD = optical density.

3. Ti^{3+} in Al_2O_3

The use of corundum as a transparent crystal host lattice for Ti^{3+} ions has been studied by McClure (233). The crystal structure of corundum was first analyzed by Pauling and Hendricks (312) and has been refined by Newnham and de Haan (287). It has space group $R\bar{3}c$, D_{3d}^6, with six formula units in the hexagonal unit cell, with lattice constants $a_0 = 4.7589$, $c_0 = 12.991$ Å. The metal ions are located on the threefold axis at $\pm(00z; 00\frac{1}{2} + z)$ and the oxygens lie on diads at $\pm(0x\frac{1}{4}; 0x\frac{1}{4}; \bar{x}\bar{x}\frac{1}{4})$ where $z = 0.3520$ and $x = 0.306$. Figure 1 shows a projection of the corundum structure on to the $(2\bar{1}.0)$ plane (287). Pairs of Al atoms are stacked along the c direction, and are surrounded by trigonally distorted octahedra of oxygen (Fig. 2) (142). The two Al atoms of the pair share the equilateral triangle of O atoms lying between them. The Al—O distance in this case is 1.97 Å (R_1—O_1 in Fig. 1). The other triangle of each octahedron is larger and lies closer to the Al atom than does the first. This Al—O distance is 1.86 Å (R_1—O_5 in Fig. 1). Thus the site symmetry of the Al atom is C_{3v} if no other distortion occurs. However the upper and lower triangles are rotated away from the σ_v planes by 2° 8.5′ in opposite directions, thus destroying the σ_v planes, Figure 2. The site symmetry is thus C_3 although it is not far from C_{3v} and the site group C_{3v} may be retained for many purposes.

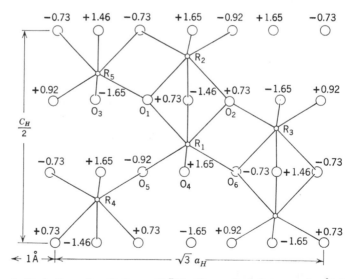

Fig. 1. Projection of α-Al_2O_3 on $(2\bar{1}.0)$. The vertical distances (in Å) from the oxygens to the plane of projection are given in parentheses. The metal ions lie at height zero (287).

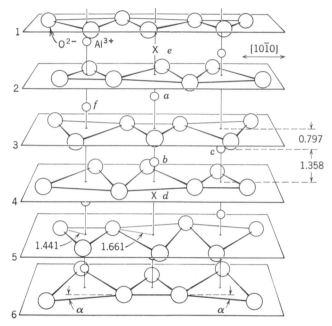

Fig. 2. Portion of the Al_2O_3 lattice. The Al ions are between equally spaced planes of oxygens. They are in octahedral oxygen coordination but the octahedron is severely distorted, the point symmetry at the Al sites being C_3. The predominant crystal field is still cubic, however, with a smaller trigonal component. The angle α is 4° 17′ (142).

The Al^{3+} ions lie on the threefold axis, but their positions on this axis are determined by a parameter (w) and not by the symmetry elements. The parameter is the distance of an Al^{3+} ion from the point midway between the two ions forming the close pair. The value of w is not expected to be the same for an ion substituting for Al^{3+} (233). The symmetry of the force system in the crystal acting on the foreign ion does not permit it to depart from the threefold axis. But the foreign ion itself may acquire an electronic dissymmetry when in a state containing uncompensated e electrons, resulting in a displacement from the threefold axis. Other ions should lie on the threefold axis, but with unknown values of w. McClure has made calculations on the potential function for an impurity ion in this lattice with these considerations in mind (233).

The splitting pattern for a single d electron in the corundum lattice is shown in Figure 3.

The major splitting is by the cubic field component and the triply degenerate $^2T_{2g}$ level is further split by a field of C_{3v} symmetry into two levels with the doubly degenerate E level lying lower for a negative value of

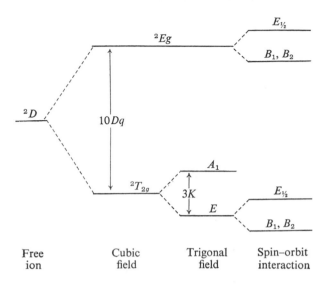

Fig. 3. The splitting pattern for a single d electron in the corundum lattice, where the trigonal field splitting parameter $K < 0$ (412).

the trigonal field parameter K. The optical spectra are expected to be determined by the selection rules for C_{3v} shown in Table III, but strictly by C_3 also shown in the table.

The polarized spectrum at 77°K of Ti^{3+} in Al_2O_3 is shown in Figure 4. The diffuse double-humped band ($\nu_{max} = 18,450 \text{ cm}^{-1}$ and $20,300 \text{ cm}^{-1}$, $Dq = 1905 \text{ cm}^{-1}$) due to $^2T_{2g} \rightarrow {}^2E_g$ is stronger when observed with light polarized parallel to the C_3 axis than with light polarized perpendicular to it. This is compatible only with the assumption that $K < 0$, and that the E component of the $^2T_{2g}$ state lies below the A_1 component. If A_1 were to lie lowest, the z-polarized component would be forbidden in either C_{3v} or C_3, see Table III; however, it is observed to appear strongly. The actual magnitude of K could be found by measuring the absorption spectrum at a series

TABLE III
Electric Dipole Selection Rules for C_3 and C_{3v} Symmetry (233)

C_3		A	A	E	$E_{1/2}$	$A_{3/2}$
	C_{3v}	A_1	A_2	E	$E_{1/2}$	$E_{3/2}$
A	A_1	$\parallel$		$\perp$		
A	A_2		$\parallel$	$\perp$		
E	E	$\perp$	$\perp$	$\parallel + \perp$		
$E_{1/2}$	$E_{1/2}$				$\parallel + \perp$	$\perp$
$A_{3/2}$	$E_{3/2}$				$\perp$	$\parallel$

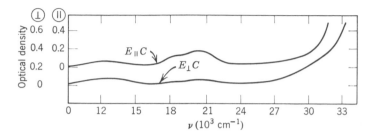

Fig. 4. Absorption spectrum of Ti^{3+} in corundum at 77°K (233).

of temperatures high enough to populate the A_1 component of $^2T_{2g}$. No effect on the polarization ratio was however observed up to 1000°K so it is concluded that $3K$ is less than about -1000 cm^{-1} (412).

The excited state ^2E is not split in a trigonal field, yet a double peak with a separation of 1850 cm^{-1} is observed. The Jahn-Teller theorem (219) states that this level is unstable when a C_3 axis is present, and the double peak could be due to transitions ending on two sheets of the potential surface. The centroid of the absorption band lies at 19,380 cm^{-1}. When this is corrected by adding K where $3K$ is guessed as -1000 cm^{-1}, one finds $Dq = 1905$ cm^{-1} (233).

4. $Ti(H_2O)_6^{3+}$ in Various Trigonal Crystal Lattices

Walker and Carlin (409b) have studied the polarized crystal spectrum of the hexaquotitanium(III) ion in the host crystals $C(NH_2)_3Al(SO_4)_26H_2O$ and $AlCl_3$. The crystal structure of guanidine aluminum sulfate hexahydrate, $C(NH_2)_3Al(SO_4)_26H_2O$, (GASH), is assumed (58b) to be similar to that of $C(NH_2)_3Ga(SO_4)_26H_2O$ (GGaSH), which has been determined by Geller and Booth (139a). It has also been examined by Varfolomeeva, Zhdanov, and Umanskii (408a) but there is considerable evidence (139b) that the trial structure of the Russian workers is inaccurate. The compound crystallizes with a hexagonal unit cell, and space group $P31m$, C_{3v}^2. The lattice constants for GASH are $a_0 = 11.745$ and $c_0 = 8.592$ Å (374c). The hexaquochromium(III) ion has been shown (57c) to replace isormorphously the hexaquoaluminum and hexaquogallium ions in, respectively, guanidinium aluminum and gallium sulfate hexahydrates. The hexaquotitanium ion is presumed to occupy a similar position in the host lattice. The gallium ions in GGaSH are surrounded by somewhat distorted octahedra of water molecules. The trivalent ions lie on threefold sites in the crystal. From symmetry considerations alone, an octahedral threefold axis must be aligned with the C_6 (hexagonal) axis of the crystal. There are three molecules in the unit cell. The placing of atoms in the cell requires that two of

TABLE IIIA

Band Maxima, cm^{-1} of $Ti(H_2O)_6^{3+}$ in Crystalline Hydrates (409b)

System	Excited state	Polarization	300°K	80°K
Ti:GASH	^2E	⊥	17,600; 20,000	17,600; 20,100
		∥	17,600; 20,000	17,600; 20,200
Ti:AlCl₃6H₂O	^2E	⊥	17,700; 19,400	17,700; 20,000
		∥	19,500	20,000

the trivalent ions have C_3 symmetry while one has C_{3v} point symmetry. This has been confirmed by ESR studies of chromic ion in this crystal.

The crystal structure of aluminum chloride has been determined by Andress and Carpenter (1a). The space group is $R\bar{3}c$, D_{3d}^6, with a bimolecular rhombohedral unit cell of dimensions; $a_0 = 7.85$ Å, $\alpha = 97°$. The trivalent Al and Ti ions, each coordinated to six water molecules, occupy the two special threefold positions lying on the long body diagonal of the unit cell. The site symmetry of the trivalent metal ion is predominantly cubic, with a small C_{3i} distortion imposed by the rest of the crystal.

One broad band is observed in the spectrum, the maxima in the two polarizations being shown in Table IIIA, and is assigned to the $^2T_2 \rightarrow {}^2E$ transition. Each band has two components, the splitting for Ti:GASH being ~ 2100 cm^{-1}. Since no splitting of the ^2E state in trigonal fields is expected, the resolution requires a tetragonal splitting in the excited state. The possibility of the lower energy component being a trigonal component of the ground state is dismissed (409b). The σ component of the spectrum is more intense than the π for both systems but lack of complete polarization prevents the sign of $v = -3K$ from being determined. Wong (421a) has reported an isotropic $g = 1.93$ for $Al(Ti)Cl_36H_2O$ which requires $v = -10,000$ cm^{-1} (i.e., $K = +3330$ cm^{-1}).

5. Ti^{3+} in $Al(acac)_3$

Piper and Carlin (330) have studied the polarized visible spectrum of aluminum acetylacetone, $Al(acac)_3$, with part of the Al^{3+} isomorphously replaced by Ti^{3+}. A complete crystal structure of $Al(C_5H_7O_2)_3$ has not been carried out, but studies have been made by Morgan and Drew (267), by Astbury (5), and more recently by Jarrett (187) whose results are quoted here. The monoclinic crystals have space group $P2_1/c, C_{2h}^5$ and lattice parameters $a_0 = 13.86$, $b_0 = 7.54$, $c_0 = 16.17$ Å; and $\beta = 98°$ 50′ with four molecules per unit cell and alternate stacking of right- and left-handed molecules. Since the unique molecular C_3 axis makes angles of $\pm 31°$ with the monoclinic b axis, confirmed from interpretation of ESR measurements (187,371), the crystal spectra had to be analyzed to obtain spectra presumed to be characteristic of the oriented gaseous molecule.

The spectrum was recorded at 77°K (Fig. 5) with light incident on the (001) face and polarized with respect to the b monoclinic axis. The maxima and shoulders at 15,400; 16,700; 18,200; 20,000; and 22,000 cm^{-1} suggest a vibrational progression with spacings of about 1500 cm^{-1}. Upon analysis into the σ and π components (referring to the molecular axis), it is found that less than 10% of the band intensity is in the π component. This is consistent with the selection rules for a $^2A_1 \rightarrow {}^2E_b$ transition, where ^2E_b is the higher of the ^2E levels. Taking into account the ground state splitting Dq is about 1300–1400 cm^{-1}. The further splittings are vibronic in origin, at least in part since Jahn-Teller splittings can give at best only two levels. The spacing of 1500 cm^{-1} suggests that the carbon–oxygen stretch may be excited. The weak lines at 25,010, 25,640, and 26,320 cm^{-1} are π polarized. This shoulder is assigned (330) to either a $\pi_3 \rightarrow \pi_4$ or $n \rightarrow \pi_4$ transition, where π_4 is an excited level of the metal–ligand π system, which would be spin forbidden in the free ligand but is allowed in the complex by interaction

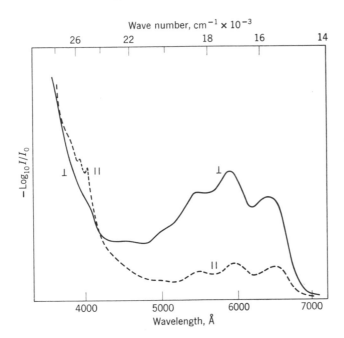

Fig. 5. Crystal spectrum of Al(Ti)(acac)$_3$ at 77°K, with light incident on the (001) face (330).

with the metal states. A trigonal field splitting parameter K of between 700 and 1500 cm^{-1} is given for Ti(acac)$_3$.

Electron spin resonance data obtained by Jarrett (187) on powdered Ti(acac)$_3$ suggested a $^2A_1$ ground state and a trigonal splitting of 13,100 cm^{-1}, $K = 4370$ cm^{-1}. More accurate work carried out by McGarvey (239) on crystals of Al(acac)$_3$ containing some Ti^{3+} were also consistent with a $^2A_1$ ground state but with a trigonal field splitting parameter $3K$ of 2000–4000 cm^{-1}. Confirming the earlier work of Jarrett (187), McGarvey (239) reports g values of $g_{\parallel} = 2.000$ and $g_{\perp} = 1.921$. An excellent review of the experimental work on the EPR spectra of Ti^{3+} in corundum, cesium titanium alum, titanium acetylacetone, and titanium-doped aluminum chloride has been given by Gladney and Swalen (150). For a single electron in a trigonal environment, the complete ligand field Hamiltonian, including the interactions with the upper states of the cubic splitting, was diagonalized in order to reexamine critically the theory of the EPR spectra of the various Ti^{3+} compounds.

6. Ti^{3+} in NaMgAl(C$_2$O$_4$)$_3$9H$_2$O

Piper and Carlin (325) have studied the polarized crystal absorption spectrum of (CH$_3$NH$_3$)$_2$NaAl(C$_2$O$_4$)$_3$H$_2$O with part of the Al^{3+} isomorphously replaced by Ti^{3+}. These crystals are monoclinic and may be grown from solutions containing a large excess of oxalate ion. Great difficulty was encountered in growing crystals of NaMgAl(C$_2$O$_4$)$_3$9H$_2$O containing small amounts of Ti^{3+}, because of air oxidation of Ti^{3+} and because it was not possible to increase the free oxalate concentration without precipitating Mg(C$_2$O$_4$). The crystal structure of NaMgAl-(C$_2$O$_4$)$_3$8H$_2$O has been determined (136), and will be discussed under d^2 configurations in connection with replacement of the Al^{3+} by V^{3+}.

Axial crystals of NaMgAl(C$_2$O$_4$)$_3$9H$_2$O containing small amounts of Ti^{3+} show a maximum at 24,000 cm^{-1}, $\Delta\nu_{1/2} = 5,600$ cm^{-1}. No crystal was grown large enough so that σ and π polarized spectra could be determined, but a smaller crystal was examined under a microscope and was found to be colorless when the light was polarized parallel to the trigonal axis. Piper and Carlin take this to be good evidence that: (1) the Ti^{3+} is present as a trisoxalate complex, (2) the transition is $^2A_1 \rightarrow {}^2E_b$, and (3) K, the trigonal field-splitting parameter, is positive. No evidence of a band in the near infrared down to 7000 cm^{-1} for $^2E_a(^2T_{2g})$ was found.

The crystal structure of (CH$_3$NH$_3$)$_2$NaAl(C$_2$O$_4$)$_3$H$_2$O has not been determined but the field is nearly trigonal and the threefold axes of all the trisoxalate complex ions are almost parallel (325). The spectra of the monoclinic crystal show the expected polarization behavior of the band at

24,000 cm^{-1}, with the σ polarization more intense than the π polarization component, leading to a ground state of $^2A_1$.

7. VOSO$_4$5H$_2$O

Ballhausen and Gray (16) have studied the polarized crystal spectrum of vanadyl sulfate, VOSO$_4$5H$_2$O, as part of a comprehensive analysis of the electronic spectrum of the vanadyl ion. The structure of VOSO$_5$5H$_2$O (299) is a distorted octahedron, which contains the VO^{2+} group situated perpendicular to a base containing the four water oxygens, but although the atoms have been placed in the general positions of $P2_1/c, C_{2h}^5$, this is incompletely supported by x-ray data. The substance is monoclinic with a tetramolecular cell of dimensions $a_0 = 7.06$, $b_0 = 9.71$, $c_0 = 13.02$ Å, and $\beta = 111°\ 23'$. This arrangement gives a V—O bond length for VO^{2+} as 1.67 Å, while the V—O bond lengths to the water ligands range from 2.03 to 2.31 Å. A sulfate oxygen completes the tetragonal structure by occupying the other axial position (V—O = 1.85 Å) (423). The V atom is coplanar with the water oxygens.

The crystal field energy level diagram for the hydrated vanadyl ion situated in a tetragonal crystal field arising from the oxide ion and five water dipoles is given in Figure 6. The axial sulfate oxygen present in the crystal is replaced with a water oxygen, but such a substitution does not affect any of the energy states of interest (16).

The parameters D_s and D_t specify the degree of tetragonality present in the field (266).

The electronic absorption spectrum for a single crystal of VOSO$_4$5H$_2$O was determined for light polarized both parallel and perpendicular to the

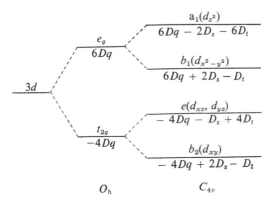

Fig. 6. d-orbital levels in crystalline fields of O_h and compressed C_{4v} symmetry, with $(-3D_s + 5D_t) > 0$ (16).

TABLE IV

Absorption Band Maxima for a Single Crystal of $VOSO_45H_2O$ (16)

	Transition	Predicted energy, cm^{-1} (polarization)	Observed energy, cm^{-1} (polarization)	$f \times 10^4$ Predicted	Obs.
	$^2B_2 \rightarrow {}^2E(I)$	12,502 ($\perp$)	13,000 ($\perp$)	3.9	1.1
	$^2B_2 \rightarrow {}^2B_1$	18,794	16,000	Vibronic	0.45
	$^2B_2 \rightarrow {}^2A_1$	44,766	(Masked)	Vibronic	—
Charge transfer bands	$^2B_2 \rightarrow {}^2E(II)$	38,800 ($\perp$)	41,700 ($\perp$)	26.4	50.3
	$^2B_2 \rightarrow {}^2B_2$	44,000 ($\parallel$)	50,000	44.7	150

molecular VO axis. The positions of the absorption maxima are tabulated in Table IV together with the calculated positions using molecular orbital theory.

The observation of the 13,000 and 41,700 bands appearing principally in the π polarization provided a convincing confirmation of the band assignments. It was not possible to obtain an accurate spectrum above 45,000 cm^{-1} so the predicted $\parallel$ polarization of the $^2B_2 \rightarrow {}^2B_2$ band could not be checked.

The position of the bands in an aqueous solution of $VOSO_45H_2O$ was almost the same and so it was concluded that the solution structure at least contains the VO^{2+} entity.

8. $VOCl_5{}^{3-}$ in $K_3TlCl_62H_2O$

The polarized crystal spectrum of the $VOCl_5{}^{3-}$ ion in the host crystal $K_3TlCl_62H_2O$ has been determined by Wentworth and Piper (413). This compound is tetragonal, space group $14/mmn, D_{4h}{}^{17}$, with a large unit cell containing 14 molecules, $a_0 = 15.841$ and $c_0 = 18.005$ Å (181). Two Tl atoms are at sites of D_{4h} symmetry, four of D_{2h}, and eight of C_{2h} symmetry. The principal sites of C_{2h} are perpendicular to the unique axis. The blue crystals containing the $VOCl_5{}^{3-}$ ion are colorless when viewed in plane-polarized light with the electric vector parallel to the unique axis. When viewed down the unique axis, the crystal is blue and not dichroic. Thus tetragonal selection rules are obeyed, and from the spectrum it is proved that the VO axis is aligned with the fourfold axis in the crystal.

The observed bands and polarizations are given in Table V and shown in Figure 7.

Although it is not known whether the $VOCl_5{}^{3-}$ ions enter the D_{4h} or the D_{2h} sites in the host crystals, the selection rules are essentially unchanged for the lower symmetry site. The ground state is undoubtedly

TABLE V
Absorption Band Maxima for the $VOCl_5^{3-}$ Ion, at 77°K (413)

Band max., cm^{-1} (polarization)	Intensity arbitrary units	
	298°K	77°K
~26,000 ($\perp$, $\parallel$)	$\begin{cases} \text{Masked } \perp \\ \text{Masked } \parallel \end{cases}$	0.4 $\perp$ 0.1 $\parallel$
~15,800 ($\perp$)		
15,500 ($\perp$)	$\begin{cases} 2.03 \perp \\ 0.25 \parallel \end{cases}$	2.02 $\perp$ 0.055 $\parallel$
~14,500 ($\perp$)		
16,200 ($\parallel$)	0.17 $\parallel$	0.090 $\parallel$

$(xy)B_2$ (see Fig. 6) and the 15,500 band $\perp$ polarized is assigned to $(xy)B_2 \rightarrow (xz,yz)E_2$. This assignment is confirmed as purely electronic in origin by the temperature independence of the $\perp$ polarized component of the band intensity. However, the intensity of the $\parallel$ component is temperature-dependent and is ascribed to vibronic intensity of the transition $xy \xrightarrow{\parallel} xz,yz$. At 77°K this intensity decreases considerably and a new absorption

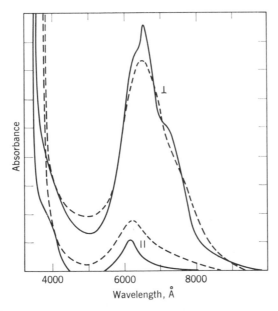

Fig. 7. Polarized crystal spectra of the $VOCl_5^{3-}$ anion in $K_3TlCl_6 2H_2O$. Incident light perpendicular to the tetragonal axis. Dashed lines: 25°C; solid lines: 77°K (413).

is evident at 16,200 cm^{-1}. This band is assigned to $B_2(xy) \to B_1(x^2 - y^2)$. The transitions $xy \xrightarrow{\shortparallel} xz$, yz, and $x^2 - y^2$ are not allowed by the electric dipole mechanism in C_{4v} symmetry and gain in intensity from the vibronic mechanism or possibly spin–orbit mixing. Dq is thus equal to 1620 cm^{-1}. Placing the $xy \to 3_{z^2-r^2}$ (A_1) transition at 38,000 cm^{-1} requires $D_s = -5330$ cm^{-1} and $D_t = -100$ cm^{-1} (413). This large difference is not expected and can only be offset by putting this transition some 50,000 cm^{-1} above the ground state. Crystal field theory is thus in reasonable accord with the MO model (16) which places this level at 45,000 cm^{-1}. The observed shoulder at 26,000 cm^{-1} is therefore considered tentatively to be a charge-transfer band but, in view of the estimated intensities of the shoulder, may be attributed to a spin-forbidden transition to an excited quartet state. Work by Ortolano, Selbin, and McGlynn (294) on KBr disks of $(NH_4)_3$-VOF_5 and $(NH_4)_3VO(NCS)_5$ is in total disagreement with these assignments for VOX_5^{3-} where all the $d \to d$ transitions are assigned to bands below 20,000 cm^{-1}. This latter assignment requires a large positive value of D_t in the region of $+1000$ cm^{-1} which is in complete contradiction with MO and crystal field theory. For example, the MO calculation (16) for $VO(H_2O)_5^{2+}$ can be interpreted in terms of the ionic model with Dq, Ds, and D_t, equal to 1880, -5670, and -900 cm^{-1}, respectively.

9. VO(acac)$_2$

Basu, Yeranos, and Belford (27) have studied the crystal spectrum of vanadyl bisacetylacetone at 90°K. The crystal structure of this compound has been determined by Dodge, Templeton, and Zalkin (100), Figure 8. The crystals are triclinic, space group $P\bar{1}$, with $a_0 = 7.53$, $b_0 = 8.23$, $c_0 = 11.24$ Å, $\alpha = 73°0'$, $\beta = 71°18'$, and $\gamma = 66°36'$. The structure consists of discrete molecules of $VO(CH_3COCHCOCH_3)_2$. Each vanadium has five oxygen neighbors at the corners of a rectangular (nearly square) pyramid with vanadium near its center of gravity. The V—O distances are 1.56 Å to the apex oxygen, and 1.96, 1.96, 1.97, and 1.98 Å for the others. Each acetylacetone is planar, and this plane makes an angle of 163° with the plane of the other acetylacetone skeleton of the same molecule. There are two orientations of the molecules, one the inverse of the other, and so polarized spectral measurements would be of interest. The molecule would have symmetry C_{2v} if the chemically equivalent angles were equal instead of being nearly equal.

The structural details of several acetylacetonate complexes of transition metals have been collected and examined by Lingafelter (213a). It was found that the metal atom may deviate from the plane of the chelate

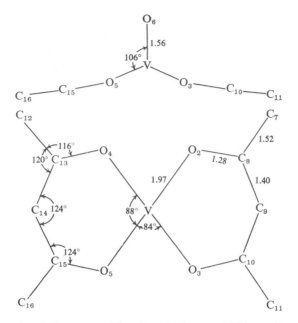

Fig. 8. Average bond distances and bond angles in vanadyl bisacetylacetone (100).

ring by as much as 0.7 Å and that the O—O distance of the acetylacetonate ring shows considerable variation, apparently decreasing with increasing oxidation state of the metal. For C_{2v} symmetry, the number of $d \rightarrow d$ bands expected is four (362) regardless of whether a simple crystal field or more detailed MO model is used, as all degeneracy is removed from the five d orbitals of the vanadium in C_{2v} giving rise to four possible $d \rightarrow d$ transitions of the lone d electron in V(IV), Figure 9.

In spite of the crystal structure being known, no polarized spectral measurements were made by Basu et al. (27). Light was shone normal to

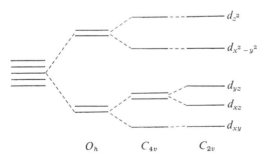

Fig. 9. Arrangement of d levels in C_{2v} symmetry (362).

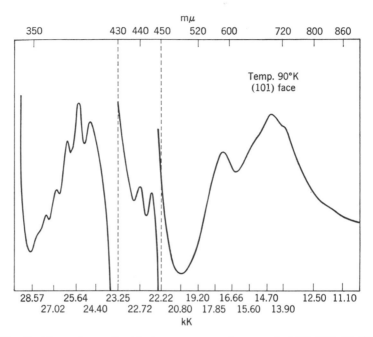

Fig. 10. Absorption at liquid air temperature of a crystal of VO(CH₃COCHCO-CH₃)₂. Spectra in the far blue and ultraviolet are arbitrarily shifted down and the interval 430–450 mμ is shown on an expanded wavelength scale to show details. Note 1 kK = 1000 cm⁻¹ = 3 × 10¹³ sec⁻¹. This face contains the b axis; otherwise its identification is tentative (27).

the (001) face (which contains the b axis). Absorption occurs in three regions of the spectrum (see Fig. 10). In the first region at about 25,000 cm⁻¹, a structure appears which is interpreted as a progression of a vibrational mode of frequency about 700 cm⁻¹, assigned to the vanadyl stretching frequency which in the ground state occurs at about 990 cm⁻¹ (281).

Yeranos (451) has studied the crystal spectrum of vanadyl bisacetylacetone at various temperatures and, assuming an electrostatic model, has assigned the three main transitions in the visible region to (1) $d_{xy} \rightarrow d_{xz}$ (12,100 cm⁻¹), (2) $d_{xy} \rightarrow d_{yz}$ (14,300 cm⁻¹), (3) $d_{xy} \rightarrow d_{z^2}$ (17,300 cm⁻¹). The position of the $d_{xy} \rightarrow d_{x^2 - y^2}$ band has been tentatively assigned to lie between 14,300 cm⁻¹ and 17,300 cm⁻¹. A semiempirical MO calculation in the C_{4v} approximation was attempted but was not found adequate to explain the observed electronic transitions. Resolution of the band near 14,000 cm⁻¹ into 3 components has been reported by Selbin, Ortolano, and Smith (361) when studying the spectrum of VO(acac)₂ in organic glasses at 77°K. This has been interpreted (27) together with the crystal spectrum as vibrational structure, as there is no unequivocal evidence of electronic splitting. A remarkable feature of the spectrum is the two sharp bands at 23,000 cm^{-1} split by about 350 cm^{-1}. The following possibilities are discussed by Basu et al. to explain this absorption: (1) components of a spin-allowed vanadyl band, (2) spin-forbidden acetylacetone to VO charge transfer, or (3) internal ligand bands.

Selbin, Maus, and Johnson (362b) have made spectral studies of six β-ketoenolate complexes of VO^{2+} and on the basis of these results now believe the relative ordering of the d orbitals to be $d_{xy} < d_{x^2-y^2} < d_{xz}$, $d_{yz} < d_{z^2}$. The observed (27) bands in VO(acac)$_2$ are reassigned as 12,100 cm^{-1}, $d_{xz} \to d_{x^2-y^2}$; 13,900 cm^{-1} sh, 14,500 cm^{-1}, 15,100 cm^{-1} sh, $d_{xy} \to d_{xz}, d_{yz}$; 17,400 cm^{-1}, $d_{xy} \to d_{yz}$.

No unambiguous assignments have been made for the electronic transitions in oxovanadium complexes but Selbin (362a) has reviewed the recent work on this species, including ESR and optical spectral data.

B. $4d^1$

1. MoOX$_5$ in Various Crystal Lattices

Although it was found that there is a unique alignment of VOCl$_5$$^{3-}$ ions in K$_3$TlCl$_6$2H$_2$O, random orientation occurred for MoOX$_5$$^{3-}$ ions in K$_2$SnF$_6$H$_2$O, (NH$_4$)$_2$In(H$_2$O)Cl$_5$, and (NH$_4$)$_2$In(H$_2$O)Br$_5$. Studies by Wentworth and Piper (413) on these crystals therefore gave no meaningful polarization data, but the spectra were recorded of the ions, shown in Table VI.

The 22,600 cm^{-1} band for MoOCl$_5$$^{3-}$ is assigned to the $B_1(xy) \to B_2(x^2 - y^2)$ transition, giving Dq a value of 2260 cm^{-1}. The spectra of molybdenyl complexes have been interpreted somewhat differently (151) from vanadyl in that the $xy \to x^2 - y^2$ transition is placed at about 23,000 cm^{-1}, while the $xy \to 3_{z^2-r^2}$ transition must occur at considerably higher energy.

TABLE VI
Absorption Band Maxima for MoOX$_5$$^{3-}$ Ions at 77°K (413)

MoOF$_5$$^{3-}$ in K$_2$SnF$_6$H$_2$O, cm^{-1}	MoOCl$_5$$^{3-}$ in (NH$_4$)$_2$In(H$_2$O)Cl$_5$, cm^{-1}	MoOBr$_5$$^{3-}$ in (NH$_4$)$_2$In(H$_2$O)Br$_5$, cm^{-1}
22,000	22,600	17,050
21,200	15,150	16,200
13,100	14,360	15,300
12,300	13,560	14,400
		13,600

IV. CONFIGURATION d^2

A. $3d^2$

The theory of the d^2 and d^8 configurations in cubic crystal fields, including spin–orbit interaction, has been worked out in detail by Liehr and Ballhausen (210). Numerical solutions were obtained for useful ranges of three parameters, namely F_2, Dq, and λ; it was assumed that $F_2 = 14F_4$. The Tanabe-Sugano (386) energy diagram for d^2 configurations in a cubic field is given in Figure 11. The ordinate in the diagram is energy in units of B (taken to be 860 cm^{-1} for V^{3+}). The abscissa is $10Dq/B$. The octahedral spin-allowed $d \rightarrow d$ transitions are then in increasing order of energy, for $10Dq/B > 13.7$, $^3T_{1g} \rightarrow {}^3T_{2g}$, $^3T_{1g} \rightarrow {}^3T_{1g}(^3P)$, $^3T_{1g} \rightarrow {}^3A_{2g}$.

Theoretical calculations for fields of symmetry lower than cubic have been made by Ballhausen and Jørgensen (9) and Maki (241) with neglect of spin–orbit interaction. The theory for d^2 in a trigonal field was worked out by Pryce and Runciman (339), and by Macfarlane (237) including spin–

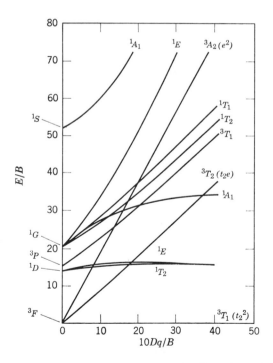

Fig. 11. Tanabe-Sugano energy diagram for the d^2 configuration (75,386).

orbit interaction, in connection with investigations of V^{3+} in corundum. These latter calculations will be discussed in more detail in the appropriate section.

1. $CsV(SO_4)_2 12H_2O$

Hartmann and Schläfer (166,167) have studied crystal spectra of the alums $NH_4V(SO_4)_2 12H_2O$ and $CsV(SO_4)_2 12H_2O$, in which the trivalent vanadium ion is surrounded by a nearly regular octahedron of water molecules (215,216). Haussühl (176) has determined the unit cell size for $NH_4V(SO_4)_2 12H_2O$ as 12.337 Å and for $CsV(SO_4)_2 12H_2O$ as 12.438 Å in the space group $Pa3, T_h^6$. The absorption band maxima and assignments (166,167) are given in Table VII.

The onset of very strong short-wavelength absorption prevents the highest crystal field band from being observed. The strong absorption undoubtedly corresponds to a charge-transfer process involving the surrounding molecules, and Hartmann (168) has pointed out that its onset decreases in a series of $V(X)_6^{3+}$ complexes with decreasing ionization potential of X.

Owen (295) in an earlier paper analyzed the available optical absorption data for octahedral and tetragonal $M(H_2O)_6$ complexes, including d^2, using an ionic model. There is fairly good agreement with the predictions of crystal field theory but in order to extend the correlation to magnetic properties, it was necessary to use a more detailed ligand field model. Only the principal features of the spectra were discussed. For further discussion of d^2 $M(H_2O)_6$ complexes, see references 17 and 191.

TABLE VII
Band Maxima for the Alums $NH_4V(SO_4)_2 12H_2O$ and $CsV(SO_4)_2 12H_2O$ (166,167)

Compound	Band max., cm^{-1}	ϵ_{max}	$\Delta\nu_{\frac{1}{2}}, cm^{-1}$	Transition
$NH_4V(SO_4)_2 12H_2O$	17,800	3.5	3200	$^3T_1 \rightarrow {}^3T_2$
	25,700	6.6	3300	$^3T_1 \rightarrow {}^3T_1(^3P)$
$CsV(SO_4)_2 12H_2O$	17,800	4.7		$^3T_1 \rightarrow {}^3T_2$
	25,700	6.6		$^3T_1 \rightarrow {}^3T_1(^3P)$

2. V^{3+} in Al_2O_3

The spectrum of V^{3+} in a corundum lattice has been studied by several groups of workers, and the problems of interpretation are typical for transition ions in crystal lattices. At low temperature, it is possible to resolve further structure; together with observed polarizations, this provides additional information about the nature of the transitions involved. The basic problem is to associate a given "line" with a particular calculated transition. The work of McClure (233) will be discussed first, as it brings out clearly many of the general features.

The crystal structure and site symmetry of a guest ion in the corundum lattice was described on p. 270. A schematic crystal field energy level diagram for the states arising from ^3F and ^3P of $(nd)^2$ (spin–orbit splitting omitted except for the lowest $^3A_2$ level) in a trigonal field is shown in Figure 12.

The ground state in an octahedral field is $^3T_{1g}$, (e^2 in a strong-field scheme) and in a corundum lattice the trigonal field forces the $^3A_2$ component below ^3E by an amount equal to the trigonal field splitting parameter $3K$. Spin–orbit coupling, mainly between the two trigonal components, causes the $^3A_2$ state to be split by 8 cm^{-1}, with the $S_z = 0$ state lying lowest. This value, calculated initially from the observed Zeeman splitting of the upper $S_z = \pm\frac{1}{2}$ level (339), has since been directly observed in the fine structure of the near-infrared lines (9741, 9748 cm^{-1}) (349). The two excited states arising from the et configuration ($^3T_{2g}$ and $^3T_{1g}$) are split in a C_{3v} field into $^3A_1 + {}^3E$ and $^3A_2 + {}^3E$, respectively. The $^3A_2 \rightarrow {}^3A_1$ transition is forbidden, but in the lower symmetry C_3 becomes an allowed $^3A \rightarrow {}^3A'$ transition. These two states should give rise to absorption bands polarized $\parallel c$ and $\perp c$, respectively, with maxima displaced

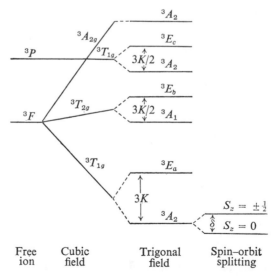

Fig. 12. Energy level diagram for two d electrons in an octahedral field with a smaller trigonal component (412).

from each other by $3K/2$ in the approximate trigonal field scheme of Figure 12, the A component lying lower when $K < 0$. The remaining triplet level from ^3F is $^3A_{2g}$, arising from the e^2 configuration. The intensity of this transition must be due to configuration interaction as it involves a two-electron jump. Originally, Ballhausen (11) considered that the transition to the $^3A_{2g}$ level would not be observed as a result of the small transition probability, but he subsequently (12) suggested that a source of configuration interaction is coupling between $^3A_2[^3T_{1g}(^3P)]$ and $^3A_2(^3A_{2g})$ in the trigonal field. The optical spectra should be determined by the selection rules for C_3 site symmetry as given under d^1 configurations, Table III, if vibronic effects are not important. In this symmetry (and also in D_3), the symmetry selection rules for electric dipole and magnetic dipole transitions are identical, but the electric dipole component will dominate. The polarized spectrum of V^{3+} in corundum at 77°K is given in Figure 13 (233), and the principal spin-allowed absorption bands are tabulated in Table VIII. The spectrum has also been discussed by Low (222) whose results are also included in Table VIII. The results of Pryce and Runciman (339) are given in more detail later.

The oscillator strength is given by $f = 1.096 \times 10^{11}\nu D$, where D is the dipole strength, the square of the dipole transition moment, and ν is the average frequency of the transition in cm^{-1} (233). f is then given by $f = 0.431 \times 10^{-8} \int \epsilon d\nu$ where ϵ is the molar decadic extinction coefficient.

From Table VIII, the splitting between the two polarizations of the $^3T_{1g}(^3F) \rightarrow {}^3T_{1g}(^3P)$ band is $380\ cm^{-1}$ (233), which gives $K = -250\ cm^{-1}$, but further work (233) on the variation of the intensity of the absorption peaks with temperature leads to a value of $-320\ cm^{-1}$. The $^3T_{2g}$ band is almost as strong in the $\parallel$ polarization as in the $\perp$ which is unexpected and there is no appreciable trigonal splitting.

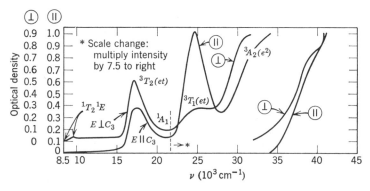

Fig. 13. Absorption spectrum of V^{3+} in corundum at 77°K (233).

TABLE VIII

Observed Absorption Bands for V^{3+} in Corundum at 77°K

($f = 0.431 \times 10^{-8} \int \epsilon d\nu$ where ϵ is the molar decadic extinction coefficient) (233)

Transition	Band max. (222), cm^{-1}	Band max. (233), cm^{-1}	$f \times 10^4$
$^3T_{1g}(^3F) \rightarrow {}^3T_{2g}(^3F)$	17,400	17,510 $\parallel$	0.279
		17,420 $\perp$	0.360
$^3T_{1g}(^3F) \rightarrow {}^3T_{1g}(^3P)$	25,200	24,930 $\parallel$	5.61
		25,310 $\perp$	1.60
$^3T_{1g}(^3F) \rightarrow {}^3A_{2g}(^3F)$	34,500	31,240	
	$Dq = 1870$	$Dq = 1750$	

Analysis of the polarized vibrational structure of this band at 5°K (Fig. 14) suggests that it is tetragonally distorted with the spin–orbit splitting quenched (233). Table IX shows the observed vibrational bands of V^{3+} in corundum.

Agreement between the observed values of the vibrational splitting on the first $^3T_{2g}$ absorption band is good as shown in Table IX. The values

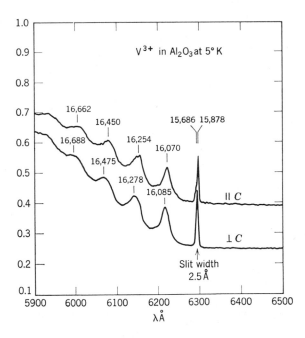

Fig. 14. Vibrational structure of the first strong absorption band of V^{3+} at ca. 5°K (233).

TABLE IX

Observed Vibrational Bands of V^{3+} in Corundum

Ref. 160	Ref. 233, 5°K		Ref. 222, 77°K	Ref. 339, 4.2°K
	‖	⊥		$\begin{cases} 15,880\ \pi \\ 15,890\ \sigma \end{cases}$
15,877	15,878	15,886	15,870	
15,895	16,070	16,085	16,050	16,060
	16,254	16,278	16,260	16,250
	16,450	16,475	16,450	16,420
	16,662	16,668	16,640	16,640
22,321			21,000	—[a]
22,340			25,400	
			29,300	
			30,150	

[a] The vibrational structure of bands in the region is not well resolved, but is qualitatively discussed.

of Low (222) were read from a graph to the nearest $m\mu$ so the small discrepancies between Low's (222) results and those of McClure (233) and Pryce and Runciman (339) are negligible. The vibrational progression of ~ 200 cm^{-1} intervals on the T_{2g} band is tentatively assigned to the 194 cm^{-1} E_u made of the corundum lattice (233).

The absorption in the region of the $^3A_{2g}(^3F)$ state also has unexpected features. It should be a weak transition appearing only in the π spectrum, but unresolved shoulders on a strongly rising background at 31,000 $\perp$ c, and 33,000 ‖ c are observed. Another shoulder occurs at 39,500 cm^{-1} and is more distinct $\perp c$ than ‖c. These bands do not arise from pure $d \rightarrow d$ transitions (233).

The value of the second-order spin–orbit splitting constant was given by Pryce and Runciman (339) as 8 cm^{-1} together with a value of -400 cm^{-1} for K. However, as pointed out in a footnote in the paper by Seed (359) a value of $K = -320$ cm^{-1} gives better agreement with their data. Seed (359) also calculates a value of 8 cm^{-1} for δ when $K = -266$ cm^{-1}.

In a trigonal field with one d electron, the lowest orbital triplet $^3T_{2g}$ is split into a doublet E_2 and singlet A_2 whose separation is $3K$, the sign being positive if the singlet level A_2 lies lowest. The effect of the trigonal field upon an energy level of an ion with more than one d electron can always be expressed in terms of this one-electron parameter, K. The trigonal potential should be fairly sensitive to the position of the metal ion in the octahedral site. For example, a downward displacement of 5% of the Al^{3+} ion in position b in Figure 2 should change the sign of the field gradient. One might therefore interpret any variation found in K for the different transition metal ions in corundum as a measure of their relative displacement from the original Al^{3+} position for which they substitute (142). Geschwind and Remeika (142) found a near constancy in the value of K for Ru^{3+}, Co^{3+}, and V^{3+} and thus assumed that they all enter in essentially the same position as the Al^{3+}. McClure (233), however, finds larger variations in the value of K and interprets this as suggesting displacements of the ions relative to each other. He further indicates that a point charge calculation gives a positive field gradient at the Al^{3+} site, corresponding to a positive K, assuming that the second-order axial potential is dominant. The validity of a point charge calculation here in determining the axial field has been questioned by Pryce and Runciman (339). Geschwind and Remeika (142) have pointed out that in determining K from the dichroism of the excited states, one must allow for theoretical uncertainties connected with spin–orbit coupling and Jahn-Teller effects in excited states which could contribute to the variations in K observed by McClure (233). However, the correct sign for K is obtained if the interstitial ion is displaced 0.1 Å along the C_3 axis toward the empty octahedral site.

A correlation between detailed crystal field calculations (omitting spin–orbit splitting) and experiment for V^{3+} in C_{3v} symmetry was made by

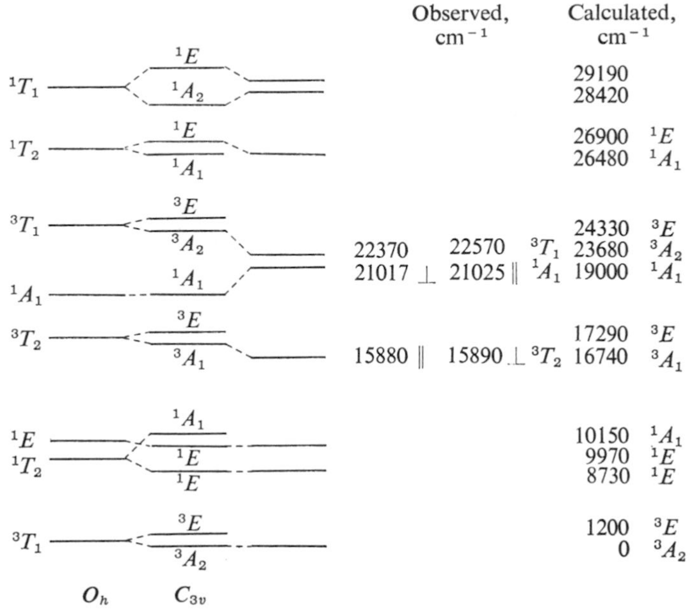

Fig. 15. Energy levels of the d^2 configuration in a trigonal field, showing observed and calculated energies (339).

TABLE X
Energy Levels for V^{3+} in Al_2O_3 at 77°K

Cubic	C_{3v}	Calculated (237), cm^{-1}	Observed, cm^{-1}	Splitting calc.	Splitting obs.
$^3T_{1g}$	$\hat{A}_1$	0			
	$\hat{E}$	7.9	8(339), 7.85 ± 0.4(128)[a]		
	$\hat{E}$	838	810–850 (128)		
	$\hat{E}$	946	960 (233)		
	$\hat{A}_2$	1,032			
	$\hat{A}_1$	1,050			
$^1E_g, {}^1T_{2g}$	$\hat{E}$	9,067	8,770 (339)	900	890
	$\hat{E}$	9,967	9,660 (339)		
	$\hat{A}_1$	10,106	15,890 (339), vib. series $\Delta\nu = 180$		
$^3T_{2g}$	$\hat{A}_2$	17,088			
	$\hat{E}$	17,094			
	$\hat{A}_1$	17,444	17,420 (233)		
	$\hat{A}_2$	17,449	17,510 (233)		
	$\hat{E}$	17,486			
	$\hat{E}$	17,520			
$^1A_{1g}$	$\hat{A}_1$	20,484	21,025 (339)		
$^3T_{1g}$	$\hat{A}_1$	25,020	24,930 (233)		
	$\hat{E}$	25,034			
	$\hat{A}_2$	25,296			
	$\hat{A}_1$	25,322	25,310 (233)	326	380
	$\hat{E}$	25,364			
	$\hat{E}$	25,407			
$^1T_{2g}$	$\hat{A}_1$	27,210			
	$\hat{E}$	27,584			
$^1T_{1g}$	$\hat{A}_2$	29,418	29,300[b]	463	850
	$\hat{E}$	29,881	30,150[b]		
$^3A_{2g}$	$\hat{A}_1$	35,376	34,500[b]		
	$\hat{E}$	35,376			
^1E_g	$\hat{E}$	45,376			

[a] Measured at 4.2°K.

[b] Temperatures at these frequencies not explicitly stated, but presumed to be 77°K as for the longer wavelength measurements as recorded in Fig. 1, ref. 222.

Pryce and Runciman (339), who carried out an extensive investigation of the polarized spectrum of V^{3+} in corundum at 77 and 4.2°K. An attempt was made to obtain electronic origins for the transitions rather than using band maxima, and some of the experimental values in Figure 15 have since been reclassified as vibrational bands (cf. Table IX).

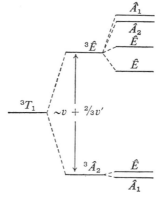

Trigonal   Spin–orbit
field   splitting

Fig. 16. Trigonal field and spin–orbit splitting of the $^3T_1$ ground term of the d^2 configuration in C_{3v} symmetry for the case trigonal splitting constant $\gg \zeta$ (237). Representations of C_{3v} indicated by caret. See text for significance of v and v'.

Higher levels taken from Low (222) give $^3A_2$, 34,640; ^1E, 44,260; and $^1A_1$ 57,170 cm^{-1}.

Macfarlane (237) has made a detailed analysis of the optical and magnetic properties of trivalent vanadium complexes on the basis of the crystal field model. Calculations have included trigonal field and spin–orbit terms. The trigonal and spin–orbit splitting of the $^3T_{1g}$ ground term is shown in Figure 16 for the case when $\Delta_T \gg \zeta$, the spin–orbit coupling constant. The trigonal splitting is calculated in terms of two parameters

TABLE XI

Summary of the Observed and Calculated g Values, Trigonal Field Splitting Parameter K and Spin–Orbit Splitting of the Ground State δ, for V^{3+} in Al_2O_3

$g_\parallel$	$g_\perp$	K, cm^{-1}	δ, cm^{-1}	Observed or calculated	Ref.
			7.85	obs./calc.	128
1.92			10.0	obs.	454
1.915			7.0	obs.	455
	1.72	−367	8.4	obs./calc.	126[a]
	1.63	93		obs./calc.	456
		−290		calc.	142
	1.720	−367	8.4	obs.	55
1.919	1.719	−315	7.9	calc.	237
		−400	8.4	obs.	57

^{a} V^{3+} in V(acac)$_3$.

$v(= -3K)$ and v', an off-diagonal term which was previously neglected.

The calculated energy levels are given in Table X for the following set of parameters: $\Delta = 18,000$, $B = 610$, $C = 2500$, $v = 800$, $v' = 200$, and $\zeta = 155$ cm^{-1} (237).

The paramagnetic resonance of trivalent vanadium complexes, especially V^{3+} in Al$_2$O, have been extensively studied (55,128,142,207,237, 454–456). The paper by Macfarlane (237) gives a crystal field analysis of the optical and magnetic properties of V^{3+} complexes. The calculation of g values, trigonal field and spin–orbit terms is outlined. Table XI gives a summary of the results obtained.

3. V(H$_2$O)$_6$$^{3+}$ in Various Trigonal Lattices

Walker and Carlin (409b) have studied the polarized crystal spectrum of V(H$_2$O)$_6$$^{3+}$ ions in the crystal lattices V:GASH and GVSH. The crystal environment of the V^{3+} ion has been discussed on p. 273. The observed bands and polarizations are listed in Table XIA.

The spectrum of pink V:GASH shows two broad bands in the visible which are assigned as $^3T_1(^3F) \rightarrow$ $^3T_2(^3F)$ at 18,500 cm^{-1} and $^3T_1 \rightarrow$ $^3T_1(^3P)$ at $\sim$26,000 cm^{-1}. Comparison of axial and polarized spectra shows that both bands are electric dipolar in origin. Magnetic susceptibility measurements (58c) of V(III) at low temperatures show that the ^3A state lies below the ^3E state (i.e., $v > 0$). The splitting between the absorption maxima of $^3T_2$ in the two polarizations is only 180 cm^{-1} at 80°K apparently giving $v = 360$ cm^{-1}. Since $\Delta(^3T_1(^3P))$ is 2100 cm^{-1}, this assignment of v requires $v' = -3800$ cm^{-1} which in turn violates the ground-state assignment (409b). A similar situation arose in the analysis of the V^{3+}:Al$_2$O$_3$ spectrum

TABLE XIA

Band Maxima (cm^{-1}) of V(H$_2$O)$_6$$^{3+}$ in Crystalline Hydrates (409b)

System	Excited state	Polarization	300°K	80°K
V:GASH	$^3T_2(^3F)$	$\perp$	18,550	19,230
		$\parallel$	18,520	19,050
	$^3T_1(^3P)$	$\perp$	27,400	27,780
		$\parallel$	25,970	25,640
GVSH	$^3T_2(^3F)$	Axial	18,590	18,960
		$\perp$	18,520	19,050
		$\parallel$	18,430	19,050
	$^3T_1(^3P)$	Axial	27,210	27,860
		$\perp$	27,590	27,780
		$\parallel$	25,970	25,640

(233) and it was concluded that the small splitting of $^3T_2$ may result from distortion into a configuration of low symmetry rather than to the trigonal splitting resulting in intensity transfer between the polarized components. The $^3T_2$ band exhibits a shoulder in parallel polarization at low temperatures, some 2000 cm^{-1} higher in energy, assigned to a tetragonal or rhombic component of the excited state.

The two-parameter scheme of Macfarlane (237) gives for the band splittings in intermediate crystal fields,

$$\Delta(^3T_1(^3F)) = v + \tfrac{2}{3}v'$$
$$\Delta(^3T_2(^3F)) = \tfrac{1}{2}v$$
$$\Delta(^3T_1(^3P)) = \tfrac{1}{2}v - \tfrac{1}{2}v'$$

and $v = +2300$ cm^{-1} and $v' = -2000$ cm^{-1} are chosen as a solution (409b). Support for the assignment of v is gained from the magnetic measurements (58c) where the zero-field splitting parameter D is 7.2 cm^{-1} in GVSH. The value calculated from the equation $D = 0.087\zeta - 0.0064\,v - 0.0036\,v' + 0.001B$ with $\zeta = 160$ cm^{-1} and $B = 650$ cm^{-1} is 6.9 cm^{-1} in fair agreement with the magnetic results (237).

4. V^{3+} in $NaMgAl(C_2O_4)_3 9H_2O$

The polarized crystal spectrum of $V(C_2O_4)_3^{3-}$ has been studied by Piper and Carlin (322,325). Single crystals of $NaMgAl(C_2O_4)_3 9H_2O$ with part of the Al^{3+} isomorphously replaced by V^{3+} were prepared. The characterization of the $NaMgCr(C_2O_4)_3 8H_2O$ was reported by Frossard (136), who established the space group as $P\bar{3}1c$, D_{3d}^2, with two formula units per unit cell. The cell is rhombohedral with $a_0 = 9.78$ and $c_0 = 12.47$ Å. Analysis of the aluminum compound by Piper and Carlin (325) has shown that the compound is a nonahydrate with the six uncoordinated water molecules per unit cell occupying the sixfold sites of symmetry 2. The aluminum ion with three bidentate ligands belongs to the site symmetry D_3. The irreducible representations of this group are A_1, A_2, E. Since a center of inversion is lacking, the $3d$ orbitals can mix with odd-parity

TABLE XII

Electric Dipole Selection Rules for D_3 Symmetry

D_3	A_1	A_2	E
A_1		$\parallel$	$\perp$
A_2	$\parallel$		$\perp$
E	$\perp$	$\perp$	$\parallel$ and $\perp$

atomic and molecular orbitals. Thus the transitions will be Laporte-allowed. The vanadium oxalate ion is assumed to replace the aluminum group and the crystal field at each vanadium ion is predominantly octahedral with a small trigonal component. The selection rules are given in Table XII.

The spectrum observed is shown in Figure 17 and the peak positions are tabulated in Table XIII.

Magnetic data (126) establish that the ground state is the $^3A_2$ component of the $^3T_2$ level. Trigonal splittings were not, however, observed in the solution spectrum and it was concluded that the ligand field is essentially octahedral in solution (169,170).

The crystal spectrum suggests that there is also trigonal splitting of the $^3T_2$ level. At 77°K the data are quite similar to those in Table XIII except that the π component occurs at 16,880 cm^{-1} and the intensities have increased by about 20%. The trigonal splitting of the $^3T_{2g}$ level is 350 cm^{-1} at 77°K (325). In the σ polarization there is an intense tail merging into the charge-transfer region without any maxima below 30,000 cm^{-1}. In the π polarization there are two weak bands centered at 23,500 and 26,300 cm^{-1}. The ratio $\epsilon_\parallel/\epsilon_\perp$ is certainly less than 10^{-2} or 10^{-3}, if $\epsilon_\perp$ is about 500

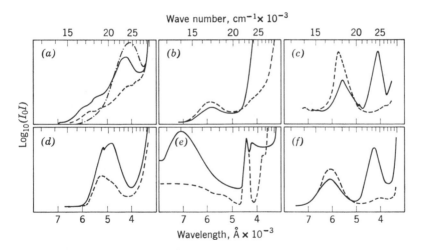

Fig. 17. Orthoaxial spectra of M^{3+} in NaMgAl$(C_2O_4)_3 \cdot 9H_2O$. The solid line is the σ polarization and the dashed line the π polarization. (a) Ti^{3+} at 25°C. In this spectrum only, the host crystal is $(CH_3NH_3)_2NaAl(C_2O_4)_3H_2O$. The dash-dotted line is an axial spectrum in NaMgAl$(C_2O_4)_3$9H$_2$O at 77°K; (b) V^{3+} at 77°K; (c) Cr^{3+} at 77°K; (d) Mn^{3+} at 77°K, calculated from the spectrum with light incident on the $(10\bar{1}1)$ face; (e) Fe^{3+} at 77°K in the pure crystal NaMgFe$(C_2O_4)_3$9H$_2$O; (f) Co^{3+} at 25°C (325).

TABLE XIII

Absorption Band Maxima and Polarization for the $V(C_2O_4)_3^{3-}$ Ion[a] (325)

Band max., cm^{-1}	$\Delta\nu_{1/2}$, cm^{-1}	Transition	Remarks
9,615 σ	250	$^3A_2 \to$ components of $^1E, \ ^1T_{2g}$	77°K
16,670 π[b]	2,500	$\left.\begin{array}{l} ^3A_2 \to {}^3A_1 \\ ^3A_2 \to {}^3E \end{array}\right\} (^3T_{2g})$	$\epsilon_{\parallel}/\epsilon_{\perp} = 1.3$ at 25°K
17,230 σ[b]	2,500		
20,480 π	30	$^3A_2 \to {}^1A_1$	77°K

[a] Axial and $\perp$ band maxima were identical.

[b] The polarization of these bands are reversed in ref. 325 owing to a misprint. We are indebted to Dr. Richard L. Carlin for confirming this point.

as in the solution spectra. The σ absorption is assigned to $^3A_2 \to {}^3E_c$. The transition $^3A_2 \to {}^3A_2$ is forbidden by symmetry but may be allowed by simultaneous excitation of vibrations. The weak absorptions in the π polarization may be due to transitions to $^3A_2(^3A_{2g})$ or possibly to $^1T_1$ and $^1T_2$.

5. V³⁺ in Al(acac)₃

Vanadium has three spin singlets in a trigonal field in the vicinity of 10,000 cm^{-1}. The single asymmetric band at 9615 cm^{-1} appearing in the σ and axial spectra is the only one observed (325). The transition $^3A_2 \to {}^1A_1$ at 20,480 cm^{-1} also only appears in the axial and σ spectra. These observations are consistent with the assumption that the spin forbidden transitions are electric dipole allowed. The trigonal field splitting parameter (K) is -230 cm^{-1}.

Piper and Carlin (330) have also studied the polarized crystal spectra of aluminum acetylacetonate with part of the Al³⁺ isomorphously replaced by V³⁺. The crystal structure and site symmetry of the vanadium ion is discussed on p. 274. The octahedral ground state $^3T_{1g}$ will again be split in the trigonal field into $^3A_2$ and ^3E components. The solution spectrum of V(acac)₃ in ethanol (23) shows no maxima but has shoulders at 18,200 and 21,700 cm^{-1} which are assigned to the $^3T_{1g} \to {}^3T_{2g}(^3F)$ and $^3T_{1g} \to {}^3T_{1g}(^3P)$ transitions. The polarized crystal spectra are rather similar and provide little improvement in the resolution of these bands. The polarization of the ultraviolet tail is similar to that of the trisoxalate. At 77°K the low-energy shoulders appear at 17,500 ($\parallel$) and 18,800 cm^{-1} ($\perp$), which is consistent with a negative trigonal splitting constant K of about -900 cm^{-1}. Thus the ground state is probably $^3A_2$. No absorption was observed in the infrared region near 10,000 cm^{-1} for the singlet levels. These data require a Dq of about 1800 cm^{-1}. The magnetic susceptibility is also only consistent with a $^3A_2$ ground state (126).

V. CONFIGURATION d^3

A. $3d^3$

The d^3 configuration in an octahedral field with a half-filled t_{2g} shell, is especially stable chemically. Reduction produces a state destabilized by electron-pairing energy while oxidation gives a state with less stabilization from the crystal field. The ground state is an orbital singlet, $^4A_{2g}$, in octahedral fields (Fig. 18).

Liehr (212) has carried out complete calculations including spin–orbit and full configuration interaction of the three-electron configuration in a cubic field. More recent calculations by Perumareddi (320a) give the eigenfunctions and energy matrices for trigonal and tetragonal ligand

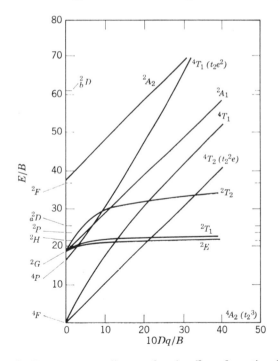

Fig. 18. Tanabe-Sugano energy diagram for the d^3 configuration (75,386).

fields not including spin–orbit interactions. Extensive optical measurements on trigonal (C_{3v}) chromium(III) complexes have been made using polarized light but not on tetragonal (C_{4v}, D_{4h}) systems. Perumareddi (320b) and Krishnamurthy, Schapp, and Perumareddi (205a) have presented complete energy-level diagrams for trigonal (205a) and tetragonal (205a,320b) ligand fields and have attempted to make assignments of the solution spectra on this basis. However, it is suggested (320b) that further polarized spectral data would be required for definitive interpretation of mono and di *trans*-substituted Cr(III) complexes. The D_σ and D_τ parameters used by Perumareddi et al. (205a,320b) are related to the trigonal splitting parameters v and v' of Pryce and Runciman (339) and to K and K' of Sugano and Tanabe (382) by the equations $v' = K' = \frac{2}{3}(3D_\sigma - 5D_\tau); v = 3K = \frac{1}{3}(9D_\sigma + 20D_\tau)$.

1. KCr(SO$_4$)$_2$12H$_2$O

The commonest ion having the d^3 configuration, Cr^{3+}, has received much study. The crystal spectrum of chrome alum, $KCr(SO_4)_2 12H_2O$, has been determined by many authors (146,182,203,286,397). In this crystal the Cr atom is surrounded by six water molecules and their grouping is very nearly octahedral. A neutron diffraction study by Bacon and Gardner (7) has completed the structure first indicated by the x-ray investigations of Lipson and Beevers (216). The angles between the axes of the group differ from 90° by only $2.2 \pm 1°$ and the Cr—O distance is 1.94 ± 0.03 Å. The space group is $Pa3$, T_h^6 and the unit cell has a side of 12.197 Å (176). Table XIV shows the positions of the observed bands in chrome alum.

The value of Dq determined for the alum from the $^4A_{2g} \rightarrow {}^4T_{2g}$ separation is 1760 cm^{-1} and the position of the $^4A_{2g} \rightarrow {}^4T_{1g}(^4F)$ and $^4A_{2g} \rightarrow {}^4T_{1g}(^4P)$ transitions are calculated to be 26,000 and 39,500 cm^{-1}, respectively, in reasonable agreement with the observations (182).

TABLE XIV

Band Maxima for Chrome Alum

Transition	Max., cm^{-1}	$\Delta\nu_{1/2}$	ϵ	$f \times 10^4$	Ref.
$^4A_{2g} \rightarrow {}^2E_g$	14,862 14,902 14,926				146
$^4A_{2g} \rightarrow {}^4T_{2g}$	17,600	3,600	7.8	1.6	182
$^4A_{2g} \rightarrow {}^4T_{1g}(^4F)$	24,700	4,400	10.5	2.2	
$^4A_{2g} \rightarrow {}^4T_{2g}$	17,700				286
$^4A_{2g} \rightarrow {}^4T_{1g}(^4F)$	24,690				
$^4A_{2g} \rightarrow {}^4T_{1g}(^4P)$	38,000				397

The transitions are highly forbidden as the Cr atom has a center of symmetry. The temperature dependence of the absorption of $KCr(SO_4)_2$-$12H_2O$ was studied by Holmes and McClure (182) to learn something of the mechanism of the transition. The band intensity was found to decrease in going from $298 \rightarrow 77°K$, the peak position shifting by 200–500 cm^{-1} toward the violet. This appeared to be greater than could be explained by freezing out ground state vibrations. Analysis of the bands showed that the transition strength must come from an admixture of p character in the predominantly d orbitals. Other factors being equal, the intensity of the bands could be taken as a measure of the p character induced in the electronic states by vibronic coupling (182). The transition strength must come solely from vibrational coupling rather than permanent distortion for electric dipole excitations.

2. Cr^{3+} in Al_2O_3

The absorption spectrum of Cr^{3+} in the corundum host lattice (ruby) has also been much studied (88,161,229,233,250,293,382,383,407). The site symmetry of the interstitial Cr^{3+} ion has been discussed on p. 270.

Macfarlane (236,238) has made an analysis of d^3 states in trigonal crystal fields with particular reference to the spectrum of Cr^{3+} in corundum. Refinement of the work of Sugano and Tanabe (382) and Sugano and Peter (385) leads to better agreement with experiment. The effect of the second trigonal field parameter v' is found to be appreciable. The trigonal splitting of the d^3 quartet levels is shown in Figure 19.

McClure (233) has determined the polarized spectrum of Cr^{3+} in corundum and this is shown in Figure 20. A comparison of the calculated

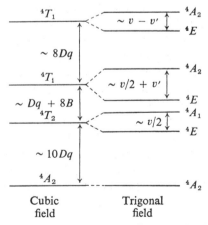

Fig. 19. Trigonal splitting of the d^3 quartet levels (236).

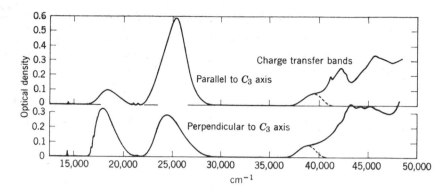

Fig. 20. Absorption spectrum of Cr^{3+} in corundum at 77°K (233).

(236) and observed bands is shown in Table XV for the set of parameters $Dq = 1810$ cm^{-1}, $B = 650$ cm^{-1}, $C = 3120$ cm^{-1}, $v = 800$ cm^{-1}, $v' = 680$ cm^{-1}, and $\zeta = 170$ cm^{-1}. As such complete data are available for the Cr^{3+} in the corundum system, the values of v and v' could be determined from splitting of the $^2T_1$ and $^2T_2$ levels. The observed spectrum is well described by taking $v' = 680$ cm^{-1}. Due to the large value of v' (to fit the lower transition) the calculated splitting of the $^4T_{1g}(^4P)$ is considerably reduced to about 65 cm^{-1}. The discrepancy with observed splittings is not considered serious in view of the difficulties of measurement (236).

The two bands of the transition from $^4A_{2g}$ to the quartets of $^4T_2(^4F)$ and $^4T_1(^4F)$ dominate the spectrum (Fig. 20). The transition to the next highest configuration $^4T_1(^4P)$ is forbidden in the approximation neglecting configuration interaction, as it involves a two-electron change. The weak band at 39,000 cm^{-1}, partially hidden by bands to the violet is assigned to the $^4A_{2g} \rightarrow {}^4E[^4T_{1g}(^4P)]$ state. The splitting of the $^4T_{1g}(^4F)$ and $^4T_{2g}(^4F)$ bands are -800 and -450 cm^{-1}, respectively. These splittings correspond to $3K/2$, where $3K$ is the trigonal field splitting parameter, and do not match theoretical expectations. From a comparison of the relative intensities of the bands under each polarization, the dipolar perturbation is unimportant relative to the octupole (233).

The ultraviolet spectrum of Cr^{3+} in Al_2O_3 has been reexamined and discussed by Naiman and Linz (279). A feature of the spectrum in the interval 28,000–30,000 cm^{-1} is the occurrence of a number of bands whose intensity varies with the square of the Cr^{3+} concentration. These "pair bands" are attributed to absorption by units of the type

$$Cr-O-Cr$$

and were first noted by Linz and Newnham (214). They show similarities

TABLE XV
Observed and Calculated Levels for Cr^{3+} in Corundum

Cubic	Trigonal	Spin–orbit	Calculated (236) No spin–orbit	Calculated (236) With spin–orbit	Observed	Splitting Calculated No spin–orbit	Splitting Calculated With spin–orbit	Observed
E_g		$E_{1/2}$	14,075	14,054	14,418 (383)		22	29
		$E_{3/2}$		14,077	14,447 (383)			
$T_{1g}(^2F)$	$^2A_2$	$E_{1/2}$	14,621	14,611	14,957 (250)		196	211
	^2E	$E_{3/2}$	14,814	14,807	15,168 (250)		10	22
		$E_{1/2}$		14,817	15,190 (250)			
$T_{2g}(^4F)$	^4E	$E_{1/2}$	17,945	17,924				
		$E_{3/2}$		17,936	$18,000 \perp (233)\,f = 4.80$	434	460	450
		$E_{3/2}$		17,966				
		$E_{1/2}$		17,985				
	$^4A_1$	$E_{3/2}$	18,379	18,411	$18,450 \parallel (233)\,f = 1.30$			
		$E_{1/2}$		18,415				
T_{2g}	^2E	$E_{3/2}$	21,557	21,509	20,993 (383)		105	75
		$E_{1/2}$		21,614	21,068 (383)			
	$^2A_1$	$E_{1/2}$	21,881	21,887	21,357 (8)		273	290
$T_{1g}(^4F)$	^4E	$E_{1/2}$	24,509	24,431				
		$E_{3/2}$		24,436	$24,400 \perp (333)\,f = 5.88$	849	1,150	800
		$E_{1/2}$		24,438				
		$E_{1/2}$		24,442				
	$^4A_2$	$E_{1/2}$	25,358	25,587	$25,200 \parallel (333)\,f = 10.16$			
		$E_{3/2}$		25,590				
$T_{1g}(^4P)$	^4E	$E_{1/2}$	39,200	39,326				
		$E_{3/2}$		39,330	$39,000 \perp (333)\,f = 1.2$	329	~0	400
		$E_{1/2}$		39,381				
		$E_{1/2}$		39,391				
	$^4A_2$	$E_{3/2}$	39,529	39,413	$39,400 \parallel (333)\,f = 1.3$			
		$E_{1/2}$		39,426				

to the solution spectrum of the ion $(NH_3)_5Cr$—O—$Cr(NH_3)_5$ (278,354).
The pair bands are thought to be due either to intervalence transfer absorption of the type

$$Cr^{3+} + Cr^{3+} \longrightarrow Cr^{2+} + Cr^{4+}$$

or to double-ion excitations. The frequencies of the pair bands are close
to twice the frequencies of individual Cr(III) bands, which makes the second
alternative a reasonable possibility. Temperature dependence and polarization data are reported (279).

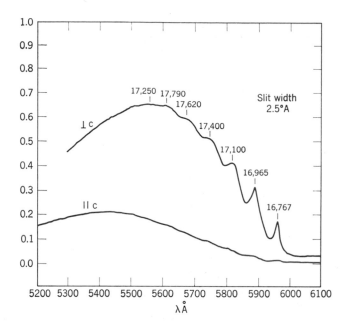

Fig. 21. Vibrational structure of the first strong absorption band of Cr^{3+} at ca. 5°K. [The suggestion of structure in the $\parallel$ spectrum is thought to be due to crystal imperfections or polarizer errors (233).]

The vibrational structure of the $^4A_{2g} \rightarrow {}^4T_{2g}(^4F)$ band at 5°K has been studied using polarized light by McClure (233), see Figure 21. Grum Grzhimailo and Brilliantov (161) observed bands at 16,760 and 16,780 cm⁻¹ at 1.7°K.

The sharp line structure is highly polarized $\perp c$ in comparison to the spectrum of V^{3+} in Al_2O_3. The lower component of $^4T_{2g}$ is ^4E in Cr^{3+} and the $^4A_{2g} \rightarrow {}^4E$ transition is polarized $xy(\perp)$, whereas in the V^{3+} case, the ^3A component of $^3T_{2g}$ is lower in the trigonal field and the polarization for $^3A \rightarrow {}^3A$ is $z(\parallel)$. This cannot be explained on the basis of electric dipole selection rules for C_3. Analysis of the bands, however, leads to the conclusion that the upper state is tetragonally distorted in both the V^{3+} and Cr^{3+} cases. This tetragonal distortion lowers the splitting of $^4T_{2g}$ state to K, giving a value of -450 cm⁻¹ to the trigonal field splitting parameter. This compares favorably with the value of -530 cm⁻¹ obtained from the $^4T_1$ band. The "best" value McClure gives as -475 cm⁻¹ (233). This may be compared with the value of -350 cm⁻¹ obtained by Tanabe and Sugano (382) who together with Sugano and Tsujikawa (383) have carried out an extensive clarification of the details of the spectrum of ruby.

The splittings of the doublet states ^2E_g, $^2T_{1g}$, $^2T_{2g}$ were calculated in terms of the trigonal field splitting parameter and the spin–orbit interaction parameter ζ. The first two of these states are possible assignments of the 14,418; 14,447 cm^{-1} lines, while the third is undoubtedly the assignment for the $\sim$21,000 cm^{-1} lines (382). The lines that correspond to the transitions to the split components of $^2T_1$ have recently been identified by Margerie (250).

The Cr^{3+} ion is one of the few trivalent ions in which all the strong crystal field bands may be observed. This is due to the particular stability of the d^3 configuration, and resulting absence of low-energy charge-transfer processes. There is some disagreement over the interpretation of spectra recorded of Cr^{3+} in a number of solid compounds and solid solutions as the $^4T_{1g}(^4P)$ state does not always appear where crystal field theory says it should. This is illustrated in Table XVI (230) for Cr^{3+} in several crystal environments.

The $^4T_2(^4F)$ and $^4T_1(^4F)$ bands change rather little as the environment is altered. Orgel (293) has shown that when Cr^{3+} is substituted for Al^{3+} it has to enter a site 0.08 Å too small for it and Dq becomes larger in consequence, thus shifting the bands to higher frequencies. In MgO there will be no compression by the crystal as the divalent ion "hole" is larger than the trivalent ion entering, and so the spectrum is similar to that of Cr_2O_3 (223).

The movement of the $^4A_{2g} \rightarrow {}^4T_1(^4P)$ band is, however, not simply explained by crystal field theory; trigonal field components do not allow for shifts as great as 10,000 cm^{-1}. The correct explanation of these data is not yet clear.

The spectrum of ruby at various temperatures has been studied by McClure (233) in the region of the $^4T_{2g}$ and $^4T_{1g}(^4F)$ bands. At 1000°K the absorption strength, measured by the areas under the bands, has increased to 1.35 times its value at 77°K. Individual bands have differing temperature dependences. The region of the $^4A_{2g} \rightarrow {}^4A_1(^4T_{2g})$ forbidden transition

TABLE XVI
The Spectrum of Cr^{3+} in Several Crystals (230)[a]

Crystal	$^4T_2(^4F)$	$^4T_1(^4F)$	$^4T_1(^4P)$
Cr_2O_3	16,800	23,000	39,500
10.9% Cr_2O_3–Al_2O_3	18,300	24,500	> 32,000 ⎰edge of
8.4% Cr_2O_3–Al_2O_3	18,350	25,000	> 36,000 ⎱band
$\sim$0% Cr_2O_3–Al_2O_3	18,150	25,730	39,100
MgO	16,200	22,700	29,700
$KCr(SO_4)_2 12H_2O$	17,500	24,700	38,000

[a] The temperature(s) at which this comparison is made is not recorded in ref. 230.

increases far more rapidly than the others. This could arise from (1) vibrational intensification of this transition or (2) depolarization of the allowed $^4E(^4T_{2g})$ transition. It is concluded that the depolarization process is less important as the $^4T_2$ band maxima remain at about the same distance apart from 300 to 1200°K.

The optical and magnetic properties of Cr^{3+} ions in noncubic sites in Al_2O_3, beryl ($Be_3Al_2Si_6O_{18}$), and rutile (TiO_2) have been studied using an LCAO—MO approach by Lohr and Lipscomb (217). It was found that the principal contribution to the zero-field splitting ($\delta = 2D$) of the $^4A_2$ state in a trigonal field arises primarily from the $^2T_2$ and not from the $^4T_{2g}(^4F)$ state. Calculations were also made for Dq, K, and D considering the symmetries C_3 and C_{3v} including and not including the $4s$ and $4p$ orbitals of Cr. The effect of shifting the Cr^{3+} ion $\pm$ 0.1 Å towards the close oxygen triad (Fig. 2) was also considered. The values of δ for ruby are more sensitive to a displacement of the Cr^{3+} ion along the threefold axis than are those of K, which are more sensitive to the inclusion of Cr $4s$ and $4p$ orbitals in the calculation.

3. $Cr(H_2O)_6{}^{3+}$ in Various Trigonal Lattices

Carlin and Walker (58a,58b) have studied the optical spectra of the hexaquochromium(III) ion in the host crystals $C(NH_2)_3Al(SO_4)_2 6H_2O$ and several isomorphs and in $AlCl_3$. The crystal structures of these compounds were discussed on p. 273. The observed absorption bands at 80°K for the series of compounds studied are shown in Table XVIA. The spectra were determined at 300, 195, 80, and 20°K, but are quoted at 80°K for comparison. The polarization refers to the molecular axis.

Figure 19 shows the energy level diagram for the d^3 configuration in octahedral and trigonal fields. The trigonal field parameters are those of Macfarlane (236). The transitions $^4A_2 \rightarrow {}^4T_2$ and $^4A_2 \rightarrow {}^4T_1$, occurring in the hydrates in the neighborhood of 17,700 and 24,500 cm^{-1}, respectively, as shown in the assignment of the bands, Table XVIA, have a molar extinction coefficient of about 10. The third transition, $^4A_2 \rightarrow {}^4T_1(^4P)$, lies at about 40,000 cm^{-1}. Macfarlane (236) has shown that for intermediate crystal fields, $\Delta(^4T_1(^4F)) = \frac{1}{2}v + v'$ and $\Delta(^4T_2(^4F)) = \frac{1}{2}v$ as shown in Figure 19. His calculations were based on ruby for which 10 $Dq/B = 25$. Since for $Cr(H_2O)_6{}^{3+}$ 10 $Dq/B = 27.5$, it is assumed (58b) that the above formulas also apply.

The $\parallel$ spectrum of Cr:GASH at 20°K is characterized by three small peaks atop the $^4T_2$ band. The first at 17,450 cm^{-1} coincides with the band maximum in the axial spectrum and hence is assigned (58b) to a magnetic dipole transition. Peaks at 18,020 and 18,690 cm^{-1} in the $\parallel$ spectrum do not

TABLE XVIA

Band Maxima for the $Cr(H_2O)_6^{3+}$ Ion at 80°K in Various Compounds (58b)

System	Polarization	Band max, cm^{-1} $^4A_2 \to$ components of $^4T_2(^4F)$	Band max, cm^{-1} $^4A_2 \to$ components of $^4T_1(^4F)$
Cr:GASH	$\perp$	18,020[a]	24,690[b]
	‖	17,540	25,480
	Axial	17,320	24,570
GCrSH	$\perp$	17,860[c]	24,690[d]
	‖	17,390	25,640
Cr:GASeH	$\perp$	17,860	24,510
	‖	17,390	25,640
Cr:AlCl$_3$6H$_2$O[e]	$\perp$	17,540(6.7)	24,390(11.4)
	‖	17,540(4.6)	24,880(7.4)

[a] $f_\perp/f_\| = 1.0$
[b] $f_\perp/f_\| = 1.1$
[c] $f_\perp/f_\| = 1.0$
[d] $f_\perp/f_\| = 1.4$
[e] $f \times 10^5$ in parentheses

occur in the axial spectrum and lying 474 and 1148 cm^{-1}, respectively, above the magnetic dipole line are assigned to vibrations. The $\perp$ spectrum has only one discernible maximum at 17,860 cm^{-1} and the trigonal field is thus taken as the difference between the 17,860 and 17,540 cm^{-1} peaks giving $v/2$ a value of -320 cm^{-1}. From the splitting of the $^4T_1(^4F)$ band the values of v and v' are calculated to be -640 and 1270 cm^{-1}, respectively. The corresponding values of v and v' in Cr:GASeH are -920 and 1800 cm^{-1}, the larger values being expected from the larger zero field splitting in the selenate. For the blue-violet crystals of Cr:AlCl$_3$6H$_2$O it was found that $v = 0$ and $v' = 610$ cm^{-1} at 20°K. The fact there is no $^4T_2$ band splitting, and yet there is ground state splitting (421a) is considered (58b) to be further evidence for Macfarlane's (236) arguments as to the origin of the ground state splitting.

The observed oscillator strengths (f) decrease as a function of temperature, the low energy side of the band suffering the greatest loss of intensity, in agreement with a vibronic mechanism.

The observation of essentially vibronic transitions and low intensities suggests that the effective electronic symmetry in the hydrates is not, in fact, either C_3 or C_{3v} but C_{3i} in which the center of inversion is preserved (58b). The broad band splittings and existence of polarized intensities are explained only by assuming that the e_u vibrations are ineffective in exciting the transitions, i.e.,

$$A_g \overset{\perp}{\longleftrightarrow} E_g \qquad A_g \overset{\|}{\longleftrightarrow} A_g \qquad E_g \overset{\perp,\|}{\longleftrightarrow} E_g$$

An alternative assumption, namely that a_u vibrations are effective only for the $A_g \leftrightarrow A_g$ and e_u vibrations for the $A_g \leftrightarrow E_g$ transitions leads to the weaker selection rules.

$$A_g \overset{\parallel}{\longleftrightarrow} A_g \qquad A_g, E_g \overset{\perp, \parallel}{\longleftrightarrow} E_g$$

Earlier analyses of the trigonal field in ruby which ignored v' were successful because ruby is anomolous in that v and v' have the same sign (381a), while the work of Carlin and Walker (58b) shows that this is not true in general.

The ^2E state is split by a trigonal field in conjunction with spin–orbit coupling into two doublets separated by an amount λ. λ (not to be confused with the spin–orbit coupling constant) is related (382,383) to the trigonal field by the formula,

$$\lambda = -\tfrac{4}{3}[v\zeta/(E(^2E) - E(^2T_2))]$$

The two components, designated R_1 and R_2, are defined such that $E(R_1) = -\tfrac{1}{2}\lambda$ and $E(R_2) = \tfrac{1}{2}\lambda$. The separation of the ^2E and $^2T_2$ levels is about 8000 cm^{-1} in the hydrates.

The value of v calculated from the $^4T_2$ splittings of GCrSH (-640 cm^{-1}) is in excellent agreement with the value of $|v|$ obtained from the ^2E line splitting (660 cm^{-1}) (58b). The two lines occur at 14,133 and 14,114 cm^{-1}, respectively, separated by 19 cm^{-1}.

The Zeeman effect of ^2E in Cr:GASH has been studied by Martin-Brunetière and Couture (57a, 256a). A value of $\lambda = 19$ cm^{-1} is reported for the dilute crystal in agreement with the optical spectra of Carlin and Walker (58b). However, Martin-Brunetière and Couture conclude that R and R' arise from different metal ion sites in the crystal, and not from the second-order term as described by Carlin and Walker.

The origin of the trigonal field has been considered in these systems. ESR studies indicate that the D values undergo a slight increase when deuterium is substituted for hydrogen in the GASH system, and a more marked decrease in the aluminum chloride system (57c,421a). The effect is thought to be due to the capacity of deuterium to form slightly longer hydrogen bonds than hydrogen thus affecting any distortion away from octahedral symmetry when hydrogen bonding is effective. The increase in the trigonal field on substituting selenate for sulfate in the GASH system is explained in a similar fashion due to distortion of the O—$\hat{\text{M}}$—O bond angle away from 90°. The bulk of the trigonal field is thought to be caused by steric influences from the sulfate and guanidinium ions in the unit cell.

The absorption spectrum of chromium(III) in corundum is markedly concentration dependent (233a), showing an increase in Dq as the concentration decreases. The energy of $^4T_2(^4F)$ in Cr_2O_3 is 16,800 cm^{-1}, while in dilute ruby it is 18,150 cm^{-1}. This change is not noted in chrome alum nor in any of the GASH isomorphs studied (1). In GASH, hydrogen bonds can relieve the strain caused by introducing a larger substituent ion into the lattice and so the octahedral and trigonal splittings do not vary greatly with concentration. With Al_2O_3, no such mechanism exists and the larger chromium ion must occupy a site which is too small by 0.08 Å. Measurements of the ^2E splitting (382,383) of very dilute chromium in corundum gave $\lambda = -29$ cm^{-1}, while in chromium oxide, $\lambda = 1$ cm^{-1} (418a) indicating that chromium–aluminum size mismatch is principally responsible for the observed trigonal field in corundum. The optical spectra, which are not in complete agreement with the line splitting, since a sizeable $^4T_2$ splitting is observed (233a), give $v = -700$ cm^{-1} in Cr_2O_3, as opposed to $v = +800$ cm^{-1} in ruby (236).

4. Cr in NaMgAl(C₂O₄)₃9H₂O

The polarized crystal spectrum of Cr^{3+} in $NaMgAl(C_2O_4)_3 9H_2O$ has been studied by Piper and Carlin (322,325). The site symmetry of the Cr^{3+} in this host lattice was discussed on p. 294 under d^2 configurations in connection with the V^{3+} analog.

The crystal structure of $K_3Cr(C_2O_4)_3 3H_2O$ has been determined by Van Niekerk and Schoening (403). Trigonal distortion occurs in this complex as the three O—$\hat{Cr}$—O bond angles about the chromium ion are reported as 81° (twice) and 74°, but these values are not accurate, as an improper space group was chosen to make the analysis simpler. The bond angle in *trans*-$KCr(C_2O_4)_2(OH_2)_2 3H_2O$ is 83° (402) while in triclinic $(NH_4)_3Cr(C_2O_4)_3 2H_2O$ the O—$\hat{Cr}$—O angle was found to be 85° (404).

The energy-level diagram is similar to that for Cr^{3+} in Al_2O_3 (Fig. 19) except that there is a positive trigonal field splitting parameter, so that A states lie lower than E states in the trigonal field.

The spectrum obtained is shown in Figure 17 with the positions and polarizations of the absorption bands tabulated in Table XVII. The extinction coefficients were measured where indicated, showing that (*1*) the σ and axial spectra are the same and (*2*) the solution spectra are more intense. At 77°K, the spin-allowed bands move to 17,620 π, 18,020 σ, and 23,950 σ cm^{-1}.

The excited doublet states of Cr^{3+} in ruby have been extensively studied (233,382,383) and analogous bands were found in the undiluted chromium oxalate spectrum (325). The transitions to ^2E were the sharpest and had the highest extinction coefficient ($\epsilon = 2.6$). The transitions to $^2T_1$ are incompletely polarized but stronger in the σ spectrum. The transitions to $^2T_2$ are completely σ polarized, except for the lowest energy component

TABLE XVII

Absorption Band Maxima and Polarizations for Cr^{3+} in $NaMgAl(C_2O_4)_3 9H_2O$ (325)

Band max., cm^{-1}	Half width $\Delta\nu$, cm^{-1}	Transition	Remarks
14,455	10	$\left.\right\}{}^4A_2 \rightarrow {}^2E$	$f = 1.2 \times 10^{-7}$
14,476	10		$f = 1.3 \times 10^{-7}$
15,216	30	$\left.\right\}{}^4A_2 \rightarrow {}^2T_1$	
15,284	50		77°K
15,323	40		
17,316 π	2,800	${}^4A_2 \rightarrow {}^4A_1({}^4T_2)$	$\epsilon_\parallel/\epsilon_\perp = 1.3$
17,620 σ	2,700	${}^4A_2 \rightarrow {}^4E_a({}^4T_2)$	25°C
20,555	30	$\left.\right\}{}^4A_2 \rightarrow {}^2T_2$	
20,704	90		
21,119	60		
23,670 σ	3,200	${}^4A_2 \rightarrow {}^4E_b({}^4T_1)$	$\epsilon_\parallel/\epsilon_\perp \leqslant 0.06$

which appears with equal intensity, $\epsilon = 0.3$, in both polarizations. Diffuse lines ascribed to vibrations were observed at 14,771 and 14,792 cm^{-1}. They are one tenth as intense as the components of ^2E, are also split by 21 cm^{-1}, and are each 316 cm^{-1} above the respective ^2E line. These lines are simultaneous electronic (^2E) and vibrational transitions, 316 cm^{-1} being assigned to a chromium–oxygen stretching vibration (325). The polarization ratios for the spin-allowed transitions of the first-row transition metal trisoxalates were calculated by Piper (328) and found to agree semiquantitatively with experiment. The trigonal field splitting parameter K is found to be 270 cm^{-1}. The change of sign of K in going from V^{3+} to Cr^{3+} seems unusual when comparison is made for the ions in Al_2O_3 (233). However, in the case of ions in Al_2O_3, the value of K probably depends more on the lattice than on the particular metal ion.

5. $Cr(C_3H_2O_4)_3^{3-}$

Hatfield (173) has determined the crystal spectrum of the trismalonatochromium(III) ion. No colorless host crystal could be found, so trismalonatochromium(III) ions were introduced into ammonium trismalonatoferrate(III) crystals and the polarized spectra recorded. The spectrum of the iron complex (see d^5 configuration, p. 345) was subtracted and resultant bands were found at 17,060 and 17,240 cm^{-1} assigned to the ${}^4A_{2g} \rightarrow {}^4E[{}^4T_{2g}({}^4F)]$ and ${}^4A_{2g} \rightarrow {}^4A_1[{}^4T_{2g}({}^4F)]$ transitions, respectively, with $\epsilon_\parallel/\epsilon_\perp = 1.4$.

The ${}^4A_2 \rightarrow {}^4T_1({}^4F)$ transition could not be observed because of intense bands in the iron compound. The close agreement of $\epsilon_\parallel/\epsilon_\perp$ of the low-energy band with that of the oxalato (325) complex indicates that the $Cr(C_3H_2O_4)_3^{3-}$ ions occupy positions which are identical with those of the $Fe(C_3H_2O_4)_3^{3-}$. The trigonal splitting parameter K was determined as -120 cm^{-1}. The magnitude is not regarded as very reliable due to inaccuracies in the subtraction of the contribution from $[Fe(C_3H_2O_4)_3]^{3-}$ and in the misalignment of the C_3 molecular axes in the crystal, but the sign of K is thought to be correct. Piper and Carlin (329) have shown that the sign of K depends on the polar angle and thereby on the O—M—O angle within the chelate ring if the radial parameters are unchanged. Since there is little change in Dq in going from $[Cr(C_2O_4)_3]^{3-}$ to $[Cr(C_3H_2O_4)_3]^{3-}$, there is presumably also little change in the radial parameters. The change of sign of K from $+270$ cm^{-1} in $[Cr(C_2O_4)_3]^{3-}$ which has a O—$\widehat{Cr}$—O angle of 83–85° (325), to -120 cm^{-1} for $[Cr(C_3H_2O_4)_3]^{3-}$ indicates that the O—$\widehat{Cr}$—O angle in the malonate chelate ring is greater than 90° (173).

6. Cr^{3+} in Al(acac)$_3$

The polarized spectrum at 77°K of chromium acetylacetone in a dilute mixed crystal with aluminum acetylacetone has been studied by Piper and Carlin (329), while earlier work in the visible region only on the pure crystal was reported by Chakravorty and Basu (65). The crystal structures and site symmetry of the Cr^{3+} ion in the host crystal has been discussed earlier under d^1 configurations, p. 274. The crystal structure of pure chromium trisacetylacetone has been determined by Shkolnikova and Shugum (368) and more recently by Morosin (275). The crystal packing of this compound is shown in Figure 22. The monoclinic unit cell, space group $P2_1/c$, C_{2h}^5 has four molecules per unit cell with $a_0 = 14.031$, $b_0 = 7.551$, $c_0 = 16.379$ Å, and $\beta = 99°4'$. The Cr—O bond distances and O—$\widehat{Cr}$—O angles are as follows:

$$
\begin{array}{ll}
Cr\text{—}O_1 = 1.943 \text{ Å} & O_1\text{—}\widehat{Cr}\text{—}O_2 = 91° \, 40' \\
Cr\text{—}O_2 = 1.942 \text{ Å} & O_4\text{—}\widehat{Cr}\text{—}O_5 = 90° \, 57 \\
Cr\text{—}O_3 = 1.951 \text{ Å} & O_3\text{—}\widehat{Cr}\text{—}O_6 = 90° \, 40' \\
Cr\text{—}O_4 = 1.958 \text{ Å} & \\
Cr\text{—}O_5 = 1.956 \text{ Å} & \\
Cr\text{—}O_6 = 1.959 \text{ Å} &
\end{array}
$$

The spectrum was recorded with light incident on the (001) face of the crystal and the polarizations refer to the crystal b axis. The spectrum is shown in Figure 23, and the bands are tabulated in Table XVIII.

The assignment of the line at 12,950 cm^{-1} to $^4A_2 \rightarrow {}^2E$ is confirmed by the observation of luminescence at 13,000 cm^{-1} by Forster and DeArmond (130). The further fine structure is ascribed to vibrations and the

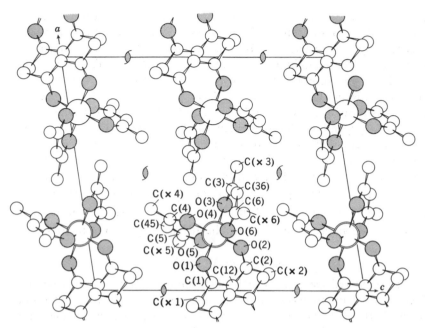

Fig. 22. Molecular arrangement in Cr(acac)$_3$ viewed along b axis. Chromium atoms near $y = \frac{1}{4}$ are denoted by double circles; those near $y = \frac{3}{4}$ by single circles (275).

state $^2T_1$ (329). This is a considerable lowering of the energy from that observed in the oxalate (14,465 cm^{-1}) (325), or in ruby (14,432 cm^{-1}) (383). This transition is confined to the $t_2{}^3$ configuration, so the interaction repulsion parameters have been considerably reduced in the acetylacetonate. This is taken as evidence of strong π covalence in the acetylacetonate as compared with the oxalate or oxide (329).

From the band in the visible region and the splitting of the σ and π components $(3K/2)$ Dq is calculated as 1810 cm^{-1} and K as $+530$ cm^{-1}. Compared with the band in Cr(C$_2$O$_4$)$^{3-}$, the near-ultraviolet band in Cr(acac)$_3$ has three peculiarities (1) it shows well-defined structure, (2) it is considerably more intense than the band in the visible region, and (3) it is about twice as intense in the π polarization as in the σ polarization.

The ultraviolet band has the wrong polarization for the expected $d \rightarrow d$ transition and is therefore ascribed to a $n \rightarrow \pi^*$ or $\pi \rightarrow \pi^*$ transition in the ring. It may have appreciable $d \rightarrow d$ character and it is possible that part of the intensity in the σ component is due to the $^4A_2 \rightarrow {}^4E_b$ transition. A similar band with the same fine structure and polarization was observed for Ti(acac)$_3$ (330), see Figure 5.

TABLE XVIII
Absorption Band Maxima and Polarizations for Cr(acac)$_3$ in Al(acac)$_3$

	Reference 329			Ref. 65 (undiluted compound)	
Band max., cm^{-1} (polarization)	$\Delta\nu_{1/2}$	$\epsilon_\parallel/\epsilon_\perp$		Band max., cm^{-1}	$\epsilon_\parallel/\epsilon_\perp$
12,950 (+5 lines with spacing 260 ± 40 cm^{-1})		2.5 (unpolarized)			
17,600 ∥	2,900	1.5		17,094 ⊥	1.0
18,400 ⊥	3,000			18,018 ∥	
25,700 (splitting of 600 cm^{-1})		2.4			

The ionic model has limited application in the case of the trisoxalates and fails to account for the spectral properties of the acetylacetonates. This failure is ascribed to greater covalence and in particular to π covalence. The acetylacetonate ring has low lying π^* levels which interact strongly with the 3d levels.

Barnum (24) has carried out a series of Hückel LCAO—MO calculations to estimate the effect of metal–ligand π bonding on the electronic

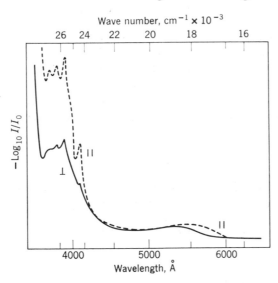

Fig. 23. Spectrum of chromium acetylacetonate in a dilute mixed crystal with aluminum acetylacetonate at 77°K. The light is incident on the (001) face and the polarizations are referred to the crystal b axis (329).

transition energies in a series of acetylacetonate complexes ($Ti^{3+} \rightarrow Co^{3+}$) with trivalent transition metal ions. Metal–ligand π bonding, increasing throughout the series, $Ti \rightarrow Co$, causes the $\pi_3 \rightarrow \pi_4$ (highest occupied $\rightarrow$ lowest unoccupied internal ligand) transition to split into four bands. The $d\epsilon \rightarrow \pi_4$ (metal–ligand charge transfer) transition is split into three bands by π bonding. These results predict K to be $+1000 \rightarrow +2000$ cm^{-1}, the magnitude of which is far too high; the π_3 level must lie considerably below the $d\epsilon$ level and/or the $d\epsilon$–π interaction is overestimated. Barnum assigns the transitions in the trisacetylacetonates near 30,000 cm^{-1} to the charge transfer transitions $t_2(d\epsilon) \rightarrow \pi_4$. The excited configuration $t_2^2\pi_4$ gives rise to quartet and doublet states of every representation of D_3 so that the assignment cannot be verified from polarizations. The similar transition in the Ti(acac)$_3$ solution spectrum was given the same assignment (23). In this case the ground state is $^2A_1$ and the excited configuration π_4^1 gives rise to only A_1 and E states. Consequently, such a transition should be $\perp$ polarized, but is found to be $\parallel$ polarized (330) so this assignment is ruled out (329). At present the most likely assignment appears to be a $\pi_3 \rightarrow \pi_4$ or $n \rightarrow \pi_4$ transition which would be spin-forbidden in the free ligand but is allowed in the complex by interaction with the metal states.

7. Cr(en)$_3^{3+}$ in Rh(en)$_3$Cl$_3$3H$_2$O

Karipides and Piper (196) have measured the crystal spectrum of Cr(en)$_3^{3+}$ in the host crystal Rh(en)$_3$Cl$_3$3H$_2$O. This paper also presents a MO model for the analysis of the experimental data. No spectra were published but the band intensities were in the ratio 1.47:1.00 for the transitions $^4A_2 \rightarrow {}^4A_1$ and $^4A_2 \rightarrow {}^4E_a$, respectively. The crystal of Rh(en)$_3$-Cl$_3$3H$_2$O absorbs strongly in the near-ultraviolet where the $^4A_2 \rightarrow {}^4T_1$ band is expected.

8. Cr(en)$_3$Cl$_3$H$_2$O

Yamada and Tsuchida (443) have measured the polarized crystal spectrum of Cr(en)$_3$Cl$_3$H$_2$O and the observed bands are tabulated below:

max., cm^{-1} $\parallel$	log α	max., cm^{-1} $\perp$	log α
2,200	1.40	22,100	1.25
28,700	0.86	28,500	1.24

The observed dichroism is opposite to that observed in ruby and is unexplained by Yamada and Tsuchida (443).

9. V^{2+} and Mn^{4+}

V^{2+} and Mn^{4+} are isoelectronic with Cr^{3+}. Absorption by V^{2+} in crystals is too weak to be observed clearly as V^{3+} usually predominates, but the sharp line fluorescence of the V^{2+} ion in MgO and Al_2O_3 has been studied by Sturge (381). A value of -140 to -160 cm^{-1} was obtained for the trigonal field splitting parameter K depending on whether a strict fit was made to the ^2E state results or an average value taken. The absorption spectrum of Mn^{4+} in Al_2O_3 shows a single band at 21,300 cm^{-1} attributed to the $^4A_2 \rightarrow {}^4T_2$ transition (143), and Dq is equal to 2170 cm^{-1}. The single crystals of α-Al_2O_3:Mn^{4+} were grown from oxide fluxes and it was found necessary to add MgO powder to provide Mg^{2+} to compensate the charge on the Mn^{4+} which enters substitutionally into the octahedral Al^{3+} site.

VI. CONFIGURATION d^4

A. $3d^4$

The d^4 configuration is chemically unstable in comparison with d^3 and d^5. Trivalent d^4 ions may lose one charge and attain the d^5 configuration which is especially stable in weak crystal fields. Alternatively, they may acquire an extra positive charge to reach the d^3 configuration, which is stable in strong crystal fields. Cr^{2+} and Mn^{3+} are the only stable d^4 ions in the first transition group and in a weak O_h field have a ^5E_a ground state (Fig. 24).

1. $Cr_2(C_2H_3O_2)_4 2H_2O$

The polarized crystal spectrum of chromous acetate has been studied by Kida, Nakashima, Morimoto, Niimi, and Yamada (201). While the crystal structure of chromous acetate (406) $Cr_2(C_2H_3O_2)_4 2H_2O$ has not been completely refined, the atomic coordinates and interatomic distances are very similar to those of cupric acetate (405). Crystals of chromous acetate are monoclinic with space group $C2/c, C_{2h}{}^6$. The unit cell of dimensions $a_0 = 13.15$, $b_0 = 8.55$, $c_0 = 13.94$ Å, and $\beta = 117°$ contains four $Cr_2(C_2H_3O_2)_4 2H_2O$ molecules (Fig. 25). Each pair of metal atoms is bridged by the four acetate groups in such a manner that four oxygen atoms arranged approximately at the corners of a square form the nearest neighbors of the Cr atom, Cr—O = 1.97 Å. A H_2O molecule at a distance of 2.20 Å and a Cr atom at a distance of 2.64 Å lie perpendicularly above and below the plane of the square, thus completing the distorted octahedral

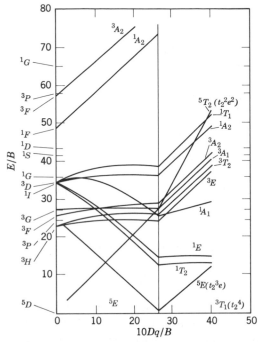

Fig. 24. Tanabe-Sugano energy diagram for the d^4 configuration (75,386).

coordination around the Cr atom. The angles between the adjacent acetate groups are 83 and 97°. Magnetic susceptibility measurements (202) have shown that this compound contains no unpaired electrons. Together with the short Cr—Cr distance of 2.64 Å this is evidence for direct interaction between the two metal ions.

Quantitative dichroism measurements were made at room temperature on the (110) face using light polarized ∥ and ⊥ to the c axis. The observed bands are given in Table XIX, the ratios of $\epsilon_\parallel/\epsilon_\perp$ being calculated from the crystal structure with ∥ referring to the Cr—Cr direction in the molecule.

TABLE XIX
Absorption Band Maxima for
Chromium(II) Acetate (201)

	Band max., cm^{-1}	$\epsilon_\parallel/\epsilon_\perp$
I	20,580	1/4.5
II	30,770	4

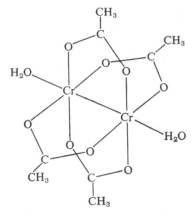

Fig. 25. Structure of chromous acetate (406).

If the molecule is considered to be made up of two units, each containing one Cr^{2+} ion, each entity has the symmetry C_{4v}, but when considered as a whole, the molecule is D_{4h}. The splitting of the orbitals in D_{4h} and C_{4v} symmetries is shown in Figure 26.

Solution spectra (201) of $Cr_2(C_2H_3O_2)_4 2H_2O$ and $Cr_2(C_2H_3O_2)_4$ in ethanol show that there is a blue shift of the first band of the anhydrate relative to that of the hydrate; but very little blue shift is observed for the second. This implies that the second band is not as sensitive to the change in the tetragonality of the ligand field as the first. The second band is therefore unlikely to be a ligand field band.

Results of the dichroism measurements support this conclusion. If both the observed bands in the visible and near-ultraviolet region were due

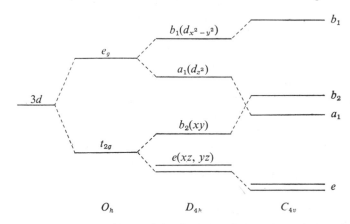

Fig. 26. d-orbital splitting in D_{4h} and C_{4v} symmetries.

to $d \rightarrow d$ transitions, the most probable assignment would be $^5B_1 \rightarrow {}^5A_1$ and $^5B_1 \rightarrow {}^5E$ for the first and second bands, respectively. Since the electric vectors transform like $E(x,y)(\perp)$ and $A_1(z)(\|)$ in C_{4v}, the second band which has been attributed to $^5B_1 \rightarrow {}^5E$ transition would be more intense in $\perp$ than $\|$. However as Table XIX shows, the second band is $\|$ polarized, so the $^5B_1 \rightarrow {}^5E$ transition is assigned to the first absorption band ($\perp$ polarized). The second band is thought to be due to an excitation of the π system of the planar 8-membered ring in the molecule, but has not been definitely characterized.

2. $CsMn(SO_4)_2 12H_2O$

The crystal spectrum of $CsMn(SO_4)_2 12H_2O$ has been determined by Hartmann and Schläfer (167) and has been analyzed together with other crystal spectra of hydrated ions of transition metals by Holmes and McClure (182). The Mn^{3+} ion is surrounded by a nearly regular octahedron of water molecules in a unit cell, space group $Pa3,T_h{}^6$, having lattice constant 12.423 Å (176). The broad single band at 21,000 cm^{-1}, $\Delta\nu_{1/2} =$ 5000 cm^{-1}, $\epsilon = 5$, and $f = 1.1 \times 10^{-4}$ is assigned to the spin-allowed $^5E_g \rightarrow {}^5T_{2g}$ transition. Owen (295) in an early paper made an extensive comparison of crystal field theory and experimental data for ions octahedrally coordinated with water molecules having a ^nD ground state. The abnormally large Dq value is explained by the large Jahn-Teller distortion in the complex similar to that observed in the Cu^{2+} ion.

3. $(NH_4)_2MnF_5$

The polarized crystal spectrum of ammonium pentafluoromanganate, $(NH_4)_2MnF_5$, has been determined by Dingle (96). Ammonium pentafluoromanganate(III) is orthorhombic, space group $Pnma$, $D_{2h}{}^{16}$ (358) with $a_0 = 6.20$, $b_0 = 7.94$, $c_0 = 10.72$ Å, and four molecules to a unit cell. The molecular symmetry is D_{4h} with an Mn(III) site symmetry of C_i. The complex is tetragonally distorted; the in-plane Mn—F bond lengths are 1.84 and 1.85 Å and the axial bond lengths 2.12 Å. The long bonds link octahedra into infinite kinked chains, the appropriate fluorines being shared by two octahedra. The molecular z axis, of the D_{4h} molecular unit, makes an angle of $\pm 20.6°$ with the b crystallographic axis. The spectra were recorded $\perp$ and $\|$ to the b axis and hence $\|b$ spectra contain 90% of the z-polarized transition and $\perp b$ spectra about 90% of the (x,y) polarized transition intensity. The observed bands are shown in Figure 27 and tabulated in Table XX.

The marked dichroism indicates that the effective symmetry is less than octahedral, although higher than rhombic. If a vibronic coupling

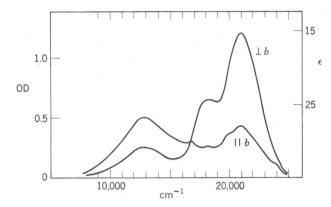

Fig. 27. Polarized crystal spectrum of ammonium pentafluoromanganate(III) taken on (001) with the electric vector ∥ and ⊥ to the b crystallographic axis. Low-temperature spectra sharpen the bands somewhat but no gross changes occur (96).

mechanism is assumed, then assignments may be made in D_{4h} symmetry. The D_{4h} splitting of the d manifold (Fig. 26), leads to the scheme for d^4 shown in Figure 28.

The possible spin-allowed transitions and associated enabling vibrations are listed in Table XXI.

On this basis, the symmetry properties of the absorption bands were assigned as in Table XX. If the odd vibrations b_{2u} and/or a_{2u} are more effective enabling vibrations than e_u, the experimental polarization ratios are in agreement with the scheme of Table XXI. The correlation between the odd vibrations of an octahedron MX_6 and those generated in a tetragonally distorted D_{4h} model are given in Figure 29 together with a pictorial representation of the vibrations.

As shown in Figure 29, the a_{2u} modes (which may be mixed) represent essentially symmetric and antisymmetric stretching motions along the

TABLE XX

Absorption Band Maxima and Polarizations for the $(NH_4)_2MnF_5$ Crystal (96)

Transition	Band max., cm^{-1}	$\epsilon_\perp/\epsilon_\parallel$	$\epsilon_\perp$ crystal, $\pm 10\%$	Type of band
$^5B_{1g} \rightarrow {}^5A_{1g}$ or $^5B_{1g}$	12,750	0.6	3.5	Strong, broad
	16,700	< 1.0?		Weak, sharp
	18,100	?		Weak, sharp
$^5B_{1g} \rightarrow {}^5B_{2g}$	18,200	≫ 1.0	8.0	Strong, broad
	19,800	< 1.0		Weak, sharp
$^5B_{1g} \rightarrow {}^5E_g$	21,000	2.3–3.0	15.0	Strong, broad
	23,800			Weak, sharp

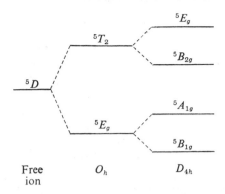

Fig. 28. The energy level scheme for d^4 configuration in D_{4h} symmetry.

····F—Mn—F—Mn—F···· chain, whereas the associated e_u components are mostly chain-bending modes. The b_{2u} mode has most amplitude along the chain, although in this case, the motion represents a bending of the F—Mn—F bonds that lie at right angles to the F—Mn—F chain. The associated e_u mode likewise represents a bending of the infinite Mn—F chain.

The assignment of the bands at 18,200 and 21,000 cm^{-1} is considered unambiguous but the 12,750 cm^{-1} band is not so easily dealt with (96). The theory (386), Figure 24, predicts a number of spin triplets, arising from the ^3H, ^3P, ^3G, and ^3F atomic terms, to lie in the visible to near-infrared region. The weak sharp lines in the region 15,000–25,000 cm^{-1} are thought to represent some of these quintet–triplet transitions, but probably not the $^3T_{1g}$ component of the ^3H term (96). This latter level has been assumed by some authors (23,189,325,330) to be responsible for the relatively intense infrared band in some tris-Mn(III) complexes. Dingle (92) has suggested an alternative assignment involving a strongly Jahn-Teller distorted ground state in these complexes, although a recent x-ray study (272) on trisacetyl-acetonate Mn(III) has indicated such an interpretation to be inappropriate, at least in that complex.

TABLE XXI
Possible Vibronic Spectral Assignments of Mn(III) in a Tetragonal Environment (96)

Transition	$2a_{2u}$	b_{2u}	$3e_u$	Polarization ratios, $\epsilon_\perp/\epsilon_\parallel$ [a]
$^5B_{1g} \rightarrow \begin{cases} ^5A_{1g} \\ ^5B_{1g} \end{cases}$	$A_{2u}(z)$	$A_{2u}(z)$	$E_u(xy)$ $E_u(xy)$	~ 3 $\sim \frac{3}{2}$
$^5B_{1g} \rightarrow {}^5B_{2g}$			$E_u(xy)$	$\gg 1$
$^5B_{1g} \rightarrow {}^5E_g$	$E_u(xy)$	$E_u(xy)$	$A_{2u}(z)$	~ 1

[a] All enabling vibrations assumed equally efficient.

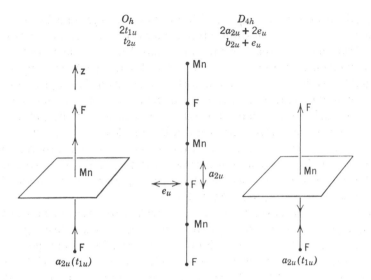

Fig. 29. Metal–ligand vibrations in the $(NH_4)_2MnF_5$ crystal (96).

In the present case of $(NH_4)_2MnF_5$, if the transition is a quintet–triplet so that mixing via spin–orbit coupling to $^5T_{2g}$ is important, the absorption has the wrong polarization properties. Very large mixing coefficients (especially for the fluoro complex) have to be invoked for the $^3T_{1g}$ state and the charge-transfer states to explain the observed polarizations. This seems unlikely (159), so the transition is probably not connected with the O_h $^5E \rightarrow {}^3T_{1g}(^3H)$ transition.

The other $d \rightarrow d$ transition that can be introduced is that arising from the tetragonal splitting of the $^5E_g(O_h)$ ground state as shown in Figure 28. This gives large tetragonal splittings: 12,750 cm^{-1} in the ground state and 2800 cm^{-1} in the excited state. The values of the tetragonal parameters D_s the D_t (cf. Fig. 6) calculated from these splittings are very large and of the same order as Dq. Tetragonal fields of this magnitude have been invoked before (16,182), but severe distortion has to occur and although a bond elongation of 0.27 Å is large, it is not certain that it is consistent with the required tetragonal parameters.

The final possibility is that the absorption band is not $d \rightarrow d$ in origin but represents a ligand-to-metal charge-transfer transition. If this is the case and if the transition is parity-forbidden but spin-allowed, then the intensity is reasonable and the excited state may be either $^5A_{1g}$ or $^5B_{1g}$.

On the basis of a later investigation (97), it seems likely that the near-infrared band where the ligands bind through oxygen or sulfur atoms can be explained not as a $d \to d$ transition but as a charge transfer involving π orbitals of the oxygen or sulfur atoms and the metal d orbitals. However, Sears (358) considered that the band near 13,000 cm^{-1} in the fluoride could not be explained on a charge-transfer model since the Mn^{3+} ion is surrounded by six fluoride ions in a distorted octahedron. The crystal used by Dingle had the same x-ray-determined characteristics as that used by Sears, although both crystals showed some evidence for the presence of O—H bonds in the infrared spectrum. It is possible (96) that in the crystal, the Mn^{3+} ion is surrounded by four fluoride ions and two shared hydroxide ions, as it is difficult to distinguish between OH$^-$ and F$^-$ in a two-dimensional x-ray determination such as that made by Sears; the formula of the complex would then be (NH$_4$)$_2$MnF$_4$OH.

If the above proposal is correct, then the hydroxides can be thought of as forming an sp^2 hybrid to take care of the σ bonding and leaving a filled nonbonding or π bonding p_z orbital to take part in the charge-transfer process (97). This arrangement also provides an explanation of the $\pm 21°$ kinking (358) of the Mn—F—Mn chain since a pure sp^2 hybrid, without strain only requires a kinking of $\pm 30°$. Further work is being carried out by Dingle to test this hypothesis which is only regarded as tentative.

4. Mn^{3+} *in* NaMgAl(C$_2$O$_4$)$_3$9H$_2$O

Piper and Carlin (325) have determined the spectrum of Mn^{3+} in the host crystal NaMgAl(C$_2$O$_4$)$_3$9H$_2$O. The D_3 symmetry of the Mn^{3+} ion was discussed under d^2 configurations, p. 294. The spectrum (Fig. 17) was recorded with light incident on the (10$\bar{1}$1) face. The spectra referring to the molecular axis were then calculated using the known inclination of this face to the trigonal axis. The observed bands are given in Table XXII. The splitting of the octahedral levels for a d^4 electron system is shown in Figure 30.

The ground state is ^5E_a, the electric dipole selection rules (Table XII) allowing transitions to the excited ^5E_b level in both polarizations, so there is no clear resolution into the two components of the excited $^5T_2$ state.

TABLE XXII

Absorption bands of Mn^{3+} in NaMgAl(C$_2$O$_4$)$_3$9H$_2$O (325)

Band max., cm^{-1} (polarization)	$\Delta\nu_{1/2}$, cm^{-1}	Transition
9,700	1500	$^5E_a \to {}^3T_1$
19,100 (π)		$^5E_a \to {}^5E_b({}^5T_{2g})$
20,000 (σ)		$^5E_a \to {}^5A_1({}^5T_{2g})$

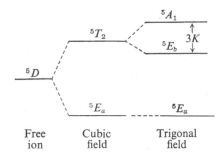

Fig. 30. Energy levels for the d^4 configuration in a trigonal field.

However, the transition $^5E_a \rightarrow {}^5A_1$ is only allowed in the σ polarization and since the σ component is found at higher energy, the $^5A_1$ state is concluded to be at higher energy, which is only compatible with a positive value of K. The total splitting is 1500 cm^{-1}, giving K a value of $+500$ cm^{-1} which is unusually large compared with the other metals in this host crystal. This is thought to be due in part to distortion from trigonal symmetry caused by Jahn-Teller splitting.

The spectra indicate Dq to be of the order of 2000 cm^{-1} which is rather large in view of the Jahn-Teller splitting of the ^5E_a ground state. A value of Dq of 1700 cm^{-1} is chosen together with a splitting of ^5E_a of 6000 cm^{-1}, placing $^5T_2$ at 20,000 cm^{-1} and $^3T_1$ at 13,000 cm^{-1} which fits the data more satisfactorily.

5. Mn^{3+} in Al_2O_3

The polarized spectrum of Mn^{3+} in a corundum host lattice has been studied by McClure (233). The site symmetry of the transition metal ion was discussed under d^1 configurations, p. 270. The observed spectrum is shown in Figure 31, and the bands are given in Table XXIII.

In the trigonal field, the upper $^5T_2$ state is split into two components, $^5A_1$ and ^5E_b. Two peaks are observed in the opposite polarizations and are

TABLE XXIII
Absorption Band Maxima for Mn^{3+} in Al_2O_3 (233)

Band max., cm^{-1} (polarization)	$f \times 10^4$
18,750 ($\perp$)	2.67
20,600 ($\parallel$)	1.77

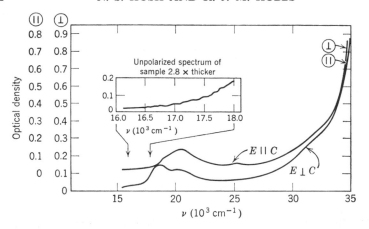

Fig. 31. Absorption spectrum of Mn^{3+} in corundum at 77°K (233).

separated by 1900 cm^{-1}. The lower energy peak at 18,750 cm^{-1} is predominantly in the $\perp$ polarization, $^5E_a \rightarrow {}^5A_1$, while the higher peak at 20,600 cm^{-1}, $^5E_a \rightarrow {}^5E_b$, has mixed polarization. These observations are consistent with the electric dipole selection rules for the $^5A_1$ state lying below ^5E_b and a negative value of the trigonal splitting parameter K of -650 cm^{-1}. The value of Dq of 1947 cm^{-1} is based on a ground-state splitting of 500 cm^{-1}.

Vibrational structure appears exclusively in the σ spectrum of Mn^{3+}. The first line at 16,760 cm^{-1} is assigned to the electronic origin but the intervals do not form a regular progression as in V^{3+} and Cr^{3+}, the most remarkable omissions being 666 and 742. A pair of narrow bands at 19,130 cm^{-1} $\parallel c$ and 19,340 cm^{-1} $\perp c$ appear exclusively in the Mn^{3+} spectrum just barely resolved in the middle of the broad absorption band. It is assumed (233) that these mark the origin of the second component of the $^5T_2$ state, ^5E_b. The origins of the $^5A_1$ and ^5E_b components are thus separated by about 2500 cm^{-1}; the separation, 210 cm^{-1}, could be assigned to spin–orbit splitting but further analysis would be required to confirm this. The extent of Jahn-Teller distortion of the upper state and possible assignments of the vibrational progression on the 16760 cm^{-1} band to e_u corundum vibrations are also discussed by McClure.

6. $Mn(H_2O)_6^{3+}$ in GGaSH

Walker and Carlin (409b) have studied the polarized crystal spectrum of the $Mn(H_2O)_6^{3+}$ ion in GGaSH. The environment of the Mn^{3+} ion was discussed on p. 273. The observed bands are listed in Table XXIIIA.

TABLE XXIIIA
Spectra of $Mn(H_2O)_6^{3+}$ in GGaSH (409b)

Excited state	Polarization	Band max., cm^{-1} 300°K	Band max., cm^{-1} 80°K
$^5T_2$	$\perp$	17,860; 21,740	17,860; 21,980
	$\parallel$	17,860; 21,500	18,860; 21,740

The ion has a ^5E ground state which may be split by a Jahn-Teller distortion, while the excited $^5T_2$ state will be split in a trigonal field, the resulting E component also being susceptible to smaller tetragonal distortions. The spectrum shows a single absorption band characteristic of the $Mn(H_2O)_6^{3+}$ ion with a main maximum at $\sim 22,000$ cm^{-1} and a well resolved shoulder with a separation of 3800 cm^{-1}. The principal band is assigned to $^5E(^5T_2)$ and the shoulder to $^5A(^5T_2)$ (409b). This analysis is similar to that of Dingle for MnF_6^{3-} (96). The small difference in the polarized maxima at 300 and 80°K is assigned to a tetragonal splitting of $^5E(^5T_2)$ (409b). These assignments are supported by the observation of large trigonal and small tetragonal distortions in the trigonal tris-(bipyridyl) copper(II) salts studied by Palmer and Piper (300).

7. Mn^{3+} in Al(acac)$_3$

The crystal spectrum of manganese(III) trisacetylacetone in the host crystal aluminum trisacetylacetone has been studied by Piper and Carlin (330). The crystal structure of the host crystal was discussed on p. 274. Recent work by Morosin and Brathovde (272) has shed more light on the structure of Mn(acac)$_3$ and its relation to the observed infrared spectra obtained by Forman and Orgel (129) of the trisacetylacetonates of Cr^{3+}, Mn^{3+}, Fe^{3+}. Mn(acac)$_3$ has the space group $P2_1/c$, C_{2h}^5, a monoclinic unit cell containing four molecular units with $a_0 = 13.87$, $b_0 = 7.467$, $c_0 = 16.20$ Å, and $\beta = 98°25'$. The Mn—O bond lengths, and O—$\widehat{Mn}$—O angles are as follows:

$$Mn—O_1 = 1.861 \text{ Å} \qquad O_1—\widehat{Mn}—O_2 = 97° 45'$$
$$Mn—O_2 = 1.860 \text{ Å} \qquad O_4—\widehat{Mn}—O_5 = 96° 43'$$
$$Mn—O_3 = 1.893 \text{ Å} \qquad O_3—\widehat{Mn}—O_6 = 96° 39'$$
$$Mn—O_4 = 1.880 \text{ Å}$$
$$Mn—O_5 = 1.875 \text{ Å}$$
$$Mn—O_6 = 1.868 \text{ Å}$$

The structure and crystal packing is almost analogous to that of Cr(acac)$_3$ (Fig. 22).

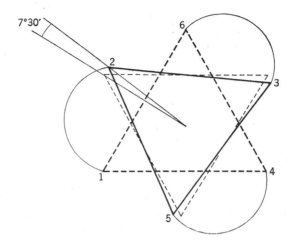

Fig. 32. Schematic arrangement of the three acetylacetone groups in Mn(acac)$_3$, viewed down the threefold axis (272).

A Jahn-Teller distortion is expected for the d^4 electron system with two opposite Mn—O bonds longer than the other four, but this is not borne out in the crystallographic structure. A distortion does however occur from the pure octahedron. Figure 32 shows a view down the only remaining threefold axis after removal of the cubic symmetry. The amount of twist of the upper triangle (2,3,5) relative to the lower (1,4,6) is approximately 7° 30′. This may not be a result of Jahn-Teller distortion; however, it still explains the infrared absorption spectra obtained by Forman and Orgel (129).

The spectra were recorded with light incident on the (001) and (100) faces at 77°K, Figures 33 and 34.

The main band observed in the crystal spectrum is assigned to the $^5E_a \rightarrow {}^5T_2$ transition. Unlike Co(acac)$_3$ (324), p. 362, the σ spectra calculated from the (001) and (100) faces are not in complete agreement. The π spectra agree, both having a maxima at 17,200 cm^{-1}. The σ components, however, have quite different band shapes and the maxima occur at 19,000 cm^{-1} in the (100) spectrum and 18,200 in the (001) spectrum. The large discrepancy cannot be accounted for by rhombic fields in the monoclinic crystal as the cobalt spectrum would also be affected. The proposed explanation is that there is a distortion from trigonal symmetry as required by the Jahn-Teller theorem. As the σ component lies at higher energies, the $^5A_1$ state is above ^5E_b, giving a positive value for K of $+500$ cm^{-1}.

The enormous intensity of the spin-forbidden transition ($\sim 10^5$ times greater than most such bands) and the general absorption (brown color)

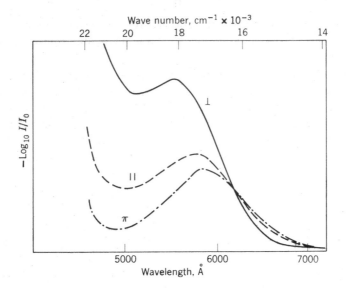

Fig. 33. Crystal spectrum of Al(Mn)(acac)$_3$ at 77°K with light incident on the (001) face. The ϵ_{max} of the parallel component is about 120 (330).

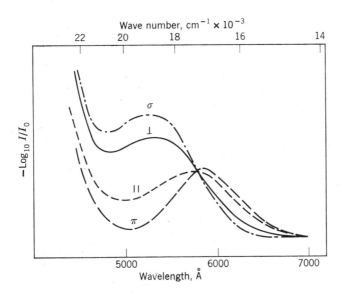

Fig. 34. Crystal spectrum of Al(Mn)(acac)$_3$ at 77°K with light incident on the (100) face (330).

indicate an almost complete mixing of all the triplet and quintet levels in this region by spin–orbit coupling, rhombic fields, and molecular orbital formation (330).

Dingle (92) determined the polarized spectrum of an undiluted crystal of Mn(acac)$_3$ and found no splitting of the near infrared band at 9000 cm^{-1} even at liquid nitrogen temperatures. This was interpreted as a result of a Jahn-Teller distortion of the ^5E_a ground state. However, an x-ray study (272) has shown that there is no evidence for such an effect. In a later study (97) of solution spectra, the analogous band at 8500 cm^{-1} (CCl$_4$ solution) is now interpreted as a ligand-to-metal charge transfer.

VII. CONFIGURATION d^5

A. $3d^5$

The half-filled d shell is especially stable when the crystal field is weak, because both the e and t_2 subshells are half-filled, and the pairing energy is zero. In strong octahedral fields the e orbitals become less available and the ion becomes more readily reduced to the stable d^6 configuration. In

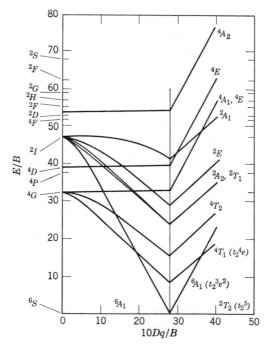

Fig. 35. Tanabe-Sugano energy diagram for the d^5 configuration (75,386).

TABLE XXIV
Absorption Band Maxima for the $Mn(H_2O)_6^{2+}$ Ion (182)

Crystal	Band max., cm^{-1}	ϵ	$\Delta\nu_{1/2}$, cm^{-1}	Transition
Aqueous solution of an Mn^{2+} salt	18,900	0.008	2,500	$^6A_1 \to {}^4T_{1g}(^4G)$
	23,000	0.006	2,500	$^6A_1 \to {}^4T_{2g}(^4G)$
	25,000	0.017	500 (doublet)	$^6A_1 \to \begin{cases} {}^4E_g(^4G) \\ {}^4A_{1g}(^4G) \end{cases}$
	28,000	0.010	1,400	$^6A_1 \to {}^4T_{2g}(^4D)$
	29,750	0.009	500	$^6A_1 \to {}^4E_g(^4D)$
	32,400	0.006	4,000	$^6A_1 \to {}^4T_{1g}(^4P)$

weak fields the ground state ($^6A_1$) is derived from the ^6S state of the free ion. Since there is only one sextet state in d^5, all excited states are either quartets or doublets. In strong fields the ground state is derived from the t_2^5 configuration, and is therefore $^2T_2$, as shown in Figure 35.

The best-known examples of the d^5 configuration are Mn^{2+} and Fe^{3+}. The spectrum of Mn^{2+} has been extensively studied in many crystal lattices. It is of special interest because of the wide use of manganese phosphors, many of which depend upon the presence of divalent Mn^{2+}. The optical spectrum is weak and spin-forbidden, so that the ion has to be present in relatively high concentrations and undiluted manganese compounds are usually employed.

1. $Mn(H_2O)_6^{2+}$

The crystal spectrum of a hydrated Mn^{2+} salt has been determined by Holmes and McClure (182). The spectrum was found to be no sharper in the crystal than in solution. Bands were observed at the frequencies given in Table XXIV for the $Mn(H_2O)_6^{2+}$ ion in solution and in the crystal lattice.

2. $RbMnF_3$

The crystal absorption spectrum of $RbMnF_3$ has been studied by Mehra and Venkateswarlu (258a). Earlier studies by Ferguson, Guggenheim, and Tanabe (122,122a) on the fine structure of the two sharp bands of Mn^{2+}, ($^4A_{1g}(^4G)$, $^4E_g(^4G)$), and $^4E_g(^4D)$ were made with the purpose of finding the effects of magnetic exchange interaction in them. Stevenson (374d,374e) has also measured the spectra of several manganese fluorides. In the perovskite structure of $RbMnF_3$, the Mn^{2+} ion is located at the body-centered position of the cubic cell while six F^- ions located at the face-centered positions form a regular octahedron around Mn^{2+}. Rubidium ions are found at the corners. The edge of the cell is 4.2396 Å, which determines all the interatomic distances (418h). The same structure is

TABLE XXIVA

Observed and Calculated Bands for $RbMnF_3$ at 77°K (258a)

Transition	Band max., cm^{-1}	$\Delta\nu_{1/2}$, cm^{-1}	$f \times 10^7$	Calculated energy[a]
$^6A_{1g} \rightarrow$				
$^4T_{1g}(^4G)$	19,150	1,100	2.0	19,060
$^4T_{2g}(^4G)$	23,106	1,300	1.6	23,257
$^4E_g(^4G)$	23,195			⎫
	25,336	130	9.2	⎬ 25,320
$^4A_{1g}(^4G)$	25,543			
	25,759			⎭
	27,917			⎫
$^4T_{2g}(^4D)$	28,129		1.1	⎬ 28,463
	28,258			
	28,385			⎭
$^4E_g(^4D)$	30,140	220		⎱ 30,136
	30,478			⎰
	32,446			⎫
$^4T_{1g}(^4D)$	32,798		1.5	⎬ 32,927
	38,000			
	39,700			⎭
$^4A_{2g}(^4F)$	41,158			40,952
$^4T_{1g}(^4F)$	41,926			41,877
$^4T_{2g}(^4F)$	43,914			44,687

[a] $B = 840$, $C = 3080$, $Dq = 780$, and $\alpha = 76\ cm^{-1}$.

retained below the Neel temperature. The absorption spectrum was recorded (258a) at room temperature and at 77°K; the results at low temperature are listed in Table XXIVA. The oscillator strengths were calculated from measuring the area under the absorption curves, and it is estimated that the error may be as great as 25%. Cubic-field calculations were made involving the parameters B, C, Dq, and α. The Trees correction factor α was fixed at 76 cm^{-1}. At 77°K, the parameters that gave the best fit to the observed absorption band positions were $B = 840$, $C = 3080$, and $Dq = 780\ cm^{-1}$. Calculations were also made on the energies of the quartet terms when spin–orbit coupling is included with $\zeta = 320\ cm^{-1}$.

3. MnF_2

The absorption spectra of the manganese halides have been studied in some detail. The polarized spectrum of a single crystal of MnF_2 has been investigated by Stout (378). Stout interpreted the spectrum in terms of a perturbation of the free-ion energy levels by an octahedral field, including the reduction of the magnitude of the e orbitals on the Mn^{2+} ion by a factor $(1 - \epsilon)^{1/2}$ in calculating the coulomb and exchange integrals. The factor ϵ is the "covalence parameter" of Koide and Pryce (204) who have discussed the energy levels of Mn^{2+} in detail. Theoretical calculations on the spectra of MnF_2 have also been carried out by Low and Rosengarten (226) who included spin–orbit interaction.

MnF_2 crystallizes with the rutile type structure. It has a tetragonal unit cell, space group $P4_2/nnm$, D_{4h}^{14}, with the Mn atoms at positions $(000; \frac{1}{2} \frac{1}{2} \frac{1}{2})$ and the F atoms at $(uuO; u + \frac{1}{2}, \frac{1}{2} - u, \frac{1}{2})$. The cell constants are $a_0 = 4.8734$, $c_0 = 3.3099$ Å, and $u = 0.305$. The six fluoride ions are arranged around the Mn ion in a distorted octahedron. Two of these at $\pm(uuO)$ are at a distance 2.14 Å and the remaining four at a distance 2.11 Å (33,377). In the crystal, the site symmetry of the Mn^{2+} ion is orthorhombic, D_{2h}.

Stout (378) made polarized measurements on a crystal of MnF_2 at 298, 77, 64, and 20.4°K. Extra bands appeared in the spectrum as the temperature was lowered. Table XXV gives a comparison of the observed

TABLE XXV
Observed and Calculated Energy Levels for MnF_2

	Ref. 378, axial spectrum				Ref. 226	
Level	Observed, cm^{-1}	$\Delta\nu_{1/2}$, cm^{-1}	ϵ_{max}	Calculated, cm^{-1}	Calculated,[a] cm^{-1}	Calculated,[b] cm^{-1}
$^6A_1$	0			0	-19.57	110 Γ_7
$^4T_1(^4G)$	19,440	1,900	0.038	19,440	19,550	19,497 Γ_6
$^4T_2(^4G)$	23,500	2,400	0.025	23,680	23,530	23,514 Γ_6
$^4A_1(^4G)$	25,190 25,300 $\big\}$[c]	290	0.248	25,190	25,438	
$^4E(^4G)$	25,500	460	0.057	25,480	25,542	25,438 Γ_6
$^4T_2(^4D)$	28,120 28,370 $\big\}$	$\sim$1,500	0.034 0.033	29,650	28,530	28,370 Γ_6
$^4E(^4D)$	30,230	850	0.074	30,240	30,115	29,955 Γ_7
$^2T_1(^2I)$	32,200				32,455	
$^4T_1(^4P)$	33,060	1,200	0.061	32,960	32,999	32,992 Γ_7
$^4A_2(^4F)$	39,000			39,000	38,824 $^2T(^2D)$	
$^4T_1(^4F)$	41,400			43,200	41,793	41,724 Γ_7
$^4T_2(^4F)$				44,440		

[a] Values are centers of gravity of multiplet using parameters $B = 820$, $C = 3150$ $\alpha = 76$, $\zeta = 320$, and $Dq = 750$ cm^{-1}.

[b] Values are for $\Gamma_6(E')$ or $\Gamma_7(E'')$ component of multiplet with parameters $B = 801 \pm 2$, $C = 3158 \pm 64$, $\alpha = 76$, $\zeta = 320$, and $Dq = 752 \pm 18$ cm^{-1}. α is the Trees' correction factor (394).

[c] This band was observed at 25,170 cm^{-1} in the $\perp c$ spectrum and at 25,270 cm^{-1} in the $\|c$ spectrum.

levels at 298°K of Stout (378), his own calculated levels, and the calculated levels of Low and Rosengarten (226). The levels have been designated by the terms for irreducible representations of the octahedral group followed by the free-ion level to which these reduce as the ligand field goes to zero.

Table XXVI shows the bands observed in the region 25,000–30,000 cm^{-1} at 20.4°K with the electric vector polarized $\perp$ and $\parallel$ to the c axis of the crystal. The very weak lines listed in the $\perp$ spectrum at 25,500, 25,530, 25,570, and 25,600 cm^{-1} have a regular spacing of about 35 cm^{-1} and are thought to represent harmonics of a crystal vibration (378).

Low and Rosengarten (226) have also carried out a calculation for the levels in Mn^{2+} including the Γ_8 (U') components of the multiplet levels. A value of 320 cm^{-1} was used for the spin–orbit coupling constant ζ. The calculated ground-state splitting is 1.0×10^{-3} cm^{-1} and the wave function of the ground state is 99.91% $^6A_1(^6S)$ + 0.07% $^4T_1(^4P)$ + 0.03% $^4T_1(^4G)$. The relativistic crystal-field splitting of the $^6S_{5/2}$ ground state of Mn^{2+} has also been discussed by Van Heuvelen (401a) using the theory of relativistic effects described by Wybourne (421b).

TABLE XXVI

Observed Energy Levels for MnF$_2$ at 20.4°K (378)

$E \perp c$ axis			$E \parallel c$ axis		
Band max., cm^{-1}	$\Delta\nu_{1/2}$	Intensity	Band max., cm^{-1}	$\Delta\nu_{1/2}$	Intensity
25,244	6	w	25,239	6	m
25,280	50	s	25,259	3	vw
25,434	10	w	25,291	6	w
25,500	15	vw	25,441	40	s
25,530	15	vw			
25,570	15	vw			
25,600	15	vw			
28,024	15	m	28,024	10	m
28,160	15	vw			
28,240	15	w	28,240	15	w
28,280	15	vw			
28,420	15	vvw	28,350	15	vw
28,480	15	vvw	28,480	15	vvw
29,930	20	vw	29,930	20	vvw

4. MnCl$_2$

The absorption spectrum of MnCl$_2$ crystals at 78°K has been determined by Pappalardo (306). The Mn^{2+} ion has a slightly distorted octahedral arrangement of Cl$^-$ ions around each Mn^{2+} ion. The unit cell space

TABLE XXVII

Observed and Calculated Energy Levels for $MnCl_2$ at 78°K

(Using the parameters (1) $B = 700$ cm^{-1}, $C = 3150$ cm^{-1}, $\alpha = 76$ cm^{-1}, $\zeta = 240$ cm^{-1}, $Dq = 740$ cm^{-1}, and (2) $B = 758 \pm 73$ cm^{-1}, $C = 3082 \pm 200$ cm^{-1}, $\alpha = 76$ cm^{-1}, $\zeta = 240$ cm^{-1} $Dq = 763 \pm 50$ cm^{-1}; values under (I) and in last column and centers of gravity of multiplet terms)

| Level | Observed at 78°K, ref. 306 | | | Calculated Ref. 226 | | Ref. 379[a] |
	Band max., cm^{-1}	$\Delta\nu_{1/2}$	α_{max}, mm^{-1}	(I)	(2)	
$^6A_1(^6S)$	0	—	—	-11.4	$-237\ \Gamma_7$	0
$^4T_1(^4G) + h\nu$	18,500	800	0.65	18,565	18,219 Γ_7	17,900
$^4T_2(^4G) + h\nu$	22,000	750	1.45	22,121	22,146 Γ_6	22,600
^4E or $^4A_1(^4G)$	23,360[b]					
^4E or $^4A_1(^4G)$ + $h\nu$	⎧ 23,590[c]	65	2.2			
	23,700[c]	90	0.9			
	23,825	90	1.25	24,255	24,266 Γ_7	24,400
	23,940	175	0.8			
	24,190	70	0.25			
	⎩ 24,250	70	0.25			
$^2A_2, ^2T_1(^2I)$	25,310					
$^4T_2(^4D)$	26,310	35	0.15			
	26,415[d]					
$^4T_2(^4D) + h\nu$	⎧ 26,540	14	2.86			
	26,558	14	2.5			
	26,650[e]	45	1.2			
	26,780	55	1.47			
	26,895	60	1.2	26,638	26,845 Γ_6	27,800
	27,030	80	1.1			
	27,120	75	0.9			
	27,250					
	⎩ 27,370					
$^4E(^4D) + h\nu$	⎧ 28,065	80	4.75			
	28,170					
	28,290	80	1.23	28,092	28,494 Γ_7	28,180
	28,375					
	⎩ 28,520					
$^4T_1(^4P)$				31,842	—	31,300
$^4A_2(^4F)$				—	—	34,200
$^4T_1(^4F)$				39,293	39,951 Γ_7	42,200
$^4T_2(^4F)$				41,934	42,477 Γ_6	42,200

[a] Energy levels calculated using free-ion values for the Racah parameters B and C, the covalence parameter (204), $\epsilon = 0.13$, and $Dq = 830$ cm^{-1}. Spin–orbit coupling and Trees correction ignored.

[b] Very weak.

[c] Two components 25 cm^{-1} apart.

[d] Weaker than 26,310 cm^{-1}.

[e] Doublet 45 cm^{-1} separation.

group $R\bar{3}m$, $D_{3d}{}^5$ is rhombohedral containing one formula unit and has dimensions $a_0 = 6.20$ and $\alpha = 34°25'$. The anions are cubic close-packed (314). The observed bands with the transition assignments are given in Table XXVII, together with the calculated values of Stout (379) and Low and Rosengarten (226) who have also considered the effect of spin–orbit coupling and have determined the Γ_6 and Γ_7 levels. Column 1 gives their calculations (averaged over multiplet components) based on the parameters $B = 700$ cm^{-1}, $C = 3150$ cm^{-1}, $\alpha = 76$ cm^{-1}, $\zeta = 240$ cm^{-1}, and $Dq = 740$ cm^{-1}. Column 2 is a further set of levels obtained by consideration of the observed bands obtained by Pappalardo (306), improving the parameters to $B = 758 \pm 73$ cm^{-1}, $C = 3082 \pm 200$ cm^{-1}, $\alpha = 76$ cm^{-1}, $\zeta = 240$ cm^{-1}, and $Dq = 763 \pm 50$ cm^{-1}, followed by a least squares analysis. Stout (379) also carried out a series of less detailed calculations following the publication of Pappalardo's spectra, the results of which are also given in the last column of Table XXVII. Free-ion values of the Racah coefficients were used, together with $Dq = 830$ cm^{-1} and covalence parameter 0.13 (204). The field is assumed to be cubic, and spin–orbit coupling is ignored.

5. MnBr₂

The absorption spectrum of $MnBr_2$ crystals at 78°K has been determined by Pappalardo (306). $MnBr_2$ is hexagonally close packed with a hexagonal unit cell, space group $P\bar{3}m1, D_{3d}{}^3$, of dimensions $a_0 = 3.81$, $c_0 = 6.26$ Å, and $u = 0.25$. The atoms are at the following special positions Mn:(000); Br: ($\frac{1}{3}$ $\frac{2}{3}$ u; $\frac{2}{3}$ $\frac{1}{3}$ $\bar{u}$) (124). Stout (379) has carried out some cubic-field calculations on the energy levels in $MnBr_2$ which are given with the observed bands of Pappalardo, Table XXVIII. The method is similar to that used in the $MnCl_2$ calculations, with $Dq = 940$ cm^{-1} and the covalence parameter 0.15 (204). Spin–orbit coupling was not taken into account.

The vibrational structure and possible trigonal splittings of the $MnBr_2$ bands are discussed by Pappalardo. An interesting feature is a simple periodicity of ~ 150 cm^{-1}, which is particularly evident for the absorption at 26,500 cm^{-1}. This, and other observed splittings, have not yet been definitely assigned.

The optical absorption spectrum of crystals of Mn^{2+} in a tetragonal field has been investigated by Goode (147). Three manganese compounds were studied in detail: $MnCl_2 2H_2O$, $2(NH_4Cl)MnCl_2 2H_2O$, and $KClMnCl_2$-$2H_2O$. The solid solutions of the form $n(NH_4Cl)MnCl_2 2H_2O$ provide approximately tetragonal sites (D_{4h}) for the Mn^{2+} ions. The crystal structure of the end member of the series when $n = 0$, $MnCl_2 2H_2O$, is monoclinic

TABLE XXVIII
Observed and Calculated Energy Levels for MnBr$_2$

Transition	Observed 78°K, ref. 306			Calculated, ref. 379[a]
	Band max., cm^{-1}	$\Delta\nu_{1/2}$ cm^{-1}	α_{max}, mm^{-1}	
$^4T_1(^4G)$	—	—	—	16,500
$^4T_2(^4G) + h\nu$	21,650[b]	~650	1.2	21,500
^4E or $^4A_1(^4G)$	22,976		0.14	
$^4E,^4A_1(^4G) + h\nu$	23,084	95	1.8	23,000 ($^4A_1$)
	23,174	55	0.5	
	23,215	50	0.8	
	23,247	45	1.0	
	23,297	50	0.98	
	23,347	55	1.1	23,600 (^4E)
	23,400	45	0.4	
	23,435	55	0.6	
	23,475	40	0.78	
	23,538	55	0.6	
	23,065	55	0.4	
$^2A_2,^2T_1(^2I)$	24,120	50	0.01	
$^4T_2(^4D)$	26,115	40	0.2	
$^4T_2(^4D) + h\nu$	26,258	45	3.5	
	26,330[c]	90	1.76	
	26,412	75	3.6	
	26,476[d]	85	2.3	
	26,565	100	3.1	27,200
	26,640[d]	100	2.35	
	26,714	80	1.83	
	26,764[d]	95	1.66	
	26,835	100		
	26,895[d]	85		
$^4E(^4D)$	27,320			
$^4E(^4D) + h\nu$	27,505	120	4.5	
	27,653	120	2.6	27,400
	27,801	90	0.8	
$^4T_1(^4P)$				31,600
$^4A_2(^4F)$				32,800
$^4T_1(^4F)$				42,100
$^4T_2(^4F)$				42,300

[a] Energy levels calculated using free-ion values for Racah parameters B and C, covalence parameter (204) $\epsilon = 0.15$ and $Dq = 940$ cm^{-1}. Spin–orbit coupling ignored.
[b] Broad band, no structure.
[c] Two components 26,295, and 26,326 cm^{-1}.
[d] Two components.

$C2/m, C_{2h}^3$ with cell dimensions $a_0 = 7.40$, $b_0 = 8.77$, and $c_0 = 3.70$ Å, and $\beta = 98°5'$ and two molecules per unit cell (401). The two manganese ions occupy identical sites which are oriented in the same direction. Two water molecules occupy the *trans* positions along the *b* crystal axis with four chlorines lying in a plane of symmetry almost at the corners of a square. The true site symmetry is in fact C_{2h} as the distances to the chlorine atoms are not quite equal (2.60 and 2.57 Å), while the Cl—$\widehat{\text{Mn}}$—Cl angles are 88° 32' and 91° 28'. However, it approximates reasonably closely to D_{4h}. The Mn—H_2O distance is 2.19 Å. Each complex shares its chlorine ions with the two complexes lying on either side of it along the *c* crystal axis, so infinitely long chains are formed in this direction.

The crystal structure of $n(NH_4Cl)MnCl_2 2H_2O$ solid solutions has been investigated by Greenberg and Walden (158), who found that they are formed in the following manner. The substitution of two adjacent NH_4^+ ions in the body centered cubic NH_4Cl lattice by two H_2O molecules occurs simultaneously with the insertion of a Mn^{2+} in the interstitial space between the H_2O molecules (Fig. 36). As the Mn^{2+} concentration increases, the solid solution becomes tetragonal with the H_2O—Mn—H_2O units aligned along the *c* axis and unit cell dimensions of $a_0 = 7.5139$ and $c_0 = 8.245$ Å.

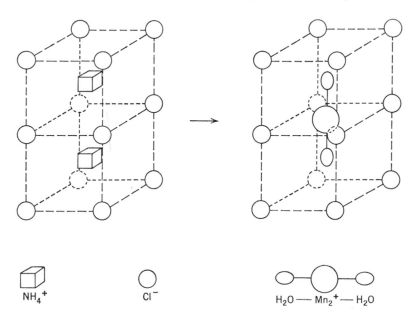

Fig. 36. Mechanism of interstitial insertion of Mn^{2+} ion in the compound cell in $n(NH_4Cl)MnCl_2 2H_2O$ (360).

Spin resonance measurements (360) have shown that when the Mn^{2+} concentration is low in $n(NH_4Cl)MnCl_22H_2O$, the Mn^{2+} ion is in a site of strong axial distortion. When all the NH_4^+ groups are replaced, the $MnCl_22H_2O$ structure is obtained.

The crystal structure of $KClMnCl_22H_2O$ has not been determined but since its optical spectrum is very similar to that of $n(NH_4Cl)MnCl_22H_2O$, it is assumed that the Mn^{2+} ion is in a similar site (D_{4h}) (147).

Spectra of these three compounds were recorded at room temperature (Fig. 37), 77°K, and 4.2°K. When the temperature was lowered to 77°K, considerable fine structure appeared in the bands associated with $^4A_1$, $^4E(^4G)$, called $^4E^a$ in Figure 37, $^4T_2(^4D)$ called $^4T_2{}^2$, $^4A_2$, and $^4T_1(^4F)$ called $^4T_1{}^3$. The energies of the most prominent bands are listed in Table XXIX. Lowering the temperature to 4.2°K had little effect except to sharpen the bands (see notes to Table XXIX).

Important similarities were found to exist between the fine line spectra of the $n(NH_4Cl)MnCl_22H_2O$ solid solutions and $KClMnCl_22H_2O$. The sharp lines near 24,000 cm^{-1}, transitions to $^4A_1$, $^4E^a(^4G)$ levels, exhibit a

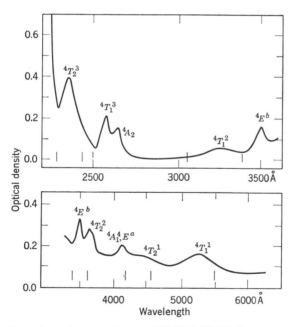

Fig. 37. The absorption spectrum of $KClMnCl_22H_2O$ at room temperature (sample thickness 0.15 cm). The band positions for $2(NH_4Cl)MnCl_22H_2O$ are identical. Also shown are the energy levels calculated for $B = 800$ cm^{-1}, $C/B = 4.00$, and $Dq = 800$ cm^{-1} (147).

TABLE XXIX
Observed Energy Levels for Mn^{2+} in D_{4h} Symmetry at 77°K (147)

Level	MnCl$_2$2H$_2$O at 77°K			2(NH$_4$Cl)MnCl$_2$2H$_2$O at 77°K			KClMnCl$_2$2H$_2$O at 77°K		
	Energy, cm^{-1}	$\Delta\nu_{1/2}$, cm^{-1}	Intensity	Energy, cm^{-1}	$\Delta\nu_{1/2}$, cm^{-1}	Intensity	Energy, cm^{-1}	$\Delta\nu_{1/2}$, cm^{-1}	Intensity
$^4T_1{}^1(^4G)$	18,940	1,900	vs	—					
$^4T_2{}^1(^4G)$	20,910	1,200	s	—					
$^4A_1(^4G)$ $^4E^a(^4D)$	23,750		w	23,856	26	w			
	23,950		w	23,917		w	23,798	50	w
	24,140		w	24,050	100	w	24,015	50	w
	24,346	25	s	24,353	16	vs	24,290	17	vs^c
	24,480		w	24,570	75	w	24,498	55	w
	24,720		w	24,659		w	24,661	37	w
				24,695		w			
$^4T_2{}^2(^4D)$	26,727		vs						
	26,807		vs						
	26,867		s	26,982	11	w	26,843	11	vs
	26,891		w	27,010	13	w	26,907	19	vs
	26,927		w	27,044	26	w	26,951	29	s
	26,969		w	27,284		vw	27,159		w
	27,020		w^a	27,346		vw	27,204		w
	27,068	24	s	27,510		vw	27,265		vw
	27,157	40	s^b	27,746		vw	27,300		vw
	27,220		w	27,815		vw	27,450		vw
	27,370		w						
	27,510		vw						
$^4E^b(^4D)$	28,210		w				28,290		w^d
	28,700	140	s	—			28,560	60	s
	28,930		w				28,790		w
$^4T_1{}^2(^4P)$	30,500	Broad	w	—			—	—	—
	37,988	28	w						
	38,191		w	37,850	Broad	vw			
	38,396		w						
$^4A_2(^4F)$	38,602		vw				37,612	28	s
							37,850		s
							38,086		w
							38,328		vw
$^4T_1{}^3(^4F)$	38,797		w				38,732		s
	38,920		w	38,060	Broad	vw	38,789		s
	39,040	75	s				38,971		w
	39,600		w				39,154		w
$^4T_2{}^3(^4F)$				—			42,550	broad	s

^{a} Increases at 4.2°K. ^{b} Appears as two lines 22 cm^{-1} apart at 4.2°K.
^{c} Faint line at 24,275 cm^{-1} at 4.2°K.
^{d} Breaks into 2 components 60 cm^{-1} apart. Also 2 faint sharp lines appear at 4.2°K at 30,300 and 30,380 cm^{-1} which could be $^4T_1{}^2(^4P)$.

distinct pattern consisting of one intense sharp line flanked on either side by several weak broader lines. The position of the central sharp line varied over a range of 40 cm^{-1} as the manganese concentration was increased, the width of the line being smallest for solutions containing a definite crystal structure corresponding to all the Mn^{2+} ion sites being similar. This pattern is quite different from that observed in other hydrates such as $MnCl_2 4H_2O$, $MnSO_4 4H_2O$, and $MnSiF_6 6H_2O$ (303), which will be discussed later.

The $^4T_2{}^2(^4D)$ absorption is composed of sharp lines superimposed on the low energy side of a broad band. At room temperature the lines become so broad that they disappear completely. This contrasts with the $^4A_1$, $^4E^a$ (^4G) and $^4E^b(^4D)$ structure, which still retains much of its sharpness. The $^4T_2{}^2(^4D)$ structure in $MnCl_2 2H_2O$ and $KClMnCl_2 2H_2O$ is similar but in $2(NH_4Cl)MnCl_2 2H_2O$, it is quite different. The $^4E^b(^4D)$ transition did not show fine structure.

The $^4A_2(^4F)$ and $^4T_1{}^3(^4F)$ spectra of $MnCl_2 2H_2O$ and $KClMnCl_2 2H_2O$ both featured a broad band with a vibrational series superimposed upon it, and a reasonably intense band at higher energies. The corresponding transitions in $2(NH_4Cl)MnCl_2 2H_2O$ were weak and showed no fine structure.

A cubic crystal field calculation was carried out by Goode (147) which included first-order spin–orbit coupling using matrix elements already published (357,386) in an attempt to explain the fine structure observed in these spectra. The values of $B = 800$ cm^{-1}, $C = 3200$ cm^{-1}, and $Dq = 800$ cm^{-1} gave the best overall fit to the quartet energy levels observed at room temperature. The high-energy bands tend to be depressed from their calculated values but this is expected when configuration interaction is neglected. The spin–orbit coupling parameter, ζ, was chosen to be 260 cm^{-1} (104). The energies of the quartet levels when $\zeta = 0$ and their splittings when $\zeta = 260$ cm^{-1} for $Dq = 800$ cm^{-1} are listed in Table XXX. Some of the energy levels are shown plotted against Dq in Figure 38. The large splitting in $^4T_2{}^2(^4D)$ is noteworthy. No first-order splitting occurs in this term either in the weak field limit (^4D) or the strong field limit because these cases represent the half-filled shells d^5 and $t_2{}^3e^2$. However, in the intermediate case, the mixing is sufficient to give this term the largest splitting of all the quartets.

The relative intensity of the transitions to the excited quartets and doublets is of interest since the electric dipole transitions from a pure sextet ground state are spin-forbidden. However, such transitions are observed as spin–orbit interaction causes the ground and excited states to be impure. The extent of this mixing, discussed by Goode (147), gives some indication of the relative intensity of the resultant transition. Englman (104) has calculated the matrix element for an electric dipole transition

N. S. HUSH AND R. J. M. HOBBS

between the $^6A_1$ ground state and a quartet excited state. This expression contains three pairs of terms involving the quartet present in the ground state, the sextet present in the excited state, and the spin mixing of the odd-parity charge transfer states. Since the absorption coefficient depends upon the square of the dipole matrix, it is more useful to use the sum of the

TABLE XXX

Energies of the d^5 Quartet Terms for Mn^{2+} and Their Splitting when Spin–Orbit Interaction is Introduced

$[B = 800, C/B = 4, \zeta = 260,$ and $Dq = 800$ cm^{-1} (147)]

Level		Energy, cm^{-1}	
		No spin–orbit	With spin–orbit
$^4T_1{}^3(^4F)$	Γ_7	$+$ 63	41,045
	$\Gamma_{8'}$	$+$ 57	41,049
	Γ_8	40,982 $-$ 72	40,910
	Γ_6	$-$ 230	40,752
$^4A_2(^4F)$	Γ_8	40,000 $+$ 117	40,117
$^4T_1{}^2(^4P)$	Γ_6	$+$ 61	32,887
	Γ_8	$+$ 22	32,848
	Γ_7	32,836 $-$ 45	32,781
	$\Gamma_{8'}$	$-$ 46	32,780
$^4E^b(^4D)$	Γ_6	$+$ 57[a]	29,657
	Γ_8	29,600 $-$ 12	29,588
	Γ_7	$-$ 39	29,561
$^4T_2{}^2(^4D)$	Γ_7	$+$ 156	27,891
	Γ_8	27,735 $+$ 67	27,802
	Γ_6	$-$ 134	27,601
	$\Gamma_{8'}$	$-$ 145	27,590
$^4E^a(^4G)$	Γ_8	$-$ 10	23,990
	Γ_6	24,000 $-$ 11	23,989
	Γ_7	$-$ 18	23,982
$^4A_1(^4G)$	Γ_8	24,000 $-$ 23	23,977
$^4T_2{}^1(^4G)$	$\Gamma_{8'}$	$+$ 32	21,968
	Γ_6	$+$ 29	21,965
	Γ_8	21,936 $-$ 56	21,880
	Γ_7	$-$ 96	21,840
$^4T_1{}^1(^4G)$	Γ_6	$+$ 43	18,236
	Γ_8	$+$ 4	18,197
	$\Gamma_{8'}$	18,193 $-$ 51	18,142
	Γ_7	$-$ 67	18,126
$^6A_1$	Γ_7, Γ_8	0–13.085	

[a] In ref. 147 this is quoted as -57, but presumably $+57$ is correct.

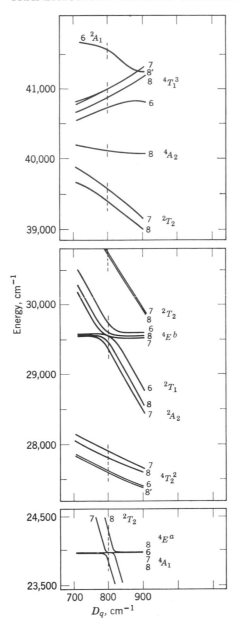

Fig. 38. Energy levels for the d^5 configuration in a cubic field including spin–orbit interaction. Only the more interesting regions of the spectrum are shown and Γ_6, Γ_7, Γ_8, $\Gamma_{8'}$, are abbreviated to 6, 7, 8, 8′. $B = 800$ cm^{-1}, $C/B = 4.00$, and $\zeta = 260$ cm^{-1} (147).

squared eigenvectors, expressed as a percentage. In this way, the ground state contains 0.0551% quartet impurity when $Dq = 800$ cm^{-1}, which is the major source of intensity.

Similar considerations may be applied to doublet states, (Table XXXI). However, Goode suggests that a more likely source of doublet intensity comes from the admixture of quartet states into the doublets. A list of total quartet impurity in the doublet states is given in Table XXXI, together with the energy difference between the doublet and quartet chiefly responsible for the mixing. Since the doublet levels contain only 0.5–0.2% quartet impurity, they are not expected to be observed.

As the Mn^{2+} ions are at sites of tetragonal symmetry, calculations were also made (147) including a tetragonal perturbation which was

TABLE XXXI

Calculated Energy of d^5 Doublet Levels and Their Total Quartet Character for Mn^{2+}

($B = 800$ cm^{-1}, $C/B = 4$, $\zeta = 260$ cm^{-1} and $Dq = 800$ cm^{-1})

The Quartet Level Interacting with Each Doublet and Their Separation in Energy is Also Shown (147)

Level	Calculated Energy, cm^{-1}	% quartet character, main impurity quartet and their separation in cm^{-1}		
		%		cm^{-1}
$^2A_1 \ \Gamma_6$	41,522	13.8[a]	$^4T_1{}^3$	800
$^2T_2 \ \Gamma_7$	39,547	0.91	$^4T_1{}^3$	1,500
$^2T_2 \ \Gamma_8$	39,379	17.2	$^4A_2$	740
$^2T_1 \ \Gamma_8$	36,698	0.54	$^4T_1{}^2$	4,120
$^2E \ \Gamma_8$	36,356	0.69 (0.25)	$^4T_1{}^2$	3,580
$^2T_2 \ \Gamma_7$	36,062	0.43	$^4T_1{}^2$	3,280
$^2E \ \Gamma_8$	32,005	2.2 (1.1)	$^4T_1{}^2$	780
$^2T_2 \ \Gamma_7$	30,658	1.4	$^4E^b$	1,090
$^2T_2 \ \Gamma_8$	30,647	0.82	$^4E^b$	1,060
$^2T_1 \ \Gamma_6$	29,699	30.5	$^4E^b$	150
$^2T_1 \ \Gamma_8$	29,428	8.3	$^4E^b$	160
$^2A_2 \ \Gamma_7$	29,310	2.4	$^4T_2{}^2$	1,420
$^2T_2 \ \Gamma_8$	24,248	1.0	$^4T_2{}^1$	2,370
$^2T_2 \ \Gamma_7$	23,829	0.95 (0.15)	$^4T_2{}^1$	1,990

[a] Figures in parentheses give indication of relative absorption due to sextet impurity ($= \%/5.51 \times 10^{-2}$).

expected to be of the same order of magnitude as that of spin–orbit coupling. The matrix elements, including spin–orbit coupling, were evaluated. Simplification was made by neglecting states where interaction was known to be small with $^4T_2{}^2(^4D)$. The splitting was determined in terms of one tetragonal variable $(4\delta-3\mu)$, where δ is the splitting of the t_2 levels, (between xy, yz, and xy) and μ is the splitting of the e levels (between z^2 and $x^2 - y^2$). Reasonable agreement was then obtained with the observed frequencies for the $^4T_2{}^2$ level in the hydrate spectra. The splittings calculated for $^4A_1(^4G)$, $^4E^a(^4G)$, $^4A_2(^4F)$ were quite small so the fine structure observed in the spectra is probably vibrational.

The crystal spectrum of $MnCl_2 2MgCl_2 2H_2O$ has been determined by Tsujikawa and Couture (400) in the visible region and by Le Paillier-Malecot (297) using polarized light in the ultraviolet region. The crystal has a rhombohedral unit cell with a threefold axis of symmetry (148). The spectra were measured at 77, 20, and 4.2°K, but are quoted only for 4.2°K (Table XXXII) as some extra bands appeared at this temperature (marked by an asterisk). There was a general sharpening of the bands in going from 77 to 4.2°K. The spectra were recorded with the electric vector $\parallel$ and $\perp$ to the threefold axis. The four bands in the region 36,610–37,250

TABLE XXXII
Observed Absorption Bands for $MnCl_2 2MgCl_2 2H_2O$ at 77°K (297,400)

Band max., cm^{-1}	$\Delta\nu_{1/2}$, cm^{-1}	Band max., cm^{-1}	$\Delta\nu_{1/2}$, cm^{-1}	Assignment
Ref. 400 23,758	4			
23,860	8			
23,928	8			
23,933	7			
23,940	7			
23,951	7			
24,043	40			
24,113	16			
$E \parallel A_3$		$E \perp A_3$		
Ref. 297 —	—	28,290[a]	50	
28,540[a]	80	28,540[a]	80	
⎡36,610	10	—	—	⎫
⎢36,812	5	36,812	5	⎬ $^6A_1 \rightarrow {}^4E(^4T_1,{}^4F)$
⎢37,030	10	37,030	10	⎪
⎢37,250[a]	50	37,250[a]	50	⎭
⎣37,000	1000	37,000	700	$^6A_1 \rightarrow {}^4A(^4T_1,{}^4F)$
⎡38,340	30	—	—	$^6A_1 \rightarrow {}^4A(^4T_2,{}^4F)$
⎣38,550	250	—	—	$^6A_1 \rightarrow {}^4E(^4T_2,{}^4F)$

[a] Extra bands appearing at 4.2°K.

cm^{-1} are attributed to the effects of spin–orbit coupling on the $^4E(^4T_1, ^4F)$ state.

6. Further Mn^{2+} Hydrated Salts

Pappalardo (303) has investigated the absorption spectra of crystals of four manganous hydrated salts, $MnSO_4 4H_2O$, $MnCl_2 4H_2O$, $MnSiF_6$-$6H_2O$, and $Mn(NH_4)_2(SO_4)_2 6H_2O$, at 77 and 20°K, while Schläfer (355) also studied the crystal spectrum of $MnCl_2 4H_2O$.

The crystal structures of these compounds are not accurately known except that $MnSO_4 4H_2O$, $MnCl_2 4H_2O$, and $Mn(NH_4)_2(SO_4)_2 6H_2O$ are monoclinic. At room temperature, a crystal of $MnSiF_6 6H_2O$ has a rhombohedral unit cell (313) with $a_0 = 6.45$ Å, $\alpha = 96° 53'$ and a trigonal arrangement of F^- ions around the Mn^{2+} ion (423). However, at 222°K, a phase change occurs which probably alters the environment of the Mn^{2+} ion (399). The bands observed at 20°K in the four compounds are listed in Table XXXIII together with an indication of the strength of the band where reported. Extra bands are observed in the spectrum of $MnSiF_6 6H_2O$ at 4°K (400).

The number of lines observed increases from the Tutton salt and fluorosilicate to the chloride and sulfate. This is compatible with the lowering of the site symmetry which is trigonal in the first two cases and is rhombic in the latter two. The number of lines observed cannot be explained on spin–orbit considerations alone so that some presumably belong to vibrational excitations.

The polarized absorption spectra of a number of crystalline hydrates at 20°K have been studied by Le Paillier-Malecot and Couture (296) and by Le Paillier-Malecot (298). Spectra were recorded for various directions of the electric and magnetic vectors of the polarized light relative to the crystalline directions for $MnSO_4 2K_2SO_4 H_2SO_4 2H_2O$ (orthorhombic), $MnSO_4 KHSO_4 2H_2O$ (triclinic), and $Mn_3Ce_2(NO_3)_{12} 24H_2O$ (rhombohedral). Since the crystal structures of these compounds are not known very precisely, especially at low temperatures, it is difficult to give any assignments to the observed bands. The region studied was 24,700–25,300 cm^{-1}.

7. Mn^{2+} in ZnS and MgO

McClure (234) has recorded the optical spectrum of pure MnS and of Mn^{2+} in the host lattice ZnS in his study of the optical spectra of exchange-coupled Mn^{2+} ion pairs. Pratt and Coehlo (336) have studied the spectrum of MnO above and below the Neel temperature and interpreted their observed bands on the theory of Tanabe and Sugano (386).

TABLE XXXIII. Observed Energy Levels for Manganous Hydrated Salts at 20°K (303)

Level	$MnSO_4 4H_2O$			$MnCl_2 4H_2O$			$MnSiF_6 6H_2O$			$Mn(NH_4)_2$-$(SO_4)_2 6H_2O$
	Band max., cm^{-1}	$\Delta\nu_{1/2}$	Intensity	Band max., cm^{-1}	$\Delta\nu_{1/2}$	Intensity	Band max., cm^{-1}	$\Delta\nu_{1/2}$	Intensity	Band max., cm^{-1}
A	24,838	30	s	24,318	26	s	24,890			24,728
	24,882	30	s	24,462	45	s	24,946			24,795
				24,500	42	w	25,002			
	24,950	30	vs	24,594	42	s	25,075	65	s	
							25,170			25,063
$A_1 \rightarrow {}^4A_1({}^4G)$	25,025									25,144
$E^a({}^4G)$	25,100	65	w							
	26,176	37	s							
B	25,221		w	24,802	45		25,330	85	s	
	27,278	68	s	24,919	60		25,420	65	s	
	25,311		w	25,300						
C	25,826	30	vw	27,137	19	vs	26,413			
	25,880			27,174	26	s	28,900			
				27,207	28	s				
	25,960									
D	27,292	27	w	27,300						
$^6A_1 \rightarrow {}^4T_2({}^4D)$	27,352	28	w	27,426						
				27,525	77	m				
	27,518	54	w	27,570		m				
	27,586			27,770						
	27,640									
	27,731		vs							
	27,778									
	27,855									
	27,910									
	27,980		s	29,070	77	s				
	28,050			29,155	95	s				
	28,090			29,330						
E	28,177									
	28,248									
	28,280									
	28,296									
	29,369									
	28,425	48	vw							
	28,465									
	28,498									
	28,670									
F										
$^6A_1 \rightarrow {}^4E({}^4D)$	29,700			31,800[a]						
$^6A_1 \rightarrow {}^4T_2({}^4D)$				40,000[a]						
$^6A_1 \rightarrow {}^4T_1({}^4F)$				42,500[a]						

[a] Ref. 355.

The ground-state splitting for Mn^{2+} has been found to be 5.7×10^{-3} cm^{-1} in an MgO environment (410). Calculations of this splitting, and of excited energy levels for a number of sets of crystal field parameters have been made by Gabriel, Johnston, and Powell (139,335).

8. $NH_4Fe(SO_4)_2 12H_2O$

Isoelectronic with the Mn^{2+} ion is the Fe^{3+} ion, but in this case charge-transfer bands permit the easy observation of only the lowest three transitions. Schläfer (355) has measured the absorption spectrum of a crystal of $NH_4Fe(SO_4)_2 12H_2O$ which has space group $Pa3, T_h^6$ and lattice constant 12.322 Å (176). The Fe^{3+} ion is surrounded by a nearly regular octahedron of H_2O molecules. The observed bands are given in Table XXXIV together with the extra weak bands obtained by Pappalardo (304) by further interpretation of the data of reference 355.

Reliable experimental data for Fe^{3+} is scarce but Low and Rosengarten (226) have made a series of calculations to predict the energy levels in an octahedral crystal field using the following parameters: $B = 730$, $C = 3150$ cm^{-1}, $\alpha = 90$ cm^{-1}, $\zeta = 420$ cm^{-1}, and $Dq = 1350$ cm^{-1}. Their results are included in Table XXXIV for comparison. The Dq value is extraordinarily small for a trivalent ion.

TABLE XXXIV

Observed and Calculated Energy Levels for Fe^{3+} in an Octahedral Environment of Water Molecules [$NH_4Fe(SO_4)_2 12H_2O$]

	Observed			Calculated	
Assignment, ref. 355	Band max., cm^{-1} (ref. 355)	Band max., cm^{-1} (ref. 304)		Band max., cm^{-1} (ref. 266)	Assignment, ref. 226
$^6A_1 \rightarrow {}^4T_1(^4G)$	12,550			13,127	$^4T_1(^4G, {}^4P)$
$^6A_1 \rightarrow {}^4T_2(^4G)$	18,500	18,200		18,176	$^4T_2(^4G, {}^4F, {}^4D)$
		24,500		24,653	$^4A_1(^4G)$
		27,500		26,749	$^4T_2(^4D, {}^4G)$
		29,000		28,701	$^4E(^4D)$
		31,500		32,358	$^2T_1(^2G, {}^2F, {}^2F)$

9. Fe^{3+} in $NaMgAl(C_2O_4)_3 9H_2O$

Piper and Carlin (325) have studied the polarized crystal spectrum of $NaMgAl(C_2O_4)_3 9H_2O$ with part of the Al^{3+} ion replaced by Fe^{3+}. The crystal structure and site symmetry of the trivalent ion has been discussed under d^2 configurations, p. 294. It is assumed that the symmetry of the Fe^{3+} ion in $NaMgFe(C_2O_4)_3 9H_2O$ is the same as in the host crystal. The crystal structure of potassium ferritrioxalate has been studied by Herpin (180). The monoclinic unit cell, space group $P2_1/c$, C_{2h}^5, with dimensions $a_0 = 7.66$, $b_0 = 19.87$, $c_0 = 10.27$ Å, $\beta = 105°6'$, has four $K_3Fe(C_2O_4)_3 3H_2O$ and formula units. The iron atoms are octahedrally surrounded by six carboxyl oxygens with Fe—O $= 2.01$–2.06 Å.

10. $(NH_4)_3Fe(C_3H_2O_4)_3$

Hatfield (173) has determined the polarized crystal spectrum of ammonium trismalonatoferrate(III), $(NH_4)_3Fe(C_3H_2O_4)_3$, and compared the results with those of Piper and Carlin (325) for the trisoxalate complex. The crystal structure of the compound was not known, so the face exhibiting maximum dichroism was determined by microscopic examination; of the other two planes perpendicular to this face, one showed similar dichroism and the other very little. The observed bands and assignments for $NaMgFe(C_2O_4)_3 9H_2O$ (325) and $(NH_4)_3Fe(C_3H_2O_4)_3$ (173) are listed in Table XXXV.

The components of ^4G were assigned for the trisoxalate complex as shown in Table XXXV. All the transitions are spin-forbidden and hence the intensities will be low. The $^4T_1$ and $^4T_2(^4G)$ transitions are $\perp$ polarized in both complexes, while the transition to $^4A_1$, ^4E is more intense in the $\parallel$ polarization, (Fig. 17). The axial and $\perp$ spectra were identical. The weak shoulder at 22,990 cm^{-1} in the trisoxalate complex is thought to be splitting of the $^4A_1$, ^4E in the trigonal field. From the assigned transitions the parameters Dq, B, C were calculated to be 1522, 609, and 3283 cm^{-1} for the trisoxalate complex and 1419, 609, and 3125 cm^{-1} for the corresponding malonate complex. The next highest state is then ^4D at 26,800 cm^{-1} in the oxalate, its $^4T_2$ component is calculated to occur at about 25,000 cm^{-1}.

TABLE XXXV

Observed Absorption Bands, Polarizations, and Assigned Transitions for
$NaMgFe(C_2O_4)_3 9H_2O$ (325) and $(NH_4)_3Fe(C_3H_2O_4)_3$ (173)

	$NaMgFe(C_2O_4)_3 9H_2O$ (325)			$(NH_4)_3Fe(C_3H_2O_4)_3$ (173)			
Assignment	Band max., cm^{-1}	$\Delta\nu_{1/2}$ cm^{-1}	ϵ	Remarks	Band max., cm^{-1}	$\Delta\nu_{1/2}$	ϵ
$^6A_1 \rightarrow {}^4T_1(^4G)$	10,000 $\perp$	1,750	0.8	77°K	11,090 $\perp$	3,400	0.7
$^6A_1 \rightarrow {}^4T_2(^4G)$	14,120 $\perp$	2,100	1.0		15,720 $\perp$	2,700	0.9
$^6A_1 \rightarrow {}^4A_1$, ^4E	22,497	1,000	2.0		22,660 $\parallel$	800	2.4
	22,990				22,880 $\parallel$		
$^6A_1 \rightarrow {}^4T_2(^4D)$	23,810		1.6		25,800[a]		

[a] Liquid N_2 temperature only.

The band at 23,810 cm^{-1} is therefore assigned to $^4T_2(^4D)$. By analogy with the spectra of $NaMgFe(C_2O_4)_3 9H_2O$, the intensities of the $^6A_1 \to {}^4T_1(^4G)$ and $^6A_1 \to {}^4T_2(^4G)$ bands in the two polarizations are taken to indicate that the face chosen for the study was parallel to the trigonal axis of the trismalonatoiron(III) ion and, furthermore, that these ions are packed with their trigonal axes parallel or nearly parallel in the crystal. The splitting of the $^4A_1$, ^4E level is caused by the lowering of the symmetry in the complex compared to an octahedral field. The splitting (220 cm^{-1}) is less than half that observed in the oxalato complex (493 cm^{-1}) (325), for which a greater trigonal distortion would not be unexpected.

11. Fe^{3+} in Al_2O_3

The spectrum of Fe^{3+} in corundum single crystals has been determined by McClure (233). The site symmetry of the metal ion was discussed under d^1 configurations, p. 270. The bands observed are listed in Table XXXVI.

TABLE XXXVI
Absorption Bands for Fe^{3+} in Al_2O_3 at 77°K (233)
($Dq = 1650$ cm^{-1})

Assignment	Band max. $\parallel$, cm^{-1}	Band max. $\perp$, cm^{-1}
$^6A_1 \to {}^4T_2(^4G)$	17,200	17,800
$^6A_1 \to {}^4A_1$, ^4E	25,770	25,830

The observed spectrum was of poor quality and was very weak. The broad band expected at 18,000 cm^{-1} could not be seen but the bands at 25,600 ($^4A_1$, ^4E) and 18,200 cm^{-1}, $^4T_2(^4G)$, were observed. The value of Dq obtained from these measurements is about 1650 cm^{-1} but the bands are so broad that this value is uncertain (233).

12. Fe^{3+} in $Al(acac)_3$

The polarized crystal spectrum of aluminum acetylacetone with part of the Al^{3+} isomorphously replaced by Fe^{3+} has been studied by Piper and Carlin (330). The crystal structure of the host crystal has been discussed on p. 274. The blood-red crystals of $Fe(acac)_3$ have an orthorhombic unit cell, space group $Pbca$, D_{2h}^{15}, of dimensions $a_0 = 15.471$, $b_0 = 13.577$, and $c_0 = 16.565$ Å (346). The structure consists of discrete molecules linked in layers by van der Waals forces. The three acetylacetonate ligands surround the central Fe^{3+} ion of each molecule with the oxygen atoms in octahedral coordination. The three ligands have planar configurations with the Fe—O distance 1.95 ± 0.015 Å. The O—$\widehat{Fe}$—O angle

TABLE XXXVII
Absorption Band Maxima for Fe(acac)₃ (330)

Band max., cm^{-1}	$\Delta\nu_{1/2}$, cm^{-1}	Remarks	Assignment
9,360		$\{$ Undiluted Fe(acac)₃	$^6A_1 \rightarrow {}^4T_1$
12,940		$\{$ 77°K	$^6A_1 \rightarrow {}^4T_2$
10,000 σ	1,700	$\{$ In host crystal	
12,820 σ	1,300	$\{$ at 77°K	
22,000 σ	6,700	$\epsilon_\perp/\epsilon_\parallel = 4$	
22,700 $\parallel$			

in the chelate ring is 89° 48′. The observed absorption bands are given in Table XXXVII.

The solution spectrum of Fe(acac)₃ in acetone shows four bands in the infrared and visible regions at 9760 (0.45), 13,160 (0.70), 23,120 (3250), and 28,410 cm^{-1} (3420) (extinction coefficient in parenthesis) (330,346). The infrared bands are strongly polarized in the crystal. The pure orthorhombic crystal at 77°K displays absorption bands at 9300 and 12,940 cm^{-1} which are assigned to the transitions $^6A_1 \rightarrow {}^4T_1$ and $^6A_1 \rightarrow {}^4T_2$, respectively. In the host monoclinic crystal Al(Fe)(acac)₃ the $\parallel$ spectrum shows negligible absorption at 77°K in this region while the $\perp$ spectrum shows broad bands at 10,000 and 12,820 cm^{-1}. These polarizations are like those in the oxalate system and confirm that the field is trigonal to a good approximation.

For the more intense absorption in the visible region, the lower energy band is found in the crystal at 77°K at 22,000 ($\perp$) and 22,700 cm^{-1} ($\parallel$) with $\Delta\nu_{1/2} = 6700$ cm^{-1}. The band is about four times as intense in the $\perp$ as in the $\parallel$ polarization. This band occurs at a wave number appropriate for a transition to the $^4A_1$ and ^4E levels, but these transitions occur in Fe(C₂O₄)₃³⁻ and hydrated Mn²⁺ salts as sharp peaks of small intensity. It is suggested (330) that either these levels are strongly mixed with nearby ligand π levels, or more probably that the $d \rightarrow d$ excitations are obscured by strong transitions involving the ligand π orbitals. Barnum (23) has assigned the higher energy band near 28,400 cm^{-1} to a $d\epsilon \rightarrow \pi_4$ transition. If the transitions to $^4A_1$ and ^4E are near to the first intense band Dq is calculated as 1640 cm^{-1} and the Racah parameters B and C as 530 and 3570 cm^{-1}, respectively (330).

B. $4d^5$

1. Ru³⁺ in Al₂O₃

Geschwind and Remeika (142) have examined the paramagnetic resonance of Ru³⁺ in Al₂O₃. Values of $g_\parallel < 0.06$ and $g_\perp = 2.430$ were found for the crystal. The trigonal field splitting parameter K was calculated to be -283 cm^{-1}. This value is in fair agreement with that obtained by Sugano and Peter of -330 cm^{-1} (385).

C. $5d^5$

1. Os(acac)$_3$

The polarized near-infrared spectrum of the trisacetylacetonate complex of osmium(III), Os(C$_5$H$_7$O$_2$)$_3$, has been determined by Dingle (94). No detailed crystal structure is available for this compound, but Jarrett (187) has reported preliminary data for a number of M(III) (acac)$_3$ complexes and found them to be near-isomorphous. The external morphology of the osmium complex crystals, plus the fact that x-ray powder patterns show the osmium, ruthenium, and cobalt complexes to be nearly isomorphous indicate that the osmium complex crystal has a structure similar to those reported by Jarrett. The detailed orientation within the crystal is, however, unknown.

The spectra were recorded in two mutually perpendicular crystal extinction directions, but bearing no significant relationship to the molecular symmetry axes. The bands observed together with the ratio OD$_\parallel$/OD$_\perp$ are given in Table XXXVIII. The bands are arranged in the vibrational progressions assigned to them. No polarized spectra were recorded below 5000 cm^{-1}, although unpolarized measurements were made in this region.

The spectrum of this complex in CCl$_4$ was also recorded and similar bands were observed. The following vibrational progressions in the crystal are identified:

1. Six or seven members based on 4020 cm^{-1} and with a spacing of 300 ± 10 cm^{-1}, with the lowest band, $\Delta\nu_{1/2} = 200$ cm^{-1}, being the most intense.

TABLE XXXVIII
Observed Absorption Bands in a Crystal of Os(C$_5$H$_7$O$_2$)$_3$ (94)

Progression 1		Progression 2		Progression 3		Progression 4	
Band max., cm^{-1}	OD$_\parallel$/OD$_\perp$	Band max., cm^{-1}	OD$_\parallel$/OD$_\perp$	Band max., cm^{-1}	OD$_\parallel$/OD$_\perp$	Band max., cm^{-1}	OD$_\parallel$/OD$_\perp$
4025 ± 20	—	6495	0.6	4025	—	4025	—
4300	—	(6800)	0.5–0.7	5400	0.75	6495	0.6
4620	—	6900	$\sim$0.5	$\sim$6900?	<1.0	$\sim$9500	<1.0
4930	—	7140	$\sim$0.55	6495	0.6		
(5240)	$\sim$0.8	7450	$\sim$0.5	8020	0.65		
(5550)	$\sim$0.8			9500	<1.0		

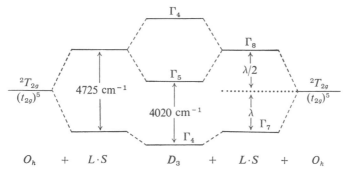

Fig. 39. Schematic diagram for the $^2T_{2g}(t_{2g})^5$ of $Os(C_5H_7O_2)_3$ illustrating the movement of the various energy levels as the trigonal field splitting Δ is increased from zero on the left-hand side to a value in the region of $\Delta = -3150$ cm^{-1} on the right side. Spin–orbit coupling constant $\zeta = 3150$ cm^{-1} (94). (Δ is here defined so that positive values correspond to ^2E higher than $^2A_1$ in absence of spin–orbit coupling.)

2. Three or four members based on 6500 cm^{-1} showing a spacing of 300 ± 10 cm^{-1}, with the lowest band, $\Delta\nu_{1/2} = 200$ cm^{-1} the most intense.

3. Broader bands at 5400, 7950, and 9400 cm^{-1} that may be considered as members of two series based on 4020 and 6500 cm^{-1}, respectively, and having common spacing 1400 cm^{-1}. It is also possible that all three bands are built on 4020 cm^{-1} and that the strong 6500 cm^{-1} absorption actually covers the member that, in these circumstances, would be expected at 6800 cm^{-1}. The crystal spectrum shows an extra band at 6900 cm^{-1}, thus indicating that the 7950 and 9400 cm^{-1} bands are based on 6500 cm^{-1}.

4. The 6500 cm^{-1} band (and perhaps 9400 cm^{-1}) may be a member of a series with upper-state spacing of 2500 cm^{-1} corresponding to the C—H stretching frequency and based on 4020 cm^{-1}. To check this, a partially deuterated sample was examined and the infrared spectrum remained unchanged, and it was therefore assumed that the 4020–6500 cm^{-1} spacing is not associated with the C—H stretching motion in the ligand.

The energy level digram for the complex is given in Figure 39 (94).

In octahedral symmetry, the ground electronic state of the osmium complex is $^2T_{2g},(t_{2g})^5$ and when spin–orbit coupling is included, the sixfold degeneracy is resolved into a twofold degenerate Γ_7 and fourfold Γ_8. Since there is a trigonal field also present, the Γ_8 level is further split into two Kramer's doublets, giving rise to three electronic states. The energy levels predicted for $\lambda = -3150$ cm^{-1} $(= -\zeta_{5d}$ for $^2T_{2g})$ and $\Delta^* = -3150$ cm^{-1}

* Δ is here defined so that positive values correspond to ^2E higher than $^2A_1$ in absence of spin–orbit coupling (94).

are 4037 and 6500 cm^{-1} (94). This is in good agreement with the observed values of 4020 and 6500 cm^{-1}, hence the transitions are assigned as:

$$\Gamma_4 \rightarrow \Gamma_5 \qquad 4020 \text{ cm}^{-1}$$
$$\Gamma_4 \rightarrow \Gamma_4 \qquad 6500 \text{ cm}^{-1}$$

K is very large when compared with the splittings of 100–1000 cm^{-1} seen in most $3d$ metal acetylacetonates (99,330), although electron spin resonance studies have indicated equally large splittings in the $3d^1$, $^2T_{2g}$ ground state of titanium(III) trisacetylacetone complex (239). The sign of K is reversed in these two cases ($3d^1, 3d^5$) as would be expected on the grounds of the electron-hole formalism (95). The large K values are probably related to the large size of the central ion and the resulting increased overlap with the nonequivalent ligand oxygen π orbitals (94).

On this interpretation, the absorption bands arise from $(t_{2g})^5 \rightarrow (t_{2g})^5$ transitions, and the upper and lower state potential surfaces will be very similar. Under these conditions, the occurrence of intense (0–0) bands with rapidly decreasing intensities of higher members of the vibrational progression, is expected and is found for the bands originating at 4020 and 6500 cm^{-1}. The 300 and 1400 cm^{-1} frequencies are assigned to symmetrical excited-state metal–oxygen and in-plane carbon–oxygen stretching modes, respectively.

VIII. CONFIGURATION d^6

A. $3d^6$

The Tanabe-Sugano energy diagram for the d^6 configuration is given in Figure 40. In a weak octahedral field, for $10Dq/B \leqslant 20$, the ground state arises from the ^5D state as it does for d^4. The Fe^{2+} ion in most of its compounds and a few compounds of Co^{3+} have the resultant $^5T_2$ ground state. In all others, the lowest $^1A_1$ state becomes the ground state, when $10Dq/B \geqslant 20$. The Jahn-Teller effect is important in the upper state of the $^5T_{2g} \rightarrow {}^5E_g$ transition. The theoretical problem of the dynamical Jahn-Teller effect has been studied by Pryce and Öpik (338).

1. FeSO$_4$7H$_2$O

The crystal absorption spectrum of FeSO$_4$7H$_2$O has been determined by Holmes and McClure (182). A preliminary crystal structure analysis has been carried out by Baur (34) who found that the monoclinic unit cell contains four formula units, has the dimensions $a_0 = 14.07$,

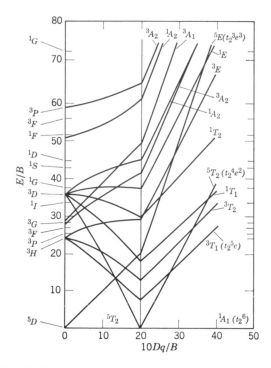

Fig. 40. Tanabe-Sugano energy diagram for the d^6 configuration (75,386).

$b_0 = 6.51$, $c_0 = 11.04$ Å, and $\beta = 105°36'$, with space group $P2_1/c, C_{2h}^5$. Six water molecules are octahedrally coordinated around the Fe^{2+} ion at a distance of 2.13 Å.

The spectrum shows a double peak centered on 10,000 cm^{-1} with a splitting of 2000 cm^{-1}. The half-width is 6000 cm^{-1} and $\epsilon = 1.6$. McClure attributes this to the complete removal of the degeneracy of the upper E_g state. From the spectrum it is impossible to tell if the splitting is tetragonal or rhombic. The ground-state splitting is less than 1000 cm^{-1}. Two maxima are predicted to appear under certain conditions as a consequence of the dynamical Jahn-Teller effect (338).

2. $Na_2Fe(CN)_5NO2H_2O$

The polarized crystal spectrum of sodium nitroprusside, $Na_2Fe(CN)_5$ $NO2H_2O$, has been determined by Gray, Manoharan, Pearlman, and Riley (155) as part of their study of the electronic structure of transition metal nitrosyl complexes. The work was extended by Gray and Manoharan (248,249). This compound has a nominal d^6 configuration with the NO

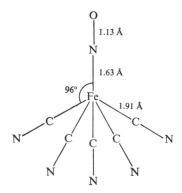

Fig. 41. Arrangement of ligands around the central iron atom in the $Fe(CN)_5NO^{2-}$ ion (247).

group being regarded as a NO^+ ion, and is thus included in this section. The crystal structure of $Na_2Fe(CN)_5NO2H_2O$ has been determined and is orthorhombic with space group $Pnnm$, D_{2h}^{12} (247). The lattice constants are $a_0 = 6.17$, $b_0 = 11.84$, and $c_0 = 15.43$ Å with four formula units per unit cell. The nitroprusside ion lies on the mirror plane and has approximate C_{4v} symmetry. The general shape of the anion is shown in Figure 41.

The ligands are colinear with the metal atom which is displaced slightly in the direction of the NO group, N—$\widehat{Fe}$—C = 96°, from the plane of the four pseudo equivalent CN groups. The chain O—N—Fe—C—N is linear. The Fe atom is 0.2 Å above the plane of the four carbon atoms. In the crystal, there are two different orientations of the NC—Fe—NO axes to the ab plane, labeled A and B in Figure 42.

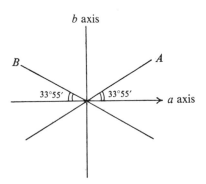

Fig. 42. Orientation of NC—Fe—NO axes in the ab plane of the $Na_2Fe(CN)_5$-$NO2H_2O$ crystal; the two different orientations are labeled A and B (249).

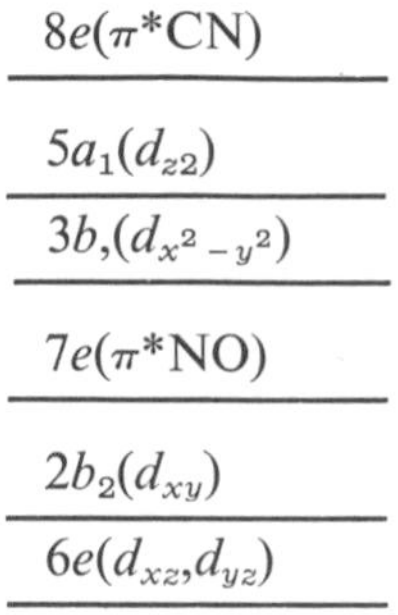

$$8e(\pi^*\text{CN})$$

$$5a_1(d_{z^2})$$

$$3b_1(d_{x^2-y^2})$$

$$7e(\pi^*\text{NO})$$

$$2b_2(d_{xy})$$

$$6e(d_{xz},d_{yz})$$

Fig. 43. Calculated (extended Hückel method), relative ordering of MO energy levels in $\text{Fe(CN)}_5\text{NO}^{2-}$ for the energy region 20,000–50,000 cm^{-1} above the ground state (248). Notation as in ref. 248.

Manoharan and Gray (248) have carried out an extended Hückel MO calculation for $\text{Fe(CN)}_5\text{NO}^{2-}$ which gives a d level ordering of $xz, yz < xy < x^2 - y^2 < z^2$. The e level derived mainly from π^* NO is the lowest empty level, lying between $x^2 - y^2$ and xy, Figure 43.

The crystal spectra were determined (247–249) with the electric vector polarized $\parallel$ and $\perp$ to the a axis of the crystal and analyzed to find the complete band polarizations of all optically equivalent molecular Fe—NO

TABLE XXXIX

Observed and Calculated Absorption Bands for Sodium Nitroprusside, $\text{Na}_2\text{Fe(CN)}_5\text{NO2H}_2\text{O}$ (249)

Observed[a] band max., cm^{-1}	ϵ	Calculated[b] band max., cm^{-1}	Assignment, notation as in Fig. 43
20,080 $\perp$	~ 8	20,540 $\perp$	$^1A_1 \rightarrow {}^1E(2b_2 \rightarrow 7e)$
25,380 $\parallel$	25	25,090 $\parallel$	$^1A_1 \rightarrow {}^1A_1(6e \rightarrow 7e)$
(30,300)[c] $\perp$	(40)[c]	30,770[d]	$^1A_1 \rightarrow {}^1A_2(2b_2 \rightarrow 3b_1)$
		(vibronic $\perp$)	
(37,800)[c]	(900)[c]	37,750[d] $\perp$	$^1A_1 \rightarrow {}^1E(6e \rightarrow 5a_1)$
(42,000)[c]	(700)[c]	40,900[d] $\perp$	$^1A_1 \rightarrow {}^1E(6e \rightarrow 3b_1)$
50,000	24,000	49,900 $\perp$	$^1A_1 \rightarrow {}^1E(2b_2 \rightarrow 8e)$

[a] Observed maxima [for an aqueous solution spectrum of $\text{Na}_2\text{Fe(CN)}_5\text{NO2H}_2\text{O}$] and polarizations (from single crystal spectra).

[b] Calculated energies and polarizations.

[c] Shoulder, ϵ_{max} values are estimated.

[d] Corrected for interelectronic repulsion energy, assuming $F_2 = 14F_4$. The Slater-Condon parameters are from ref. 38. The value of $F_2 - 5F_4 = 400$ cm^{-1} from the Fe(CN)_6^{4-} spectrum (154) was assumed for $\text{Fe(CN)}_5\text{NO}^{2-}$. Interaction between the two closely spaced ^1E states was included.

groups. The observed bands are listed in Table XXXIX together with their assignments, and calculated levels (248,249).

The band listed at 30,300 cm^{-1} separates at 77°K into two well-resolved peaks at 28818 and 30960 cm^{-1}, i.e., with a separation of 2142 cm^{-1}. The assigned transition $^1A_1 \rightarrow {}^1A_2$ is orbitally forbidden, but if the electronic transition occurs simultaneously with excitation of a vibration of e symmetry, the total wave function will be of e symmetry and allowed in xy polarization. The value of 2142 cm^{-1} for the excited electronic state is not far from the reported frequencies for the e stretching mode in the ground state of $Fe(CN)_5NO^{2-}$ ion, 2161.6 and 2156.7 cm^{-1} (45). Thus it is concluded that the $^1A_1 \rightarrow {}^1A_2$ transition is allowed by simultaneous excitation of a vibration of e symmetry.

The polarized infrared spectrum of single crystals of $Na_2Fe(CN)_5$-$NO2H_2O$ has been determined by Sabatini (350a). The spectrum shows that the absorption band at 650 cm^{-1} is absent when the plane of polarization is $\|c$ while the band at 662 cm^{-1} appears $\|a,b,c$. The 650 cm^{-1} band is assigned to the Fe—N stretching vibration and the 662 cm^{-1} band to the Fe—NO rocking vibration. In the free ion, the C≡N stretching vibrations divide among the symmetry species $2A_1 + E + B_1$. From the observed polarizations, the following assignments are made 2175, 2163 cm^{-1}, A_1 polarized $\|a,b$; the band at 2145 cm^{-1} corresponds to the vibration of species E, polarized $\| a,b,c$; and the band at 2158 cm^{-1} corresponds to the vibration B_1, polarized $\| c$ only.

The Mössbauer hyperfine interactions in sodium nitroprusside single crystals has been studied by Danon and Iannarella (83a).

3. BaFeSiO$_4$O$_{10}$

Burns, Clark, and Stone (57d) have studied the polarized spectrum of gillespite, $BaFeSiO_4O_{10}$, a mineral with Fe(II) in a square-planar configuration using the technique of Burns (57c). A single crystal of the mineral was placed at the center of a three-axis universal stage which enabled it to be oriented accurately. This was placed in a polarizing microscope and, in turn placed in the sample beam of a recording spectrophotometer.

Gillespite is uniaxial, space group P_4/ncc, $D_{4h}{}^8$, with the pleochroic scheme $E \perp c$, pale pink; $E \| c$, deep rose red. The unit cell with $a_0 = 7.495$ and $c_0 = 16.050$ Å contains four molecules. The structure (295a) contains Si_8O_{20} sheets consisting of two linked sets of square groups of four (SiO_4) tetrahedra at two levels along the c axis. Oxygen atoms occur in three kinds of positions. Two types, O_I and O_{II}, bridge SiO_4 tetrahedra, and the third, O_{III} forms unshared vertices. Only the O_{III} type are coordinated to cations. Fe^{2+} is in square-planar coordination with 4 O_{III} oxygens (Fe—O = 1.97 Å) in one sheet of Si_4O_{10}. The square–planar (FeO_4) group is perpendicular to the c axis. The nearest atoms vertically above and below Fe^{2+} are other Fe^{2+} ions at 8.025 Å along the c axis.

The magnetic moment of gillespite shows that the Fe^{2+} ion has four unpaired electrons. Two energy level schemes are considered by Burns et al., one $d^1_{x^2-y^2} > d^1_{xy} > d^1_{xz},d^1_{yz} > d^2_{z^2}$ and the other $d^1_{x^2-y^2} > d^2_{z^2} > (d_{xz},d_{yz})^3$. The former has a $^5A_{1g}$ ground state and correlates best with the observed spectra. If E_g were the ground state, the allowed transitions $^5E_g \rightarrow {}^5B_{2g}$ and $^5E_g \rightarrow {}^5B_{1g}$ should result in two absorption bands in each polarization, whereas the π spectrum only contains one band. An E_g ground state would also be unstable due to Jahn-Teller distortion. A large contribution to the magnetic moment would be expected from the orbital angular momentum, whereas no first-order contribution is expected for the $^5A_{1g}$ state. The observed magnetic moment is only slightly greater than the "spin only" value, the small difference being attributed to spin–orbit coupling effects. Simple $d \rightarrow d$ transitions are forbidden by the Laporte selection rule as they are of the $g \rightarrow g$ type but may become allowed by simultaneous excitation of a u vibrational mode. The observed bands are assigned as in Table XXXIXA taking account of the normal modes allowing vibronic coupling. The axial and σ spectra are coincident showing that the spectra are due to electric dipole transitions (57d).

The band assigned to the $^5A_{1g} \rightarrow {}^5B_{1g}$ transition differs in intensity and frequency in the two polarized spectra. The shift represents the difference in energy of the fundamental frequencies of the vibrational modes involved, i.e., assuming a single vibrational quantum is involved in each case

$$\nu_{E_u} - \nu_{B_{2u}} = 20{,}160 - 19{,}650 = 510 \text{ cm}^{-1}$$

Since E_u vibrations are in-plane distortions of the (FeO_4) square, while the B_{2u} mode is a bending of the square, it is considered reasonable that $\nu_{E_u} > \nu_{B_{2u}}$. If the absorption peaks corresponding to the $^5A_{1g} \rightarrow {}^5E_g$ transition could be identified in the polarized spectra, similar wavelength shifts would be expected for the different vibrational modes involved in the vibronic coupling (57d). Direct experimental observation of this band, predicted to occur (67a) at ~ 2850 cm^{-1} is unlikely to be achieved as the energy denominator in the expression, for the intensity of a band allowed by vibronic coupling (159) is larger in this case than for the other transitions of gillespite. Also, calculations (418f) show that on inclusion of spin–orbit interaction, the fine structure arising from ^5E_g spreads over a

TABLE XXXIXA

Assignment of Observed Bands in the Polarized Spectrum of Gillespite, $BaFeSiO_4$, Taking Account of the Normal Modes Allowing Vibronic Coupling (57d)

Ground state	Polarization of electric vector	Transition	Normal modes allowing vibronic coupling	Assignment band max., cm^{-1}	ϵ
$^5A_{1g}$	$E \parallel c$	$^5A_{1g} \rightarrow {^5E_g}$	E_u	—	—
		$^5A_{1g} \rightarrow B_{2g}$	None	Forbidden	—
		$^5A_{1g} \rightarrow B_{1g}$	B_{2u}	19,650	9.8
$^5A_{1g}$	$E \perp c$	$^5A_{1g} \rightarrow {^5E_g}$	$A_{2u} + B_{2u}$	—	—
		$^5A_{1g} \rightarrow {^5B_{2g}}$	E_u	8,300	0.6
		$^5A_{1g} \rightarrow {^5B_{1g}}$	E_u	20,160	2.55

much wider range than for the other excited states, thus reducing the maximum peak height of the $^5A_{1g} \rightarrow {^5E_g}$ transition.

4. Fe(bipy)$_3^{2+}$

Palmer and Piper (300) have studied the electronic spectra of the 2,2′ bipyridyl (bipy) complexes of the type $M(bipy)_3^{2+}$ where M = Cu(II), Ni(II), Co(II), Fe(II), Ru(II). Polarized optical spectra of single crystals were obtained using the hexagonal crystal $M(bipy)_3Br_26H_2O$ and the monoclinic crystal $M(bipy)_3SO_47H_2O$. Both undiluted crystals and those diluted with Zn(II) were used. The crystal structures and orientation of the crystals with respect to the light beam will be discussed here while the spectral results will be discussed under the appropriate electronic configuration for each ion.

Single-crystal x-ray and optical spectral data indicate that in the bromide crystals the $M(bipy)_3^{2+}$ ions are aligned with their C_3 axes parallel. The bromides grow as pseudo-hexagonal plates which alter to truly hexagonal after removal from solution. The space group of $Zn(bipy)_3Br_26H_2O$ is hexagonal, P622, D_6^1, or $P6, C_6^1$, with two molecules per unit cell, $a_0 = 13.54$ and $c_0 = 10.71$ Å. The C_3 axes are parallel to the crystallographic c axis (301).

The tabular crystals of $Zn(bipy)_3SO_47H_2O$ have been described morphologically by Jaegar and Van Dijk (186) and the space group of the isomorphous nickel compound has been determined by Jacobs and Speeke (185). The crystal is monoclinic, space group $C2/c, C_{2h}^6$ and has eight molecules per unit cell, with $a_0 = 22.90$, $b_0 = 14.19$, $c_0 = 24.80$ Å and $\beta = 117°\,3'$. The $M(bipy)_3^{2+}$ ions are aligned with their C_3 axes perpendicular to the (001) plane.

The sulfate crystals were aligned so that the beam was either perpendicular to the (001) plane (axial spectrum, unpolarized) or parallel to this plane (orthoaxial, polarized ⊥ or ∥). The bromide crystals were oriented in the same manner with respect to the (001) plane. The axial spectra of each of the $M(bipy)_3^{2+}$ ions (M = Cu, Ni, Co, Fe, Ru) were identical in both the pure $M(bipy)_3Br_26H_2O$ crystals and in dilute solid solutions in the corresponding zinc crystal. In the region 8000–3333 cm^{-1} the absorption of a $Zn(bipy)_3Br_26H_2O$ crystal of the same thickness was subtracted from the total absorption of a diluted crystal to obtain the net absorption of the metal chromophore. The orthoaxial spectra of dilute solutions of the $M(bipy)_3^{2+}$ ions in single crystals of $Zn(bipy)_3SO_47H_2O$ show that the ⊥ and axial spectra coincide in each case. The results in the region below 8000 cm^{-1} in the sulfate spectra are not as accurate as those in the bromide crystals owing to the overlapping of the vibrational bands, but there is agreement of the general features of the spectrum.

TABLE XL

Band Maxima of $Fe(bipy)_3^{2+}$ Spectra (300)

Matrix	Spectrum	Excited state[a]	Band max., cm^{-1} 300°K	80°K	$\Delta\nu_{1/2}$, cm^{-1} (at 300°K)	ϵ
$Fe(bipy)_3Br_26H_2O$	Axial	^3T	11,500	11,500	3,800	3.9
	σ	^3T	11,500	11,500	3,800	4.4
$Zn(bipy)_3Br_26H_2O$	Axial	CT-1	25,500	25,500		
		CT-2	24,400	24,400		
		CT-3	—	22,900		
		CT-4	22,000	21,900		
		CT-5	20,200	20,370		
		CT-6	18,760	18,760		
$Zn(bipy)_3SO_47H_2O$	σ	CT-1	—	25,500		
		CT-2	24,500	24,200		
		CT-3	—	—		
		CT-4	22,000	21,600		
		CT-5	20,200	20,280		
		CT-6	18,870	18,690		
	π	CT-7	26,300	25,500		
		CT-8	23,500	$\begin{cases} 23,900 \\ 23,260 \end{cases}$		
		CT-9	—	22,000		
		CT-10	20,410	20,200		
		CT-11	18,870	18,800		

[a] CT = charge transfer.

The observed absorption bands for Fe(bipy)$_3^{2+}$ are listed in Table XL, and the orthoaxial spectrum of Zn(Fe)(bipy)$_3$SO$_4$7H$_2$O at 80°K is shown in Figure 44. Because of the extremely high intensity of the visible absorption, only the spectrum below 15,000 cm^{-1} could be measured in the undiluted crystal. The polarization ratio $I_\perp/I_\parallel$ for the predominant visible bands is 11.4 and 10.5 at 300 and 80°K, respectively.

The assignment of the extremely intense bands in the visible region to Laporte-allowed transitions in which the electron is excited from orbitals primarily localized on the metal to π^* antibonding orbitals primarily localized on the ligands ($t_2 \to \pi^*$) (190) is supported by the crystal spectra (300). The half-widths of these bands are only ~ 1100 cm^{-1} at 80°K and ~ 1500 cm^{-1} at 300°K, being considerably narrower than $d \to d$ transitions which have half-widths of 3000–4000 cm^{-1}. The increase in the intensity of the bands on lowering the temperature is consistent with their assignment to charge-transfer bands as the transitions would be enhanced by a decrease in the Fe—N distance which occurs with decrease in temperature. This phenomenon is atypical of crystal field bands. It is concluded that the crystal field spectrum of Fe(bipy)$_3^{2+}$ is not observed in the visible region (300).

The ground state of Fe(bipy)$_3^{2+}$ is $^1A_1$ (58) from magnetic susceptibility measurements. The $d \to d$ transitions expected in the near-infrared visible region in order of increasing energy are $^1A_{1g} \to {}^3T_{1g}$, $^3T_{2g}$, $^1T_{1g}$,

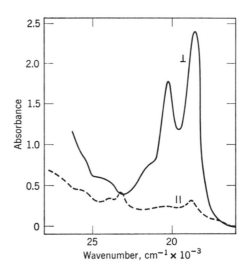

Fig. 44. Polarized orthoaxial spectra of Zn(Fe)(bipy)$_3$SO$_4$7H$_2$O at 80°K. Crystal thickness, 1.9 mm; approximately 4×10^{-3} mole % iron (300).

$^1T_{2g}$ in an octahedral field (386,387) (Fig. 40). The symmetry of the ion is no lower than D_3 and in this symmetry, intensity is predicted for $^1A_1 \rightarrow {}^1T_1$ in both $\parallel$ and $\perp$ polarization, but only in $\perp$ for $^1A_1 \rightarrow {}^1T_2$. Since the observed band at 11,500 cm^{-1} is $\perp$ polarized, the transition would be assigned to $^1A_1 \rightarrow {}^1T_2$, but then the $^1A_1 \rightarrow {}^1T_1$ should be observed near 6000 cm^{-1}, and no absorption was observed of comparable intensity. This is therefore not assigned to a spin-allowed transition.

A rough estimate of the intensity that might be gained by the spin-forbidden band at 11,500 cm^{-1} through coupling with the $(t_2 \rightarrow \pi^*)$ transition at 18,000 cm^{-1} gives an integrated band intensity ≈ 1.2 compared with an experimental value of 1.4 for the $\perp$ polarization (300). Calculations on the variation of the Racah parameter B with change in $10Dq$ show that for the assignment $^1A_1 \rightarrow {}^3T_1$ reasonable values of B are obtained for $16,500 < 10Dq < 19,000$ cm^{-1} and for the assignment $^1A_1 \rightarrow {}^3T_2$, for $14,000 < 10Dq < 14,500$ cm^{-1}. As no electronic transitions were observed below 11,500 cm^{-1}, the $^1A_1 \rightarrow {}^3T_1$ is favored, giving Dq a value $1650 < Dq < 1900$ cm^{-1}.

Polarized intervalence transfer spectra due to transitions of the type

$$Fe(II) + Fe(III) \longrightarrow Fe(III) + Fe(II)$$

in some naturally occurring crystals are discussed in reference 184.

5. FeF$_2$

The absorption spectrum of a polycrystalline sample of FeF$_2$ was determined by Hatfield and Piper (175) in their study of the distortions about six-coordinate ferrous ion. FeF$_2$ has the rutile type structure, space group $P4_2/mnm, D_{4h}{}^{14}$, with cell constants $a_0 = 4.6966$, $c_0 = 3.3091$ Å, and $u = 0.300$ (32,377). Two well-resolved bands appear in the spectrum at 10,660 ($\Delta\nu_{1/2} = 2500$ cm^{-1}) and 6990 cm^{-1} ($\Delta\nu_{1/2} = 1000$ cm^{-1}). These are assigned to the transitions $B_{1g} \rightarrow A_g$ and $B_{1g} \rightarrow B_{1g}$ (in D_{2h}), although the ground state symmetry is not definitely established. The spectrum is used as a basis for calculations on the tetragonal distortion in this type of complex (175).

6. Co(en)$_3{}^{3+}$

Polarized single-crystal data of the cobalt trisethylenediamine, (Coen$_3$)$^{3+}$, ion have been reported by Yamada and Tsuchida (443), Karipides and Piper (197), and by Dingle (93). dl-(Coen$_3$Cl$_3$)3H$_2$O has space group $P\bar{3}c1, D_{3d}{}^4$, with four molecules per trigonal unit cell (282). The cell constants are $a_0 = 11.50$ and $c_0 = 15.52$ Å. The six nitrogen atoms are arranged in a distorted octahedron around the cobalt atom, with Co—N distance = 2.00 Å. The c axis in the crystal is coincident with the threefold axis of the $(Coen_3)^{3+}$ ions. The crystal structure of $(Coen_3Br_3)$-$3H_2O$ is assumed to be similar (443). Crystals of the optically active d-$(Coen_3Cl_3)NaCl6H_2O$ have space group $P3,C_3^1$ with one molecule per unit cell and cell constants $a_0 = 11.47$, $c_0 = 8.06$ Å (283).

The bands observed by Yamada and Tsuchida (443) for the three compounds dl-$(Coen_3Cl_3)3H_2O$, dl-$(Coen_3Br_3)3H_2O$, and d-$(Coen_3Br_3)$-$2H_2O$ are listed in Table XLI. The spectra were recorded parallel and perpendicular to the c axis in the crystal (the threefold molecular axis).

The first absorption band at $\sim 21,300$ cm^{-1} is ascribed to the $^1A_1 \rightarrow {}^1T_1$ transition and the second to the $^1A_1 \rightarrow {}^1T_2$. The small dichroism is ascribed to splitting of the T_1 level into A_2 and E levels, this being very small. Piper and Carlin (329) have carried out some crystal field calculations on the variation of K with the angles a and b where a is the polar angle of each ligand from the (pseudo) threefold axis, and b is the ligand–metal–ligand angle. Their results for $(Coen_3)^{3+}$ ions are given in Table XLII. The agreement for $(Coen_3Cl_3)3H_2O$ is poor.

Karipides and Piper (197) measured the crystal spectrum of $(Coen_3)^{3+}$ in the host crystal $(Rhen_3Cl_3)3H_2O$ as part of a study of the optical activity of coordination compounds. The relative band intensities for the transitions $^1A_1 \rightarrow {}^1A_2$ and $^1A_1 \rightarrow {}^1E$ were found to be 1.38:1.00, with the observed splitting of the $^1T_1$ band very small although the data from four different crystals indicated K to be $+30 \pm 15$ cm^{-1}. The host crystal absorbed in the ultraviolet where the $^1A_1 \rightarrow {}^1T_2$ band was expected and so this could not be observed.

Dingle (93) has studied the polarized crystal spectrum of dl-$(Coen_3Cl_3)$-$NaCl3H_2O$ at 300, 77, and 5°K. The $\perp$ and axial spectra were found to coincide from which it is concluded that the spectra are almost entirely

TABLE XLI

Absorption Band Maxima and Polarizations Referring to the Threefold Axis for
dl-$(Coen_3Cl_3)3H_2O$, dl-$(Coen_3Br_3)3H_2O$, d-$(Coen_3Br_3)2H_2O$ (443)
[α = absorption coefficient (mm^{-1})]

dl-$(Coen_3Cl_3)3H_2O$		dl-$(Coen_3Br_3)3H_2O$		d-$(Coen_3Br_3)2H_2O$	
Band max., cm^{-1}	log α	Band max., cm^{-1}	log α	Band max., cm^{-1}	log α
21,270 $\parallel$	1.49	21,400 $\parallel$	1.48	21,300 $\parallel$	1.54
21,330 $\perp$	1.33	21,400 $\perp$	1.39	21,370 $\perp$	1.43
29,300 $\parallel$	1.22	29,300 $\parallel$	1.30	29,300 $\parallel$	1.27
29,600 $\perp$	1.61	29,300 $\perp$	1.63	29,300 $\perp$	1.67

TABLE XLII

Point Charge Calculation of Variation of Trigonal Splitting Constant K with the Polar Angle a and the Ligand–Metal–Ligand Angle b for the $Coen_3{}^{3+}$ Ion (329)

[Values of b are for (1) d-$(Coen_3Cl_3)NaCl6H_2O$ (283), and (2) dl-$(Coen_3Cl_3)3H_2O$ (282)]

b	a	Obs. K, cm^{-1}	Calc. K, cm^{-1}
87° 24′ (1)	53° 54′	—	−750
90° 18′ (2)	54° 30′	+45(34)	−280

electric dipole in nature. There was considerable absorption in the $\parallel$ spectrum in the region of the $^1A_1 \rightarrow {}^1T_2$ transition, while the temperature dependence showed that the D_3 selection rules are not obeyed rigorously. Vibronic contributions to the intensity must be of considerable importance. In the crystals it was found that in the region of the $^1T_1$ transition the ^1E state lies at higher energy than the $^1A_2$ state but that the splitting was only 0 ± 2 cm^{-1}. Vibrational fine structure was observed on the low energy side of this band at 5°K, with the important nontotally symmetric vibrations 185, 345, and $\sim$400 cm^{-1}. These were thought to be reasonable values for the t_{1u} and t_{2u} vibrations of the CoN_6 octahedron. The only totally symmetric mode observed had a frequency 255 cm^{-1}. Dingle (93), on the basis of this very small value of K, questions the treatment of Karipides and Piper (197,327) and also suggests that the ionic model (327,384) can no longer be used to explain the dichroism of the ^1E_b band or ^1E_a and $^1A_2$ bands since with $K \simeq 0$, the model predicts zero optical activity in solution in these regions.

Denning (87) obtained a small value of $K = +3.5$ cm^{-1} from his studies of the circular dichroism of d-$(Coen_3)^{3+}$ in the host crystal dl-$(Rhen_3Cl_3)NaCl6H_2O$. Vibronic structure was observed in the circular dichroism spectrum on the low energy side of the $^1T_1$ band. The two components of the $^1T_1$ band are attributed to Jahn-Teller components rather than trigonal components.

7. Co^{3+} in $NaMgAl(C_2O_4)_3 9H_2O$

Piper and Carlin (322,325) have studied the polarized single-crystal spectrum of Co^{3+} in the host crystal $NaMgAl(C_2O_4)_3 9H_2O$. The site symmetry of the trivalent ion has been discussed under d^2 configurations for V^{3+}, p. 294. Spectra were recorded at 25°C $\parallel$ and $\perp$ to the threefold axis and are shown in Figure 17. The spectra at 77°K are essentially the same except that the bands shift by about 150 cm^{-1} to higher energy and

TABLE XLIII

Observed Absorption Bands for Co^{3+} in the Host Crystal
$NaMgAl(C_2O_4)_3 9H_2O$ (325)

Band max., cm^{-1}	$\Delta\nu_{1/2}$, cm^{-1}	Transition	Remarks
16,350 σ	2,700	$\left.\begin{array}{l}{}^1A_1 \rightarrow {}^1E_a \\ {}^1A_1 \rightarrow {}^1A_2\end{array}\right\} {}^1T_{1g}$	$\epsilon_\parallel/\epsilon_\perp = 1.16$
16,500 π	2,600		at 25°K
23,400 σ	3,400	$\left.\begin{array}{l}{}^1A_1 \rightarrow {}^1E_b \\ {}^1A_1 \rightarrow {}^1A_1\end{array}\right\} {}^1T_{2g}$	$\epsilon_\parallel/\epsilon_\perp = 0.05$

increase in ϵ by about 20. The observed bands, polarizations, and assignments are given in Table XLIII. From the observed maxima, the trigonal field splitting parameter K is calculated to be -100 cm^{-1}, and $Dq = 2000$ cm^{-1}.

8. Co^{3+} in Al(acac)$_3$

Piper (324) has analyzed the polarized crystal spectrum of cobalt acetylacetone in the host crystal aluminum acetylacetone. The structure of the host crystal was discussed under d^1 configurations, p. 274. The crystal structure of $Co(C_5H_7O_2)_3$ has been determined (369). The unit cell contains four molecular units, has space group $P2_1/c$, C_{2h}^5, and lattice parameters $a_0 = 14.16$, $b_0 = 7.48$, $c_0 = 16.43$ Å, and $\beta = 98°41'$. The spectra were recorded perpendicular to the (001) and (100) faces and polarized $\perp$ or $\parallel$ to the b axis. Figure 45 shows the observed spectra and the bands are listed in Table XLIV.

Analysis into components π and σ ($\parallel$ and $\perp$ to molecular threefold axis), Figure 45, gives results agreeing quite well for the two orientations, indicating little breakdown of the trigonal symmetry within the crystal. The first excited state for Co(III) is $^1T_1$, split by the trigonal components into $^1A_2$ and ^1E components. The π and σ bands are assigned to these two transitions, respectively. The splitting of the two levels is $-K$ ($^1A_2$) and $+\frac{1}{2}K(^1E)$ (382) and by correlation with the π and σ maxima, K is found to be $+600$ cm^{-1}. The sign of K is opposite to that found for the trisoxalate complex (325) ($K = \sim -100$ cm^{-1}). The ionic model (323) predicts that if the radial parameters are unchanged, then the sign of K depends upon the angle $O\widehat{-M}-O$ in the chelate ring. K will have one sign if the angle is greater than 90° and the opposite sign if less than 90°. The radial parameters are assumed to be similar since the $^1A_1 \rightarrow {}^1T_1$ transition occurs at nearly the same energy in both compounds. Comparison of the angles in acetylacetonate and oxalate complexes of chromium, 93° (368)

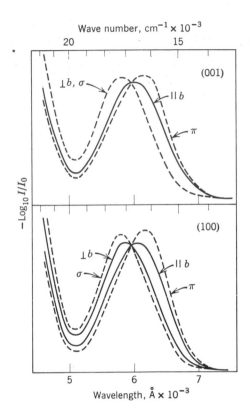

Fig. 45. Spectra of cobalt acetylacetonate ($\sim 1\%$) in aluminum acetylacetonate, crystals ca. 2 mm thick. The measured spectra are the solid lines and the calculated spectra dashed, except that there is only one dashed curved for both the $\perp$ and σ (001) spectra (324).

and 83° (403,404), respectively, shows that these angles are in good agreement with the signs of the trigonal field parameters. It is assumed that although cobalt is slightly smaller than chromium and its bond angle may be slightly larger, the difference cannot be great (291).

TABLE XLIV
Observed Absorption Bands for Co^{3+} in Al(acac)$_3$ (324)

Crystal face	Band max., cm^{-1} $\parallel b$	Band max., cm^{-1} $\perp b$	$\epsilon_\perp/\epsilon_\parallel$
(001)	16,250	17,100	0.99
(100)	16,200	17,200	0.97

9. Co^{3+} in Al_2O_3

McClure (52) has studied the polarized optical absorption spectrum of Co^{3+} in corundum at 77°K. The site symmetry of the trivalent ion in this lattice was discussed under d^1 configurations, p. 270. The observed bands are shown in Figure 46, and listed in Table XLV.

The observed spectrum consists of two well-separated bands and a number of very weak ones. The strong bands are assigned to singlets of configuration t_2^5e ($^1T_1$) 15,560 and $^1T_2$ 22,980 cm^{-1}. They are split by the trigonal field by 360 and 370 cm^{-1} with $A > E$, assuming C_3 electric dipole selection rules. The first-order splitting of each band is $3K/2$, so that the trigonal field splitting parameter $K = -240$ cm^{-1} which is comparable with the value for V^{3+}. The suggestion of a band at $\sim 30,000$ cm^{-1} near the sharply rising charge-transfer absorption is in the correct position for a transition to $^1T_2$ ($t_2^4e^2$). The two weak bands at 18,880 and 19,800 cm^{-1} are assigned to the transitions to $^3T_2$ of ($t_2^4e^2$) and the accidentally degenerate pair ($^3T_1 + {}^3T_2$) of ($t_2^4e^2$). The $^1T_2(t_2^5e)$ band is doubled. It is believed (233) that the low energy component in the $\parallel$ spectrum is a triplet intensified by coupling to the singlet.

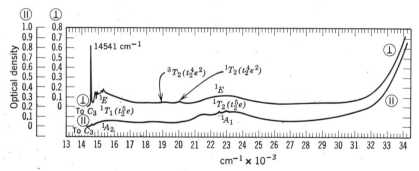

Fig. 46. Absorption spectrum of Co^{3+} in corundum at 77°K (233).

TABLE XLV
Observed Absorption Bands for Co^{3+} in Al_2O_3 at 77°K (233)
($Dq = 1830$ cm^{-1})

Band max., cm^{-1}	Oscillator strength, f	Assignment	
15,380 $\perp$	0.87	$^1A_1 \rightarrow {}^1E$	$\Big\}$ $^1T_{1g}(O_h)$
15,740 $\parallel$	0.27	$^1A_1 \rightarrow {}^1A_2$	
22,800 $\perp$	0.60	$^1A_1 \rightarrow {}^1E$	$\Big\}$ $^1T_{2g}(O_h)$
23,170 $\parallel$	1.05	$^1A_1 \rightarrow {}^1A_1$	

TABLE XLVI

Correlation of Vibrational Modes Observed for Co^{3+} in Al_2O_3
at 77°K on the $^1T_1(t_2{}^5e)$ Band with the Corundum Vibrations

Co^{3+} in Al_2O_3 (233), band max., cm^{-1}			Pure Al_2O_3 (206)	
	$\parallel c$	$\perp c$		
14,535	−53			
	+80	+235	244	a_{2u}
	+253	259		
		369		
		393	375	e_g
		419	417	e_g
		497		
		571	578	a_{1g}
		666	642	e_g
		742	751	a_{1g}
		915	909	e_u
		1025		

A good deal of vibrational structure was observed on the $^1T_1(t_2{}^5e)$ peak. Most of the vibrations can be correlated with the corundum lattice vibrations as shown in Table XLVI.

According to the assignments made in Table XLVI, not only non-degenerate vibrations, which become totally symmetric in C_3, but also degenerate vibrations are excited. The degenerate modes are nontotally symmetric in C_3, and their appearance in the spectrum can be ascribed either to vibrational electronic interaction or to a distorted equilibrium configuration in the upper state. The intensity of the $E(T_{2g})$ band increases by a factor of 1.36 from 300 to 1300°K, hence vibrational–electronic interaction is no more important here than for Cr^{3+}. The $14,535 - 53$ cm^{-1} band in the $\parallel c$ spectrum is assigned tentatively to the origin of the $\parallel$ transition. The next band at $14,535 + 80$ cm^{-1} is 133 cm^{-1} higher, similar to the 100 cm^{-1} interval observed in α-Al_2O_3:Ni^{3+}, and assigned (233) to the inactive twisting motion of the octahedron around the C_3 axis. This mode might be excited because the transition is made allowed by a twist of the octahedron from the ideal C_{3v} geometry.

10. $Co(H_2O)_6{}^{3+}$ in GASH

Walker and Carlin (409b) have determined the polarized crystal spectrum of Co:GASH. The crystal environment of the $Co(H_2O)_6{}^{3+}$ ion was discussed on p. 273. The observed bands are listed in Table XLVIA.

Excited state	Polarization	300°K	80°K
$^1T_1$	$\perp$	16,450	16,560
	$\parallel$	16,130	16,260
	Axial	16,130	16,200
$^1T_2$	$\perp$	25,000	25,580
	$\parallel$	25,320	25,440
	Axial	25,000	25,400

The two maxima are assigned as $^1A_1 \rightarrow {}^1T_2$ at $\sim 16,000$ cm^{-1} and $^1A_1 \rightarrow {}^1T_2$ at $\sim 25,000$ cm^{-1}. Comparison of axial and polarized spectra reveals that the transition to $^1T_1$ has considerable magnetic dipole character while the $^1T_2$ transition is electric dipolar in origin. Trigonal splittings were observed in both bands and are opposite in sign showing the importance of the off-diagonal term v'.

11. trans [Co(amine)$_2$X$_2$]$^+$ Complexes

The polarized crystal spectra of tetragonal amine complexes of cobalt (III) have been extensively investigated. In this section we shall discuss the spectra of the following:

trans-dichloro-bisethylenediamine cobalt(III), (Coen$_2$Cl$_2$)$^+$

(98,388,418,427,435)

trans-dibromobisethylenediamine cobalt(III), (Coen$_2$Br$_2$)$^+$(388,418,435)

trans-dichloro bis-1-propylenediamine cobalt(III), (Co(l-*pn*)$_2$Cl$_2$)$^+$(98).

The crystal structure of *trans*-(Coen$_2$Cl$_2$)ClHCl2H$_2$O has been determined by x-ray analysis (280,289). The compound is monoclinic, space group $P2_1/c, C_{2h}^5$, with cell constants $a_0 = 10.68$ Å, $b_0 = 7.89$, $c_0 = 9.09$ Å, and $\beta = 110° 26'$. There are two molecules per unit cell, with molecular symmetry C_{2h} for the (Coen$_2$Cl$_2$)$^+$ ion and C_i for the cobalt site symmetry. Two ethylene diamine molecules and two *trans* chloride ions coordinate to a central Co^{3+} ion, so that four nitrogen atoms of the ethylenediamine ligand and the cobalt atom are coplanar. The two coordinating ethylene-diamine molecules have the gauche form, kk', with one ligand related to the other by inversion. The ethylenediamine metal rings are kinked in a manner that preserves the center of symmetry at the cobalt site. The Cl—Co—Cl axis is nearly parallel to the c axis in the crystal, and on (100), it projects almost entirely normal to b (Fig. 47). The CoN$_4$ group is almost in the ab plane. In polarized light, the crystal appears yellow with the electric vector $\parallel$ to the b axis, and blue $\parallel$ to the c axis.

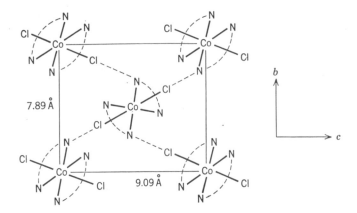

Fig. 47. Arrangement of *trans*-(Coen$_2$Cl$_2$)$^+$ in *trans*-(Coen$_2$Cl$_2$)ClHCl2H$_2$O. Projection of single layer upon (100) plane (280).

There is no detailed crystal structure of *trans*-(Coen$_2$Cl$_2$)ClO$_4$, the salt studied by Dingle (98) in the visible and ultraviolet regions and by Tanaka, Sato, and Fujita (388) in the infrared region. However, from the crystal habit, optical properties of the crystal and the absorption spectrum (435), it is concluded that the (Coen$_2$Cl$_2$)$^+$ units are arranged in the same way as in the (Coen$_2$Cl$_2$)ClHCl2H$_2$O structure. This is confirmed by measurements of the infrared spectra (388) showing that the Cl—Co—Cl axis nearly coincides with the *c* axis. Since (Coen$_2$Cl$_2$)ClHCl2H$_2$O decomposes easily in air, losing HCl and H$_2$O, experiments were carried out by Dingle (98) on the perchlorate. Wentworth (418) overcame this difficulty by immersing the (Coen$_2$Cl$_2$)ClHCl2H$_2$O crystal in a rubber cement diluted in acetone, this medium being transparent between 5000 and 30,000 cm^{-1} and also preventing any decomposition.

The crystal structure of *trans*-dibromobisethylenediamine cobalt(III) ion in *trans*-(Coen$_2$Br$_2$)BrHBr2H$_2$O is the same as that of the chloride (289).

The structure of *trans*-[Co(1-*pn*)$_2$Cl$_2$]ClHCl2H$_2$O has been determined by Saito and Iwasaki (351). The unit cell was found to be monoclinic, space group $C2, C_2{}^3, a_0 = 22.09, b_0 = 8.406, c_0 = 9.373$ Å, and $\beta = 99°39'$ with four molecules per unit cell. The molecular symmetry of the [Co(1-*pn*)$_2$Cl$_2$]$^+$ unit is D_2 if the CH$_3$ group of 1-propylenediamine is neglected, but C_2, if it is included. The cobalt site symmetry is C_1. The arrangement of the atoms follows closely to that found in the ethylenediamine analog, Figure 47. On (100), Cl—Co—Cl makes an angle of $\sim 10°$ with the *c* axes while the CoN$_4$ group is approximately in the *ab* plane.

TABLE XLVII

Absorption Band Maxima for $(Coen_2X_2)^+$ Complexes

	$(Coen_2Cl_2)ClHCl2H_2O$				D_{4h} Assignment (418)	$(Coen_2Cl_2)ClO_4$			$[Co(l\text{-}pn)_2Cl_2]ClHCl2H_2O$		
	Band max., cm^{-1} (435)	ϵ	Band max., cm^{-1} (418)	ϵ		Band max., cm^{-1} (98)	ϵ	$\Delta\nu_{1/2}$, cm^{-1}	Band max., cm^{-1} (98)	ϵ	$\Delta\nu_{1/2}$, cm^{-1}
300°K	16.1 ∥c	20	15.9 ∥c	20	$^1A_1 \to {}^1E_g$	16,150 ⊥b	51	2,550	16,100 ⊥b	56	2,500
	16.1 ⊥c	7	16.1 ⊥c	8	$^1A_1 \to {}^1E_g$	16,300 ∥b	12	2,100	16,200 ∥b	18.5	2,200
	23.3 ⊥c	15	23.1 ⊥c	17	$^1A_1 \to {}^1A_{2g}$	22,000 ∥b	30		22,400 ∥b	~35	
	—^{c}		~24.0 ∥c^a		$^1A_1 \to {}^1A_{2g}$	24,400 ∥b	35		24,000 ∥b	43	
						>27,000 ⊥,∥b	>100				
80°K (418, 435)			16.0 ∥c	14	$^1A_1 \to {}^1E_g$	16,350 ⊥b	22	2,100	16,350 ⊥b	32.5	1,900
			16.4 ⊥c	6	$^1A_1 \to {}^1E_g$	16,600 ∥b	10	1,750	16,600 ∥b	17.5	2,100
25°K (98)			23.3 ⊥c	15	$^1A_1 \to {}^1A_{2g}$	22,500 ∥b	~23	2,000^g	22,700 ∥b	29	2,000^g
			23.9 ∥c	5	$^1A_1 \to {}^1A_{2g}$	24,250 ∥b	30	2,100^g	24,200 ∥b	37.5	2,100^g
						27,100 ⊥,∥b	25 ∥ / 15 ⊥	} 2,400^g	27,300 ⊥,∥b	33 ∥ / 20 ⊥	2,300^g
						>28,500 ⊥,∥b	>100				
300°K	trans-$(Coen_2Br_2)BrHBr2H_2O$										
	14.9 ∥c	30	14.9 ∥c	31	$^1A_1 \to {}^1E_g$						
	15.0 ⊥c	6	15.2 ⊥c	8	$^1A_1 \to {}^1E_g$						
	—^{c}		—^{d}		$^1A_1 \to {}^1A_{2g}$						
	21.7 ∥c	14	21.9 ∥c	16	$^1A_1 \to {}^1A_{2g}$						
	—^{de}		—^{d}		$^1A_1 \to {}^1E_g$						
	—^{de}		—^{d}		$^1A_1 \to {}^1E_g$						
	25.6 ∥c	152	—^{de}		$^1A_1 \to {}^1B_{2g}$						
	29.0 ⊥c	135	—^{de}		—^{e}						
					—^{e}						

| 80°K | 15.1 $\parallel c$ | 23 | $^1A_1 \rightarrow {}^1E_g$ |
| | 15.4 $\perp c$ | 6 | $^1A_1 \rightarrow {}^1E_g$ |

80°K	~22.0 $\parallel c^a$	~12(~2)^b	$^1A_1 \rightarrow {}^1A_{2g}$
	22.3 $\perp c$	15(10)^b	$^1A_1 \rightarrow {}^1A_{2g}$
	25.4 $\parallel c$	40	$^1A_1 \rightarrow {}^1E_g$
	26.8 $\perp c$	42	$^1A_1 \rightarrow {}^1B_{2g}$
	—^{de}		
	—^{de}		

[a] Not clearly defined.
[b] Result of curve analysis.
[c] Not reported.
[d] Not observed.
[e] See text.
[g] From resolved curves.

The Co—Cl bond distance is 2.29 Å, exactly the same as found in (Coen$_2$-Cl$_2$)ClHCl2H$_2$O. The four Co—N distances are 1.99, 1.97, and 2.02, 1.94 Å for the two 1-propylenediamine rings, giving a distorted octahedral environment to the Co^{3+} ion. The complex has a twofold symmetry about the Cl—Co—Cl axis (351), the ligands thus having a kk form. The C—CH$_3$ bonds on the 1-propylenediamine rings are equatorial with respect to the Cl—Co—Cl axis.

Wentworth (418) recorded the spectra in the same manner as Yamada et al (427,435) of (Coen$_2$Cl$_2$)ClHCl2H$_2$O and (Coen$_2$Br$_2$)BrHBr2H$_2$O with the incident light polarized $\perp$ and $\parallel$ to the c axis. To ensure the best alignment of the electric vector with this axis, the spectrum of the region between 14,000 and 17,000 cm^{-1} was recorded (418) at 2° angle intervals until the absorbancy reached a maximum, where it is assumed that the electric vector is accurately parallel to the c axis. The spectra were then obtained by rotating the polarizer through 90°. Dingle (98) recorded the spectra of (Coen$_2$Cl$_2$)ClO$_4$ and [Co(1-pn)$_2$Cl$_2$]ClHCl2H$_2$O at temperatures 300, 80, and 25°K with the incident light $\perp$ and $\parallel$ to the b axis.

Table XLVII lists the results obtained in these studies. The spectra shown in Figure 48 of *trans*-(Coen$_2$Cl$_2$)ClO$_4$ is a typical example of this series (98).

The results of Yamada et al. (435) have been interpreted by Ballhausen and Moffitt (10) in a crystal field model using D_{4h} molecular sym-

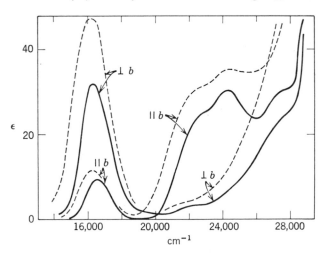

Fig. 48. The polarized single-crystal spectrum of *trans*-(Coen$_2$Cl$_2$)ClO$_4$ taken on (100) at room temperature (broken line) and at 80°K (solid line). The 25°K spectrum is very similar to the latter. The polarizations refer to the b axis of the crystal. On this face this axis is almost normal to the Cl—Co—Cl direction (98).

metry and a vibronic coupling mechanism. McClure (231) questioned this assignment and suggested an alternative D_4 nonvibronic analysis. Dingle (98), however, having determined the temperature dependence of the absorption band intensities, concluded that a vibronic intensity-gaining model could explain the observed intensities in the different polarizations. In $trans$-$(Coen_2Cl_2)^+$, the C_i site symmetry cannot explain the results, while the molecular symmetry is C_{2h} with z passing through the center of the C—C bonds in the ligands. Approximate molecular symmetries D_{2h} and D_{4h} (z parallel with Cl—Co—Cl) which neglect the kinking of the ethylenediamine rings are also possible. As the spectra of $trans$-$(Coen_2Cl_2)^+$ and $trans$-$[Co(1$-$pn)_2Cl_2]^+$ are similar, while the essential difference between the two molecules is the conformation of the chelate rings, it is concluded that this is unimportant in the determination of the absorption spectrum. On this basis the spectra were assigned (98) on a vibronic D_{4h} model, Table XLVIII.

From these measurements (98) it was not possible to distinguish between $^1A_{2g}$ and $^1B_{2g}$. Wentworth and Piper (416,417) conclude from crystal field and Hückel MO calculations generally similar to those of Yamatera (450) and McClure (232) on the splittings in tetragonal complexes of cobalt(III) that the $^1A_{2g}$ state lies below the $^1B_{2g}$. Dingle (98) has pointed out that these two states will probably be mixed via a vibronic coupling mechanism, while the $^1E_g{}^a$ and $^1E_g{}^b$ states will mix under the influence of the tetragonal field. Using the matrix element $\langle {}^1E_g{}^a \mid V_T \mid {}^1E_g{}^b \rangle = \sqrt{3}(D_s + \frac{5}{4}D_t)$ (10), it is found that $^1E_g{}^b$ contains an admixture of about 20% $^1E_g{}^a$ (98). With allowance for π bonding and interaction of configurations, a reasonable fit with the theory can be obtained. The polarized infrared spectrum (388) has been partially interpreted on the basis of C_{2h} symmetry in $trans$-$(Coen_2Cl_2)^+$, but these results were not available at the time of Dingle's (98) publication.

The circular dichroism and optical rotation of these compounds has been determined (177,415) and Dingle has argued from these results that

TABLE XLVIII

Assignment of Bands in $trans$-$(Coen_2Cl_2)^+$ on a Vibronic D_{4h} Model (98)[a]

Assignment	Band max., cm^{-1}	Term level in O_h	$\Delta\nu$, tetragonal
$^1A_{1g} \rightarrow {}^1E_g{}^a \times \epsilon_u$	16,150	$^1T_{1g}$	6,300 cm^{-1}
$\rightarrow {}^1E_g{}^a \times \alpha_{2u}, \beta_{2u}$	16,300		
$\rightarrow {}^1A_{2g} \times \epsilon_u$	22,500		
$\rightarrow {}^1B_{2g} \times \epsilon_u$	24,250	$^1T_{2g}$	2,750 cm^{-1}
$\rightarrow {}^1E_g{}^b \times \epsilon_u, \alpha_{2u}, \beta_{2u}$	27,100		

[a] $Dq = 2175$ cm^{-1}, $D_s \sim 600$ cm^{-1}, and $D_t \sim 700$ cm^{-1}.

TABLE XLIX

Assignments of Bands in *trans*-[Co(1-*pn*)$_2$Cl$_2$]$^+$ on a D_2 Vibronic Model (98)

Assignment			Band max., cm^{-1} (300°K)	Polarization
D_{4h}	D_2(electronic)	D_2(vibronic)		
$^1A_{1g} \rightarrow \, ^1E_g{}^a(xy,z)$	$\left\{ \begin{array}{c} ^1B_2(y) \\ ^1B_3(x) \end{array} \right\}$ $\alpha_1,\beta_1,\beta_2,\beta_3,(x,y,z)$		16,000, 16,200	$xy;z$
$^1A_{2g}(xy)$	$^1B_1(z)$	$\left. \begin{array}{c} \\ \end{array} \right\}\beta_2,\beta_3(x,y)$	22,400	xy
$^1B_{2g}(xy)$	^1A(forbidden)		24,200^a	xy
$^1E_g{}^b(xy,z)$	$\left\{ \begin{array}{c} ^1B_2(y) \\ ^1B_3(x) \end{array} \right\}$ $\alpha_1,\alpha_1,\beta_2,\beta_3(x,y,z)$		27,300^a	$xy;z$

^{a} Maximum at 80°K.

the spectrum of *trans*-[Co(1-*pn*)$_2$Cl$_2$]$^+$ is more appropriately assigned on the basis of a D_2 vibronic model. The D_{4h} representations in D_2 transform as shown in Table XLIX while the assignments for [Co(1-pn)$_2$Cl$_2$]$^+$ are also included in Table XLIX.

The polarization of the bands at 22,400 and 24,200 cm^{-1} is consistent with the source of the intensity being almost entirely vibronic rather than electronic in origin for the crystal spectra. In accord with Wentworth and Piper (415), Dingle (98) proposes that there is no vibronic contribution to the circular dichroism.

Wentworth (418) has given a somewhat different interpretation of the spectrum of *trans*-(Coen$_2$Br$_2$)BrHBr2H$_2$O. The splitting of the $^1T_1$ band in the tetragonal field is interpreted (418) in terms of $\delta\Delta$, the difference in the crystal field stabilization energies due to the axial and in-plane ligands. The appearance of bands at 25,400 and 26,800 cm^{-1} at 80°K in opposite polarizations is attributed to splitting of the $^1T_2$ band, when $\delta\Delta$ is $-12,600$ cm^{-1}. The ordering of the $^1E_g{}^b$ and $^1B_{2g}$ components is determined by the amount of π bonding from the bromide ion. With small π bonding, the $^1E_g{}^b$ state should lie higher than the $^1B_{2g}$ state, but with increasing π bonding, the energy separation decreases until the states cross, which then leads to inversion of ^1E_g and $^1B_{2g}$ levels. It is assumed that the $^1B_{2g}$ state is in fact highest, as shown in the assignments (Table XLVII). The two bands observed by Yamada et al. (435) at 25,600 and 29,000 cm^{-1} may correspond to the two bands observed by Wentworth (418), with higher values of ϵ due to the higher temperature. In the chloro complex, the onset of charge-transfer at 27,000 cm^{-1} masks any bands of this type.

12. Co^{3+}en$_2$X$_2$ Complexes

Yamada (445) has reported the polarized absorption spectra of *trans*-(Co^{3+}en$_2$X$_2$) complexes where X = Cl$^-$, Br$^-$, ONO$_2{}^-$, OH$_2$, NCS$^-$, and

TABLE L

Assignment of Observed Bands in *trans*-($Co^{3+}en_2X_2$) Complexes (445)

X	Transition $^1A_{1g} \rightarrow {}^1E_g$		Transition $^1A_{1g} \rightarrow {}^1A_{2g}$	
	Band max., cm^{-1}	Polarization	Band max., cm^{-1}	Polarization
Br^-	14,800	z	22,000	xy
Cl^-	16,000	z	22,700	xy
ONO_2^-	18,600	z	23,000	xy
OH_2	18,700	z	21,200	xy

NO_2^-. The observed maxima and assignments are tabulated in Table L. The splitting of the bands is interpreted in terms of the magnitude of σ and π bonding in each complex.

13. Other Co^{3+} Complexes

The infrared spectra of single crystals of $[Co(NH_3)_4CO_3]Br$ have been measured using polarized light incident on the (001) face of the crystal (105). Definite polarization data were obtained for four carbonate modes, from which it is concluded that the carbonate group has effective C_{2v} symmetry (as distinct from free-carbonate D_{3h}). This is consistent with carbonate acting as a bidentate ligand in this complex.

Tsuchida and Kobayashi (395) measured the absorption spectrum of crystals of *cis*-$[Co(NH_3)_4Cl_2]Cl$. Maxima were obtained at 19,300 and 18,300 cm^{-1}, for the two components $\parallel$ and $\perp$ to the c axis. The dichroism of $[Co(bipy)_2Cl_2]Cl$ has also been determined (374).

The dichroism of some complexes of the type *cis*-$(Coen_2X_2)^+$ and $[Co(NH_3)_5X]^{2+}$ have been investigated (445) and the results discussed in comparison with those for the *trans* compounds. The splitting of the $^1A_{1g} \rightarrow {}^1E_g$ band is found to be larger in the *trans* than in the *cis* series, but the magnitude of the splitting is greater than that predicted by Yamada's theoretical estimate. In the more recent work by Stanko (374a) on the polarized optical spectra of $[Co(NH_3)_5X]SiF_6$, where X = Cl,Br,CN, a vibronic intensity mechanism is invoked to account for the main polarization features of the components of the two spin-allowed absorption bands. The crystal structure of $[Co(NH_3)_5Cl]SiF_6$ is monoclinic with $a_0 = 6.26$, $b_0 = 8.22$, $c_0 = 10.18$ Å, and $\beta = 99° 40'$.

Dichroism of $[Co(NH_3)_3Cl_2H_2O]Cl$ has been observed and the crystal structure determined (438). The hexagonal unit cell contains two $[Co(NH_3)_3Cl_2H_2O]Cl$ molecules with $a_0 = 7.37$ and $c_0 = 8.75$ Å, space group $C6/mmc, D_{6h}^4$. Two chlorine atoms are coordinated to a cobalt atom in *trans* positions at 2.33 Å. The Cl—Co—Cl chain is parallel to the c axis, which accounts for the marked red–blue dichroism of the crystals. This is in qualitative agreement with the results for $(Coen_2Cl_2)ClHCl2H_2O$ crystals (280).

The crystal spectrum of $Co(NH_3)_6Co(CN)_6$ has been studied by Wentworth (414). Earlier studies were made by Tsuchida and Kobayashi (397) and by Kondo (205). The crystal is hexagonal with space group $P\bar{3}, C_{3i}^2$ and the site symmetry of the Co^{3+} atom of $[Co(NH_3)_6]^{3+}$ is either C_3 or S_6 (172). The $\perp$ and $\parallel$ spectra are identical indicating small distortions from O_h symmetry. At 80°K, progressions of at least 12 vibrational bands superimposed on the $^1A_1 \rightarrow {}^1T_1$ transition at 21,190 cm^{-1} extending from $\sim$24,390–18,520 cm^{-1} are observed. Substitution of hydrogen by deuterium shifted the band maximum to 21,500 cm^{-1} with the first component discernible between 19,230 and 18,870 cm^{-1}. The vibrational structure of $Co(NH_3)_6Co(CN)_6$ can be explained in terms of a uniquantal progression of a 420 $\pm$ 10 cm^{-1} vibration and its combinations with a 390 $\pm$ 10 cm^{-1} vibration. Deuteration lowers these to 380 $\pm$ 10 cm^{-1} and 340 $\pm$ 15 cm^{-1}, respectively. These frequencies are too small to accommodate internal ligand motions. Hence, only skeletal vibrations of the CoN_6 octahedron appear to contribute to the intensity of the $^1T_{1g}$ absorption, as was found in dl-$(Coen_3Cl_3)_2NaCl6H_2O$ by Dingle (93). This is in contradiction to the suggestion by Mason and Norman (257) that in $(Coen_3)^{3+}$, the N—H and C—H vibrations are responsible for a significant share of the intensity of the spectroscopically accessible $d \rightarrow d$ transitions.

Polarized crystal spectra studies have been carried out on the binuclear compounds $[(NH_3)_5Co(III)—O_2—Co(IV)(NH_3)_5](NO_3)_5$ and $[(NH_3)_5Co(III)—O_2—Co(III)(NH_3)_5](NO_3)_42H_2O$ (428). The spectra were recorded $\perp$ and $\parallel$ to the Co—O—O—Co direction and the observed bands are listed in Table LI.

TABLE LI
Observed Absorption Bands for $Co(NH_3)_5—O_2—Co(NH_3)_5$ (428)

	Band max. $\parallel$, cm^{-1}	log ϵ	Band max. $\perp$, cm^{-1}	log ϵ
Co(III)–Co(IV) cpd.	15,000	1.88		
	21,500	1.32	21,000	1.13
	27,500	1.38	29,300	1.38
Co(III)–Co(III) cpd.	16,000	0.85	15,300	1.48
	21,000	1.2		
	27,700	1.53		

Paramagnetic resonance absorption measurements (103,245) have shown that $[(NH_3)_5Co—O_2—Co(NH_3)_5]^{5+}$ has one unpaired electron which interacts equally with both cobalt nuclei. Any formulation of the structure must be made under the assumption that both the cobalt atoms are equivalent. Two structures have been suggested, one with a single Co—O—O—Co linkage (103) and the other with a

$$
\begin{array}{ccc}
 & O & \\
\diagup & & \diagdown \\
Co & & Co \\
\diagdown & & \diagup \\
 & O & \\
\end{array}
$$

link (54), giving each cobalt atom seven-coordination. However, due to these uncertainties, no assignments of the bands are possible at this stage.

B. $4d^6$

1. Ru(bipy)$_3^{2+}$

The polarized visible spectrum of Ru(bipy)$_3^{2+}$ has been studied by Palmer and Piper (300) in the host lattice Zn(bipy)$_3SO_4 7H_2O$. The site symmetry of the divalent ion has been discussed earlier (see p. 356). The observed spectrum is shown in Figure 49. It is very similar to that of Fe(bipy)$_3^{2+}$ in both crystal and solution. The entire spectrum is shifted

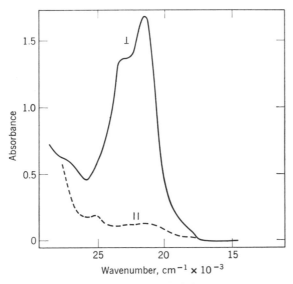

Fig. 49. Polarized orthoaxial spectra of Zn(Ru)(bipy)$_3SO_4 7H_2O$ at 300°K. Crystal thickness, 2.0 mm; approximately 2×10^{-4} mole % ruthenium (300).

about 2700 cm^{-1} to higher energy, but there is no shoulder at 11,500 cm^{-1} as in the Fe(bipy)$_3^{2+}$ spectrum. The observed shoulder at 18,000 cm^{-1} is not assigned to a $d \rightarrow d$ transition as it is too intense, in disagreement with Crosby, Perkins, and Klassen (81) who assigned it to the $^1A_1 \rightarrow {}^1T_1$ transition when studying the spectrum in EMPA rigid glass at 82°K (81). The band observed by Crosby et al. at 15,050 cm^{-1} has been acknowledged (ref. 88 p. 300) to be due to an impurity.

The polarization of the intense visible bands is almost complete, with $I_\perp/I_\parallel$ being 26.5 at 300°K. The integrated intensity of these bands shows no temperature dependence betweeen 80 and 300°K.

IX. CONFIGURATION d^7

A. $3d^7$

The Tanabe-Sugano diagram for the d^7 configuration in an octahedral field is shown in Figure 50. Liehr (212) has given a complete d-level scheme for the three-electron and three-electron hole (i.e., d^7), cubic field states

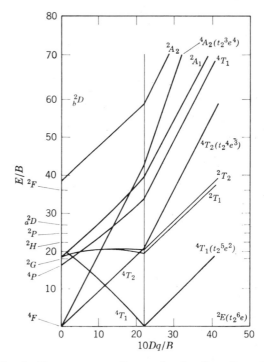

Fig. 50. Tanabe-Sugano energy diagram for the d^7 configuration (75,386).

including spin–orbit coupling. The commonest representative of the d^7 configuration is Co^{2+}, which has been studied in both octahedral and tetrahedral coordination.

1. Co^{2+} in MgO

Pappalardo, Wood, and Linares (310) have studied the optical absorption spectra of several oxide systems—MgO, ZnO, aluminum spinel, yttrium gallium garnet, and germanium garnets doped with Co^{2+} at room temperature, 78, and 4.2°K. Wood and Remeika (421) have studied the optical spectra of garnets containing Co^{2+} and Co^{3+} ions, and compared their results for the substitution of the interstitial ion into octahedral, tetrahedral, and dodecahedral sites. These have been analyzed using a crystal field model. No polarized data were recorded, so the results will not be discussed in detail. The absorption spectrum of a Co^{2+}-doped MgO crystal has also been studied and interpreted by Low (225), who found good agreement of observed frequencies with crystal field predictions.

Pappalardo et al. (310) used the absorption spectra of Co in MgO and ZnO as reference spectra for octahedral and tetrahedral coordination of Co^{2+} to determine the site symmetry of the Co^{2+} ion in the Co doped spinels and garnets. A comparison of the absorption bands of MgO:Co and ZnO:Co shows that the ZnO:Co spectrum is shifted towards lower energies and that the band shape in the infrared region is markedly different for the two systems. The shape of the bands in the visible region does not however show the drastic difference found for Ni^{2+} in MgO and ZnO (309).

Magnesium oxide crystallizes in a NaCl-type lattice. The unit cube contains four molecules with a lattice constant of 4.2112 Å at 21°C (422). The Mg atoms are in positions 000; $\frac{1}{2}\,\frac{1}{2}0$; $\frac{1}{2}0\frac{1}{2}$; $0\frac{1}{2}\,\frac{1}{2}$ and the oxygens at $\frac{1}{2}\,\frac{1}{2}\,\frac{1}{2}$; $00\frac{1}{2}$; $0\frac{1}{2}0$; $\frac{1}{2}00$. Each Mg atom has six equidistant oxygen atoms as nearest neighbors and vice versa. The neighbors are at the corners of a regular octahedron and hence any substitution of Mg^{2+} by a divalent cation will be in an octahedral site.

The energy level scheme for Co^{2+} in MgO was calculated (310) assuming $Dq = 927$ cm^{-1}, $B = 945$ cm^{-1}, and $C = 4.5B$. The $^4T_1(^4F) \rightarrow$ $^4T_2(^4F)$, and $^4T_1(^4F) \rightarrow {}^4T_1(^4P)$ transitions were fitted to the experimental values. The first-order spin–orbit splittings of the quartet terms were also calculated. The experimental results are shown in Table LII. No unequivocal assignments of individual lines have been made, and vibrational bands are not yet identified: it is possible, for instance (310) that the line systems A_1 and A_2 may be vibrational replicas of A_0 (Table LII). Possible multiplet component assignments are also discussed in (310).

TABLE LII
MgO:Co (0.20%); Sample 0.37 cm thick; Temperature 4.2°K (310)

Principal assignments	Absorption peaks	Wave number, cm^{-1}	ϵ_{max}, cm^{-1}	Half-width $(\Delta\nu)_{\frac{1}{2}}$ at $\alpha=\frac{1}{2}\,\alpha_{max}$, cm^{-1}	Oscillator strength
$^4T_1(^4F) \rightarrow {}^4T_2(^4F)$	A_0	8,146.6	0.43	6	1.0×10^{-7}
		8,163.2	0.38	11	1.6×10^{-7}
		8,203	0.32	11	1.3×10^{-7}
		8,319.4	0.24		
	A_1	8,350.7	0.18		4.0×10^{-7}
		8,403.3	0.17		
		8,547	0.17		
	A_2	8,598.4	0.21		$2.3 \times 10^{-6\,a}$
		8,633.3	0.17		
	B	18,559⎱ 18,797⎰	0.32		
$^4T_1(^4F) \rightarrow {}^4T_1(^4P)$	C	19,723	0.6		1.6×10^{-5}
		20,555⎱ 20,833⎰	0.6		

[a] Inclusive of long tail to high energies.

2. Co^{2+} in $CoCl_2$, $CoBr_2$, $KCoF_3$, and $CoWO_4$

Ferguson, Wood, and Knox (118) have studied the crystal spectra of Co^{2+} in the lattices $KCoF_3$, $CoCl_2$, $CoBr_2$, and $CoWO_4$. No polarization data were recorded. In all these compounds, the Co^{2+} ion is in a position more or less distorted from a pure octahedral site. $KCoF_3$ at room temperature has a cubic perovskite structure in which each Co^{2+} ion lies in a perfect octahedral field of six F^- ions (291). Below 114°K, a small tetragonal distortion occurs (291). $CoCl_2$ crystallizes in a rhombohedral unit cell, containing one molecule, space group $R\bar{3}m, D_{3d}^5$, with lattice constants for the corresponding hexagonal cell of $a_0 = 3.544$ and $c_0 = 17.430$ Å (352). The anions are cubic close-packed, with a slight trigonal distortion. At $\sim 90°K$ the color of $CoCl_2$ is dull pink rather than the blue observed at room temperature but no change in the crystal structure could be detected (352). $CoBr_2$ crystallizes in a hexagonal unit cell, space group $C\bar{3}m, D_{3d}^3$ with one molecule per unit cell and lattice constants $a_0 = 3.68$ and $c_0 = 6.12$ Å, with $u = 0.25$ (123). When $c_0/a_0 = 1.63$ and $u = 0.25$, the Br atoms are in perfect hexagonal close packing; here $c_0/a_0 = 1.663$, so it is slightly trigonally distorted. $CoWO_4$ belongs to an isomorphous series of tungstates which includes $MgWO_4$, $MnWO_4$, $FeWO_4$, $NiWO_4$, and $ZnWO_4$. Keeling (198) has determined the structure of $NiWO_4$ and has found that

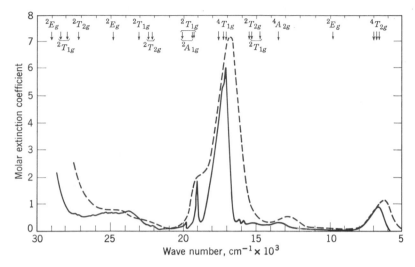

Fig. 51. Absorption spectrum of $CoCl_2$. The broken curve is the room temperature spectrum and the full line that taken at 20°K. The computed energies are shown by vertical lines (118).

the crystal has space group $P2/c, C_{2h}^4$. The structure is found to contain WO_6 groups rather than WO_4 groups as previously reported (380). It is based on hexagonal close-packing of the O atoms with Co and W atoms occupying one fourth of the octahedral interstices. The Co atom lies off-center of the distorted octahedron of oxygen atoms in a direction parallel to the b axis with symmetry C_2—2.

The spectrum of $CoCl_2$ at 20°K together with the proposed band assignments (118) is shown in Figure 51 as an example of this series. Calculations were carried out on a four parameter (Dq, ζ, B, and C) crystal field model and fair agreement was found for the doublet levels in $CoCl_2$ and $KCoF_2$. An attempt was made to correlate the different band shapes observed in the crystals with the size of the octahedral hole that the cobalt ion occupies. Assignments are similarly proposed for $KCoF_3$, $CoBr_2$, and more tentatively for $CoWO_4$ on the basis of fitting to the calculated levels.

3. Co^{2+} in ZnO

The absorption spectrum of Co^{2+} in a tetrahedral lattice (ZnO) has been studied by Pappalardo et al. (310) and also by Weakliem and McClure (412). The d-orbital splitting diagram for Co^{2+} in a tetrahedral field is the same as for d^3 in an octahedral coordination (Fig. 18).

Zinc oxide crystallizes in a hexagonal unit cell, space group $C6mc$, C_{6v}^4, with dimensions $a_0 = 3.2495$ Å, $c_0 = 5.2069$ Å, and $u = 0.345$ with two molecules per unit cell. The Zn atoms are at positions 000; $\frac{1}{3}\,\frac{2}{3}\,\frac{1}{2}$ and the oxygens at $00u$; $\frac{1}{3}\,\frac{2}{3}$, $u + \frac{1}{2}$, which are derived from two sets of special positions of $C_{6v}{}^4$, $\frac{1}{3}\,\frac{2}{3}\,v$; $\frac{2}{3}\,\frac{1}{3}\,v + \frac{1}{2}$ by a change of origin to the point $\frac{2}{3}\,\frac{1}{3}0$ (178,380). The wurzite ZnO structure may be visualized as two interpenetrating hexagonal close-packed lattices, one an anion and the other a cation lattice. The two lattices are displaced by approximately $3c/8$ and each atom is tetrahedrally surrounded by four atoms of the opposite type. If the lattice were ideal, the c_0/a_0 ratio would be $(8/3)^{1/2} = 1.633$, however, the c_0/a_0 ratio is less than this for ZnO (1.6024). This distortion means that the Zn—O distance parallel to c is slightly shorter than the other three equivalent distances. Taking δ to be the difference and d equal to the three equivalent Zn—O distances, $\delta/d = 0.0154$ (412,422). The symmetry at a cation site is C_{3v}, but because of the small magnitude

TABLE LIII
ZnO:Co. 0.17% (0.02 cm thick) at 4.2°K (310)

Transition		Absorption peaks	λ, Å	ν, cm^{-1}	ϵ_{max}, cm^{-1}	$(\Delta\nu)_{1/2}$ at $\alpha = \frac{1}{2}\,\alpha_{max}$, cm^{-1}	Oscillator strength
				4,140[a]	2.3		
				4,336[a]	3.3		
$^4A_2 \rightarrow {}^4T_1(^4F)$	T_1	16,540		6,045.9	11	12.6	0.6×10^{-5}
		16,390		6,101.3	17.5	34	1.7×10^{-5}
		16,160		6,188.1	18.5	47	2.5×10^{-5}
	T_2	15,900		6,289.3	12.5		
		15,300		6,535.9	9		
		15,000		6,666.6	17		0.53×10^{-3}
		14,730		6,788.8	24		
		14,400		6,944.4			
	T_3	13,950		7,168.4	60		
		13,665		7,317.9			
		13,510		7,401.9	31.5		
		13,335		7,499	34		
		13,000		7,692.3	32.5		
		12,500		8,000	15.5		
		12,200		8,196	10		
$^4A_2 \rightarrow {}^4T_1(^4P)$	D	6,500		15,384	138		
		6,380		15,674	115		0.68×10^{-2}
		6,270		15,949	107		
	E	6,150		16,260	120		
	F	5,665		17,652	40		
	G	4,950		20,202	4		

[a] In a sample 0.33 cm thick, and containing a nominal 5% Co, recorded at 77°K.

of δ/d, the approximate symmetry T_d is predominantly reflected in the absorption spectra.

Since a center of symmetry is absent, the extinction coefficient is greater than for octahedral coordination by a factor of roughly 100. The spectral results at 4.2°K, together with overall assignments, are shown in Table LIII. There is a good deal of structure, and an attempt has been made (310) to fit the predicted spin–orbit splitting of the $^4T_1(^4F)$ band (overall splitting calculated at 1019 cm^{-1}) to the data. With the lowest level (Γ_6) identified with the line at 6009 cm^{-1} (78°K) and the (Γ_7,Γ_8) pair identified with the 7189 cm^{-1} (78°K) line, there is approximate agreement. However, the remaining assignments are undecided, and in detailed calculations a symmetry lower than T_d should be considered (see below).

The close similarity of the absorption spectra of Co-doped spinels and garnets with the spectra of ZnO:Co suggests that Co^{2+} is present in such systems in sites possessing predominantly tetrahedral coordination (310). This is in disagreement with the assignments of Geller, Miller, and Treuting (140) who believed that the Co^{2+} ion preferentially takes up an octahedral site. This discrepancy may be resolved if it is assumed that the strong absorption of a minority of Co^{2+} ions in tetrahedral coordination swamp out the weaker spectrum of the majority of ions in other coordinations (310). The earlier assignments (140) were based on considerations of electronic configuration, ionic radii, and relative size of tetrahedral and dodecahedral sites in garnets containing Co.

Weakliem and McClure (412), in studying the polarized absorption spectrum of Co^{2+} in ZnO, took into account the fact that there is a small trigonal (C_{3v}) component in the ZnO field. Figure 52 shows the spectrum at 298 and 77°K. The observed bands at 4°K in the two polarizations in the region 14,900–19,200 cm^{-1} are shown in Figure 53.

The ground state in T_d is $^4A_2$ which becomes a G state when spin–orbit coupling is included. The spin–orbit components of $^4T_1(^4F)$ are

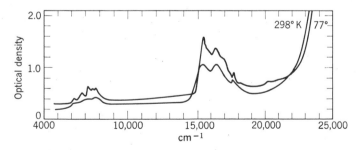

Fig. 52. Absorption spectrum of Co^{2+} in ZnO at 298 and 77°K (412).

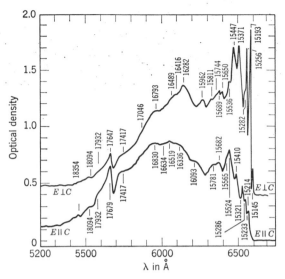

Fig. 53. Absorption spectrum of the visible bands of Co^{2+} in ZnO at 4°K (412).

clearly visible (Fig. 52) and the visible band is a mixture of the ^2E and $^2T_1(^4G)$ components and the $^4T_1(^4P)$ state. In C_{3v}, the ground state becomes $E_{1/2} + E_{3/2}$ and these levels are separated by only about 1 cm^{-1} (260). The two states are equally populated and since transitions are allowed to both $E_{1/2}$ and $E_{3/2}$ excited states in both polarizations, the selection rules based on the orbital symmetries are most important:

$$(e^4t_2{}^3)A_2 \rightarrow (e^3t_2{}^4)A_2 \quad \parallel \text{ polarized}$$
$$(e^4t_2{}^3)A_2 \rightarrow (e^3t_2{}^4)E \quad \perp \text{ polarized}$$

Transitions from the quartet ground state to the doublets are made allowed through spin–orbit mixing when $^2E(^2E)$ couples with $^4E(^4T_2)$ and $^4A_1(^4T_2)$, whereas $^2A_2(^2T_1) + {}^2E(^2T_1)$ is mixed with $^4A_2(^4T_1)$, $^4E(^4T_1)$, $^4A_1(^4T_2)$, and $^4E(^4T_2)$. The trigonal potential can cause further mixing of the states with the same symmetry in C_{3v}. Transitions to the excited states $^2E(^2E)$ and $^2E(^2T_1)$ are only expected for the $\perp$ spectrum and transitions only to $^2A_2(^2T_1)$ for the $\parallel$ spectrum. This is qualitatively confirmed by experiment (Fig. 53), but the large number of narrow bands which have small energy separations remain unexplained. They probably result from coupling between the electronic and vibrational motions of the complex.

Pappalardo and Dietz (308) have discussed the absorption spectra of first transition series metals (V to Cu) in the near ideal tetrahedral CdS lattice. There are many interesting features in the recorded spectra but they will not be discussed here as the oxidation state in some cases is not precisely known.

4. Cs_2CoCl_4 and Cs_3CoCl_5

The polarized crystal spectrum of Co^{2+} in a tetrahedral Cl^- environment has been the subject of several studies. However, assignments of the crystal spectra present many difficulties. We will summarize the experimental work with brief comments on the problems of interpretation. The

TABLE LIV

Observed Bands and Transitions for Cs_2CoCl_4, at $4°K$ (318,319)

Transition	Band max., cm^{-1} [a]	$\Delta\nu_{\frac{1}{2}}$, cm^{-1}		
$^4A_2 \rightarrow\ ^4T_2(^4F)$	3,100	2,500		
$^4A_2 \rightarrow\ ^4T_1(^4F)$	5,400	2,500		
$^4A_2 \rightarrow\ ^4T_1(^4P)$	15,300	2,500		
$^4A_2 \rightarrow$ terms from (^2G)	17,700			
			$E \parallel A_2$ [c,d]	
	Band max., cm^{-1} [b]	Strength	Width	
$^4A_2 \rightarrow$ terms from $(^2G,^2D)(^2H,^2P)$	26,342	m		
	26,395	s		
	26,465	m	n	
	26,624	m	n	
	26,674	m	b	
	26,747	w		
	26,914	s	b	
	26,962	w		
$^4A_2 \rightarrow$ terms from ^2F	30,437	s	n	
	30,475	s	b	
	30,630	w	b	
	30,730	w	n	
	30,820	w		
	30,910			
	31,110	m	b	
	32,390	m	n	
	32,470	s	b	
	32,570		b	
	32,670	s	b	
	32,710	s	b	

[a] Ref. 318.

[b] Ref. 319.

[c] Measured parallel to twofold axis A_2. The same bands are obtained with light polarized parallel to either of the remaining twofold axes, but with different qualitative intensities and line widths.

[d] s = strong; m = medium; w = weak; vw = very weak; n = narrow; b = broad; vb = very broad.

compounds Cs_2CoCl_4 and Cs_3CoCl_5 will be considered together, for while they have different crystal structures, both contain the near-tetrahedral $CoCl_4{}^{2-}$ complex. Cs_2CoCl_4 has a K_2SO_4-type structure (332), space group $Pnam, D_{2h}{}^{16}$, and an orthorhombic unit cell having lattice constants $a_0 = 9.737$, $b_0 = 12.972$, $c_0 = 7.392$ Å with four molecules per unit. There is a small distortion around the cobalt atom from a perfect tetrahedron, but the average Co—Cl distance is 2.26 Å. The crystal structure of Cs_3CoCl_5, was first determined by Powell and Wells (334) and a least squares refinement has recently been carried out by Figgis, Gerloch, and Mason (127). The compound has a tetramolecular tetragonal unit cell, $a_0 = 9.219$, $c_0 = 14.554$ Å of space group $I4/mcm, D_{4h}{}^{18}$. The Co—Cl bond length is 2.252 Å. The slight deviation from strict tetrahedral symmetry of the ligand arrangement about the cobalt ion involves angular distortions, (Cl—$\widehat{\text{Co}}$—Cl $= 106°$ twice, $111°\ 12'$ twice) rather than bond stretching or contraction. This situation is entirely similar to that of $CoCl_4{}^{2-}$ in Cs_2CoCl_4 (332) and of $Co(CNS)_4{}^{2-}$ in $K_2Co(CNS)_4 4H_2O$ (453).

Pelletier-Allard has studied the absorption spectrum of Cs_2CoCl_4 in the near-infrared and visible regions (318) and also in the near-ultraviolet at 4°K (319). The light was polarized with the electric vector parallel to each of the twofold axes in turn. The observed bands and suggested assignments are listed in Table LIV. The combination of spin–orbit splitting and the low symmetry field for the ^2F term would produce seven lines, yet many more are observed for Cs_2CoCl_4; the remainder are attributed to vibrational bands (319).

5. R_2CoCl_4; R $= \frac{1}{2}$Zn, Cs, or $N(CH_3)_4$

Ferguson (117) has studied the polarized spectra at temperatures down to 4.2°K of single crystals of the type R_2CoCl_4, where R is $\frac{1}{2}$Zn, Cs, $N(CH_3)_4$, or quinolinium ion. These have been measured from the near-infrared to the near ultraviolet. Cs_2ZnCl_4 is known to be isostructural with Cs_2CoCl_4 (52). The Cl^- tetrahedron is again slightly angularly distorted, with Cl—$\widehat{\text{Co}}$—Cl angles $107°\ 20'$, $108°\ 50'$, $109°\ 20'$, and $116°\ 20'$. The cobalt site belongs to the point group C_s (mirror-plane symmetry only). The structure of the crystals $[N(CH_3)_4]_2 Zn(Co)Cl_4$ has been determined by Morosin and Lingafelter (268); they belong to the same space group as the cesium compounds, but have space-group orientation $Pnma$.

The internal Cl—$\widehat{\text{Co}}$—Cl "tetrahedral" angles are $108°\ 18'$, $109°\ 6'$, $110°24'$, and $111°42'$. The structure of the quinolinium compound is not known.

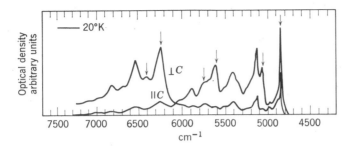

Fig. 54. The $^4A_2 \rightarrow {}^4T_1(^4F)$ absorption band of $Cs_2Zn(Co)Cl_4$ at 20°K. The six electronic origins are shown by arrows (117).

Three-parameter crystal-field theory leads to prediction of a $^4A_2$ ground state in T_d symmetry, and $^4A_2 \rightarrow {}^4T_2(^4F)$ and $^4A_2 \rightarrow {}^4T_1(^4F)$ as the lowest two spin-allowed transitions. The first of these (observed in the work recorded in Table LIV; note that the electric dipole transition is formally forbidden in regular T_d symmetry) is outside the range of Ferguson's experiments (117); the second is studied in detail. The structure assigned to this transition at 20°K in $Cs_2Zn(Co)Cl_4$ is shown in Figure 54. Six origins are identified, and each has a progression of the same vibrational mode (usually two-quantum, 270–290 cm^{-1}). The splitting pattern cannot be interpreted for pure T_d symmetry, but instead corresponds to the case in which the C_s splitting is greater than spin–orbit splitting as shown in Figure 55. In the $[N(CH_3)_4]_2Zn(Co)Cl_4$ crystals, the splitting pattern resembles the left-hand side of Figure 55, in which the influence of the low-symmetry field component is very small. The structure of this compound is more open than that of $Cs_2Zn(Co)Cl_4$, with a larger Co^{2+}—monovalent cation distance.

It is estimated that the $^4A_2 \rightarrow {}^4T_1(^4F)$ transition is approximately $(e^4t_2{}^3) \rightarrow 0.66\ e^3t_2{}^4 + 0.33\ e^2t_2{}^5$; owing to transfer of electrons from e to t_2, the excited-state equilibrium configuration will be larger for the ground

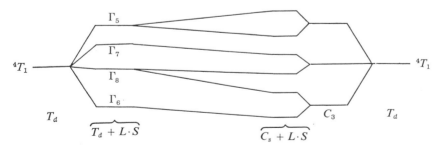

Fig. 55. Spin–orbit coupling in the $^4T_1$ state of $CoCl_4{}^{2-}$ in T_d or C_s symmetry (117).

state, so that transitions other than the pure electronic ones are probable. From the polarization observed with the cesium compound (Fig. 54), the excited-state vibration must be totally symmetric within the site group C_s, which implies that the $CoCl_4{}^{2-}$ ion retains the same site symmetry in the excited state. The vibration is not positively identified, but the totally symmetric stretching frequency of $CoCl_4{}^{2-}$ ion has been measured as 276 cm^{-1} (68). The lowest energy bands for the transition to $^4T_1(^4F)$ are within 120 cm^{-1} of each other for the three crystals studied, but the splitting patterns and overall widths are different, reflecting different relative importance of the low-symmetry field component (117). Measurements of integrated absorption intensity for Cs_2CoCl_4 show that the x, y, and z transition moments are all approximately equal, and that admixture of odd character into the otherwise even d functions is isotropic, as would be expected.

The remaining bands [$^4A_2 \rightarrow {}^4T_1(^4P)$ and doublet states] are not so easily analyzed. The complications of structure are attributed to breakdown of Born-Oppenheimer separability, arising from strong interaction of different electronic configurations.

Yamada and Tsuchida (433,434) observed fine structure in the polarized crystal absorption spectrum of Cs_3CoCl_5. No anisotropy was observed. Ferguson (109) observed two strong transitions in the crystal spectrum of Cs_3CoCl_5 at 5500 and 16,000 cm^{-1} which he attributed to the $^4A_2 \rightarrow {}^4T_1(^4F)$ and $^4A_2 \rightarrow {}^4T_1$ (^4P) transitions, respectively.

The polarized crystal spectrum of Cs_3CoCl_5 has been determined by Pelletier-Allard in the infrared (318), visible (315), and ultraviolet regions (319). The bands observed in the infrared are the same as for Cs_2CoCl_4, showing the similar environment of the Co^{2+} ion. Dq was calculated to be 300 $\pm$ 20 cm^{-1} and the Racah parameters B and C to be 780 and 3680 cm^{-1}, respectively. For Cs_3CoCl_5 the spin–orbit coupling constant (ζ) was calculated as 405 $\pm$ 45 cm^{-1}. The polarized spectra were determined with the electric vector $\parallel$ and $\perp$ to the fourfold axis in the crystal. A great deal of detail ($>$ 140 lines) is revealed at 4.2°K in the region 15,390–32,870 cm^{-1}. A difficulty in the interpretation of this spectrum is lack of knowledge of the low temperature environment of the Co^{2+} ion. Studies of the Zeeman effect in Cs_3CoCl_5 (316, 317) have produced evidence for the existence of two crystalline phases at low temperature. In slow cooling from 80 to 20°K, the initial fourfold symmetry is not retained. There is a phase change in the lattice with a decrease in symmetry of the crystal field around the Co^{2+} ion. The spectrum of Cs_3CoCl_5 is also discussed by Clark, Dunn, and Stoneman (69), and by McClure (228).

Yamada and Tsuchida have determined the dichroism of $K_2Co(NCS)_4 \cdot 4H_2O$ (433) and of $HgCo(NCS)_4$ (436).

6. $CoSO_4 6H_2O$

Holmes and McClure (182) have determined the crystal spectrum of cobalt sulfate hexahydrate, $CoSO_4 6H_2O$, at various temperatures down to 90°K. The visible band centered at 19,800 cm^{-1} shows a good deal of fine structure. Three components are present at room temperature and four at 90°K, with peak positions at 19,000, 20,700, 21,500, and 22,250 cm^{-1}. The infrared band lies at 8600 cm^{-1}. The visible band is assigned to superposed $^4T_1(^4F) \rightarrow {}^4T_1(^4P)$ and $^4T_1(^4F) \rightarrow {}^4A_2(^4F)$ transitions, four components being expected at low temperature since the field at the Co^{2+} ion has appreciable rhombic character. The change in shape of this band with temperature may be interpreted by considering the effect of lattice contraction on the crystal field splitting.

No structure is observed on the infrared band. Since it is very broad ($\Delta\nu_{1/2} \approx 3000$ cm^{-1}), some of the breadth is probably contributed by the rhombic field, and each component transition must be broad. The spin-orbit coupling is not strong enough in Co^{2+} to account for the splitting of over 3000 cm^{-1} in the visible band. It is thought more likely that the environment is distorted from a regular octahedral shape, and that this gives the observed splitting in the upper $^4T_1(^4P)$ state. The component at 20,700 cm^{-1} is tentatively assigned to the A_2 state as it shifts most markedly to the red with increasing temperature.

7. $CoPy_2Cl_2$

Ferguson (109,110) has measured the crystal absorption spectra of several complexes having the general formula CoX_2Y_2. Cobalt(II) complexes of the form $CoPy_2X_2$ where X = Cl, Br, I, NCS, or NCO can exist in two forms, the relatively stable tetrahedral and the bridged octahedral ones. The polarized crystal spectrum of the unstable violet form of $CoPy_2$-Cl_2 which has the cobalt atom octahedrally coordinated has been investigated (110). The structure (101) consists of polymeric chains running parallel to the c needle axis (Fig. 56), each chlorine being shared by two cobalt ions.

It is monoclinic, space group $P2/b, C_{2h}^4$, eight molecules per unit cell having dimensions $a_0 = 34.42$, $b_0 = 17.38$, $c_0 = 3.66$ Å, and $\gamma = 90°$. The arrangement of the four chloride ions surrounding each cobalt is rectangular–planar so that the octahedral unit, $CoPy_2Cl_4$, has the point symmetry D_{2h}, not D_{4h}. The y axis was taken as bisecting the smaller $Cl—\hat{Co}—Cl$ angle (85° 30') and the z axis coincident with the N—Co—N direction.

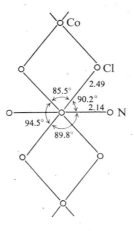

Fig. 56. Interatomic distances and angles in solid cobalt dipyridine dichloride (110).

The observed spectrum is shown in Figure 57, and consists of four bands, three of which overlap. At low temperature a narrow band at 18,700 cm^{-1} appears.

The slight rhombic distortion of the tetragonal symmetry in the complex removes the triple orbital degeneracy of the $^4T_{1g}$ ground state in octahedral symmetry. The correlations between the irreducible representations of O_h, D_{4h}, D_{2h}, and C_{2h} are given in Table LV. The band at 16,200 cm^{-1} is assigned to the $^4B_{3g} \rightarrow {}^4A_g$ ($^4A_{2g}$ in O_h) transition. The upper state is chosen since the $^4A_{2g}$ state should lie below $^4T_{1g}(^4P)$ for Dq less than 1000 cm^{-1} (as in CoPy$_2$Cl$_2$) and because the spectra of CoPy$_2$Cl$_2$ and CoPy$_2$-(NCS)$_2$ (110) show no pronounced splitting of the absorption band. The

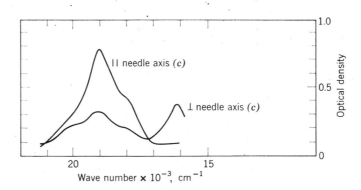

Fig. 57. Polarized crystal spectrum of cobalt dipyridine dichloride at room temperature (110).

TABLE LV

Correlations between Irreducible Representations of O_h, D_{4h}, D_{2h}, C_{2h} (110)

O_h	D_{4h}	D_{2h}	C_{2h}
A_{1g}	A_{1g}	A_g	A_g
A_{2g}	B_{1g}	A_g	A_g
E_g	$A_{1g} + B_{1g}$	$2A_g$ [a]	$A_g + A_g$
T_{1g}	$A_{2g} + E_g$	$B_{1g} + B_{2g} + B_{3g}$	$B_g + A_g + B_g$
T_{2g}	$B_{2g} + E_g$	$B_{1g} + B_{2g} + B_{3g}$	$B_g + A_g + B_g$

[a] This replaces $B_{2g} + B_{3g}$ in Ferguson's table. Note that correlations for the lower symmetry groups are not all unique; see, e.g., *Molecular Vibrations* by E. B. Wilson, Jr., J. C. Decius, and P. C. Cross, McGraw Hill, New York, 1955, p. 333.

components are grouped close together so that a simple tetragonal distortion is unlikely to produce a splitting as large as 3000 cm^{-1} in $CoPy_2Cl_2$.

The electronic transitions are forbidden for electric dipole absorption and excitation of internal odd-parity vibrations is assumed to provide the mechanism whereby intensity is stolen from more strongly allowed transitions. The u fundamentals belong to the representations B_{1u}, B_{2u}, B_{3u} of D_{2h} and can be combined with all the possible direct products of the ground and excited states as shown in Table LVI.

Only those vibronic combinations with symmetry B_{1u}, B_{2u}, B_{3u} are allowed, (A_{1u} is inactive), and correspond to transition moments along the z, x, and y axes, respectively. The assignment of the 16,200 band to an upper ^4A_g state which is polarized $\perp$ to the c axis ($\parallel$ to b) means that the ground state must be $^4B_{3g}$ since $B_{2g} \times A_g \times (\beta_{1u}, \beta_{2u}, \beta_{3u}) = B_{2u}, B_{1u}, A_u$; with zero component on the c axis. Either a B_{3g} or B_{2g} ground state would be consistent with the interpretation of the rhombic distortion as arising from the Jahn-Teller effect. The highest energy component of the three other bands has nearly equal intensity along the b and c crystal axes and is assigned to the upper $^4B_{3g}$ state because $B_{3g} \times B_{3g} \equiv A_g$ which results in

TABLE LVI

Combination of Symmetry Representation of Ground and Excited Electronic Wave Functions with Perturbing Vibration Representation in D_{2h} (110)

Symmetry representation of perturbing vibration	Symmetry representation of the product of ground and excited electronic wave functions			
	A_g	B_{1g}	B_{2g}	B_{3g}
β_{1u}	B_{1u}	A_u	B_{3u}	B_{2u}
β_{2u}	B_{2u}	B_{3u}	A_u	B_{1u}
β_{3u}	B_{3u}	B_{2u}	B_{1u}	A_u

three active components B_{1u}, B_{2u}, B_{3u}. The other two bands have the same polarization ratio 2/1 (*c/b*) and this is expected for both B_{1g} and B_{2g} upper states.

8. Co(*p*-toluidine)$_2$Cl$_2$

Crystal absorption spectra of several complexes having the general formula CoX_2Y_2, with a tetrahedral environment for the Co^{2+} ion, have also been measured (109,110). In a tetrahedral field, only the transitions $^4A_2 \rightarrow {}^4T_1(^4F)$ and $^4A_2 \rightarrow {}^4T_1(^4P)$ are allowed by the electric dipole selection rules (assuming there is some admixture of odd parity orbitals into the *d* manifold). These correspond to the bands observed in $CoCl_4{}^{2-}$ at about 5500 and 16,000 cm^{-1}. When complexes of the type CoX_2Y_2 are considered, the near infrared band corresponding to the transition $^4A_2 \rightarrow {}^4T_1(^4F)$ in $CoX_4{}^{2-}$ should split under the reduced symmetry C_{2v} into three transitions $^4A_2 \rightarrow {}^4A_2$, $^4A_2 \rightarrow {}^4B_1$, and $^4A_2 \rightarrow {}^4B_2$. Each of these is allowed under C_{2v} symmetry and they correspond to polarizations parallel to the *z*, *y*, and *x* molecular axes. The visible band corresponding to the $^4A_2 \rightarrow {}^4T_1(^4P)$ transition should be split in a similar manner.

The crystal structure of cobalt di-*p*-toluidine dichloride has been reported (42) and if only the nitrogen of the *p*-toluidine is considered, the complex belongs to the $C_{2v}{}^6$ point group. The compound crystallizes

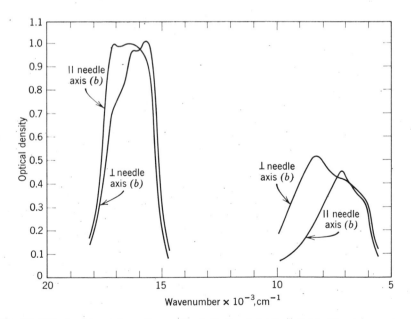

Fig. 58. Polarized crystal spectrum of cobalt di-*p*-toluidine dichloride at room temperature (109).

with a monoclinic space group $C2/c, C_{2h}^6$ having cell constants $a_0 =$ 12.3, $b_0 = 4.59$, $c_0 = 26.1$ Å, $\beta = 93° 45'$. The relevant bond angles are $Cl—\widehat{Co}—N$, $109°$; $Cl—\widehat{Co}—Cl$, $111°$; and $N—\widehat{Co}—N$, $111°$. The crystals have the (001) face well developed and the light was incident on this face in the spectral measurements (109) which are shown in Figure 58.

The two absorption bands were recorded (109) on different crystals as the visible band is much stronger than the one in the near infrared. The visible band has three components, but these do not appear at the same frequency in each polarization. This suggests that intercomplex interaction has split the energy levels, and that the absorption band in the crystal may be an exciton band. This is not considered unreasonable as the absorption band is strong for a $d \rightarrow d$ transition and resonance interaction is therefore likely to be significant. The near infrared band has three components, one being completely polarized $\perp$ to the b crystal axis. This is assigned to either the $^4A_2 \rightarrow {}^4B_1$ or $^4A_2 \rightarrow {}^4B_2$ transition. As the other two bands have roughly equal absorption in the two crystal directions, it is assumed that they are mixed, either in the crystal or in the complex. The symmetry of the ligand field is thus thought to be higher than C_{2v} but below that of T_d.

9. $CoPy_2Br_2$ and $CoPy_2I_2$

The polarized crystal spectra of the stable tetrahedral forms of $CoPy_2Br_2$ and $CoPy_2I_2$ have also been studied by Ferguson (109). Although the crystals are monoclinic, the developed face and cell dimensions are different from those of cobalt di-p-toluidine dichloride (258). The b crystal axis is perpendicular to the well-developed face and the extinction directions make a small angle with the needle axis. The spectra are shown in Figure 59.

The replacement of p-toluidine by pyridine alters the strength of the ligand field. Measurements on $CoPy_2Br_2$ showed no crystal splitting of the visible band, and as the developed face lies perpendicular to the symmetry axis, no splitting of exciton bands is expected in this face. Figure 59 shows that the near infrared band is split in $CoPy_2Br_2$ and $CoPyI_2$ more than in cobalt di-p-toluidine dichloride, Figure 58. This may be due to resonance interaction rather than to the effect of the ligand field. The degree of polarization of the bands is higher which could also indicate a further approach of the field to C_{2v} symmetry.

From the observed spectra, it was concluded that formal reduction of the symmetry of the ligand field modifies the structure and energy of the transition $^4A_2 \rightarrow {}^4T_1(^4F)$ more than that of $^4A_2 \rightarrow {}^4T_1(^4P)$. The unassociated forms of $CoPy_2X_2$ show the same type of band structure in the band corresponding to the transition $^4A_2 \rightarrow {}^4T_1(^4P)$ as in $CoCl_4^{2-}$ (53),

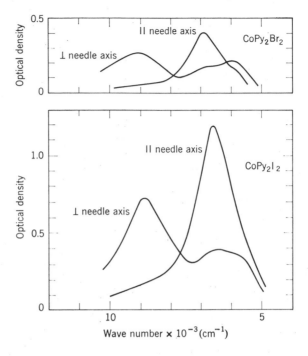

Fig. 59. Polarized crystal spectra of cobalt dipyridine dibromide and cobalt dipyridine diiodide at room temperature (109).

and they appear at nearly the same position as in the spectrum of $CoCl_4^{2-}$. On the other hand, the near infrared band $^4A_2 \rightarrow {}^4T_1(^4F)$ is split and shifted to higher energy in the $CoPy_2X_2$ complexes. As the upper states belong to the same representation of the point group symmetry of the complex and mixing of the two must occur, it is surprising that perturbation of the tetrahedral ligand field affects the two states differently. It is thought that mixing of the ^2G and ^4P states through spin–orbit interaction leads to the peculiar structure of the $^4A_2 \rightarrow {}^4T_1(^4P)$ transition and its insensitivity to changes of formal symmetry of the ligand field. Because there are no doublet levels close to the $^4T_1(^4P)$ state, it is perturbed only by the noncubic components of the ligand field and is split and shifted to higher energy (109).

Measurement of the spectra in bromoform solution (109) has shown how intercomplex interaction modifies the spectra of the compounds of the type $CoPy_2X_2$. The infrared band is still split but there is a large shift of the components to higher energy. The interpretation of the polarization of the bands in the crystal is complicated, as it is not clear to what extent the association has modified the energy levels. Interaction between the next nearest neighbors of the cobalt atom, a second pair of chloride ions and nitrogens, must modify the energy level which is responsible for the change. Simple crystal field theory neglects the interaction between next nearest neighbors, and this may be a poor approximation here.

10. $CoCl_2 6H_2O$

Ferguson (110) has determined the polarized crystal spectrum of cobalt chloride hexahydrate, $CoCl_2 6H_2O$, following the earlier work of Gielessen (144) and Pappalardo (305). The analysis of the spectrum was

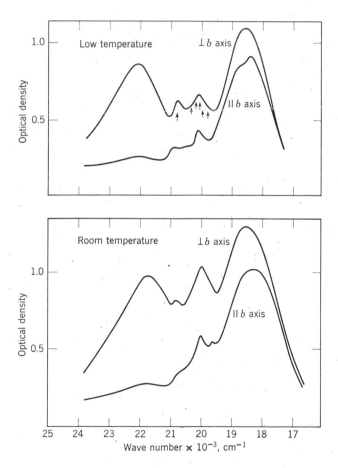

Fig. 60. Polarized crystal spectrum of cobalt chloride hexahydrate at room temperature and low temperature. The arrows show the position of bands recorded in Table LVII (110).

made on the basis of the crystal structure determination by Mizuno, Ukei, and Sugawara (265). It is monoclinic with unit cell dimensions $a_0 = 10.34$, $b_0 = 7.06$, $c_0 = 6.67$ Å, and $\beta = 122°\ 19'$, having two molecules per unit cell. The basic octahedral unit is $CoCl_2(H_2O)_4$, the arrangement of the water molecules being square–planar, so the complex unit belongs to the point group D_{4h}. The b crystal axis is parallel to the needle axis and the Co—Cl bonds lie in the ac plane at right angles to the b axis. The spectra were recorded on the (001) face and are shown in Figure 60. Extra bands were observed (110) at low temperature and are listed in Table LVII.

The earlier work of Gielessen (144) and Pappalardo (305) is summarized in Table LVIII. Pappalardo determined the crystal spectrum of $CoCl_26H_2O$ and $CoBr_26H_2O$ at room temperature, 90, and, 20°K. He also made simple crystal field calculations of the energy levels, not including spin–orbit interaction of noncubic components of the crystal field. The proposed assignments of Pappalardo are also included.

A treatment similar to that for $CoPy_2Cl_2$ does not explain the observed polarizations of the bands in $CoCl_26H_2O$ (110). Inclusion of next nearest neighbors to the central cobalt atom, four Cl^- ions lying towards the corners of the unit cell at 4.5 Å, gives a symmetry of C_{2h} instead of D_{4h} around the Co^{2+} ion. The correlations between D_{4h} and C_{2h} were given in Table LV. The twofold axis of C_{2h} is taken to be identical with the twofold axis which bisects the H_2O—$\widehat{Co}$—H_2O bond angle in D_{4h}. The ground state is now ^4B_g while the excited states, in the region studied, are $^4A_g(^4F)$, $^4B_g(^4P)$, $^4A_g(^4P)$, and $^4B_g(^4P)$. Only the α_u and β_u vibration fundamentals are capable of removing the symmetry prohibition of the electronic transition. The combinations of these with the direct products of the ground state and possible excited states are given in Table LIX.

The spectrum has been analyzed (110) on the assumption that the Co—Cl electron-transfer transition is the most important from which the intensity is stolen for the forbidden $d \to d$ transitions. The important vibrations will be those which are essentially Co—Cl stretching modes,

TABLE LVII
Positions of the Narrow Bands in the
Spectrum of $CoCl_26H_2O$ at Low Temperature (± 20 cm^{-1}) (110)

19,850 cm^{-1}	20,230 cm^{-1} (a pol.)
20,030	20,250 (b pol.)
20,120 (a pol.)	20,410
20,130 (b pol.)	20,580
	20,850

TABLE LVIII
Observed Band Maxima and Possible Assignments (305) in $CoCl_26H_2O$ and
$CoBr_26H_2O$

$CoCl_26H_2O$, ref. 144, 84°K	$CoCl_26H_2O$, ref. 305, 20°K	$CoBr_26H_2O$, ref. 305, 20°K	$CoCl_26H_2O$, ref. 305, room temperature		
Band max., cm^{-1}	Band max., cm^{-1}	Band max., cm^{-1}	Observed, cm^{-1}	Calculated, cm^{-1}	Assignment
				0	$^4T_1(^4F)$
				7,300	^2E
				8,000	$^4T_2(^4F)$
15,992			16,400	14,700	$^2T_1(^2G)$
17,852	17,800	16,200?		15,200	$^2T_2(^2G)$
18,340		16,490		17,100	$^4A_2(^4F)$
18,940	18,370	18,120	18,500	18,500	$^4T_1(^4P)$
19,687		18,600			
19,820	18,900	19,745	19,700⎫ 21,000⎭	19,000	$^2T_1$
20,051					
20,110	20,175	20,040			
20,180	20,260	20,245			
20,247	20,420	20,325			
20,420	20,620	20,515			
20,497					
20,623	20,880	20,805			
20,741					
20,927	21,525	20,980			
21,418	21,710	21,750	21,700	21,700	$^2A_1$
21,655	21,960	22,770			
21,867	22,150	22,940			
23,423	23,420	23,070			
23,787	23,590	23,250			
	23,890	23,535			
23,900	24,150				
24,130					
24,505	24,785	23,820		23,400	$^2T_2$
24,725					
24,942	25,000	24,060			
25,155					
25,365	25,400	24,375	24,400	24,700	$^2T_1$
	26,300	26,250	26,200	26,700	^2E
				32,000	^2E
				33,000	$^2T_2$
				33,300	$^2T_1$
				34,700	$^2T_1$
				37,100	$^2A_2$
				39,100	$^2T_2$
				46,400	$^2T_2$
				54,300	E

TABLE LIX
Combination of Symmetry Representations of Ground and Excited Electronic
Wave Functions with Perturbing Vibration Representation in C_{2h} (110)

Symmetry representation of perturbing vibration	Symmetry representation of the product of ground and excited electronic wave functions	
	A_g	B_g
α_u	A_u	B_u
β_u	B_u	A_u

α_{2u} in D_{4h}, β_u in C_{2h}. With β_u as enabling vibration, the transitions have the following polarizations: $^4B_g \rightarrow {}^4A_g(^4F)$, b polarized; $^4B_g \rightarrow {}^4B_g(^4P)$, a polarized; $^4B_g \rightarrow {}^4A_g(^4P)$, b polarized; $^4B_g \rightarrow {}^4B_g(^4P)$, a polarized.

The strongly polarized band at 22,000 cm^{-1} is interpreted as $^4B_g \rightarrow {}^4B_g(^4P)$, being a split component of the octahedral band and the one at 18,500 cm^{-1} as representing the accidentally degenerate pair $^4B_g \rightarrow {}^4A_g(^4P)$ and $^4B_g \rightarrow {}^4B_g(^4P)$, the other components of the octahedral $^4T_{1g} \rightarrow {}^4T_{1g}$ transition. The weaker $^4B_g \rightarrow {}^4A_g(^4F)$ transition is assumed to lie at about 18,000 cm^{-1}, masked by the stronger $^4B_g \rightarrow {}^4A_g(^4P)$ band. The structure in the latter band at low temperatures is thought to be a component of the quartet–doublet transition which also accounts for the narrow bands in the region 19,000–21,000 cm^{-1}, Table LVII. It will be noted that these assignments differ in a number of respects from those in Table LVIII.

11. Co(bipy)$_3{}^{2+}$

Palmer and Piper (300) have studied the polarized crystal spectra of the tris-(2,2′ bipyridyl) cobalt(II) ion in Co(bipy)$_3$Br$_2$6H$_2$O and Zn(Co)(bipy)$_3$SO$_4$7H$_2$O. The crystal structures and site symmetry of the Co^{2+} ion have been discussed under d^6 configurations, p. 356. The observed bands are listed in Table LX and the orthoaxial spectrum at 80°K of Zn(Co)-(bipy)$_3$SO$_4$7H$_2$O is shown in Figure 61.

The spectrum of Co(bipy)$_3{}^{2+}$ reveals a band at 11,300 (± 300) cm^{-1} which is assigned (300) to the octahedral field transition $^4T_1(^4F) \rightarrow {}^4T_2(^4F)$. The second spin-allowed transition $^4T_1(^4F) \rightarrow {}^4T_1(^4P)$ is assigned to the shoulder on the charge-transfer band at 22,000 cm^{-1} in the crystal spectrum. From the above transitions, Dq and B are calculated to be 1267 and 791 cm^{-1}, respectively. On this basis, the transition $^4T_1(^4F) \rightarrow {}^4A_2(^4F)$ is predicted to occur at 23,000 cm^{-1}, but owing to the very weak absorption due to the two-electron transition and its location between $^4T_1(^4P)$ and the charge transfer region, it is not expected to be seen. The weak

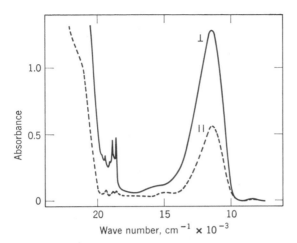

Fig. 61. Polarized orthoaxial spectra of Zn(Co)(bipy)$_3$SO$_4$7H$_2$O at 80°K. Crystal thickness, 3.8 mm; approximately 20 mole % cobalt (300).

shoulder at 18,500 cm^{-1} is resolved at 80°K into at least eleven sharp peaks between 18,450 and 19,604 cm^{-1}. This coincides closely with the energy calculated (212) for the $^2T_1$ level arising from the ^2P and ^2H free ion states and is assigned to this transition. The broad, low absorption between 14,000 and 18,000 cm^{-1} is likely to be due to transitions to the $^2T_1(^2G)$ and $^2T_2(^2G)$ (212) states which should also be split by the combination of spin–orbit coupling and the trigonal field, and may in addition be affected by vibronic interaction. No explanation for the absence of the sharp lines and progressions similar to those observed for the $^2T_1(^2P,^2H)$ levels has been proposed (300).

The close agreement of the axial and $\perp$ spectra and the small temperature dependence of the intensities suggest an electric dipole interpretation of the spectrum. Although strong polarization effects are observed, splitting of the $^4T_2(^4F)$ levels is not seen even at 80°K. This may mean that the trigonal splitting is smaller than or of the same magnitude as the spin–orbit splitting. The total first-order spin–orbit splitting of the $^4T_2$ level should be $\frac{2}{3}\zeta$ or 360 cm^{-1} (310), but the trigonal field may be large enough to completely confuse the six-component pattern predicted by the combination of spin–orbit and trigonal splitting.

In an octahedral field, the $^2T_1(^2P,^2H)$ level should give rise to two levels with a separation of the order of magnitude of ζ (212). In a trigonal field, second-order effects will lift the degeneracy further to produce three Kramers doublets. Two lines in the region of 19,000 cm^{-1} which stand out because of their intensity occur at 18,554 and 18,814 cm^{-1} and are split by

TABLE LX

Band Maxima of Co(bipy)$_3$$^{2+}$ Spectra (300)

Matrix	Spec-trum	Excited state	Band max., cm^{-1}		$\Delta\nu_{1/2}$, cm^{-1}	ϵ
			300°K	80°K		
Co(bipy)$_3$SO$_4$6H$_2$O	Axial	$^4T_1(^4P)$	22,000			50 at 80°K
		$^2T_1(^2P,^2H)$	18,500	19,330	80°K, 50–100	0.5–1.5
				19,000		
				18,800		
				18,600		
		$^2T_1,^2T_2(^2G)$	16,000	16,000		
		$^4T_2(^4F)$	11,000	11,350	$\begin{cases} 300°K\ 3,200 \\ 80°K\ 2,500 \end{cases}$	9.2 11.36
Zn(bipy)$_3$SO$_4$7H$_2$O	σ	$^4T_1(^4P)$	22,000	—		
		$^2T_1(^2P,^2H)$	18,700	19,750		
				19,604		
				19,474		
				19,331		
				19,220		
				18,964		
				18,814		
				18,699		
				18,554		
				18,450		
		$^2T_1,^2T_2(^2G)$	16,000	16,000		
		$^4T_2(^4F)$	11,100	11,490		
	π	$^4T_1(^4P)$	21,500	—		
				19,327		
		$^2T_1(^2P,^2H)$	18,700	18,939		
				18,808		
				18,692		
				18,563		
		$^2T_1,^2T_2(^2G)$	16,000	16,000		
		$^4T_2(^4F)$	10,950	11,490		

260 cm^{-1}. These two lines are thought to arise from (0,0) transitions, but it is uncertain whether there is another (0,0) line in this region. If not, the trigonal field must be quite small. In this interpretation, the weak shoulder at 18,450 cm^{-1} would have to be assigned to a hot line and then each of the two strong lines would have associated with it a vibrational component at ∼150 cm^{-1} above the origin. The whole pattern seems to be repeated at about 780 cm^{-1} higher in energy, similar to that of the out-of-plane C—H deformation at 778 cm^{-1}. If such a vibration were excited, marked delocalization of metal electrons onto the ligand would be implied.

12. Ni^{3+} in Al_2O_3

A study of the polarized spectrum of Ni^{3+} in Al_2O_3 has been made by McClure (233). The site symmetry of the trivalent ion has been discussed under d^1 configurations, p. 270. Absorption bands were observed at 16,300 ($\perp$) and 16,800 ($\parallel$) cm^{-1} with f (the oscillator strength) = 4.5 and 3.3, respectively. The spectrum shown in Figure 62 consists of a weak sharp band series originating at 12,300 cm^{-1}, two broad bands in the visible and some bands in the ultraviolet which are too strong to be crystal field excitations. The assignment of the ground state is difficult from the spectrum alone.

Geschwind and Remeika (142) have measured the ESR spectrum of Ni^{3+} in the Al_2O_3 lattice and find a single isotropic line with $g = 2.146$ from room temperature down to about 50°K. At 4.2°K the spectrum splits, owing to a static Jahn–Teller distortion. The ground state of the ion is ^2E, $S = \frac{1}{2}$ (in contrast to Co^{2+}, which is $^4T_2$, $S = \frac{3}{2}$). The larger value of Dq for the trivalent ion evidently results in a low-spin ground state. The Ni^{3+} ground state thus contains a single unpaired e electron, with a two-fold orbital degeneracy which is not removed either by spin–orbit coupling or by the trigonal field.

In the spin paired state ^2E, the spin–orbit representation in a cubic field is G which splits in a trigonal field into $E_{\frac{1}{2}} + E_{\frac{3}{2}}$, the splitting of these components being only about 10 cm^{-1}. The relevant selection rules for transitions in C_3 were given in Table III. The observed polarizations of the bands are given in Table LXI.

The occurrence of features polarized exclusively along c is compatible only with a $E_{\frac{3}{2}}$ ground state. This can only arise from a ^2E ground state,

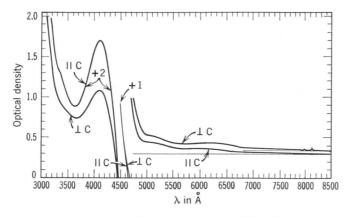

Fig. 62. Absorption spectrum of Ni^{3+} in corundum at 35°K. The high-energy band of this spectrum may be due to a charge transfer transition (233).

TABLE LXI
Observed Band Maxima and Polarizations for Ni^{3+} in the Corundum Lattice at 35°K (233)

Band position, cm^{-1}	Description	Polarization
12,300	Line	$\perp$
16,300	Broad band maximum	$\perp$
16,350	Origin of group of lines	$\parallel$
16,700	Broad band maximum	$\parallel$
17,750	Line	$\perp$
17,850	Line	$\parallel$
19,600	Broad band	$\perp + \parallel$

but in order for the polarization to be strongly marked, the ground state must be split and $E_{\frac{1}{2}}$ must not contribute to the absorption. If there is a distortion, however, the symmetry C_3 is no longer correct and the selection rules have to be reexamined. The new symmetry is C_1 and only intensity calculations can be used, but it is probable that these would give results closely resembling the symmetry rules of C_3. The assignments of these absorption bands are not clear. The trigonal field-splitting parameter K might be measurable from the first band, with a splitting of $3K/2$, with K having a value of -330 cm^{-1}.

B. $4d^7$

1. Rh^{2+} in ZnWO$_4$

The ESR and electronic spectra of crystals of $ZnWO_4$ doped with Rh have been studied by Townsend (393a); these are attributed to Rh^{2+} ions substituted at Zn^{2+} sites in the lattice. Zinc tungstate is monoclinic, space group $P2/c, C_{2h}^4$ with two magnetically equivalent Zn sites in each unit cell, and is isostructural with $MgWO_4$ and $NiWO_4$ (198). In $NiWO_4$, both Ni and W are surrounded by six oxygen ions which form distorted octahedra. Although the point site symmetry of Zn in $ZnWO_4$ is strictly no higher than C_2, the noncubic components of the crystal potential are

TABLE LXIA
Spectrum of Rh^{2+} in ZnWO$_4$, $E \parallel b$ axis (393a)

Band max., cm^{-1}	ϵ	$f \times 10^4$	Assignment in O_h
12,800	16	2×10^4	$^2E \rightarrow {}^2T_1(t_2{}^5e^2, {}^3A_2)$
15,400	34	3×10^4	$^2E \rightarrow {}^2T_2(t_2{}^5e^2, {}^1E)$
18,900	49	4×10^4	$^2E \rightarrow {}^2T_1(t_2{}^5e^2, {}^1E)$
22,200	63	6×10^4	$^2E \rightarrow {}^2T_2(t_2{}^5e^2, {}^1A_1)$

probably small, and in first approximation the potential can be regarded as due to a regular octahedral field. Rh^{2+} has a doublet ground state $^2E(t_2^6e)$ in the strong-field scheme. There are four doublets of the first excited configuration $(t_2^5e^2)$ and neglecting higher configuration (e.g., $t_2^4e^3$) four spin-allowed transitions are expected. The spectrum with electric vector parallel to the b axis in fact shows four bands, which have been assigned as in Table LXIA. The approximate value of Dq calculated from these assignments in 1600 cm^{-1}. Dichroism is observed but has not yet been interpreted.

X. CONFIGURATION d^8

A. $3d^8$

The crystal-field splitting of the d^8 configuration is the same as that for d^2 if the sign of Dq is changed. The comprehensive paper of Liehr and Ballhausen (210) discusses this configuration in detail for ranges of parameter values appropriate to Ni^{2+}. The Tanabe-Sugano energy diagram for d^8 ions in an octahedral field is given in Figure 63.

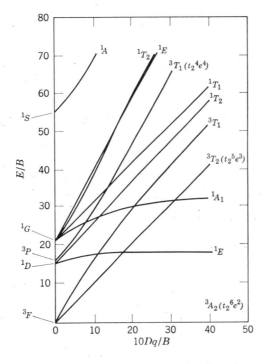

Fig. 63. Tanabe-Sugano energy diagram for the d^8 configuration (75,386).

More crystal spectra have been recorded for Ni^{2+} than for any other ion over a wide range of crystal field symmetries. The study of Ni^{2+} in an oxide or silicate lattice has been the subject of many investigations. A complete analysis of the Ni^{2+} line spectrum has been given by Shenstone (367).

1. Ni^{2+} in MgO

Pappalardo, Wood, and Linares (309) have measured the optical absorption spectrum of single crystals of MgO, ZnO, $MgAl_2O_4$, and yttrium garnet doped with Ni^{2+}.

This work was carried out at 4°K and considerable fine structure was observed. In the earlier work of Low (244) on Ni^{2+} dissolved in single crystals of MgO measured at room temperature and at 77°K, the positions of almost all the excited singlets as well as triplets were reported, but no fine structure was observed.

A comparison between the observed and predicted bands for Ni^{2+} in MgO is given in Figure 64. The Ni^{2+} is at a site of octahedral symmetry in this lattice. In order to simplify the analysis, it is assumed that, at least for the leading lines of each absorption group, the *same* odd-parity vibration modulates the electronic transition from the ground level to the excited term multiplet components. The calculated levels are mostly for first-order spin–orbit interaction; some further comparison is also made with the more detailed calculations of reference 210.

2. $NiSiF_66H_2O$

The vibronic absorption spectrum of the Ni^{2+} ion in nickel fluorosilicate crystals at temperatures down to 4°K has been reported by Pryce, Agnetta, Garofano, Palma-Vittorelli, and Palma (340). Earlier work on this crystal was carried out by Pappalardo (302). The crystal structure (423) of $NiSiF_6$-$6H_2O$ shows that the Ni^{2+} ion is subject to a predominantly octahedral, trigonally distorted field. The energy level diagram is shown in Figure 65, with the small trigonal component neglected (i.e., for O_h symmetry). The observed band maxima and oscillator strengths (4.2°K) are shown in Table LXII. The spin–orbit fine structure is calculated for a number of sets of parameters and possible assignments of the multiplets (not shown in Table LXII) are discussed in some detail in reference 340. The main bands correspond to those observed by Holmes and McClure (182) for Ni^{2+} ions in solution and in the cubic field of MgO by Low (224), and by Pappalardo et al. at 4°K (309). The assignments are given in Table LXII except for the C band which will be discussed later. The complex structure of the B band

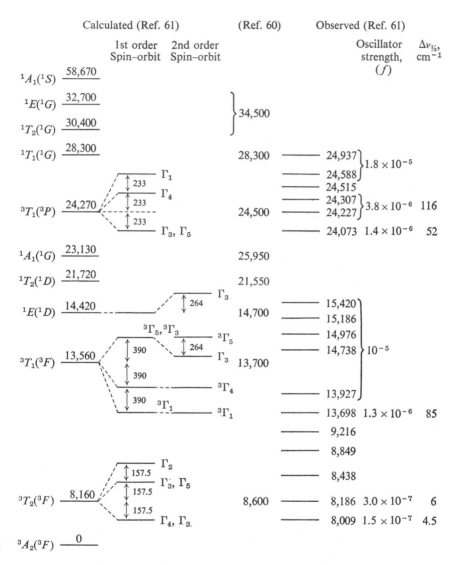

Fig. 64. Comparison of calculated and observed levels for Ni^{2+} in MgO (309). The energy levels were calculated assuming $Dq = 815 \text{ cm}^{-1}$, $B = 890 \text{ cm}^{-1}$, $C = 4.4B$, and $\zeta = 630 \text{ cm}^{-1}$. Values in last two columns are oscillator strength (f) and the band half-width (cm^{-1}). Further second-order spin–orbit splittings are discussed in ref. 309.

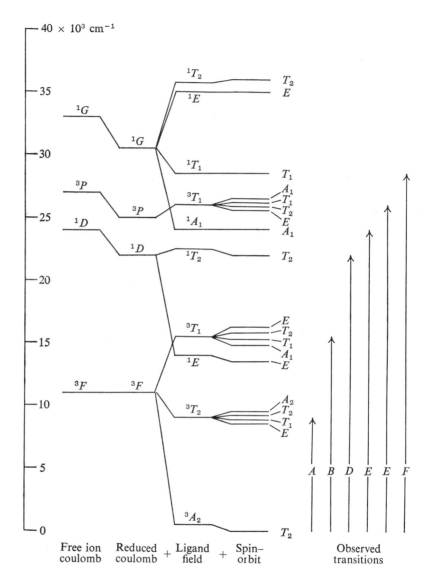

Fig. 65. Energy level diagram of Ni^{2+} in a predominantly octahedral field, calculated using the parameters $Dq = 920$ cm^{-1}, $B = 960$ cm^{-1}, $C = 3600$ cm^{-1}, α (Trees constant) $= 70$ cm^{-1}, and $\zeta = 600$ cm^{-1}. Successively indicated are the centers of the multiplets of the free ion (367); the corresponding energies with the reduced Racah, Trees, and spin–orbit parameters; the same plus the ligand field; and the energies including the spin–orbit interaction. Vertical arrows indicate the observed transitions (340).

TABLE LXII
Observed Band Maxima at Liquid Helium Temperature (Nominal) with
Proposed Assignments, for $NiSiF_6 6H_2O$ (340)

Observed band	Positions ΔE, cm^{-1}	Assignment	Observed width, cm^{-1}	Oscillator strength observed (approx.)
A	9,150	$^3A_2 \rightarrow {}^3T_2(^3F)$	2,000	2×10^{-5}
B	15,400	$^3A_2 \rightarrow {}^3T_1(^3F)$	2,600	2×10^{-5}
C	19,085		130	1×10^{-7}
	19,570		200	0.7×10^{-7}
	20,040		350	0.5×10^{-7}
D	22,935	$^3A_2 \rightarrow {}^1T_2(^1D)$	1,600	2×10^{-6}
E	26,100	$^3A_2 \rightarrow {}^3T_1(^3P)$	2,200	3.5×10^{-5}
F	29,070	$^3A_2 \rightarrow {}^1T_1(^1G)$	—	—

is explained in terms of the mutual perturbation of the ^1E and $^3T_1(^3F)$ levels which strongly modifies the vibronic coupling and results in two fairly narrow bands, each showing resolved vibrational structure, superimposed on a broad band. The following parameters are found to give good agreement: $Dq = 910$, $B = 955$, $C = 3750$, $\alpha = 80$, and $\zeta = 600$ cm^{-1}, where α is the proportionality factor in the expression $\alpha L(L + 1)$, the Trees correction term (340).

The peaks C_1, C_2, C_3 cannot be explained by the ligand field theory predicted in Figure 63. However, it is likely that they correspond to combinations of the transitions causing the B band with either one quantum of one of the fundamental frequencies 3652, 3756 cm^{-1} of the water molecule, or two quanta of the frequency 1595 cm^{-1}. Piper and Koertage (321) have observed the spectrum of a number of nickel salts at 77°K and find a series of weak bands in the region 18,000–21,000 cm^{-1}. The salts studied were $Ni(BrO_3)_2 6H_2O$, $Ni(BrO_3)_2 6D_2O$, $NiSeO_4 6H_2O$, $NiSeO_4 6D_2O$, $NiSO_4$-$7H_2O$, $NiSiF_6 6H_2O$, and $NiCl_2 xH_2O$. The weak bands of the hydrated nickel bromate and selenate shift, on average, by 790 and 890 cm^{-1}, respectively, upon deuteration to lower wave numbers. The isotope effect rules out any assignment of the weak bands as pure electronic transitions to singlet levels (302), and verifies the hypothesis of transitions to the electronic level $^3T_{1g}$ with simultaneous excitation of the O—H stretching mode in agreement with Pryce et al.

The observed widths of the bands in $NiSiF_6 6H_2O$ are explained as the combined result of spin–orbit splitting and excitation of lattice vibrations (340). The significance of the vibrations is illustrated in Figure 66, which plots the energy associated with various electronic levels as a function of

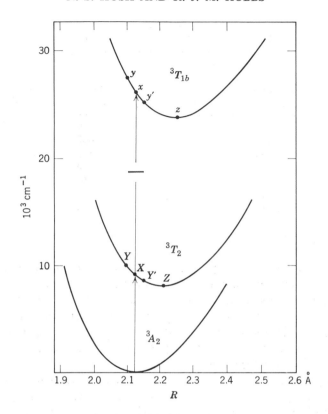

Fig. 66. Energy of $(Ni6H_2O)^{2+}$ as a function of metal–ligand distance R for the electronic ground state $^3A_2$ and two typical excited states from the multiplets $^3T_2$, $^3T_{1b}$. The vertical arrows indicate electronic transitions without change of lattice configuration. The short horizontal line is a measure of zero-point vibration in the ground state (340).

the Ni—H_2O distance R. Application of the Franck-Condon principle indicates that for the transitions of $^3T_2$, (the A band), on the average about 1000 cm^{-1} of energy associated with octahedrally symmetrical vibrations around the ion will be excited. This is indicated by the vertical transition, which meets the $^3T_2$ curve at X, whereas the minimum energy is at Z. The zero-point fluctuation in the ligand distance R (spanning 50% points in the probability distribution of the lowest vibrational state) is indicated by the horizontal line, and corresponds to a probable spread between Y and Y' on the $^3T_2$ curve. The vertical spread between Y and Y' ($\sim$1500 cm^{-1}) together with the overall spin–orbit splitting of about 630 cm^{-1}, leads to a good agreement with the observed overall width of 2000 cm^{-1}.

The completely vibrationless transition which will be a magnetic dipole excitation, leads to Z on the curve, some 1070 cm^{-1} below X. This represents the threshold of the band, but will be extremely weak. The actual threshold expected, if the pure magnetic dipole strength is very small, will be a group of at least three sharp peaks, in each of which the transition is assisted by one quantum of an odd-parity lattice vibration. Such peaks (containing no symmetric quanta) are termed the 'precursors' of a band. The lowest calculated electronic level is at 8827 cm^{-1} (with $Dq = 910$ cm^{-1}, $B = 955$ cm^{-1}, $C = 3750$ cm^{-1}, $\alpha = 80$ cm^{-1}, and $\zeta = 600$ cm^{-1}). The completely vibrationless (magnetic dipole) transition would therefore be expected around $8827-1070 = 7747$ cm^{-1}; and with the help of an odd vibration of 150 cm^{-1}, the next precursor would be at ~ 7900 cm^{-1}, in fair agreement with band A_2 at 7925 cm^{-1} (340).

Jahn-Teller distortion will cause further broadening of the bands, by excitation of corresponding vibrations with a consequent reduction in the intensity of the precursor. In the spectrum of Ni^{2+} in MgO of Pappalardo (309), only two sharp lines are seen in the corresponding band. This was attributed to the fact that only one electronic level is involved [$^3T_2(A_2)$] but the work of Pryce et al. shows that a more complete analysis is required (340).

Various estimates of the frequency of the odd-parity vibrations (408) that have to be excited in an assumed centrosymmetric lattice to give the necessary electric moment for an electric dipole transition have been made (204,209,302). The vibrational energy that has to be supplied might be between 100 and 300 cm^{-1}. This should be subtracted from the experimental band position to give the electronic energy difference. As it is small compared with the electronic separations, the uncertainty in its value is not serious although it does limit the accuracy to which the ligand field and Racah parameters may be determined.

3. NiSO$_4$6H$_2$O and NiSO$_4$7H$_2$O

Hartmann and Müller (171) determined the crystal spectrum of NiSO$_4$6H$_2$O and NiSO$_4$7H$_2$O at room temperature and at 68°K. Distortion from octahedral symmetry was shown by splitting of the main absorption bands. Polarized spectra were recorded by Holmes and McClure (182) on NiSiF$_6$6H$_2$O and on NiSO$_4$7H$_2$O. Spectra were recorded along the three principal axes of the orthorhombic crystal of NiSO$_4$7H$_2$O and for the $^3A_2 \rightarrow {}^3T_1(^3F)$ and $^3A_2 \rightarrow {}^3T_1(^3P)$ bands the order of increasing absorption was $b < a < c$ while for the $^3A_2 \rightarrow {}^3T_2(^3F)$ band, the order was $a < c < b$. The differences were accounted for by the tetragonal distortion of the field, the splitting being of the order of 200 cm^{-1} for each band.

4. Ni^{2+} in $KMgF_3$ and $KNiF_3$

The electronic absorption spectrum of Ni(II) in perovskite fluorides has been determined in detail by Ferguson, Guggenheim, and Wood (120). The fine structure of the spectrum of Ni^{2+} in $KNiF_3$ and $KMgF_3$:Ni is analyzed in terms of the spin–orbit splitting and vibrational intervals using a four-parameter crystal field approach. $KNiF_3$ and $KMgF_3$ are cubic crystals with the perovskite structure (284,291). The fluorine atoms are located at sites with D_{4h} symmetry and the metal ion sites have O_h symmetry.

Much fine structure is observed, and the vibrational structure is analyzed for the $^3A_{2g} \rightarrow {}^1T_{2g}(^1D)$ and $^3A_{2g} \rightarrow {}^3T_{1g}(^3F)$ bands. The structural features of the $^3A_{2g} \rightarrow {}^3T_{2g}(^3F)$ band were found to be quite different from those of the other spin-allowed bands or the $^3A_{2g} \rightarrow {}^1T_{2g}(^1D)$ band. This suggests a fundamental difference in the mechanism of the electronic transition for this band. Ferguson, Guggenheim, Johnson, and Kamimura (116) had previously studied the $^3A_{2g} \rightarrow {}^3T_{2g}$ transition in $KMgF_3$ and MgF_2 doped with Ni^{2+}, in which the Ni^{2+} ion lies at the center of an orthorhombic field. They concluded that this transition is predominantly magnetic dipole in character, in contrast to the forbidden

TABLE LXIII

Comparison of Observed and Calculated Energies of the Spin–Orbit States in $KMgF_3$:Ni^{2+}, for the Following Values of the Parameters, $Dq = 698$, $B = 950$, $C = 3990$, and $\zeta = 620$ cm^{-1} (120)

State	Spin–orbit designation	Calculated energy, cm^{-1}	Observed energy, cm^{-1} (20°K)	Comments
$^3T_{2g}(^3F)$	Γ_3	6,713	6,699 ± 2	Purely electronic
	Γ_4	6,863	6,866 ± 2	Purely electronic
	Γ_5	7,265	7,265 ± 5	Purely electronic
	Γ_2	7,412	7,422 ± 5	Purely electronic
$^3T_{1g}(^3F)$	Γ_1	11,084	11,324	False origin
	$\Gamma_4{}^a$	11,605	11,814	False origin
	Γ_3	12,287	—	
	Γ_5	12,428	12,663	False origin
$^1E_g(^1D)$	Γ_3	15,247	15,156	Lowest energy line
$^1T_{2g}(^1D)$	Γ_5	21,102	20,998	False origin
$^3T_{1g}(^3P)$	Γ_3	23,256	23,186	False origin
	Γ_4	23,638	—	
	Γ_5	23,732	—	
	Γ_1	24,027	—	

[a] There is an obvious misprint (Γ_2 instead of Γ_4) in Table IX in ref. 120.

electric dipole character of the other bands. On this assumption, the positions of the pure electronic spin–orbit transitions can be determined for this band. These are shown in Table LXIII, together with the calculated values based on the parameters $Dq = 698$, $B = 950$, $C = 3990$, and $\zeta = 620$ cm^{-1}. For the remaining bands in Table LXIII [cf. the 'precursor bands' of Pryce et al. (340)], several false origin bands are tentatively identified (corresponding to $00 + 1$ quantum of a different perturbing vibration) for $^3A_{2g} \to {}^1T_{2g}$. For this crystal symmetry, four perturbing vibrations are predicted. The oscillator strengths of the transitions (20°K) are (labeling by excited state): $^3T_{2g}(^3F)$, 6.7; $^3T_{1g}(^3F)$, 8.4; $^1E_g(^1D)$, 1.5; $^1T_{2g}(^1D)$, 1.9; $^3T_{1g}(^3P)$, 21, all $\times 10^{-6}$. The unusual structure of the $^3T_{2g}$ band is discussed in terms of the nature of the excited-state wave function (almost entirely $t_2{}^5e^3$, in contrast to the others, which are mixtures of $t_2{}^4e^4$ and $t_2{}^5e^3$).

A review of the fine structure in these compounds has been given by Ferguson (119). The absorption spectrum of Ni^{2+} in MgWO$_4$ has been determined by Ferguson, Knox, and Wood (114).

5. Ni^{2+} in ZnO and CdS

Pappalardo, Wood, and Linares (309) have determined the absorption spectrum of Ni^{2+} in ZnO. The site symmetry of the Ni^{2+} ion in this lattice has been discussed under d^7 configurations, p. 397. The calculated and observed levels are given in Table LXIV (309). The energy levels were calculated using the parameters $B = 795$, $Dq = 405$ cm^{-1}, $C = 4.36$ B, and $\zeta = 630$ cm^{-1}, and a spin–orbit ground state stabilization of 900 cm^{-1}. In the region near 8600 cm^{-1}, the structure is attributed to vibrational fine structure rather than spin–orbit coupling. Although spin–orbit coupling was considered in the assignment of the bands, no account was taken of the trigonal component in the tetrahedral field. Brumage and Lin (56) have determined the magnetic susceptibilities of Ni^{2+} in ZnO and CdS. CdS also has the wurtzite structure with $c_0/a_0 = 1.6358$ and so the Cd^{2+} ion is in a near ideal tetrahedral coordination. Thus in Ni^{2+}:CdS the spin–orbit splittings are much greater than the trigonal terms, while in Ni^{2+}:ZnO they are of comparable magnitude. Figure 67 shows the energy level diagram for Ni^{2+} in a tetrahedral site with a trigonal component.

Magnetic susceptibility measurements (56) show that the spin–orbit coupling constant ζ for Ni^{2+}:ZnO is 350 ± 50 cm^{-1} and the trigonal field splitting of the $^3T(^3F)$ state is 100 ± 10 cm^{-1}. The corresponding quantities for Ni^{2+}:CdS are 340 ± 20 cm^{-1} and 10 ± 4 cm^{-1}. In both crystals, the A_2 component of $^3T(^3F)$ lies below the E component. The large

TABLE LXIV
ZnO : Ni (0.02%) at 4.2°K (309)

Level	Calculated level, cm^{-1}	ν_1, cm^{-1}	Half-width at $\alpha = \frac{1}{2}\alpha_{max}$, cm^{-1}	Oscillator strength, $f \times 10^5$
$^3T_2(^3F)$	4,192 Γ_3,Γ_5	4,221.2	2.7	0.16
	4,508 Γ_4	4,252.6	9	1.3
	4,665 Γ_2	4,329		
		4,345		
		4,424.7		2.1
		4,524.9		
		8,350	5.5	0.85
$^3A_2(^3F)$	8,380	8,406.2	12	0.34
		8,445.9		
		8,583.7		1.0
		8,711		
$^1T_2(^1D)$	12,800	8,818		
		8,882.6	45	1.0
$^1E(^1D)$	13,450	12,682		
		12,923		7.3
		13,422		
$^3T_1(^3P)$	15,345 Γ_3,Γ_5	15,332		
	15,955 Γ_4	15,468		
	16,260 Γ_1	15,643		
		15,772.8		81
		16,286		
		17,300		
		(17,860)[a]		

[a] At this frequency $I_0/I = 10$, namely optical density $D = 1$.

reduction of the spin–orbit coupling constant from the free-ion value indicates rather strong covalent bonding between the Ni^{2+} ion and the ligands.

Weakliem and McClure (412) have determined the absorption spectrum of Ni^{2+} in ZnO at 298 and 77°K, and the polarized absorption spectrum of the visible region at 4°K, shown in Figure 68.

The detailed features of the $^3T_1(^3P)$ state are shown in Figure 68. The splitting of the lowest energy sharp band, which is the T_2 spin–orbit component can be accounted for by the trigonal field of ZnO. The observed spectrum together with the electric dipole selection rules gives A_1 lowest. (The lowest spin–orbit component of the ground state is A_1, see Fig. 67.) The T_1 level should be split by the same amount, however, with E lower than A_2. Now $A_1 \to A_2$ is forbidden and taking the weak band at 15,500

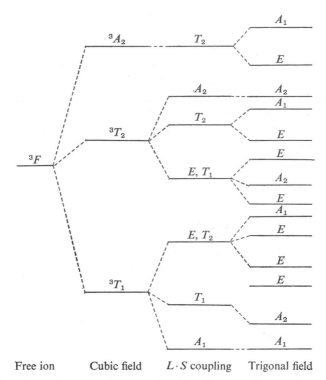

Free ion	Cubic field	$L \cdot S$ coupling	Trigonal field

Fig. 67. Splitting of the energy levels for the d^8 configuration in a tetrahedral field with a trigonal component (56).

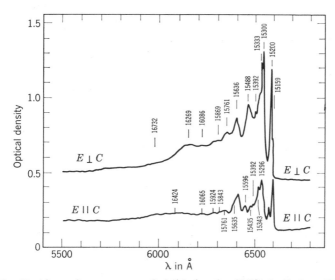

Fig. 68. Absorption spectrum of visible bands of Ni^{2+} in ZnO at 4°K (412).

cm^{-1} to be $A_1 \rightarrow A_2$ and the band at 15,488 cm^{-1} to be $A_1 \rightarrow E_1$ it is seen that the order of levels is correct although the splitting is smaller than for the T_2 component. The band at 15,300 cm^{-1} is identified as $E \rightarrow {}^3T_1({}^3P)$ and the band at 15,635 cm^{-1} identified as $A_1 \rightarrow {}^3T_1({}^3P)$. The identification fits well with the tetrahedral symmetry having a small trigonal component. The spectra of Ni^{2+} in Cs$_2$ZnCl$_4$ and Cs$_2$ZnBr$_4$ were also recorded by Weakliem and McClure (412).

6. Ni(CN)$_4{}^{2-}$

The polarized crystal spectra of tetracyano nickelate crystals have been studied by Ballhausen, Bjerrum, Dingle, Eriks, and Hare (19). Included in the investigation were the following crystals: K$_2$Ni(CN)$_4$H$_2$O, Na$_2$Ni(CN)$_4$ 3H$_2$O, CaNi(CN)$_4$5H$_2$O, SrNi(CN)$_4$5H$_2$O, and BaNi(CN)$_4$4H$_2$O. The crystal structures (48,49,208) of a number of these complexes are summarized in Table LXV. However the R values calculated from some of the data leave much to be desired, and in some cases they are so high that the entire structure determination becomes doubtful.

The site symmetry of the Ni^{2+} ion is D_{4h} in all the complexes, the nickel atoms being stacked along the c axis. Studies on the tetracyano nickelate(II) ion have been carried out by other authors (50,70,71,425), but these results will not be discussed further.

The polarized spectra were measured (19) with the electric vector $\parallel$ and $\perp$ to the c axis of the crystals, this axis being nearly but not exactly perpendicular to the planes of the stacked Ni(CN)$_4{}^{2-}$ groups. It is thus nearly but not quite coincident with the fourfold symmetry axis of the complex ion. The overall electronic spectrum for all but the barium salt shows a very weak band at 20,000 cm^{-1}, a broad $\parallel$ band at 23,000 cm^{-1}, and a somewhat sharper band seen in both $\parallel$ and $\perp$ polarizations at 27,000 cm^{-1}. In addition, the Ba complex exhibits a rather narrow, temperature dependent $\perp$ band at 22,000 cm^{-1}. Above 30,000 cm^{-1} the absorption rises steeply. In addition, two very weak bands are observed at 5600 cm^{-1} (three peaks with a separation of $\sim$50 cm^{-1}), and at 7000 cm^{-1} (two peaks separated

TABLE LXV
Crystal Structures of Tetracyano–Nickelate(II) Complexes

Ref.	Compound	Space group	Z	a_0	b_0	c_0	α	β	γ
48	Na$_2$Ni(CN)$_4$3H$_2$O	$P\bar{1}, C_i^1$	4	15.01	8.84	7.34	95° 42′	92° 16′	89° 20′
208	SrNi(CN)$_4$5H$_2$O	$C2/m, C_{2h}^3$	4	10.33	15.18	7.28		98° 47′	
49	BaNi(CN)$_4$4H$_2$O	$C2/c, C_{2h}^6$	4	11.71	13.48	6.63		104° 50′	

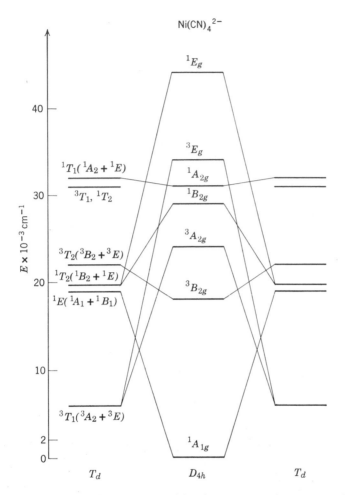

Fig. 69. Energy level diagram for $Ni(CN)_4^{2-}$ for D_{4h} and T_d symmetries. The diagram does not imply anything about the amount of distortion, and the connecting lines between the D_{4h} and T_d configurations are drawn as straight lines only for convenience (19).

by ~ 300 cm^{-1}). However, these two bands are not electronic transitions. With the O—H stretching frequency in water being ~ 3500 cm^{-1} and the C—N stretching frequency in cyanide being 2050 cm^{-1}, the first band is assigned as the combination ($3500 + 2050 = 5550$ cm^{-1}) and the second one an overtone of the O—H stretching frequency ($2 \times 3500 = 7000$ cm^{-1}). These predictions are substantiated by measurements on deuterated crystals (19).

The electronic structure of the square–planar nickel(II) cyanide complex has been considered by several authors (152,200,320). If the complex has a regular square–planar configuration in both the ground and excited states, a vibronic intensity-gaining mechanism has to be invoked for an electric dipole transition. The $Ni(CN)_4^{2-}$ unit has sixteen different normal vibrations: $2a_{1g}, a_{2g}, 2b_{1g}, 2b_{2g}, e_g, 2a_{2u}, 2b_u, 4e_u$. The electronic ground state in all proposed level schemes is $^1A_{1g}$. A closer analysis then shows that with a vibronic mechanism, some transitions are forbidden $\parallel$ but all possible bands should be seen in the $\perp$ polarization. However, the band at 23,000 cm^{-1} is seen $\parallel$ but not $\perp$.

The absence of this band in the $\perp$ polarization indicates that the D_{4h} vibronic selection rules may not be appropriate. Figure 69 shows the energy level diagram for $Ni(CN)_4^{2-}$ in D_{4h} and T_d (19). A comparison of the states in T_d shows that identifying the observed transition at 23,000 cm^{-1} with $^1A_{1g} \rightarrow {}^1B_2$, the ordering of the states is as in Figure 69. The intense allowed $^1A_{1g} \rightarrow {}^1B_2$ transition is rather broad since several quanta of the b_{1u} are excited. This is the vibration which distorts D_{4h} towards T_d; in D_{2d} it transforms as a_1. From Figure 69, ^1E is expected to have a very distorted equilibrium configuration. The transition $^1A_{1g} \rightarrow {}^1E$ in D_{2d} is $\perp$ allowed, but distortion of the Jahn-Teller type means the transition to the split components of ^1E may be observed in both $\parallel$ and $\perp$ polarizations. It is thus possible that the sharp peak at 26,000 cm^{-1} is a vibrational level of ^1E. The weak absorptions near 20,000 cm^{-1} are assigned (19) to spin forbidden transitions but not in agreement with those reported previously (420).

The specific band at 22,000 cm^{-1} found in $BaNi(CN)_4 4H_2O$ appears sharp $\perp$ and as a weak shoulder in the $\parallel$ polarization. In $BaNi(CN)_4 4D_2O$, it is very sharp $\parallel$ ($\Delta\nu_{1/2} = 800$ cm^{-1}) although in $\perp$ it is seen at almost the same position and with almost the same intensity as in $\parallel$, but with a shoulder towards the red. No trace of the band was found in either crystal in the axial spectrum. The assignment of this band is uncertain.

7. Ni(DMG)$_2$ and Other Ni(II) Glyoximates

The polarized crystal spectrum of nickel bisdimethylglyoxime $Ni(DMG)_2$ has been determined by Basu, Cook, and Belford (28), and more recently by Anex and Krist (4). Solid $Ni(DMG)_2$ has a strong green absorption (21,22,431,432) that is not reported in the solution spectrum. This was seen in single crystals by Yamada and Tsuchida (431,432) and by Banks and Barnum (21,22) as a rather featureless absorption edge. But it has been observed as a sharp well-defined peak in the absorption spectra

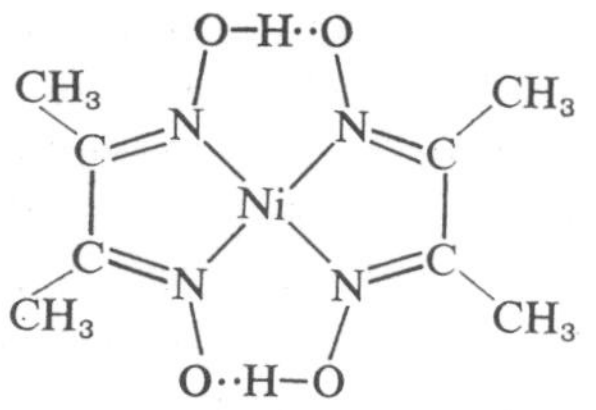

Fig. 70. The nickel(II) bisdimethylglyoxime molecule.

of powders, mulls, and suspensions (21,22,373). Zahner and Drickamer (452) found that this sharp band showed very strong red shifts with pressure and accordingly assigned it to $3d^8 \rightarrow 3d^7 4p$.

The crystal structure of Ni(DMG)$_2$ has been determined by Godychi and Rundle (145). The orthorhombic crystals contain four molecules in a unit cell with lattice constants $a_0 = 16.68$, $b_0 = 10.44$, $c_0 = 6.49$ Å, and space group $Ibam, V_h{}^{26}$. The planar molecules, Figure 70, stack one above the other along the c axis with a rotation of 90° for alternate layers. The nickel atoms line up to form a chain having a Ni—Ni distance of 3.25 Å.

The crystal spectra were recorded parallel and perpendicular to the molecular plane (28). The observed bands are given in Table LXVI, being slightly more intense in the ∥ than in the ⊥ polarization.

The $d \rightarrow p$ assignment (452) does not fit the observed polarization of the red tail which is stronger in the direction perpendicular to the molecular planes. The observed band in the infrared at 10,300 cm^{-1} is thought to contain one or more of the expected d-like nickel ion transitions. The visible absorption then probably consists of the higher energy d-like transitions and some of the partially allowed charge transfer bands. Due to the several n and π ligand orbitals crowded into a fairly small energy range, the broad nearly featureless character of the visible absorption is understandable.

TABLE LXVI

Observed Absorption Band Maxima and Polarizations in Ni(DMG)$_2$

Ref. 28, 77°K	Band max., cm^{-1}, ref. 431	Ref. 4
10,300 Unpolarized		18,600 ⊥
		24,400 ⊥
18,020 ⊥	20,000 ⊥	24,500 ∥
18,870 ∥ $\epsilon = 320$	17,700 ∥	25,300 ⊥
27,780 ⊥	24,500 ⊥	25,200 ∥
27,780 ∥	29,000 ∥ and ⊥	29,700 ⊥
		32,000 ⊥

An extensive comparative study of the solution and solid state spectra of a series of nickel(II) glyoximates has been undertaken by Anex and Krist (4). This work has shed considerable light upon the nature of the color band which is apparent only in the crystalline state. Included in the study were Ni(DMG)$_2$ itself and the Ni(II) complexes of α and β ethylmethylglyoxime and heptoxime (cycloheptanedione dioxime). These three compounds have identical solution spectra but the saturated ligand substituents give rise to varying interplanar spacings in the crystal so that the effects of changing this parameter may be studied. The α modification of nickel ethylmethylglyoxime has a crystal structure in which the nickel atoms do not form a chain structure (135a), and hence was used to ascertain the nature of the solid state effects observed in the other compounds. The polarized crystal spectra were recorded in two directions A and B, the A direction corresponding to almost completely in-plane absorption ($\parallel$) while the B direction is largely out-of-plane ($\perp$).

The nearest nonbonded atom to the nickel atom in α-Ni(EMG)$_2$ is an oxygen lying 3.44 Å away and the shortest Ni—Ni distance is 4.75 Å, as compared to 3.25, 3.40, and 3.60 Å, respectively, for Ni(DMG)$_2$ (145), β-Ni(EMG)$_2$ (366a), and Ni(Hept)$_2$ (22). This situation together with the similarity of the crystal and solution spectra, suggests (4) that the crystal spectrum may reasonably be taken to characterize the absorption of the isolated complex. From the recorded spectra it is assumed that the isolated molecules of the complexes studied possess an out-of-plane absorption at $\sim$23,000 cm^{-1} indicating that this band is the single molecule origin of the band that in Ni(DMG)$_2$ appears in the solid. This out-of-plane absorption shifts to the red and increases by a factor of $\sim$8 in intensity as the interplanar distance decreases to 3.25 in Ni(DMG)$_2$. This band is assigned (4) as the $d_{x^2-y^2} \rightarrow p_z, \pi^*$, metal $\rightarrow$ ligand charge transfer band. This is based upon interpretation of a one-electron molecular orbital level scheme modified from that used by Gray and Ballhausen (152) for square-planar complexes, as the glyoximates do not have D_{4h} symmetry and their ligands do not involve in-plane π orbitals. Various alternative assignments are discussed for the other observed bands.

8. trans-Ni(NH$_3$)$_4$(NCS)$_2$ and trans-Ni(NH$_3$)$_4$(NO$_2$)$_2$

The polarized crystal spectra of the isostructural trans-Ni(NH$_3$)$_4$-(NCS)$_2$ and trans-Ni(NH$_3$)$_4$(NO$_2$)$_2$ have been determined at room temperature by Hare and Ballhausen (164). These spectra are particularly interesting in that they are quite different in appearance (three prominent bands in the first and two prominent bands in the second). The x-ray structures have been determined by Porai-Koshits and Dikareva (333). The two compounds are isomorphous and crystallize as monoclinic plates. The space group is $C_2/m, C_{2h}{}^3$ and there are two molecules per unit cell. The lattice parameters for the isothiocyanate are $a_0 = 11.46$, $b_0 = 8.19$, $c_0 = 5.68$ Å, $\beta = 105°$; and for the nitrite $a_0 = 10.77$, $b_0 = 6.58$, $c_0 = 6.12$ Å, and $\beta = 128°$. The molecular symmetry is D_{4h} and the nickel ion occupies a site of symmetry C_{2h}. In the isothiocyanate the Ni—N (Ni—NCS) distance is 2.07 Å and the Ni—N(NH$_3$) distance is 2.15 Å; in the nitrite the Ni—N(NO$_2$) distance is 2.15 Å and the Ni—N(NH$_3$) distance is 2.07 Å. The crystal structure in the (001) projection is shown in Figure 71; from this it is seen that the a crystal axis corresponds to the fourfold molecular axis and the b axis roughly to the plane of the four ammonia molecules. The spectra were recorded with the electric vector parallel to these axes and conform to the $\parallel$ and $\perp$ spectra of the complex, respectively.

The transformation of the energy levels in O_h to those in D_{4h} together with the tetragonal splitting parameters D_s and D_t are shown in Figure 72.

The absorption bands and assignments based on D_{4h} symmetry are given in Table LXVII and the spectra shown in Figures 73 and 74. The dominant feature of the Ni(NH$_3$)$_4$(NCS)$_2$ spectrum is the similarity of the $\parallel$ and $\perp$ bands compared with the spectrum of Ni(NH$_3$)$_4$(NO$_2$)$_2$ where the bands are highly polarized and the splittings due to the tetragonal field are higher than in the isothiocyanate.

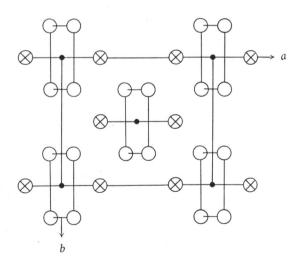

Fig. 71. Crystal structure of *trans*-Ni(NH$_3$)$_4$X$_2$ (X = NCS$^-$ or NO$_2{}^-$) in the (001) projection. (●) Ni, (○) NH$_3$, (⊗) NCS or NO$_2$ (164).

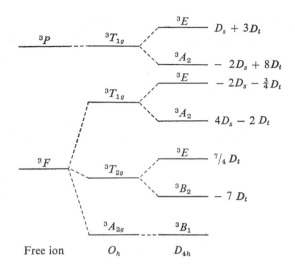

Fig. 72. First-order O_h and D_{4h} level schemes for d^8. For C_4 symmetry, the right-hand
subscripts are omitted (163).

The magnitude of the tetragonal splitting in *trans*-NiY_4X_2 complexes
is dependent on the difference of the positions of the ligands X and Y in the
spectrochemical series (26). The assignments of $Ni(NH_3)_4(NCS)_2$ are based
largely on the relative positions of the two ligands in the series:

$$SCN^- < NH_3 \ll NO_2^- \ll CN^-$$

From the observed splittings $D_s = 34$ and $D_t = 37$ cm^{-1} were calculated
(164). The uncertainty in the splitting of the band is 150 cm^{-1}; this is in-
sufficient to have consistency in the splitting parameters. The positive
sign for these parameters was chosen so that the tetragonal field gives rise

TABLE LXVII

Polarized Crystal Spectra at Room Temperature of *trans*-
$Ni(NH_3)_4(NCS)_2$ and *trans*-$Ni(NH_3)_4(NO_2)_2$ (assignments
based on D_{4h} symmetry) (164)

O_h	$Ni(NH_3)_4(NCS)_2$	$Ni(NH_3)_4(NO_2)_2$
$^3T_{2g}(^3F)$	$^3B_{2g}$; 10,750, $\perp$	^3E_g; 11,200, $\parallel$
	^3E_g; 10,750, $\parallel$	$^3B_{2g}$; 12,000, $\perp$
$^1E_g(^1D)$	$^1A_{1g}, ^1B_{1g}$; 13,000, $\perp, \parallel$	$^1A_{1g}, ^1B_{1g}$; 12,750, $\perp$
	^3E_g; 17,350, $\parallel$	$^3A_{2g}$; 19,900, $\perp$
$^3T_{1g}(^3F)$	$^3A_{2g}$; 17,500, $\perp$	^3E_g; 20,350, $\parallel$
	^3E_g; 27,900, $\parallel$	
$^3T_{1g}(^3P)$	$^3A_{2g}$; 28,000, $\perp$	

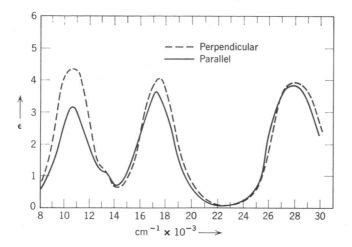

Fig. 73. Crystal spectrum of *trans*-Ni(NH$_3$)$_4$(NCS)$_2$ at room temperature; ϵ is extinction coefficient in units, liters mole-cm^{-1} (164).

to a slight axial stabilization. That is, the nature of the ligands is more important than the internuclear distance. The opposite effect would give parameters with a negative sign and the assignment for the tetragonal compounds would be reversed. In the case of Ni(NH$_3$)$_4$(NCS)$_2$, it is difficult to say which is more important as it is a cooperative phenomenon.

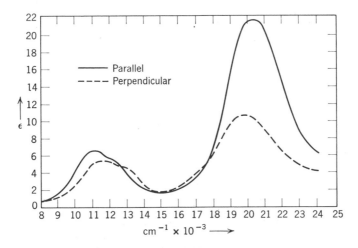

Fig. 74. Crystal spectrum of *trans*-Ni(NH$_3$)$_4$(NO$_2$)$_2$ at room temperature; ϵ is extinction coefficient in units, liters mole-cm^{-1} (164).

In $Ni(NH_3)_4(NO_2)_2$ the problem is more clearly resolved as NO_2^- is a very much "stronger" ligand than NH_3. The values of the tetragonal parameters are calculated to be $D_s = -100$, $D_t = -91$ cm^{-1}. The intensity of the observed transitions arises from vibronic interactions since both the molecule and the site symmetry are centrosymmetric.

The position of the bands in $Ni(NH_3)_4(NCS)_2$ agrees very well with the "complete" level diagram for nickel complexes given by Liehr and Ballhausen (210) ($Dq = -1075$ cm^{-1}), and the agreement is satisfactory for $Ni(NH_3)_4(NO_2)_2$ ($Dq = -1200$ cm^{-1}). (The hole formalism is used here.)

9. Ni^{2+} Salicylaldehyde Complexes

Basu and Belford (29) have carried out a survey of the temperature dependence of the intensities of polarized transitions in the crystal spectra of nickel(II) complexes of salicylaldehyde derivatives. The structural formulas of the compounds are shown in Figure 75.

The crystal structure of diaquo bissalicylaldehydato nickel(II) has been determined by Stewart, Lingafelter, and Breazedale (376). It has space group $A_2/m, C_{2h}^3$ with two molecules per unit cell and lattice constants

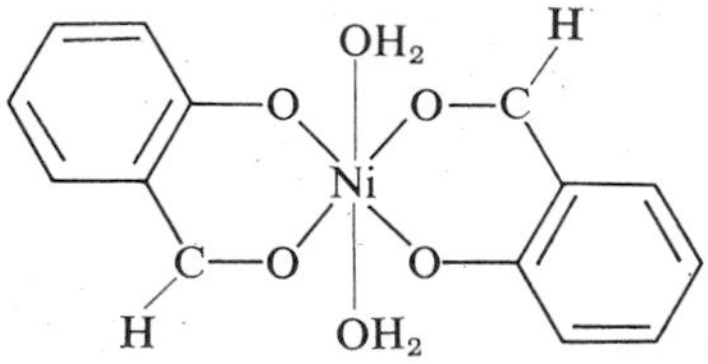

Diaquo bissalicylaldehydato Ni(II)

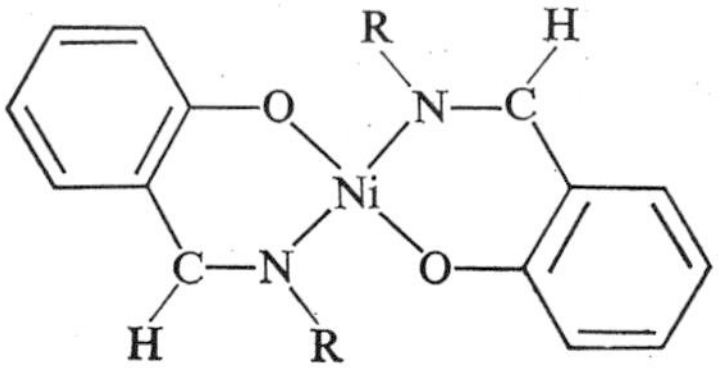

where R = H   Bissalicylaldiminato Ni(II)
R = CH$_3$ Bis-N-methylsalicylaldiminato Ni (II)
R = C$_2$H$_5$ Bis-N-ethylsalicylaldiminato Ni(II)
R = C$_3$H$_7$ Bis-N-propylsalicylaldiminato Ni(II)
R = iso-C$_3$H$_7$ Bis-N-isopropylsalicylaldiminato Ni(II)
R = OH   Bissalicylaldoximinato Ni(II)

Fig. 75. Structural formulas of bissalicylaldiminato complexes of Ni(II).

$a_0 = 12.93$, $b_0 = 7.32$, $c_0 = 7.41$ Å, and $\beta = 90° 15'$. The well-developed (100) face of the crystal has the shape of a square with the crystal axes along the diagonals. The salicyldehyde molecules and the Ni^{2+} ion lie exactly in the (010) plane and the nickel–water bonds lie exactly along the b axis.

Bissalicylaldiminato-Ni(II), [Ni(Salim)$_2$] crystallizes with a monoclinic space group $P2_1/c, C_{2h}^5$ with unit cell dimensions $a_0 = 12.96$, $b_0 = 5.83$, $c_0 = 8.11$ Å, $\beta = 95° 35'$, and two molecules per unit cell (375). The molecule is planar and has the *trans* configuration. The planar molecules are stacked in essentially the same way as in naphthalene. Although the set of four ligand atoms bonded to the Ni^{2+} ion is exactly planar, and the bond distances are equal, the configuration is not exactly square, the angle O—$\widehat{Ni}$—N being 93° 48'. The shortest distance between the nickel ions is 4.99 Å. Intermolecular interaction effects are unlikely in the solid state, as the closest atom to the nickel atom of the neighboring molecule is an oxygen at 3.85 Å.

Bis-(*N*-methylsalicylaldiminato)-Ni(II),[Ni(MeSalim)$_2$] exists in several crystalline modifications. In the monoclinic form, space group $P2_1/c, C_{2h}^5$, $a_0 = 11.94$, $b_0 = 7.00$, $c_0 = 8.30$ Å, and $\beta = 92° 30'$; these are two molecules per unit cell (135). The well-developed face is (100). The shortest Ni—Ni distance is 5.32 Å. The bonds around each nickel ion are in a transplanar arrangement.

In the orthorhombic form, space group *Ibam*, D_{2h}^{26}, $a_0 = 24.46$, $b_0 = 9.25$, $c_0 = 6.60$ Å; there are four molecules per unit cell (213). The crystal is isomorphous with the corresponding copper compound (132), in which the molecular planes lie exactly parallel to the (001) plane, and are packed with copper atoms spaced 3.30 Å apart in a chain along the c axis. Alternate molecules are rotated by 90° about the c axis. The last three compounds have also been studied by Ferguson (111).

Bis-(*N*-isopropylsalicylaldiminato)-Ni(II),[Ni(*i*-PrSalim)$_2$] crystallizes in the orthorhombic modification, space group *Pbca*, D_{2h}^{15}, with lattice constants $a_0 = 13.22$, $b_0 = 19.70$, $c_0 = 15.14$ Å, and eight molecules per unit cell (132). The structure consists of discrete molecules in which there is a distorted tetrahedral disposition of ligand atoms about Ni, the angle between the two ligand planes being 82°. The Ni—O bond distance of 1.90 Å is intermediate between the 1.80 Å found in planar bis-*N*-methyl derivative (213) and 2.02 Å found in tetragonal octahedral diaquo bissalicylaldehydato-Ni(II) (376). The Ni—N distance of 1.97 Å is also greater than that in the *N*-methyl compound.

The structure of bissalicylaldoximato Ni(II) was investigated by Merritt, Guare, and Lessor (259). The unit cell is monoclinic, space group $P2_1/n, C_{2h}^2$, with two molecules and $a_0 = 13.83$, $b_0 = 4.89$, $c_0 = 10.20$ Å, and $\beta = 110°\,26'$. The bonds around each nickel atom have a transplanar arrangement.

Lingafelter (213a) has collected the structural details of a number of substituted salicyldimines of Ni, Pd, and Cu, and discussed the observed bond lengths and angles in relation to the acetylacetonate complexes.

TABLE LXVIII

Observed Absorption Bands for Bissalicylaldiminato Complexes of Ni(II) (29,111)[a]

Compound	Band max., cm^{-1} (29)	Comments	Band max., cm^{-1} (111)	Comments
Bissalicylaldoximin-ato-Ni(II)	16,100 20,300			
Diaquo bissalicylalde-hydato-Ni(II)			16,100	$\|c/\|b = 2$
Bissalicylaldiminato-Ni(II), (100) face	6500 9000 14,700 18,000 21,000	$\|/\perp \simeq 2$ $\perp$	18,800	$\|b/\|c \simeq 2$
Bis-N-methyl salicyl-aldiminato-Ni(II), orthorhombic (100) face	16,660 19,500 20,768 21,008 21,231 21,561 21,930 22,204 24,000	$\|b$ $\|c$ $\left.\rule{0pt}{36pt}\right\}\|c$ 80°K only	16,500 $\|b$ 19,600 $\|c$	
Bis-N-methyl salicyl-aldiminato-Ni(II), monoclinic	12,800 17,300 20,900	$\left.\rule{0pt}{18pt}\right\}$ (001) face $\|/\perp = 2.1/1.7$ 80°K	16,800 $\left.\rule{0pt}{12pt}\right\}$ (100) 20,800 $\}$ face	$\|c/\|b = 4$
Bis-N-ethyl salicyl-aldiminato-Ni(II)	16,300 16,700 20,900 22,400	$\perp$ $\|$ $\|$ $\|$		
Bis-N-propylsalicyl-aldiminato-Ni(II)	16,400 20,500	$\|$ 90°K		
Bis-N-isopropyl-salicylaldiminato-Ni(II)	$\sim 6,000$ 11,000 14,280 17,500 19,000 22,400	Very weak $\|b/\|c \simeq 5$		

[a] The polarization ratios are optical density ratios.

The absorption bands are given in Table LXVIII. The spectra obtained by Basu and Belford (29) are in reasonable agreement with those of Ferguson (111) where there is overlap of results. The main interest in Basu and Belford's investigation was the interpretation of the temperature dependence of the intensities. For the series of centrosymmetric compounds studied Ni(Salim)$_2$, Ni(PrSalim)$_2$, and Ni(MeSalim)$_2$, each band becomes less intense at low temperature as would be expected for a vibronic electric dipole mechanism. However, the integrated intensity of each band of the noncentric compound Ni(i-PrSalim)$_2$ is nearly constant between room temperature and liquid nitrogen temperature. Figure 76 shows the temperature dependence of the bands in the 17,000 to 22,000 cm^{-1} region for

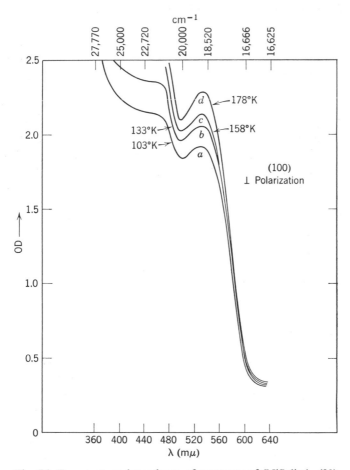

Fig. 76. Temperature dependence of spectrum of (NiSalim)$_2$ (29).

bissalicylaldiminato-Ni(II). The strikingly different temperature dependence of the two bands in Ni(Salim)$_2$ suggests that very different enabling vibrations are involved.

In Ni(Salim)$_2$, the 6500 cm^{-1} band is tentatively attributed to the first harmonic of the imine N—H stretch. The fundamental was found at 3320 cm^{-1} and the same sharp 6500 cm^{-1} band was found in the corresponding crystal spectrum of Cu(Salim)$_2$(29). The low-temperature satellites on the high-energy side of the 19,500 cm^{-1} absorption of orthorhombic Ni(MeSalim)$_2$ might be associated with a spin-forbidden Ni^{2+} transition.

Ferguson has discussed the band assignments in the compounds he studied (111). The band at 16,500 cm^{-1} in Ni(MeSalim)$_2$ is polarized in the molecular plane (vibronic symmetry B_{2u} and B_{3u}). The vibronic representations in D_{2h}, for the combination of one quantum of a u vibration and the direct product of the ground and excited electronic representations were shown in Table LVI. The electronic transition is assigned to $^1A_g \rightarrow$ $^1B_{1g}$,($b_{3u}(x)$ and $b_{2u}(y)$ enabling vibrations). The higher-energy band at 19,600 cm^{-1} is observed with the electric vector $\parallel$ to the a and c axes and not along the b axis, i.e., along the molecular y and z axes (vibronic species B_{2u} and B_{1u}) and the electronic transition is therefore assigned to $^1A_g \rightarrow$ $^1B_{3g}$. For the monoclinic modification, the band at 16,800 cm^{-1} is also assigned to the $^1A_g \rightarrow {}^1B_{1g}$ transition while the higher energy band appears to be polarized $\parallel$ and $\perp$ to the b axis and this would be true for either of the assignments $^1A_g \rightarrow {}^1B_{3g}$ and $^1A_g \rightarrow {}^1A_g$. Maki (244) assigned the transition at 16,800 to $^1A_g \rightarrow {}^1B_{3g}$ and the shoulder at about 20,500 cm^{-1} to $^1A_g \rightarrow$ $^1B_{1g}$, while Ballhausen and Liehr (14) have used the strong field formalism to assign the bands to $^1A_g \rightarrow {}^1B_{1g}$ and $^1A_g \rightarrow {}^1A_g$, respectively.

The analysis (111) of the crystal spectrum of diaquo bissalicylaldehydato-Ni(II) also gives results differing from a previous assignment. Maki (243) has reported the spectrum of the powder and assigned the absorption band at 16,100 cm^{-1} to the transition $^3B_{3g} \rightarrow {}^3B_{2g}$ in D_{2h} symmetry. If this is correct, the allowed vibronic states belong to the representations B_{2u} and B_{3u} and the transition should be polarized in the molecular (xy) plane (243). However, the absorption band appears in both $\parallel$ and $\perp$ components to the c axis, with the polarization ratio c/b of about 2. The only possible assignment within D_{2h} is $^3B_{3g} \rightarrow {}^3B_{1g}$.

The electronic absorption spectrum of a single crystal of bisacetylacetonate iminato Ni(II) at 80°K has been determined by Basu (30). The compound crystallizes with a monoclinic unit cell having dimensions $a_0 = 8.40$, $b_0 = 3.36$, $c_0 = 3.745$ Å, and $\beta = 98°\,28'$. The main band in the spectrum at 18,520 cm^{-1} is assigned to the $^1A_{1g} \rightarrow {}^1B_{1g}$ transition while the much weaker band at 9090 cm^{-1} is assigned to $^1A_{1g} \rightarrow {}^3B_{1g}$.

10. $[(NH_2)_2CS]_4NiCl_2$

The polarized crystal spectrum of tetrakisthiourea nickel chloride has been determined by Hare and Ballhausen (163). The crystal structure of $[(NH_2)_2CS]_4NiCl_2$ has been reported by Cavalca, Nardelli, and Braibant (64) and more recently by Lopez-Castro and Truter (221) whose data are quoted. The space group is $I4, C_4^5$ and the unit cell dimensions are $a_0 = 9.558$ and $c_0 = 8.981$ Å with two molecules per unit cell at 110°K. The crystal is composed of hexacoordinated nickel units in which four sulfur atoms (thiourea) are in a plane about the nickel atom and the two chlorine atoms occupy the two axial positions (*trans*). The four sulfur atoms are related by a fourfold axis, Ni—S = 2.464 Å, while the two chlorine atoms occupy nonequivalent positions at 2.40 and 2.52 Å. The nickel atom is 0.3 Å above the plane of the sulfur atoms. The molecular symmetry is C_{4v} and the site symmetry C_4. The Cl—Ni—Cl axis of the molecular units corresponds to the tetragonal axis (c) of the crystal.

The observed absorption bands with their polarizations and assignments in C_4 are given in Table LXIX (163), together with the theoretical octahedral levels predicted by Tanabe and Sugano (386), using $Dq = 798$, $B = 750$, $C = 2175$ cm^{-1}; and by Liehr and Ballhausen (210) using $B = 675$, $C = 2625$, and $\zeta = 750$ cm^{-1}. Agreement is better when spin–orbit coupling is included (210) and it is suggested this is important for the prediction of energy levels in this complex (163).

TABLE LXIX

Observed Absorption Bands at Room Temperature and Polarizations in
$[(NH_2)_2CS]_4NiCl_2$, (163) and Comparison with Calculated Levels in O_h

Absorption and polarization, cm^{-1}	Assignment, C_4	Calculated levels in O_h, cm^{-1}	
		Tanabe and Sugano[a] (386)	Liehr and Ballhausen[b] (210) including spin–orbit coupling
7,100 ∥	^3B	7,980	7,980
8,200 ⊥	^3E		
10,350 ⊥	$^1A, ^1B(^1E_g)$	10,350	10,374
13,400 ∥	^3A		13,315 (Γ_5)
		13,080	
13,700 ⊥	^3E		13,500 (Γ_3)
17,800 ⊥	$^1B, ^1E(^1T_{2g})$	18,340	18,020
22,300 ∥	^3A	22,170	21,410 (Γ_5)

[a] Octahedral levels, $Dq = 798$, $B = 750$, $C = 2175$ cm^{-1}, and $C/B = 2.9$.
[b] Octahedral levels, $B = 675$, $C = 2625$, and $\zeta = 750$ cm^{-1}.

The transformation of the energy levels in O_h to those in C_4 was shown in Figure 72 together with the tetragonal matrix elements, and the selection rules for the possible transitions in C_{4v}, C_4, and S_4 are shown in Table LXX.

TABLE LXX

Electric Dipole Selection Rules for Vibronically Allowed Transitions in C_{4v}, C_4, and S_4 Arising from the Excited States $^3T_{1g}$, $^3T_{2g}$ in O_h (163)

Octahedral states	C_{4v}	C_4	S_4
$^3T_{2g}$	$^3B_1 \nrightarrow {}^3B_2$	$^3B \rightarrow {}^3B, \parallel$	$^3B \nrightarrow {}^3B$
	$^3B_1 \rightarrow {}^3E, \perp$	$^3B \rightarrow {}^3E, \perp$	$^3B \rightarrow {}^3E, \perp$
$^3T_{1g}$	$^3B_1 \nrightarrow {}^3A_2$	$^3B \nrightarrow {}^3A$	$^3B \rightarrow {}^3A, \parallel$
	$^3B_1 \rightarrow {}^3E, \perp$	$^3B \rightarrow {}^3E, \perp$	$^3B \rightarrow {}^3E, \perp$

It is clear from the experimental results that the selection rules for either C_{4v}, C_4, or S_4 symmetry are not obeyed since intense $\parallel$ components are observed for all the spin-allowed bands. The polarization ratios ($\epsilon_\perp / \epsilon_\parallel$) for the first and second spin-allowed bands are 1.5 and 1.6, respectively. In the group C_{4v} there are no vibrations of symmetry a_2 which would allow the first $\parallel$ band to steal intensity from the allowed states, but vibrations of symmetry b_2 do exist to account for $\parallel$ absorption in the second band. However, under the group C_4, the site symmetry, there are appropriate vibrations of symmetry a and b to couple allowed states and give all the $\parallel$ bands intensity. Perpendicular absorption is allowed under C_{4v} and C_4 symmetry and appropriate e vibrations also exist. The same situation would also apply to a site symmetry of S_4 as suggested by Cavalca et al. (64). The preferred site symmetry is however C_4 because the experiments indicate that the polarization ratio of the first band is less than that of the second (163). However, since the two polarization ratios are nearly the same, it is concluded that the main intensity-giving mechanism for both bands is vibronic. The situation is therefore interesting in that it offers an example in which a vibronic intensity mechanism seems to be operative in a complex with an inherent lack of a center of symmetry. Factor group interaction is also suggested to be partly responsible for the intensity.

Configurational mixing in the octahedral field scheme leads to a splitting of $-4.9D_s + 0.5D_t$ for the $^3T_{1g}(^3F)$ state and $-1.9D_s + 4.35D_t$ for the $^3T_{1g}(^3P)$ state. These modify the levels of Figure 72. With these splittings, D_t and D_s are calculated to be 126 and -48 cm^{-1}, respectively, for this complex.

11. Ni(DPM)₂

Cotton and Wise (80) have determined the polarized crystal spectrum of bis(2,2,6,6 tetramethylheptane-3,5 dionato)Ni(II),Ni(DPM)₂, also called bis(dipivaloylmethanido)-Ni(II). The crystal structure of this compound has been determined by Cotton and Wise (76). The unit cell is monoclinic, space group $P2_1/a, C_{2h}^5$ with dimensions $a_0 = 10.70$, $b_0 = 10.98$, $c_0 = 10.39$ Å, $\beta = 113° 16'$, and contains two formula units. The chelate rings and indeed the whole molecule were found to be essentially planar (Fig. 77). The Ni atom is surrounded by a plane of four oxygen atoms with Ni—O = 1.836 Å, while the O—$\widehat{\text{Ni}}$—O ring angle is 94° 36'. The projections of the molecules onto the (010), (001), (100), and (110) crystal planes are shown in Figure 77. These are included here for reference when the polarized crystal spectrum of Cu(DPM)₂ is discussed, p. 463, as the Ni(DPM)₂ and Cu(DPM)₂ compounds are isomorphous (76). The two molecules within the unit cell are related by a twofold screw axis parallel to b; they have identical projections on the (010) plane as is apparent from Figure 77. The dihedral angle between the mean planes of the two molecules is 44° 54' ± 54' while the line passing through Ni and the carbon atoms C(2) and C(2)' makes an angle of 24° 12' ± 48' with the b axis of the unit cell. This is most clearly shown on the (001) projection, Figure 77. The observed bands are tabulated in Table LXXI.

MO calculations (78) suggest that the d_{xy} orbital is some 16,700 cm⁻¹ higher than the other four, which lie relatively close together in an order which the calculation suggests should be $d_{z^2} > d_{x^2-y^2} > d_{xz} > d_{yz}$. These results are taken to show only that the four transitions should occur close

TABLE LXXI
Observed Absorption Bands (25°C), Polarizations, and Proposed Assignments for Ni(DPM)₂ (80)

	Observed			Calculated		
Transition	Band max., cm⁻¹	Polarization	Oscillator strength	Band max., cm⁻¹	Polarization	Oscillator strength
	~16,000	y and/or z				
	~18,500	Mainly x		See ref. 80		
	~20,000	Mainly x				
$\sigma_L \rightarrow 3d_{xy}$	~29,000 (shoulder)			27,370	x	0.41
$\pi_L \rightarrow \pi_L^*$	37,300		~0.4	32,230 / 32,910	y / y	0.44 / 0.35
$\sigma_L \rightarrow 3d_{xy}$	42,700		~0.3	48,220	y	0.31

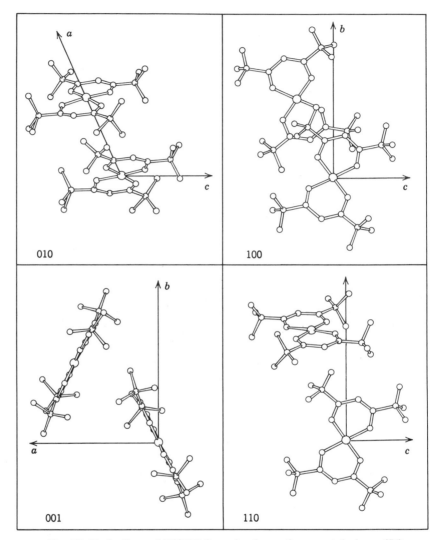

Fig. 77. Projections of Ni(DPM)$_2$ molecules on four crystal planes (76).

together and in approximately the indicated energy range. It is suggested that three or all four transitions lie within the envelope of the band at $\sim 19,000\ cm^{-1}$.

The tentative assignments of the higher-energy bands were made on the basis of a comparison of the observed energies, polarizations, and oscillator strengths (from solution spectra) with the calculated values (78).

12. Ni(bipy)$_3^{2+}$

Palmer and Piper (300) have studied the polarized crystal spectrum of the tris-(2,2' bipyridyl)nickel(II) ion in the matrices Ni(bipy)$_3$Br$_2$6H$_2$O, Zn(Ni)(bipy)$_3$Br$_2$6H$_2$O, and Zn(Ni)(bipy)$_3$SO$_4$7H$_2$O. The crystal structures and site symmetry of the Ni^{2+} ion were discussed on p. 356 in connection with the Fe^{2+} analog. The crystal spectrum of Zn,(Ni)(bipy)$_3$SO$_4$7H$_2$O is shown in Figure 78 and the $d \rightarrow d$ transition band maxima and absolute intensities are tabulated in Table LXXII.

The temperature dependence of the intensities is quite low; $I^3[E(^4T_1)]$ decreases 16% and $I^3[E(^3T_2)]$ remains unchanged as the temperature decreases from 300 to 80°K. The crystal spectra reveal the presence of a weak spike at 20,960 cm^{-1} and a narrow shoulder at about 23,000 cm^{-1}, neither of which is observed in solution (188).

The octahedral ground state is $^3A_2$ (58) and the spin-allowed $d \rightarrow d$ transitions are then in increasing order of energy $^3A_2 \rightarrow {}^3T_2(^3F)$, $^3T_1(^3F)$, and $^3T_1(^3P)$ (386,387). Each of the triply degenerate octahedral levels will be split into an A and an E component with D_3 symmetry about Ni^{2+}. The pertinent electric dipole selection rules in D_3 symmetry are

$$A_2 \;\not\!\longrightarrow\; A_2$$
$$A_2 \;\overset{\parallel}{\longrightarrow}\; A_1$$
$$A_2 \;\overset{\perp}{\longrightarrow}\; E_1$$

The assignment of electric dipole character to the $d \rightarrow d$ transitions is supported by the strong polarization of the $^3T_1$ band in the crystal spectra.

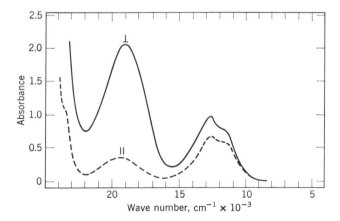

Fig. 78. Polarized orthoaxial spectra of Zn(Ni)(bipy)$_3$SO$_4$7H$_2$O at 300°K. Crystal thickness, 3.4 mm; approximately 25 mole % nickel (300).

TABLE LXXII

Observed Absorption Bands for Ni(bipy)$_3{}^{2+}$ Ion, Together with Proposed Assignments (300)

| | | | Band max., cm^{-1} [a] | | | | | Temp., |
| | | Excited | | | | | | |
Matrix	Spectrum	state	300°K	80°K	$\Delta\nu_{1/2}$, cm^{-1}	ϵ	I[b]	°K
Ni(bipy)$_3$Br$_2$6H$_2$O	Axial	$^1A_1$	(21,000)	20,960	400	0.2	0.01	300
					100	0.8	0.01	80
		$^3E(^3T_1)$	18,870	19,120	3,000	20.7	3.7	300
					2,300	23.9	3.1	80
		$^3E(^3T_2)$	12,740	12,820	2,500	6.7	1.57	300
					2,100	7.8	1.6	80
		^1E	(11,400)	(11,700)	1,700	2.1	0.4	300
					1,000	1.7	0.2	80
Zn(bipy)$_3$Br$_2$6H$_2$O	Axial	$^1A_1$	—	20,920				
		$^3E(^3T_1)$	18,870	19,100				
		$^3E(^3T_2)$	12,690	12,850				
		^1E	(11,690)	(11,760)				
Zn(bipy)$_3$SO$_4$7H$_2$O	σ	$^1A_1$	—	(20,800)				
		$^3E(^3T_1)$	19,040	19,410				
		$^3E(^3T_2)$	12,740	12,800				
		^1E	(11,400)	(11,800)				
	π	$^1A_1$	—	(21,000)				
		$^3A_2(^3T_1)$	19,600	20,100				
		$^3A_1(^3T_2)$	12,660	12,750				
		^1E	(11,400)	(11,900)				

[a] Bands appearing as unresolved shoulders are enclosed in parentheses.
[b] Integrated band intensity $I = \int (\epsilon/\nu)d\nu \approx (1/\nu_{max}) \int \epsilon d\nu$ (300).

However, the intensity of the ∥ component $^3A_2$, represents partial breakdown, presumably vibronic, of the point group selection rules by which the electronic transition is strictly forbidden. The polarization of the $^3T_2$ band is very slight.

Energy level calculations for the d^8 configuration (210,386,387) indicate that two lower singlet states $^1E(^1D)$ and $^1A_1(^1G)$ and two higher, $^1T_2(^1D)$ and $^1T_1(^1G)$, should lie low enough to give rise to absorptions in the optical spectrum. The band at 20,960 cm^{-1} is assigned to the $^3A_2 \rightarrow {}^1A_1$ transition. The partially resolved peak which appears in the low-temperature spectra at 23,040 cm^{-1}, is not assigned to a $d \rightarrow d$ transition because of its small half-width. The spin-forbidden transition $^3A_2 \rightarrow {}^1T_1$ is a $t_2 \rightarrow e$ transition and must have a half-width of 2000–4000 cm^{-1}. The peak at 23,040 cm^{-1} is thought to be an internal ligand transition.

The shoulder on the low-energy side of the infrared band is assigned to the spin-forbidden transition $^3A_2 \to {}^1E$, in agreement with Jørgensen (188). However, Liehr and Ballhausen (210) assigned this double band to the spin–orbit components of the $^3T_2$ level. Adopting this assignment ($^3A_2 \to {}^1E$), Palmer and Piper (300) arrive at a value of 1.8 for Dq/B (386), and using the value of $Dq = 1279$ cm^{-1} from the low-temperature spectra, B is 710 cm^{-1}. Consistency of agreement for the energies of ^1E, $^3T_2$, $^3T_1$, and $^1A_1$ with theory (386) is good.

The trigonal splitting of both $^3T_2$ and $^3T_1$ is $3K/2$. However, owing to the apparent partial vibronic character of the bands, particularly the $^3T_2$, the small differences in band maxima shown by the $\parallel$ and $\perp$ components cannot be interpreted in terms of a trigonal splitting, and it is concluded that K is very small.

13. $Ni_2{}^+$ in Al_2O_3

Geschwind and Remeika (142) have determined the electron spin resonance of Ni^{2+} and Cu^{3+} in Al_2O_3. When the cubic field ground state is an orbital singlet, the magnetic properties of the ground state are determined by perturbations from higher cubic field states and a determination of K in this case is more complex and may be less reliable with regard to sign. However, values of $K = 370$ cm^{-1} for Ni^{2+} and 330 cm^{-1} for Cu^{3+} were obtained. The change of sign of K from Ru^{3+}, Co^{2+}, and V^{3+} could be interpreted as meaning that the Ni^{2+} and Cu^{3+} are sufficiently displaced from the positions occupied by the former ions as to change the sign of K. The optical spectrum of Ni^{2+} in Al_2O_3 was determined by McClure (233) who obtained a value of 930 cm^{-1} for Dq.

14. $K_2Ni(COS)_4$

The polarized crystal spectrum of $K_2Ni(COS)_4$ has been reported (199). The absorption bands are discussed in terms of the electronic nature of the complex ions.

B. $4d^8$

1. K_2PdCl_4

A few measurements have been made of square–planar Pd(II) complexes. The polarized crystal spectrum of K_2PdCl_4 has been determined by Yamada (425) and more recently by Day, Orchard, Thompson, and Williams (84), both using a microspectrophotometric method. The compound has a tetragonal structure, with the atoms in special positions of

$P_4/mmm, D_{4h}{}^1$ and the unimolecular prism has dimensions $a_0 = 7.04$, $c_0 = 4.20$ Å (89,392). The spectra were recorded by Day et al. with light incident on the (100) or (010) face with the electric vector first parallel to the c axis ($\|$) and then parallel to the b or a axis ($\perp$). Table LXXIII lists the observed absorption bands with the three possible assignments of Day et al. Basch and Gray (25) have proposed assignments for the bands on the basis of an SCCC-MO one-electron calculation (d-energy ordering $x^2 - y^2 > xy > xz, yz > z^2$) and a separate crystal field calculation. The possible ordering of the d orbitals in an initially octahedral field with increasing tetragonal distortion is shown in Figure 79 (252). There are three qualitatively different level schemes, referred to as A, B, and C. Basch and Gray's ordering for $PdCl_4{}^{2-}$ corresponds to C, and the significance of this is discussed. The proposed MO assignments of Basch and Gray are also included in Table LXXIII for comparison.

The three distinct symmetry assignments of Day et al. are all compatible with the observed polarizations, if it is assumed that all the odd normal modes are vibronically active. In assignments 1 and 2, the absorption near 17,000 cm^{-1} is considered to be the spin-forbidden counterpart of that near 23,000 cm^{-1}. The transition $^1A_1 \rightarrow {}^1A_2$ must always occur in

TABLE LXXIII

Observed Room Temperature Band Maxima (Molar Decadic Extinction Coefficients in Parentheses) and Polarizations in $K_2PdCl_4{}^a$

Band max.,[b] cm^{-1} (425)	Band max., cm^{-1} (84)	Symmetry assignments			MO assignments
		1	2	3	
16,700 (1.5) $\|$					
17,000 (4)[c] $\perp$	17,000 (7) $\|$ ⎫ 18,000 (19)[c] $\perp$ ⎭	^3E_g	$^3B_{1g}$	$^1B_{1g}$	$^3B_{1g}$
21,300 (56) $\perp$	20,000 (67)[c] $\perp$	$^1A_{2g}$	$^1A_{2g}$	$^1A_{2g}$	$^1A_{2g}$
23,300 (30) $\|$	22,600 (128) $\perp$ ⎫ 23,000 (80) $\|$ ⎭	^1E_g	$^1B_{1g}$	^1E_g	^1E_g
	29,500 (67)[c] $\perp$				$^1B_{1g}$

[a] Possible assignments on the basis of symmetry arguments alone are those of Day et al. (84). Assignments on basis of MO calculation of Basch and Gray (25) in last column.

[b] Yamada's polarization convention (425) has been reversed here so that $\|$ and $\perp$ also refer to polarization with respect to the fourfold axis rather than to the xy plane. The extinction coefficients in parentheses have been calculated from the $\log_{10} \alpha$ data of ref. 425.

[c] Unresolved shoulder.

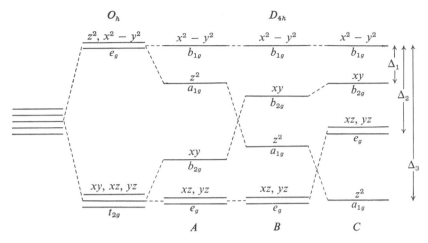

Fig. 79. Possible ordering of the five d orbitals in octahedral coordination with a tetragonal distortion (252).

$\perp$ polarization, and in each of the three assignments it is identified with the inflexion at 20,000 cm^{-1}.

Assignment *3* neglects the inflexion at 29,500 cm^{-1}, though the intensity is of the order expected for a ligand field band. This high-frequency band cannot be identified with the spin-forbidden charge-transfer transition $^1A_1 \rightarrow {}^3A_2$, since this may occur only in the $\|$ polarization, but it could be due to an orbitally forbidden charge transfer. The remaining possibilities (*1* and *2*) each identify the 23,000 ($\|$) and 22,600 cm^{-1} ($\perp$) bands with a single transition, but differ in the relative ordering of the d_{z^2} and d_{xz}, d_{yz} orbitals. In each case it is assumed that the $\|$ component of the band indicated by 29,500 cm^{-1} inflexion is lost beneath the first charge transfer band (itself $\|$ polarized), only the $\perp$ component being sufficiently intense to appear as a shoulder. The spin-forbidden counterpart of this high-frequency absorption is presumed to lie under the main absorption in the 23,000 cm^{-1} region, while the excitation $^1A_1 \rightarrow {}^3A_2$ must be either too weak to be detected or else is contained in the band occurring at about 16,500 cm^{-1} in $\perp$ polarization.

2. K_2PdI_4 and $K_2Pd(SCN)_4$

Yamada and Tsuchida (430) have examined the polarized spectra of crystals of K_2PdI_4 and $K_2Pd(SCN)_4$ by the microscopic method (395, 396), while Yamada (425) has studied the dichroism of $K_2PdBr_4 2H_2O$, $Pd(NH_3)_4Cl_2H_2O$, and $K_2Pd(CN)_4 3H_2O$. This latter compound has also

TABLE LXXIV

Observed Room-Temperature Absorption Band Maxima and Intensities in Tetragonal Pd(II) Complexes (α = extinction per mm)

Ref.[a]	Compound	Band max., cm^{-1}	log α	Band max., cm^{-1}	log α
430	K_2PdI_4			13,400 ∥	1.88
		16,100 ⊥	1.93	20,000 ∥	1.86
430	$K_2Pd(SCN)_4$	19,300 ⊥	1.54	20,700 ∥	1.14
		24,300 ⊥	1.44	25,300 ∥	1.48
425	$K_2PdBr_4 2H_2O$	19,000 ⊥	1.80	21,000 ∥	1.50
425	$Pd(NH_3)_4Cl_2H_2O$	33,000 ⊥	1.40	30,700 ∥	1.35
425	$K_2Pd(CN)_4 3H_2O$	Absorption edge near 34,000 cm^{-1}			

[a] Listed data are interpolated from figures in refs. 430 and 425. Polarization is ∥ when electric vector is parallel to fourfold axis. (This reverses the convention of refs. 430 and 425.)

been studied by Brasseur and Colin (70). Table LXXIV lists the absorption bands observed in these complexes. The symbols ∥ and ⊥ refer to orientation with respect to the (pseudo) fourfold molecular axis.

3. Pd(DMG)₂

Basu, Cook, and Belford (28) studied the absorption spectrum of single crystals of palladium dimethyl glyoxime Pd(DMG)₂. While NaCl pellets of Pd(DMG)₂ show a series of sharp bands in the visible region (22,452), the spectrum of a single crystal gives a fairly steep rise in absorption in the region of the sharp band, followed by a plateau which forms a shoulder on the side of the ultraviolet absorption. No explanation of this observation has been given. The crystal spectrum of $K_2Pd(COS)_4$ has been reported (199).

C. 5d⁸

The possible ordering of the metal d orbitals in the $5d^8$ tetrahalide square-planar complexes (D_{4h} symmetry) of Pt(II) and Au(III) [as well as Pd(II) ($4d^8$)] has been the subject of numerous papers (67,84,85,107,152, 156,251–254,264,425). Interest has been focussed on the PtCl$_4^{2-}$ ion where attempts have been made to match the bands in the visible and near-ultraviolet regions of the spectrum to one-electron excitations from the occupied d orbitals [$b_{2g}(xy)$, $e_g(xz, yz)$, and $a_{1g}(z^2)$] to the empty $b_{1g}(x^2 - y^2)$, Figure 79. The spectrum has been assigned variously depending on the author's choice of ordering for the occupied d orbitals and on whether a low-intensity band occurring at 20,300 cm^{-1} is considered to be a transition to a spin singlet or to a spin triplet excited state (25). However experimental studies by several groups (84,251,252,254,276), including circular dichroism measurements (46,253), have considerably narrowed the possibilities. In particular, the work of Martin and co-workers (251–254), reviewed in references 255 and 256, is particularly useful here.

1. K_2PtCl_4

Potassium chloroplatinate, K_2PtCl_4, has a structure built up of alkali ions and planar tetrahedral complex ions. The crystal structure (89) has space group $P4/mmm, D_{4h}^{1}$, with lattice constants of the unimolecular prism $a_0 = 6.99$, and $c_0 = 4.13$ Å. Atoms are in special positions at:

Pt   (000)

K $0\frac{1}{2}\frac{1}{2}; \frac{1}{2}0\frac{1}{2}$

Cl $\pm(uuO; u\bar{u}O)$ with $u = 0.233$–0.238

In this structure, platinum has its usual square coordination with Pt—Cl = 2.32 Å; each potassium atom has eight chlorine neighbors with K—Cl = 3.28 Å. The fourfold symmetric axis (z axis) is aligned with the crystallographic c axis, the Pt^{2+} ion having D_{4h} symmetry. The ions lie directly over one another with an eclipsed arrangement of chlorides and with Pt atoms separated along the z axis by 4.13 Å.

Table LXXV lists the observed band maxima and polarizations together with the proposed assignments for the three independent studies (84,254,276). The last column shows the assignments of Basch and Gray (25) on the basis of an SCCC–MO calculation and a separate crystal field calculation. Strong vibrational structure was observed on the 2,400 and 26,300 cm^{-1} bands (254,276) and also on the 20,700 cm^{-1} band by Mortenson at 5°K (276). The progression reaching from 23,000 to 28,000 cm^{-1} contains some 19 members separated by 270–290 cm^{-1}, the progression around 23,940 cm^{-1} ($\|$), 8 members separated by 275–295 cm^{-1}, and the progression around 20,700 cm^{-1} ($\|$), 3 members separated by 260–280 cm^{-1} (276).

The differences in assignments have arisen from the different assumed orderings of the d orbitals in the square–planar ions (Fig. 79). The qualitative arrangement of orbitals shown in Figure 79c is that first proposed by Chatt, Gamlen, and Orgel (67), corresponding to a d-level ordering of $x^2 - y^2 > xy > xy,xz > z^2$. This was also the conclusion of Basch and Gray (25) who estimated the level ordering by the SCCC–MO method and also by Cotton and Harris (77) who used a somewhat different extended Hückel approach. Cotton and Harris obtained values of 20,165, 21,778, and 29,037 cm^{-1} for $\Delta_1, \Delta_2, \Delta_3$, respectively (cf. Fig. 79). Basch and

TABLE LXXV

Observed Band Maxima in cm^{-1} for Crystals of K_2PtCl_4 (Ground State $^1A_{1g}$ in D_{4h}), Together with the Various Proposed Assignments

| Assignments[a] | | Observed | | | | | | | | | | | | | | | Calculated |
| A | B | Ref. 254, 15°K | | | | | Ref. 276, 5°K | | | | Assignment | | Ref. 84, 300°K | | | | Ref. 25 |
		$z(\parallel)$	ϵ	$xy(\perp)$	ϵ	Assignment[a]	$z(\parallel)$	$f\times10^4$	$xy(\perp)$	$f\times10^4$	(1)	(2)	$z(\parallel)$	ϵ	$xy(\perp)$	ϵ	Assignment
							16,930	0.005									
$^3E_g[\Gamma_5(1)]$	$^3B_{1g}[\Gamma_5(1)]$	17,000	<1	18,000	2	$^3A_{2g},{}^3E_g(\Gamma_1)$	17,190	0.03			$^3A_{2g}$	$^3A_{2g}$	17,300				^3E_g
$^3E_g[\Gamma_4(1)]$	$^3A_{2g},{}^3E_g\ [\Gamma_1(2)]$	19,000		18,000		$^3E_g(\Gamma_2)$	17,190	0.03	18,100	0.09							
$^3A_{2g}[\Gamma_5(2)]$	$^3A_{2g},{}^3E_g\ [\Gamma_5(2)]$	20,600	10	20,900	9	$^3E_g(\Gamma_3)$	20,660	0.77	20,810	0.73	^3E_g	$^3B_{1g}$	20,200	20	20,400	17.5	$^3A_{2g}$
$^3B_{1g}[\Gamma_5(3)]$	$^3A_{2g},{}^3E_g\ [\Gamma_5(3)]$	24,100	3	24,000	7	$^3B_{1g}(\Gamma_5)$	23,940	0.13	24,000	0.35	$^1A_{2g}$	$^1A_{2g}$	$\sim$24,000[b]		24,000[b]		$^3B_{1g}$
$^3B_{1g}[\Gamma_4(2)]$	$^1A_{2g}[\Gamma_2(2)]$			26,300	28	$^1A_{2g}$			26,050	1.6	$^1A_{2g}$	$^1A_{2g}$			($\sim$26,000)[c]		$^1A_{2g}$
$^1E_g[\Gamma_5(4)]$	$^1E_g[\Gamma_5(4)]$	29,800	55	29,200	37	$^1B_{1g}$	29,900	4.3	29,250	3.6	$^1A_{2g}$	$^1A_{2g}$	29,300	70	28,500	57	^1E_g
$^1B_{1g}[\Gamma_3(2)]$						^1E_g									($\sim$36,500)[b]		$^1B_{1g}$

[a] The Bethe classification Γ_1, Γ_2, Γ_3, Γ_4, Γ_5 is retained here for the spin–orbit coupled states in D_{4h}. In Mulliken notation, these are A_1, A_2, B_1, B_2, E, respectively.

[b] The existence of this band was confirmed (84) after publication of ref. 251.

[c] From reflectance spectrum of K_2PtCl_4 (84). The numbers in parentheses for assignments A and B are the ordinal number of the Γ_n term arranged in order of increasing energy.

Gray (25) found Δ_1, Δ_2, Δ_3 = 28,700, 34,300, 42,200 cm^{-1}, respectively. Figure 79b shows the arrangement of orbitals where $x^2 - y^2 > xy > z^2 > xz, yz$ found by Gray and Ballhausen (152).

The assignments of Day et al. (84) were based upon symmetry considerations alone and they were not able to distinguish between the two arrangements listed without further information on the enabling vibrations. Martin and co-workers (252,254), who have carried out an extensive series of calculations, including spin–orbit coupling, arrive at the two possible sets of assignments A and B shown in Table LXXV. Alternative A assigns the $^1A_{1g} \rightarrow {}^1B_{1g}$ transition to a band observed (84) at 36,500 cm^{-1} in the reflectance spectrum of K_2PtCl_4. This leads to the ordering of Figure 79c. Alternative B places this transition in the 29,000 cm^{-1} region, bringing us back to the ordering of Figure 79b; however in this case the one-electron energy of the z^2 and (xz, yz) levels is about the same. Mortenson (276), whose experimental results are very similar to those of Martin et al. (254), qualitatively considered spin–orbit splitting and assumed the ordering of Gray and Ballhausen (152) i.e., that of Figure 79b, in proposing assignments. The various assignments (25,254) have in each case been fitted quite closely to the calculated levels.

A complete crystal field treatment of the D_{4h} d^8 problem including spin–orbit coupling was made by Fenske et al. (107). The states with the lowest energy and those which will make the greatest contribution to the intensities are considered to be the one-electron transfer states in which one electron has been excited from a lower orbital to b_{1g}. In the absence of spin–orbit perturbation, the ground state for $PtCl_4^{2-}$ can be designated as $^1(b_{1g}b_{1g})-{}^1A_{1g}$. The one-electron transfer states are then: $^3(b_{1g}b_{2g})-{}^3A_{2g}$; $^1(b_{1g}b_{2g})-{}^1A_{2g}$; $^3(b_{1g}a_{1g})-{}^3B_{1g}$; $^1(b_{1g}a_{1g})-{}^1B_{1g}$; $^3(b_{1g}e_g)-{}^3E_g$; and $^1(b_{1g}e_g)-{}^1E_g$ (252). The ground state is $^1A_{1g}$ and the predicted polarizations for a vibronic electric dipole mechanism are shown in Table LXXVI for the three odd-parity vibrations of the planar complex. From Table LXXVI

TABLE LXXVI

Polarizations for Electric Dipole Transitions of d^8 Configuration in D_{4h} Symmetry (e.g., $PtCl_4^{2-}$) from the Ground State $^1A_{1g}$ to the Various Excited States Induced by the Asymmetric Vibrations of an MX_4 Complex in the Vibronic Model (252)

Vibration	Excited state				
	A_{1g}	A_{2g}	B_{1g}	B_{2g}	E_g
a_{2u}	z, ‖	—	—	—	xy, ⊥
b_{2u}	—	—	z, ‖	—	xy, ⊥
e_u	xy, ⊥	xy, ⊥	xy, ⊥	xy, ⊥	z, ‖

it can be seen that transitions to A_{2g} and B_{2g} states are polarized $\perp$. Transitions to A_{1g} and B_{1g} have $\parallel$ polarizations excited by single out-of-plane bending vibrations and $\perp$ polarizations excited by the e_d (stretching and in plane bending), whereas E_g transitions the $\parallel$ polarization is excited by the e_u vibrations and the $\perp$ polarization by the two out-of-plane bendings (252). There is no one-electron transfer singlet state mixed into the B_{2g} state and hence transitions to B_{2g} are expected to be low in intensity (254).

Because of its polarization and intensity, the band at 26,300 cm^{-1} is generally assumed to be a transition to an A_{2g} state (cf. Table LXXV), comprising mainly $^1A_{2g}$, except in arrangement A of Martin et al. (254). However, here the $^1A_{2g}[\Gamma_2(2)]$ level is very close to the $^3B_{1g}[\Gamma_4(2)]$ level and accounts for the observed polarization as the transition to the Γ_4 multiplet level may be very weak (254).

The assignment schemes are all tentative. The critical point of difference between the two arrangements of Martin et al. (254) is the assignment of the 36,500 cm^{-1} band either to $^1B_{1g}$ or to a spin-forbidden charge-transfer band. The assignment of the lower-energy bands is also open to question. A deeper knowledge of the vibronic mechanism, and consistent experimental results for allied compounds are evidently necessary.

2. $Pt(NH_3)_4PtCl_4$ and Related Complexes

The electronic spectra of crystals of planar Magnus Green salt (MGS), $Pt(NH_3)_4PtCl_4$, and related substances have been the subject of considerable interest due mainly to the unusual colors of such crystals compared with those of their solutions (3,84,85,263,264,426) [MGS is deep green, while $PtCl_4^{2-}$ and $Pt(NH_3)_4^{2+}$ are red and colorless, respectively, in water solution (6)]. The spectrum of the solid appears to show an extra "color band." As the crystals involved have a chain structure, this has led to speculations that direct metal–metal interaction is important in the solid. Such a view was reinforced by the fact that the apparent new absorption is strongest when the incident light is polarized with its electric vector parallel to the metal chains. According to recent work by Anex and Krist (4) on nickel dimethylglyoxime, $Ni(DMG)_2$, the "color band" in this substance is actually a highly perturbed single-molecule transition that has moved from the ultraviolet into the visible and increases sharply in intensity under the influence of the crystal perturbation. Day, Orchard, and Williams (84) had previously demonstrated that the color of MGS also does not result from a new solid-state transition but rather arises from an intensification and red shift of $d \rightarrow d$ transitions that are observed in the solution and single-crystal spectra of K_2PtCl_4. The green body color of MGS can be regarded

(4) as the result of a "window" developing in the crystal at around 20,000 cm^{-1} as the bands on either side of this point increase in intensity, just as the red color of solid Ni(DMG)$_2$, in contrast to the pale yellow of its solution, is related to sharply increased green absorption in the solid.

The Magnus Green salt structure belongs to space group $P4/mnc, D_{4h}^6$, the lattice containing a chain system of alternate planes of Pt(NH$_3$)$_4$ and PtCl$_4$ groups (6). Adjacent cations and anions are staggered by some 28°, while in the tetragonal lattice of K$_2$PtCl$_4$ the anions are perfectly eclipsed. There are two molecules in the tetragonal unit cell, with lattice parameters $a_0 = 9.03$ and $c_0 = 6.49$ Å. The atoms are in the positions:

Pt(1)   (000; ½½½)

Pt(2)   (00½; ½½0)

Cl $\pm (uv0; v\bar{u}0; u + \frac{1}{2}, \frac{1}{2} - v, \frac{1}{2}; v + \frac{1}{2}, u + \frac{1}{2}, \frac{1}{2})$
    with $u = 0.0563, v = 0.2528$

NH$_3$ $u = 0.5627, v = 0.7197$

Both kinds of Pt atom have a square–planar coordination, the Pt(1) atoms having four chlorines as nearest neighbors with Pt—Cl = 2.34 Å and the Pt(2) atoms four ammonia molecules with Pt—N = 2.06 Å. The interplatinum distance is 3.25 Å. The infrared spectrum of MGS does not indicate any significant contribution from hydrogen bonding between the ammonia and chlorine ligands, so that the partial staggering is probably due to the packing requirements of the hydrogen atoms. Miller (262,263) has shown that Pt(CH$_3$NH$_2$)$_4$PtCl$_4$, (Me—MGS), has the stoichiometrically analogous structure with $a_0 = 10.35$ and $c_0 = 6.49$ Å. Pt(C$_2$H$_5$NH$_2$)$_4$-PtCl$_4$ does not, however, crystallize with a tetragonal structure, but this does not preclude a lattice in which Pt—Pt chains are present, the Pt—Pt distance being 3.40 $\pm$ 0.02 Å (337). The structure of Pt(CH$_3$NH$_2$)$_4$PtBr$_4$ is also tetragonal with $a_0 = 10.55$ and $c_0 = 6.61$ Å.

The observed bands are given in Table LXXVII and the polarized spectrum of MGS is shown in Figure 80 (85). (Ultraviolet bands are discussed below.)

Anex, Ross, and Hedgcook (3) have determined the specular reflection spectra of MGS Me—MGS, and Et—MGS by Kramers-Krönig analysis of the ‖ polarized crystals in the ultraviolet spectra; an absorption band was calculated to lie at 34,500 ($\epsilon = 58,000$); 34,400 ($\epsilon = 49,000$); and 39,900 cm^{-1} ($\epsilon = 70,000$) for the three compounds, respectively. The ⊥ spectra shows little absorption in this region. In the ‖ polarization, another apparent band occurred near 20,000 cm^{-1} whereas Day et al. (84) show only weak absorption in this region; in fact, Figure 80 shows a minimum at 20,000 cm^{-1}. This is interpreted as a spurious effect resulting from back-

TABLE LXXVII

Observed Absorption Band Maxima, (cm^{-1}) and Polarizations for MGS and MeMGS

| Assignments | | MGS, refs. 84,85 | | | | MGS, ref. 426 | | | | MeMGS, ref. 84 | | | |
(1)	(2)	∥	ε	⊥	ε	∥	α	⊥	α	∥	ε	⊥	ε
^3E_g	$^3B_{1g}$	7,500	1.75	7,500	0.6								
$^3B_{1g}$	^3E_g	16,500	150	16,500	20					17,300	100	17,300	35
		(23,000)sh	~190			22,700	37			(23,000)sh	~120		
^1E_g	$^1B_{1g}$	(24,900)	305	24,900	170			24,700	28			25,200	190

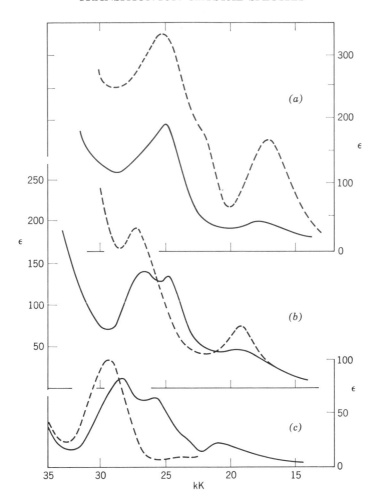

Fig. 80. The polarized absorption spectra at room temperature of single crystals of (a) $Pt(CH_3NH_2)_4PtCl_4$, (b) $Pt(C_2H_5NH_2)_4PtCl_4$, and (c) K_2PtCl_4. The dotted lines indicate spectra measured with the incident electric vector parallel to a line through the platinum atoms, i.e., the needle axis. The continuous lines are the spectra measured with the vector perpendicular to this direction. ϵ denotes extinction coefficients, in units liters, mole-cm^{-1}; 1 kK = 1000 cm^{-1} (85).

scattering of light near an intense absorption band, and similar effects have been observed elsewhere (1b,2). Day et al. (84) also observed a band at $\sim 34{,}500$ cm^{-1} in the diffuse reflectance spectra of both MGS and MeMGS.

Day et al. (84) assign the spectra of MGS and MeMGS on the same basis as that of K_2PtCl_4, (Table LXXV), the assignments *1* and *2* being related to *1* and *2* of Table LXXVII for the MGS salts. The transitions $^1A_{1g} \rightarrow {}^{1,3}A_{2g}$ are assumed to lie beneath the xy components of $^1A_{1g} \rightarrow {}^{1,3}E_g$, or $^{1,3}B_{2g}$.

According to these assignments, the $PtCl_4{}^{2-}$ anion excitation $^1A_{1g} \rightarrow$ ^1E_g or $^1B_{1g}$ requires about 5000 cm^{-1} less energy than in K_2PtCl_4, and diffuse reflectance data (84) also show that the third ligand field band has shifted appreciably to the red. Thus (apart from the new infrared band at 7500 cm^{-1} in MGS), the MGS and MeMGS spectra are attributed to transitions of a perturbed $PtCl_4{}^{2-}$ anion. Specific metal–metal interactions need not be involved to explain the observations. It is suggested that the allowed transition from which the ∥ polarized $d \rightarrow d$ transitions gain intensity through a vibronic mechanism is a $PtCl_4{}^{2-}$ charge-transfer excitation, which can plausibly be identified with the UV band observed by Anex et al.

This UV band can in turn be related to the intense band at 42,500 cm^{-1} in the diffuse reflectance spectrum of solid K_2PtCl_4 (84) and at 46,000 cm^{-1} for $PtCl_4{}^{2-}$ in water solution (67). The $PtCl_4{}^{2-}$ band has been interpreted by Gray and Ballhausen (152) as a $Cl^-(3p_z) \rightarrow Pt^{2+}(5d_{x^2-y^2})$ ligand-to-metal charge transfer; Anex et al. (3) consider an alternative assignment to a $Pt^{2+}5d_{z^2} \rightarrow 6p_z$ excitation. The changes of frequency and intensity of the $PtCl_4{}^{2-}$ ∥ absorption on changing from $K_2PtCl_4{}^{2-}$ to MGS and its analogs are related to variation of the ultraviolet transition by both Day et al. (84) and Anex et al. (3). Similar smaller band shifts in related crystals containing $PtCl_4{}^{2-}$ ions are discussed in detail in reference 84, and the intensity-gaining mechanism is also discussed in detail in reference 3. The effects being studied here are evidently of great importance for the understanding of the interplay between allowed and forbidden transitions in metal complexes.

The near-infrared band of MGS at 7500 cm^{-1}, ∥ polarized, is interpreted differently, as an electron-transfer from anion to cation. Three different models are discussed: (*1*) $5d_{z^2}(a_{1g})$ orbitals and empty $6p_z(a_{2u})$ orbitals of both cations and anions may interact separately to give molecular orbitals. (*2*) The $5d_{z^2}$ orbitals of $PtCl_4$ interact with the $6p_z$ orbitals of adjacent cations. (*3*) Extended interaction resulting from δ overlap of the filled $5d_{xy}$ orbitals of $PtCl_4$ and the empty $5d_{x^2-y^2}$ orbitals of the neighboring $Pt(NH_3)_4{}^{2+}$. No final assignment is suggested, however.

3. $Pt(NH_3)_4PtBr_4$

Yamada (426) has determined the dichroism of crystals of $Pt(NH_3)_4$-$PtBr_4$ and found bands at 20,600 (∥) and 21,500 cm^{-1} (⊥), the polarizations being quoted here with respect to the fourfold axis. He later (446) studied qualitatively the dichroism of the series of complexes [Pt(amine)$_4$] (PtX$_4$) where X = Cl$^-$, Br$^-$, I$^-$ to investigate the presence of a Pt—Pt bond.

4. Pt(DMG)$_2$

Basu, Cook, and Belford (28) have studied the absorption spectrum of single crystals of platinum dimethylglyoxime, Pt(DMG)$_2$, and in binary mixed crystals with Ni(DMG)$_2$ and Pd(DMG)$_2$. The absorption is similar to that in Pd(DMG)$_2$. Yamada and Tsuchida (432) observed a band at 20,000 cm^{-1} in the ∥ absorption and below 16,700 cm^{-1} in the ⊥ absorption. A direct interaction between the Pt atoms in the crystal lattice was postulated.

5. *Other Square–Planar* Pt *Complexes*

Yamada (425) has determined the dichroism of Pt(NH$_3$)$_4$Cl$_2$H$_2$O and K$_2$PtBr$_4$2H$_2$O. The spectrum of Pt(NH$_3$)$_4$Cl$_2$H$_2$O shows a featureless absorption edge at ∼34,000 cm^{-1} in both the ∥ and ⊥ polarizations. The polarized absorption spectrum of K$_2$Pt(COS)$_4$ has been determined by Kida, Nakamoto, and Tsuchida (199). Yamada and Tsuchida (430) have studied the dichroism of K$_2$Pt(SCN)$_4$. The observed absorption bands are given in Table LXXVIII for K$_2$PtBr$_4$2H$_2$O and K$_2$Pt(SCN)$_4$.

Yamada (424) has studied the dichroism of the salts BaPt(CN)$_4$4H$_2$O, CaPt(CN)$_4$5H$_2$O, and MgPt(CN)$_4$7H$_2$O. In these compounds the molecular planes are parallel to each other and perpendicular to the c axis. The interaction of the Pt ions was found stronger as the distance between the planes decreased in order of Ca, Ba, and Mg. This was accompanied by the main absorption band shifts towards longer wavelengths in the order Ca, Ba, Mg (ν_{max} = 22,900, 21,900, and 18,000 cm^{-1} for Ca, Ba, Mg, respectively). This suggests that the red shift is more pronounced as the lattice becomes more compact. These compounds have also been studied by

TABLE LXXVIII

Observed Absorption Bands and Polarizations with Respect to the Fourfold Axis for
K$_2$PtBr$_4$2H$_2$O and K$_2$Pt(SCN)$_4$ (α = extinction per mm)

Ref.	Compound	Band max., cm^{-1}	log α	Band max., cm^{-1}	log α
425	K$_2$PtBr$_4$2H$_2$O	18,300 ⊥	1.25	17,700 ∥	1.25
		24,300 ⊥	1.55	27,700 ∥	1.55
430	K$_2$Pt(SCN)$_4$	22,300 ⊥	1.31	23,700 ∥	1.25
		26,700	1.44	27,700 ∥	1.42

Brasseur (50) and Colin (70). The dichroism of Wolffram's Red salt has been studied by Yamada and Tsuchida (440).

XI. CONFIGURATION d^9

A. $3d^9$

In this configuration the octahedral field d levels are reversed compared with d^1 with the ^2E_g state lying lowest. This level is particularly susceptible to Jahn-Teller splitting. Orgel and Dunitz (292) pointed out that the copper(II) ion in crystals is rarely if ever found in a regular octahedral environment, and concluded that the Jahn-Teller effect is responsible for this. The usual distortion involves the two polar ligands moving away from the Cu^{2+} ion and leaving a near square-equatorial coordination. This is in agreement with crystal-field calculations by Belford (36) who studied the effects of ligand fields of varying symmetries around the Cu^{2+} ion. According to these results, the E_g state in a cubic field is usually split to such an extent that its upper component falls among the components of T_{2g}. Thus the ligand field is more nearly square planar than octahedral.

1. $CuSO_4 5H_2O$

Holmes and McClure (182) have studied the crystal spectrum of copper sulfate pentahydrate, $CuSO_4 5H_2O$. The detailed atomic arrangement in this crystal has been determined by Lipson and Beevers (35), while a more recent neutron diffraction study (8) has confirmed the earlier work and established the positions of the hydrogen atoms. The space group is $P\bar{1}$ C_i^1, and the bimolecular triclinic unit cell has dimensions $a_0 = 6.1130$, $b_0 = 10.7121$, $c_0 = 5.9576$ Å, $\alpha = 82° 17.7'$, $\beta = 107° 17.5'$, and $\gamma = 102° 34.3'$ (8). Each copper atom is surrounded by a near octahedron of oxygen atoms, four of which belong to water molecules and two to sulfate ions. This is a higher coordination than prevails in many copper complexes where it is usually fourfold and square. The field experienced by the ion has a strong tetragonal component along the axis joining the sulfate oxygens and smaller rhombic components in the equatorial plane.

A single band was observed at 13,000 cm^{-1}, $\epsilon = 5$, $\Delta\nu_{1/2} = 5300$ cm^{-1}, with f (oscillator strength) $= 1.4$. The doubly degenerate ground state ^2E_g in O_h and the triply degenerate excited state $^2T_{2g}$ in O_h should split into two and three components, respectively, in the field experienced by the Cu^{2+} ion in this lattice. Holmes and McClure (182) analyzed the 13,000 cm^{-1} band into three Gaussian curves centered at 10,500, 13,000, and

14,500 cm^{-1}, each having a half-width of 3000 cm^{-1}. There is a certain amount of disagreement in the 14,500 cm^{-1} region probably caused by uncertainties in the base line of zero absorption, but the resolution into two components seems definite. Bjerrum, Ballhausen, and Jørgensen have reported a similar resolution for the aqueous copper ion (40). The two upper bands are assigned to the close-lying pair of upper levels $^2E(^2T_{2g})$ in a tetragonal field and the 10,500 cm^{-1} band to a third component, predicting a large splitting of the ground state.

Roos (346a) has made a semiempirical open-shell SCF—MO calculation on $Cu(NH_3)_6{}^{2+}$ and $Cu(H_2O)_6{}^{2+}$. The tetragonal distortion from octahedral symmetry is discussed and compared to the experimental observations (40,182). Bjerrum et al. (40) resolved the spectrum of $Cu(NH_3)_6{}^{2+}$ into two bands at 11,700 and 15,600 cm^{-1}, identified (346a) with the A_{1g} $(3a_{1g} \rightarrow b_{2g}{}^*)$ and the almost degenerate $E_g(a_g \rightarrow b_{2g}{}^*)$, $B_{1g}(b_{1g} \rightarrow b_{2g}{}^*)$ levels, respectively. The calculations showed that configurational interaction is important for the A_{1g} state; the interaction between this state and the A_{1g} state $2a_{1g} \rightarrow b_{2g}$ is strong leading to a considerable decrease in energy for A_{1g}. Calculations for $Cu(H_2O)_6{}^{2+}$ show that the B_{1g} and E_g states are expected to absorb at 12,010 and 13,760 cm^{-1} when the A_{1g} state is fitted to the observed peak at 9400 cm^{-1} (40). The mean value of these energies is 12,890 cm^{-1} in agreement with the observed peak at 12,700 cm^{-1}. However, the tetragonality is unexpectedly slight. Support for this is found in the spectrum of $CuSO_45H_2O$ discussed by Holmes and McClure (182). Bands are calculated (346a) to occur at 10,500, 12,200, and 14,000 cm^{-1} for a Cu—O axial distance of 2.5 Å in fair agreement with the analyzed bands but again requiring a large tetragonal parameter.

2. Cu^{2+} in ZnO

The absorption spectrum of Cu^{2+} in ZnO has been determined at 77°K by Pappalardo (307) and at 4°K by Weakliem and McClure (412). The site symmetry of the Cu^{2+} ion in the wurtzite lattice has been discussed under d^7 configurations, p. 379. Pappalardo (307) showed that there is agreement between crystal field theory and experiment for Cu^{2+} in a (pseudo) tetrahedral site. The site symmetry of Cu^{2+} in yttrium garnet was inferred from the recorded spectra and showed that at least a sizeable number of Cu^{2+} ions are in tetrahedral sites. The energy level diagram for d^9 in a tetrahedral field including spin–orbit coupling is shown in Figure 81.

The lowest energy transition between the spin–orbit components of $^2T_2(E_{5/2}$ and $G)$ is expected to lie in the region 1000–1200 cm^{-1}. This energy is close to very strong absorption due to lattice vibrations in ZnO and

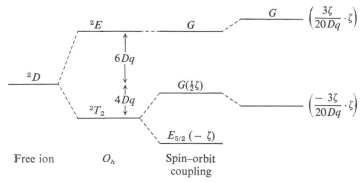

Fig. 81. Energy level diagram for the d^9 configuration in O_h symmetry including spin–orbit coupling (412).

hence has not been directly observed. The combined effect of the trigonal field and spin–orbit interaction gives rise to a splitting of the upper G level. Two sharp lines on the low-energy side of the spectrum (near 5800 cm^{-1}) are separated by 38 cm^{-1} and are tentatively assigned to the two trigonal components. The electric dipole selection rules are such that the transitions to both components are allowed in the $\perp$ polarization which was studied. The ground state is $E_{1/2}$ in C_{3v} and the trigonal field splits the upper G state into $E_{1/2} + E_{3/2}$. The $E_{1/2} \rightarrow E_{3/2}$ transition is predicted to be forbidden in the $\parallel$ polarization but this has not yet been verified. A pattern of lines between 5780 and 6000 cm^{-1} is repeated twice with a separation of 430 cm^{-1}, the second repeat being a broad structureless band. The interval of 430 cm^{-1} is close to the reported value of 414 cm^{-1} for the transverse optical mode frequency of ZnO (72). A series of several weak bands observed in the region 5860–6071 cm^{-1} remains unexplained.

3. $CuCl_4{}^{2-}$

The electronic energy levels of the $CuCl_4{}^{2-}$ complex ion have been much discussed (121,138,174,179,218,273,274,365,366,372). The initial definitive work was done by Helmholz and Kruh (179) who determined the crystal structure of Cs_2CuCl_4 and discussed the d-level ordering. Felsenfield (106) then showed that the observed departure from tetrahedral bond angles could be understood on the basis of an electrostatic model (symmetry D_{2d}). Lohr and Lipscomb (218) made an LCAO—MO calculation to estimate the relative ordering of the levels, while Furlani and Morpurgo (137) had interpreted their observations on the solution spectrum of $CuCl_4{}^{2-}$ and $CuBr_4{}^{2-}$ in terms of a ligand field model. Morosin and Lawson (273,274) measured the polarized crystal spectrum of Cs_2CuCl_4 and $(NH_4)_2CuCl_4$, and an extensive investigation on Cs_2CuCl_4 was carried out by Ferguson (121), whose observations agreed very well with the theoretical predictions of Lohr and Lipscomb (218). Hatfield and Piper (174) measured the optical spectra of a $Cs_2Zn(Cu)Cl_4$ crystal (and also reflectance and solution spectra of a series of 4, 5, and 6 coordinated chlorocuprate(II) complexes) and interpreted the results in terms of a crystal field model. Sharnoff and Reimann (366) determined the polarized ultraviolet absorption spectra of the tetrahedral $CuCl_4^{2-}$ ion oriented in single crystals of Cs_2CuCl_4 and $Cs_2Zn(Cu)Cl_4$. Their band assignments differ from those proposed by Ferguson (121) and will be discussed later. Furlani, Cervone, Calzona, and Baldanza (138) have determined the polarized absorption spectrum of bistrimethylbenzyl-ammonium tetrachlorocuprate(II), $(TMBA)_2CuCl_4$.

The crystal structure of Cs_2CuCl_4 was determined by Helmholz and Kruh (179) and was later refined by Morosin and Lingafelter (271). The crystal space group is $Pnma$, D_{2h}^{16}, with unit cell dimensions $a_0 = 9.719$, $b_0 = 7.658$, and $c_0 = 12.358$ Å with four molecules per unit cell. In the crystal the configuration found for the ion may be considered as a tetrahedron distorted by a flattening that removes the threefold symmetry of the tetrahedron but still preserves the fourfold inversion axis. The site symmetry of the Cu^{2+} ion is thus C_s by the requirements of the structure but the approximate site symmetry is D_{2d}. The z axis of $CuCl_4^{2-}$ is taken (121,366) as the approximate axis, of rotary reflection symmetry of the molecule. The x and y axes lie along the long edges, (3.80 Å), of the tetrahedron (179). The molecules are arranged with one long edge parallel to the b axis; this is defined as the y axis. The x and z axes of each molecule lie in the ac plane of the crystals.

Cs_2ZnCl_4 belongs to the same space group as Cs_2CuCl_4 (52) and has lattice constants $a_0 = 9.737$, $b_0 = 7.392$, and $c_0 = 12.972$ Å (52). Spectra were taken by Sharnoff and Reimann (366) containing 0.01–0.1% substituted $CuCl_4^{2-}$ and of thin layers of Cs_2CuCl_4 deposited on the Cs_2ZnCl_4 substrate. The influence of the environment on the shape of the complex ion is reflected in the g values (363) and in the $d \rightarrow d$ spectra (364) of dissolved $CuCl_4^{2-}$ as well as in the crystal spectra (366).

The x-ray structure of $(TMBA)_2CuCl_4$ was carried out by Bonamico, Dessy, and Vaciago (44). The compound has space group $P2_1/a, C_{2h}^5$, and the tetramolecular unit cell has dimensions $a_0 = 9.5844$, $b_0 = 9.104$, $c_0 = 28.434$ Å, and $\beta = 92° 50'$. The $CuCl_4^{2-}$ ion has a flattened tetrahedral structure of D_{2d} symmetry. It was further shown that the four S_4 axes of the four $CuCl_4^{2-}$ ions in the unit cell are practically parallel, because they lie almost parallel to the glide planes of symmetry. This results in clearer assignment of the spectral bands (138).

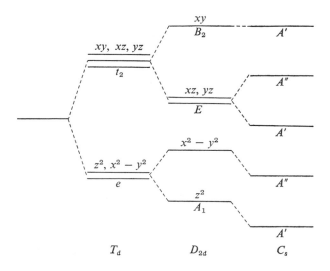

Fig. 82. Schematic d-orbital splitting in D_{2d} and C_s symmetry (after ref. 121).

The relatively weak copper $d \rightarrow d$ transitions occur in the near infrared region (3500–10,000 cm^{-1}) and are unlikely to be confused with the charge-transfer bands ($>20,000$ cm^{-1}). The splitting of the $3d$ orbitals under D_{2d} and C_s symmetries is shown in Figure 82. The observed (121, 138,174,273,274,365) bands and assignments for the near-infrared region are given in Table LXXIX.

TABLE LXXIX
Observed Bands and Assignments for $CuCl_4^{2-}$ in the Region 3500–10,000 cm^{-1}

Cs$_2$CuCl$_4$				Cs$_2$Zn(Cu)Cl$_4$		(TMBA)$_2$CuCl$_4$
Ref. 121			Refs. 173 and 274	Ref. 174	Ref. 365a	Ref. 138
	Assignment					
Band max., cm^{-1} (20°K)	D_{2d}	C_s	Band max., cm^{-1}	Band max., cm^{-1} (77°K)	Band max., cm^{-1} (77°K)	Band max., cm^{-1}
			~9500b			
9050	$^2B_2 \rightarrow {}^2A_1$	$^2A' \rightarrow {}^2A'$	6135	8300	8685 ± 15	~8850 ∥, z
7900	$^2B_2 \rightarrow {}^2B_1$	$^2A' \rightarrow {}^2A''$			7500 ± 100	
5550⎱ ~4800⎰	$^2B_2 \rightarrow {}^2E$	$\begin{cases} {}^2A' \rightarrow {}^2A' \\ {}^2A' \rightarrow {}^2A'' \end{cases}$	~3500		5210 ± 20 4450 ± 50	~5920 ⊥, xy

a Assignments same as in ref. 121 under D_{2d} (Γ_6 and Γ_7 E components 5210 and 4450).
b Interpolated from Fig. 2c of ref. 274.

The selection rules for allowed electric dipole transitions between orbitals transforming as the irreducible representations of the pseudo-group D_{2d} are given in Table LXXX.

Electronic transitions are polarized either along the z axis or in the xy plane of the group of operations D_{2d}, when spin–orbit interactions are neglected. In relating the selection rules for the ionic transitions to the axes of the crystal, use is made of the fact that the z axis of each $CuCl_4^{2-}$ ion lies in the ac plane of the host lattice, making angles of 55° 15′ and 34° 42′ with the a and c axes in the case of Cs_2CuCl_4 (179,271) and 50° and 40° in the case of Cs_2ZnCl_4 (52).

In the point group D_{2d}, only xy and z transitions are allowed (Table LXXX), i.e., from a B_2 ground state to E or A_1 excited states, transitions to A_2 and B_1 being electric dipole forbidden. The relative ordering of the $3d$ energy levels in $CuCl_4^{2-}$ is now most generally considered to be that of Figure 82. Purely electrostatic calculations (106,137,273) place $x^2 - y^2$ above xz,yz rather than below in contrast with the results of single-crystal polarized spectra by Ferguson (121) and by Karipides and Piper for $CuBr_4^{2-}$ (195). The system depicted in Figure 82 is, however, supported by an improved electrostatic model (138) and by MO calculations of Lohr and Lipscomb (218). On the other hand Morosin and Lawson (273) present a rather different assignment with $d_{x^2-y^2}$ above d_{xz}, d_{yz} and also with the peak at ~ 9500 cm^{-1} not assigned to a $d \rightarrow d$ transition.

The measurements of Ferguson (121) at 20°K reveal the presence of a band at 7900 cm^{-1}. If the field were exactly D_{2d} and spin–orbit coupling was zero, only two bands would be observed $^2B_2 \rightarrow {}^2E$, and $^2B_2 \rightarrow {}^2A_1$. This does not yet take into account spin–orbit coupling. However, $\zeta = 829$ cm^{-1} for the free Cu^{2+} ion so that spin–orbit effects cannot be neglected. Ferguson (121) using the theory of Liehr (211) assigned the

TABLE LXXX

Selection Rules for Allowed Electric Dipole Transitions for the Group D_{2d} [a] (121)

	A_1	A_2	B_1	B_2	E	Γ_6	Γ_7
A_1	—	—	—	$z(a)$	$xy(ab)$	—	—
A_2	—	—	$z(a)$	—	$xy(ab)$	—	—
B_1	—	$z(a)$	—	—	$xy(ab)$	—	—
B_2	$z(a)$	—	—	—	$xy(ab)$	—	—
E	$xy(ab)$	$xy(ab)$	$xy(ab)$	$xy(ab)$	—	—	—
Γ_6	—	—	—	—	—	$xy(ab)$	$xyz(aba)$
Γ_7	—	—	—	—	—	$xyz(aba)$	$xy(ab)$

[a] x,y,z refer to the group D_{2d}, and a,b to the corresponding crystal directions. Γ_6 and Γ_7 are the spin–orbit representations.

transitions as shown in Table LXXIX as a combination of spin–orbit coupling and the C_s component of the ligand field. The forbidden transition $^2B_2 \rightarrow {}^2B_1$ appears as a result of the combined effects.

Detailed consideration of the observed polarizations shows that the assignments in Table LXXIX are consistent with experiment. Sharnoff and Reimann (365) obtain very similar data in this region for $Cs_2Zn(Cu)Cl_4$ (77°K), and make the same assignments as Ferguson (121). The average decrease of band maxima in the Zn diluted crystal is ~ 360 cm^{-1}. A marked sharpening of the bands on cooling from room temperature was observed in references 121 and 365 but not in ref. 273.

Sharnoff and Reimann (365), in an effort to determine how much of the distortion of the tetrahedral complex $CuCl_4{}^{2-}$ may be ascribed to an intrinsic Jahn-Teller effect rather than environmental distortion, studied the changes in optical and ESR spectra which occur when the ion is embedded in several crystal lattices.

Large differences between the g tensors of $CuCl_4{}^{2-}$ in Cs_2CuCl_4 and $Cs_2Zn(Cu)Cl_4$ are found, in spite of the similar (3% difference) lattice constants. These are interpreted as resulting from a change in the amount of ligand character in the e orbitals, the composition of the t_2 orbitals remaining almost unchanged. This is consistent with the optical data. In spite of this, the low symmetry of the ion (C_s) is considered to result essentially from an intrinsic Jahn-Teller effect.

The observed electron transfer bands and their polarization are given in Table LXXXI while the spectrum of Cs_2CuCl_4 obtained with light incident upon the (001) face of the Cs_2ZnCl_4 substrate crystal is shown in Figure 83.

TABLE LXXXI

Electron Transfer Bands and Assignments for $CuCl_4{}^{2-}$ in Cs_2CuCl_4 and Cs_2ZnCl_4

Cs$_2$CuCl$_4$ at 77°K					Cs$_2$ZnCl$_4$	(TMBA)$_2$CuCl$_4$
Ref. 366, band max., cm^{-1}	Polarization ref. to crystal axes	Assignment	Ref. 121, band max., cm^{-1}	Ref. 121, assignment	Ref. 366, band max., cm^{-1}	Ref. 138, band max., cm^{-1}
$22{,}700 \pm 200$	$E \parallel b,(a)$		$23{,}000$	$^2B_2 \rightarrow {}^2A_2$	$22{,}800 \pm 200$	$22{,}700$
$24{,}600 \pm 30$	$E \parallel c,a$	$^2B_2 \rightarrow {}^2E$	$24{,}800$	$^2B_2 \rightarrow {}^2E$	$24{,}570 \pm 30$	$\sim 25{,}000$
$24{,}850 \pm 30$	$E \parallel b$	$^2B_2 \rightarrow {}^2E$			$24{,}940 \pm 30$	
$28{,}400 \pm 200$	$E \parallel b$	$^2B_2 \rightarrow {}^2E$	$\sim 29{,}000$	$^2B_2 \rightarrow {}^2E$	$28{,}600 \pm 200$	
$29{,}250 \pm 200$	$E \parallel c,a$	$^2B_2 \rightarrow {}^2E$			$29{,}400 \pm 200$	
$33{,}060 \pm 40$	$E \parallel b$	$^2B_2 \rightarrow {}^2E$	$34{,}000$	$^2B_2 \rightarrow {}^2E$	$33{,}100 \pm 40$	
$33{,}900 \pm 40$	$E \parallel c,a$	$^2B_2 \rightarrow {}^2E$	$34{,}500$	$^2B_2 \rightarrow {}^2A_1$	$33{,}750 \pm 40$	
$43{,}000 \pm 100$	$E \parallel a,c$	$^2B_2 \rightarrow {}^2A_1$	$42{,}400$	$^2B_2 \rightarrow {}^2A_1$	$42{,}900 \pm 100$	

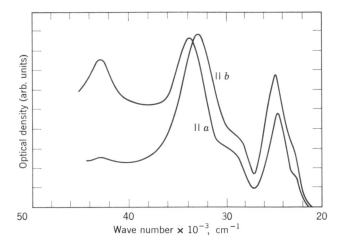

Fig. 83. Polarized ultraviolet absorption spectrum at liquid air temperature of Cs_2CuCl_4 obtained with light incident on the (001) face of the substrate Cs_2ZnCl_4 crystal. The notations on the curves refer to the orientation of the electric vector of the incident radiation with respect to a and b crystal axes (366).

In the region between 20,000 and 27,000 cm^{-1}, the Cs_2CuCl_4 spectra are in agreement in the two investigations (121,366) but the twin peaks near 34,000 cm^{-1} lie nearly 800 cm^{-1} lower in the spectra of Sharnoff and Reimann (366) than in those of Ferguson (121). Sharnoff and Reimann attribute these differences to the presence of a small amount of impurity of $CsCuCl_3$ (red) in the orange crystals of Cs_2CuCl_4 used by Ferguson in his determination. Sharnoff and Reimann also consider their results to represent the Cs_2CuCl_4 charge-transfer spectra more closely than Ferguson's because they agree very closely in energy with the charge-transfer spectra of $CuCl_4^{2-}$ in Cs_2ZnCl_4, Table LXXXI, and because they exhibit bands which are generally narrower and better resolved than Ferguson's.

The relative intensities of the peaks at 24,600, 29,400, and 33,750 cm^{-1} are very nearly the same in c and a polarizations. The intensity of the peak at 42,900 cm^{-1} is about 50% higher than these three in both polarizations, indicating that it is polarized along a different ionic axis: the z axis. According to the D_{2d} selection rules, Table LXXX, this transition is assigned to $^2B_2 \rightarrow {}^2A_1$. The rest of the spectrum cannot be interpreted on the basis of Table LXXX. Neither of the two peaks near 33,500 m^{-1} is z polarized nor isotropically polarized within ionic xy planes, as the selection rules for D_{2d} require. The 33,100 cm^{-1} peak is polarized only along the y axis and the 33,750 cm^{-1} component is polarized along the z axis.

The degeneracy of the E level of the $CuCl_4{}^{2-}$ ion is removed under the action of spin–orbit interaction and of a rhombic component in the coulomb Hamiltonian of the ion. Calculations on these effects show that charge-transfer transitions of the type $^2B_2 \rightarrow {}^2E$ in D_{2d} will occur in pairs containing two oppositely polarized bands of roughly equal oscillator strength lying as much as 1000 cm^{-1} apart. In Cs_2CuCl_4, the band pairs lying at 33,060 and 33,900 cm^{-1} and at 24,600 and 24,850 cm^{-1} and the band shoulders lying at 28,400 and 29,250 cm^{-1} are tentatively assigned to this transition. The earlier assignment of Ferguson (121) of the 33,800 cm^{-1} band to $^2B_2 \rightarrow {}^2A_1$ is unlikely as the transition is not z polarized. The MO interpretation of the transitions is discussed in both (121) and (366). Spectral data for the electron-transfer bands of Cs_2CuCl_4 are also given in references 273 and 274; no interpretation is suggested, however.

Willett, Liles, and Michelson (418f) have reexamined the electronic absorption spectrum of the monomeric copper(II) chloride species in the light of further crystal spectra and ESR studies. The compounds included in the study were $(CH_3NH_3)_2CuCl_4$, $(C_2H_5NH_3)_2\,CuCl_4$, $[(CH_3)_2NH_2]_2$-$CuCl_4$, $[(CH_3)_2CHNH_3]_2CuCl_4$, and $[(CH_3)_4N]_2CuCl_4$.

The compounds $[(CH_3)_4N]_2CuCl_4$ and $[(CH_3)_2NH_2]_2CuCl_4$ are assumed (418f) to have a distorted tetrahedral anion as in Cs_2CuCl_4 from the relative positioning of the large band centered in the 6000–7000 cm^{-1} region. Table LXXXIA gives the observed band maxima in these com-

TABLE LXXXIA

Band Maxima (cm^{-1}) in the Spectra of the Tetrahedral $CuCl_4{}^{2-}$ Species (418f)[a]

		Cs_2CuCl_4[b]	$[(CH_3)_4N]_2CuCl_4$	$[(CH_3)_2NH_2]CuCl_4$
Charge transfer				
	$a_1 \rightarrow b_2$	42,400	43,500 sm	44,500 sm
σ	$a_1(\pi) \rightarrow b_2$	34,500		
	$e(\sigma,\pi) \rightarrow b_2$	34,000	36,000 b	34,500 b
	$e(\pi,\sigma) \rightarrow b_2$	29,000		
π	$e(N) \rightarrow b_2$	24,800		
	$a_2(N) \rightarrow b_2$	23,000	25,000 sm	22,000 sm
$d \rightarrow d$ transitions				
$d_{z^2} \rightarrow d_{xy}$		9,050	9,000 b	9,000 b
$d_{x^2-y^2} \rightarrow d_{xy}$		7,900		
$d_{xz} \rightarrow d_{xy}$		5,550	6,000 b	6,000 b
$d_{yz} \rightarrow d_{xy}$		~4,800		

[a] sm indicates a sharp distinct maximum, b indicates a broad band—probably containing more than one transition; the number given is the maximum point and not necessarily the center of the band; sh indicates a shoulder.

[b] Ref. 121.

pounds, together with the results of Ferguson (121) for Cs_2CuCl_4 for comparison.

The existence of a square–planar $CuCl_4^{2-}$ ion has been shown in the yellow complexes $(NH_4)_2CuCl_4$, $(C_2H_5NH_3)_2CuCl_4$, and $[(CH_3)_2CHNH_3]_2$ $CuCl_4$ (374b,418c). These compounds are thermochromic, changing to a pale green at low temperature. The compounds crystallize as flat, yellow sheets above the transition temperature and as long, light-green needles in the low-temperature phase (418f). The transition temperature is near $-40°C$ for $(CH_3NH_3)_2CuCl_4$ and $(C_2H_5NH_3)_2CuCl_4$ although covering a temperature range of 50°C in the latter compound (374b), but occurs at about $+60°C$ for $[(CH_3)_2CHNH_3]_2CuCl_4$ (345b). The structural properties of both phases of all the compounds are not known in detail, but the crystal structure of $(NH_4)_2CuCl_4$ at room temperature has been determined by Willett (418c), and it is thought that $(CH_3NH_3)_2CuCl_4$ and $(C_2H_5NH_2)_2$-$CuCl_4$ probably have related structures. $(NH_4)_2CuCl_4$ crystallizes in an orthorhombic unit cell, space group $Cmca$, D_{2h}^{18}, with parameters $a_0 = 15.46$, $b_0 = 7.20$, and $c_0 = 7.20$ Å. The structure contains discrete planar $CuCl_4^{2-}$ ions with Cu—Cl distances of 2.30 and 2.33 Å. These ions are then bonded together by larger Cu—Cl bonds of 2.79 Å [2.98 Å in $(C_2H_5NH_3)_2CuCl_4$ (374b)] to form two-dimensional sheets. This gives each Cu atom a pseudo-octahedral coordination but the 2.79 Å bonds are so long that no covalent bonding can be considered to be present. Below the transition temperature it is proposed that for $2/3$ of the $CuCl_4^{2-}$ ions, the phase change rearranges the immediate neighbors so that there is only one other $CuCl_4^{2-}$ ion in contact with it. This gives these copper

TABLE LXXXIB

Band Positions (cm^{-1}) of the Square–Planar $CuCl_4^{2-}$ Species at Room Temperature (418f)

	$Pt(NH_3)_4CuCl_4^a$	$(CH_3NH_3)_2CuCl_4$	$(C_2H_5NH_3)_2CuCl_4$
Charge transfer			
$\sigma \rightarrow d_{x^2-y^2}$		49,000 sm	49,000 sm
		38,500 b, sh	38,500 b, sh
$\pi \rightarrow d_{x^2-y^2}$		33,300 b	33,300 b
$\pi(N) \rightarrow d_{x^2-y^2}$	24,900	24,000 sm	24,000 sm
$d \rightarrow d$ transitions			
$d_{xz,yz} \rightarrow d_{x^2-y^2}$	14,300		12,800 b
$d_{xy} \rightarrow d_{x^2-y^2}$	13,100	13,000 b	
$d_{z^2} \rightarrow d_{x^2-y^2}$	10,900	10,700 sh	10,500 sh

^{a} Values quoted from ref. 174. As pointed out by W. E. Hatfield (quoted in reference 418f), the order reported in reference 174 contained a typographical error.

TABLE LXXXIC
Band Positions of the Low-Temperature Thermochromic Form (418f)

	$[(CH_3)_2CHNH_3]_2CuCl_4$	$(CH_3NH_3)_2CuCl_4$	$(C_2H_5NH_3)_2CuCl_4$
	48,000 sm	50,000 sm	51,000 sm
σ transition	44,500 sh	41,500 b, sh	41,500 b, sh
	33,300 b	30,000 b	29,500 b
π transition	24,500 sm	24,500 sm	24,500 sm
$d \rightarrow d$ transition	13,000 b	13,100 b	13,100 b

atoms a square–pyramidal configuration but again with one atom outside the coordination sphere.

The observed band positions and assignments are given in Table LXXXIB for the square–planar $CuCl_4{}^{2-}$ species at room temperature and in Table LXXXIC for the low-temperature thermochromic form.

The assignments of the more intense charge transfer bands were made by analogy to the spectrum of the distorted tetrahedral $CuCl_4{}^{2-}$ ion. The only major difference between the spectra of the tetrahedral and square–planar species is the position of the $d \rightarrow d$ band. As expected, this band appears at lower energy (6000–9000 cm^{-1}) in the tetrahedral species than in the square–planar species (10,000–13,000 cm^{-1}). In general, there are very slight changes in the positions of the bands between the spectra of the two phases of the square–planar species. In particular, there is very little variation in the positions of the $d \rightarrow d$ transition, a small shift to higher energy. Further studies of these properties are clearly required before definitive assignments can be made.

The electronic absorption spectrum and bonding in the $Cu_2Cl_6{}^{2-}$ dimer has been studied by Willett and Liles (418e). The compounds studied included $KCuCl_3$ (418b), $LiCuCl_3 2H_2O$ (409a), $(CH_3)_2NH_2CuCl_3$ (418d), and $(CH_3)_2CHNH_3CuCl_3$ (345a). The crystal structures of some of these compounds are known (409a,418b,418d), but no detailed polarized measurements were made and so will not be discussed in detail. The spectra closely resemble the spectra obtained for the square–planar $CuCl_4{}^{2-}$ ion. With the exception of an extra band at $\sim 19,000$ cm^{-1} characteristic of the dimer species, very little difference is noted. This extra band has been observed in other bibridged copper(II) dimers (418g) and is used as a criterion for the identification of dimeric species.

4. $CuBr_4{}^{2-}$

The polarized crystal spectrum of $CuBr_4{}^{2-}$ in Cs_2CuBr_4 and $Cs_2Zn(Cu)$-Br_4 has been determined by Karipides and Piper (195) and by Morosin and

Lawson (273). (No difference was observed in reference 195 between the spectra of $Cs_2Zn(Cu)Cl_4$ mixed crystals and Cs_2CuBr_4 plated on the Zn substrate.) The crystal structures of Cs_2CuBr_4 and Cs_2ZnBr_4 have been determined by Morosin and Lingafelter (269,270). They are isostructural with Cs_2CuCl_4 and have the lattice constants: Cs_2CuBr_4; $a_0 = 10.2$, $b_0 = 7.97$, $c_0 = 12.94$ Å; and Cs_2ZnBr_4; $a_0 = 10.196$, $b_0 = 7.770$, $c_0 = 13.517$ Å. In Cs_2CuBr_4, the Cu—Br bond distances are nearly all equal. In the flattened tetrahedron, the two obtuse angles are 130° 24′, and 126° 24′ from which the average polar angle is 64°; in a tetrahedron $\alpha = 54°$ 42′ and in Cs_2ZnBr_4 α is 56° (269). The observed bands and assignments are tabulated in Table LXXXII. In the work of Karipides and Piper (195) light was incident on the (001) face and polarized ∥ and ⊥ to the plane of symmetry.

The assignments of Morosin and Lawson (273) reverse the ordering of the $3d$ energy levels given previously by Karipides and Piper (195) whose energy level diagram indicates the $^2B_1$ state higher than ^2E. Morosin and Lawson state that the $CuBr_4{}^{2-}$ ion approaches the square–planar configuration and the $d_{x^2-y^2}$ energy level thus lies above that of the d_{xz},d_{yz} pair (contrary to the scheme of Fig. 82) while their results are in agreement with the calculated levels of Felsenfield (106). Unfortunately the complete low-temperature infrared spectrum is not available, so that these assignments cannot yet be considered in any detail.

Braterman (51) determined the visible and near-ultraviolet spectrum of $CuBr_4{}^{2-}$ in $[(n\text{-}C_4H_9)_4N]_2CuBr_4$. It is again assumed that there is an electron hole in the xy orbital in the ground state. The bands in the visible were assigned to forbidden or weakly allowed charge transfer and the ultraviolet bands to allowed charge transfer. Calculations on the allowed $d \rightarrow d$ transitions on both models with either B_1 or E lying lower favored the latter arrangement in agreement with the results of Karipides and Piper (195).

TABLE LXXXII

Observed Bands and Proposed Assignments for $CuBr_4{}^{2-}$ in Cs_2CuBr_4 and $Cs_2Zn(Cu)Br_4$

Cs_2CuBr_4, ref. 195			Ref. 273	
	Assignment			Assignment
Band max., cm^{-1} (77°K)	Cs	D_{2d}	Band max., cm^{-1}	D_{2d}
8000 ∥	$\epsilon_\parallel/\epsilon_\perp = 4.6$	$A' \rightarrow A'$ $\Big\}$ $B_2 \rightarrow E$	8475	$^2B_2 \rightarrow A_1$
7570 ⊥		$A_1 \rightarrow A''$	5404	$^2B_2 \rightarrow {}^2E$
			4545	$^2B_2 \rightarrow {}^2B_1$

5. $Cu_2(C_2H_3O_2)_4 2H_2O$

The polarized crystal spectrum of copper(II) acetate monohydrate, $Cu_2(C_2H_3O_2)_4 2H_2O$, was first determined by Yamada, Nakahara, and Tsuchida (441) and was reinvestigated at room temperature and at 77°K by Tonnet, Yamada, and Ross (393). An additional band was observed in the spectrum at 77°K by Reimann, Kokoszka, and Gordon (345).

Copper(II) acetate has the same binuclear structure as chromium(II) acetate, Figure 25. The crystal structure has been determined by Van Niekerk and Shoening (405). The crystal has a space group $C2/c, C_{2h}^6$ with four dimer molecules in the monoclinic unit cell of dimensions $a_0 = 13.15$, $b_0 = 8.52$, $c_0 = 13.90$ Å, and $\beta = 117° 0'$. Each copper atom has another copper atom at 2.64 Å, a water molecule at 2.20 Å, and four carboxy oxygens ~ 1.97 Å away as its nearest neighbors. The dimer shows a strongly temperature-dependent magnetism, being paramagnetic at room temperature but diamagnetic below 20°K, indicating that spin pairing occurs between the copper atoms (41). Bleaney and Bowers (41) explained that this was due to a low-lying triplet state of the molecule, which is partly populated at room temperature. The singlet–triplet splitting is found to be near 300 cm^{-1}.

The observed absorption bands are listed in Table LXXXIII for the three independent investigations, and are shown in Figure 84.

The data in reference 393 are a reinvestigation of that of reference 441; the later work of Reimann, Kokoszka, and Gordon (345) established the presence of a new ‖ band at 11,000 cm^{-1}. The qualitative features are thus

TABLE LXXXIII
Observed Absorption Bands in $Cu_2(C_2H_3O_2)_4 2H_2O$[a]

Face	Polarization direction [b]	Ref. 393, band max., cm^{-1} (77°K)	Ref. 345, band max., cm^{-1} (77°K)	Ref. 441, band max., cm^{-1}
(110)			11,000 ‖	
(110)	Green ‖	14,400	80 $\sim$14,400 ⊥	
	Blue ⊥	14,400	180	
(20$\bar{1}$)	Green ‖	14,700	140	
	Blue ⊥	14,300	190	14,300 ⊥
(110)	Green ‖	27,500	$\sim$30[c] 27,000 ‖	
	Blue ⊥	29,000	$\sim$20[c]	
(20$\bar{1}$)	Green ‖	27,500	$\sim$20[c]	26,700 ‖
	Blue ⊥	Absent	$\sim$0[c]	

[a] Spectra in solutions in which the complex remains dimeric show an additional strong band ($\epsilon > 10^3$) at 40,000 cm^{-1} (393).

[b] Polarization ‖ and ⊥ refer to z and xy directions, respectively.

[c] After subtracting an estimate of the background absorption.

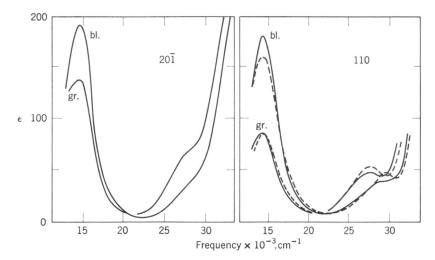

Fig. 84. Absorption spectra of crystalline copper acetate monohydrate. (————)
room temperature; (– – – –), 77°K. bl., blue polarization direction; gr., green
polarization direction. For the (201) face the absorption continues towards higher
frequencies without change of slope to $\epsilon = 500$, the limit of the measurements (393).

fairly weak bands at $\sim 11,000$, $\sim 14,000$, and $\sim 28,000$ (particularly low
intensity) and a very strong absorption at 40,000 cm^{-1}. The band near
28,000 does *not* occur in monomeric Cu(II) analogs and is considered to
be a consequence of the dimeric structure. Three models of the Cu(II)....
Cu(II) interaction have been proposed. Figgis and Martin (125) proposed
that this weak bond was a δ bond formed by overlap of electrons in the
half-filled $3d_{x^2 - y^2}$ orbitals on each atom. Ross et al., in a series of
publications (347,348,393), have supported this proposal by energy level
calculations and interpretation of the ESR splitting parameters. However,
Forster and Ballhausen (131) on the basis of MO calculations, proposed
a σ copper–copper bond. Further one-electron MO calculations by
Boudreaux (47) gave results essentially analogous to those of reference 131,
although it is suggested that a more detailed calculation would favor a
δ Cu····Cu bond. [Different *d*-level orderings are assumed in (47) and
(131).] Hansen and Ballhausen (162) point out that a more appropriate
theoretical approach may be one using the weakly coupled chromophore
model used for interacting conjugated π systems (220); this method allows
a maximum use of empirical data. Hansen and Ballhausen propose that
the singlet character of the ground state results from greater stabilization
of the singlet relative to the triplet by configuration interaction. (Experimentally the energy difference is 300 cm^{-1}, which is too small to calculate

with any accuracy.) The excited states are divided into (*1*) locally excited (13,000–15,000 cm^{-1}), (*2*) charge transfer (90,000 and 100,000 cm^{-1}), and (*3*) doubly excited states (23,000–25,000 cm^{-1}), with the three types being so separated energetically that they will not interact appreciably. The experimental band at 14,000 cm^{-1} is then assigned to a transition of type *1*. The 27,000 cm^{-1} has ∥ polarization which could correspond to an allowed charge-transfer transition but the observed energy is far too low. The absence of the band is monomeric copper complexes makes an assignment as a copper ligand transition unlikely, as also recognized by Tonnet et al. (393). This transition is assigned by Hansen and Ballhausen as one to a doubly excited state (i.e., type *3*). This transition can gain intensity via configuration interaction and vibronic coupling. The third band near 40,000 cm^{-1} observed in solutions where the dimeric structure of the complex is retained is presumed to be associated with the ligands. The ESR data are also interpreted in this scheme.

Reimann, Kokoszka and Gordon (345) have considered the various possible one-electron transition assignments for the 14,000 and 27,000 cm^{-1} bands. The band at 11,000 cm^{-1} observed as a shoulder on the side of the 14,000 cm^{-1} band is assigned by them either to a $d_{x^2-y^2} \rightarrow d_{z^2}$ transition or to a $d_{x^2-y^2} \rightarrow d_{xy}$ transition.

6. [C$_5$H$_5$NO)$_2$CuCl$_2$]$_2$

In order to obtain additional experimental data on the nature of the metal–metal interaction in binuclear copper complexes, a study of the magnetic and optical spectra of [(C$_5$H$_5$NO)$_2$CuCl$_2$]$_2$, Cu(pno)$_2$, and [(C$_5$H$_5$NO)CuCl$_2$H$_2$O]$_2$, Cu(pno), has been made by Kokoszka, Allen, and Gordon (204a). Cu(pno)$_2$ crystallizes (354a) in a monoclinic space group $P2_1/b$, C_{2h}^5 with four copper atoms per unit cell of dimensions $a_0 = 5.844$, $b_0 = 10.049$, $c_0 = 13.643$ Å, and $\gamma = 104° 52'$. The copper atom is in a distorted tetrahedral environment. The x-ray crystal structure of Cu(pno) is not available. Thin crystals of Cu(pno)$_2$ with a large (001) face were studied using polarized light. Any pronounced polarization effects should have been observed in this plane as the z axes are only about 20° out of the (001) face. The effects were found to be relatively slight, and considered to be due to low effective symmetry at the copper ion site. Three bands were observed at 8750; 11,750, and 21,000 cm^{-1} with extinction coefficients 50, 100, and 25, respectively. In Cu(pno), the main band at 12,300 cm^{-1}, $\epsilon = 200$ severely masked the other two bands at 9500 and 22,000 cm^{-1} with extinction coefficients of ~ 50. The main band is assigned to the transition $d_{x^2-y^2} \rightarrow d_{xz,yz}$.

7. Cu(acac)$_2$

The polarized crystal spectrum of copper bisacetylacetone, Cu(acac)$_2$, was first reported by Yamada and Tsuchida (439) and subsequently redetermined by Ferguson (112). Ferguson's results were later reinterpreted by Piper and Belford (326)* in the light of an accurate crystal structure determination by Dahl (82). Ferguson, Belford, and Piper (115) then presented further data where better resolution was obtained in the spectrum. Subsequently, a series of papers appeared concerning the possible exciton interpretation of the crystal spectrum (37,90,122). Allen (1) reinterpreted the spectrum by describing the ligand field symmetry by the molecular point group C_{2h}. Comparison was made with the work of Reimann, Kokoszka, and Allen (345a) in other copper(II) chelates. La Mar (206a) has performed a Wolfsberg-Helmholz MO calculation on Cu(acac)$_2$, but the results were not considered sufficiently reliable to make definitive band assignments.

Cu(acac)$_2$ crystallizes in the monoclinic system, space group $P2_1/n$, C_{2h}^5, with two molecules per unit cell and lattice constants $a_0 = 10.30$, $b_0 = 4.71$, $c_0 = 11.34$ Å, and $\beta = 92°\ 12'$ (82). The molecule is not completely planar, the carbon atoms lying slightly above the copper–oxygen framework but is symmetrical within experimental error as required by enolate resonance. The Cu—O distance is 1.92 Å and the angle O_1—$\hat{Cu}$—O_2 is $93°\ 30'$. The copper atoms lie at centers of inversion. The x molecular axis is taken as a line passing through the copper atom and parallel to the O_1—O_2 internuclear distance, this axis being inclined to its projection on the ac plane by only $42'$. The y axis is taken as the axis passing through the copper atom and the midpoint of the O_1—O_2 distance. The y axis is inclined at $44°\ 54'$ to its projection on (010) and hence y and z axes make the same angle to the b crystallographic axis. The angle of $\sim 45°$ is in excellent agreement with the ESR results of Maki and McGarvey (242).

The observed absorption bands and polarizations are listed in Table LXXXIV and shown in Figure 85 (115).

The unusual molecular orientation in this crystal makes the analysis of the spectrum more difficult than for many different metal chelates, and definitive assignments have not yet been made. Assuming a vibronic mechanism, independent knowledge of the symmetries of the most effective enabling vibrations (for example, from a general comparison of analogous spectra) seems to be necessary in this case to arrive at definite

* The z (∥) column in the transformation table (iv) in ref. 326 is incorrect, and the (∥) in the α_{2u} column should be moved to the β_{1u} column. (Private communication from Dr. R. Linn Belford.)

TABLE LXXXIV
Observed Crystal Absorption Bands in Cu(acac)$_2$

Refs. 326,115, 30°K		Ref. 90, 300°K	
Band max., cm^{-1}[a]	Polarization[b]	Band max., cm^{-1}	Polarization
26,000	x; y and z		
18,000	y and z	$\begin{cases} 18,500 \\ 18,000 \end{cases}$	$\perp b$ $\| b$
16,300	x; y and z	16,100	$\perp b$
14,500	y and z	15,600	$\| b$

[a] The results in refs. 326 and 115 were obtained by analysis of spectra $\| b$ and $\perp b$.

[b] See text for relation of x, y, and z molecular axes to crystal axis b.

predictions (115,326). Possible assignment schemes are discussed in references 1, 112, 115, and 326. An alternative suggestion by Dijkgraaf (90) is that intercomplex resonance interaction is important, leading to exciton excited states, rather than the single-molecule "oriented-gas" states usually assumed. This is also thought to lead to Davydov-type splitting of the bands. However, Ferguson (122) and Belford and Belford (37) have shown that the observed band separations and polarizations are not consistent with Davydov splitting. The importance of resonance interaction could be checked experimentally by using a diluted crystal. This has not yet been reported.

In assigning the transitions for D_{2h} symmetry (or pseudosymmetry), care has to be taken to label the axes consistently, as the choice is arbitrary. Figure 86 shows the two conventions that have been employed. In scheme a, the x and y (or y and x) axes are coincident with molecular twofold

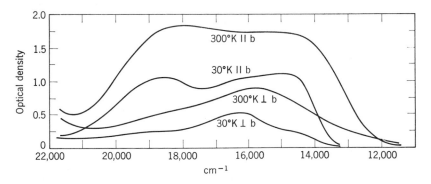

Fig. 85. The absorption spectra with polarized light incident on the (101) face of copper acetylacetonate, for a crystal ca. 1 mm thick (115).

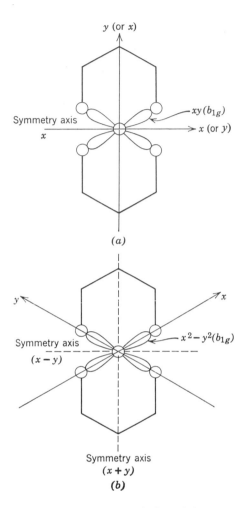

Fig. 86. Two conventions used in the labeling of the symmetry axes in copper bisacetylacetone. (*a*) According to Piper and Belford, *Mol. Phys.*, *5*, 169 (1962), ref. 326 and (*b*) after Dijkgraaf, *Theoret. Chim. Acta*, *3*, 38 (1965), ref. 90. (After figure kindly provided by Dr. R. Linn Belford.)

axes whereas in scheme *b*, they are rotated by $\pi/4$ with respect to these axes. The *z* axis in both cases is normal to the plane of chelate rings. Thus, in *a*, the *x*,*y* axes are directed at 45° to the nearest Cu ligands, whereas in *b*, they point toward the nearest ligands. Convention *a* is used by Piper and Belford (326), and by Allen (1). Convention *b* is used by Dijkgraaf (90) [and also for D_{4h} by Ballhausen and Gray (152)]. This has led to some confusion in Dijkgraaf's work, since the highest-energy orbital pointing to the ligands is labeled a_g, whereas it should be b_{1g}, because it changes sign on twofold rotation (37).

Gersmann and Swalen (141) have determined the ESR spectrum of Cu(acac)$_2$ in the vitreous state and found $g_{\parallel} = 2.264$, $g_{\perp} = 2.036$, in fair agreement with the values of $g_{\parallel} = 2.266$ and $g_{\perp} = 2.053$ reported by Maki and McGarvey (242).

8. Cu[3φ(acac)]$_2$

Belford and Carmichael (37a) have studied in detail the polarized crystal spectrum of bis(3-phenyl-2,4-pentanedionato) copper(II), Cu(3φ-acac)$_2$ following the earlier work of Basu, Belford, and Dickerson (26a). The crystal structure has been determined by Carmichael, Steinrauf, and Belford (58d). The crystal belongs to the monoclinic space group $P2_1/c$ with two molecules per unit cell and parameters $a_0 = 10.250$ Å, $b_0 = 6.778$ Å, $c_0 = 13.763$ Å; $\beta = 93° 33'$. The acetyl carbon skeleton is tilted by 14° from the plane defined by the copper and oxygen atoms. The phenyl ring is bent slightly down from the acetyl carbon skeletal plane, and is twisted through a torsional angle of 70° from the near plane of the chelate ring. The O—Cu—O bond angle is 91° 24' while the Cu—O bond distances are 1.911 and 1.902 Å in comparison to 1.92 Å for Cu(acac)$_2$. Cu(3φacac)$_2$ differs from Cu(acac)$_2$ only by the replacement of the hydrogen atom on the 3-carbon atom by a phenyl group. This also alters the molecular packing but not the bonding, thus removing the problem encountered in interpreting the crystal spectra of Cu(acac)$_2$ in terms of an oriented gas model.

The polarized spectra of the (100) and (001) faces at 6°K and the spectra of (100) parallel and perpendicular to b at 299, 195, 142, 77, and 6°K were determined. Molecular axes were chosen so that the y axis passes through the origin and the midpoint of the internal copper oxygen angle; the x axis is in the same plane as the oxygen and copper atoms and the z axis is perpendicular to the main molecular plane. The observed bands and transitions assigned to them are listed in Table LXXXIVA. From the variation of the intensity in the bands for each face, an electric dipole mechanism is assumed. The mechanism whereby the forbidden transitions obtain intensity is assumed to be a vibronic one. Using D_{2h} pseudosymmetry for the molecule and considering the product of the representations of the d orbitals and perturbing vibrations, the proposed assignments are as given in Table LXXXIVA. The spectra at 6°K were analyzed to obtain the spectra along the molecular axes x, y, and z. The entire spectrum is y polarized, the x spectrum not showing the well-defined broad maxima which characterizes the y and z curves, but rather gradually ascending to

TABLE LXXXIVA
Band Maxima at 6°K and Assignments for
$Cu(3\phi acac)_2$ (37a)

Band max., cm^{-1}	Polarization	Transition
20,600	$y \gg x > z$	$d_{xy} \rightarrow d_{yz}$
19,000	$y \gg x$	$d_{xy} \rightarrow d_{xz}$
16,900	$y \gg x,z$	$d_{xy} \rightarrow d_{x^2-y^2}$
15,400	$y \gg x,z$	$d_{xy} \rightarrow d_{z^2}$

higher frequency. It is thought that there is some x contribution to the visible portion of the spectrum; but it is not certain that all the visible bands have significant x components. The z spectrum clearly shows complete absence only of the 19,000 cm^{-1} band. From the polarization it is inferred that the bulk of the intensity is borrowed from one charge transfer transition, of an electron from the nonbonding oxygen $B_{3u}(\sigma)$ orbital to the antibonding copper $B_{1g}(xy)$ orbital in D_{2h}. The temperature dependence implies that in the y spectrum the lower energy bands are promoted by coupling to stiff stretching modes, which can only be β_{2u} and β_{3u}. This sets the lower transitions as $B_{1g} \rightarrow A_g$ and by elimination leaves the other two as $B_{1g} \rightarrow B_{2h}, B_{3g}$; y allowed by out-of-plane bending motions β_{1u} and α_u (37a).

9. Cu(DPM)$_2$

Cotton and Wise (80) have determined the polarized crystal spectrum of Cu(DPM)$_2$ and have compared the results with those obtained for Cu(acac)$_2$ (90,115,326). The crystal structure of Cu(DPM)$_2$ is isomorphous with Ni(DPM)$_2$ (76) and has been discussed under d^8 configurations, p. 427, Figure 77.

The observed bands and assignments are given in Table LXXXV.

An extended Hückel calculation by Cotton and Wise (78) for Cu(DPM)$_2$ gives a d-level ordering in which d_{xy} lies some 20,000 cm^{-1} above the other four, which are relatively close together in the order $d_{z^2} > d_{xz} > d_{x^2-y^2} > d_{yz}$. An assignment of observed transitions is made on this basis (Table LXXXV).

The assignment of the transitions in the ultraviolet was based upon consideration of the calculated symmetry-allowed transitions (78) and on the calculated and observed oscillator strengths. The $d \rightarrow d$ transitions are considered to be vibronically allowed rather than magnetic dipole excitations. For magnetic transitions, $d_{z^2} \rightarrow d_{xy}$ and $d_{x^2-y^2}$ should be z dipole polarized which is contrary to the assignments of reference 78. The observed reduction in the band intensity in Cu(acac)$_2$ on cooling (90,115) also points to a vibronic coupling mechanism.

TABLE LXXXV

Possible Assignment of Visible and Ultraviolet Transitions in $Cu(DPM)_2$ (80)

Transition	Calc. energy[a]	Band max., cm^{-1}	Polarization[b]
$d_{z^2} \rightarrow d_{xy}$	19,180	15,600	x
$d_{xz} \rightarrow d_{xy}$	19,490	16,400	y
$d_{x^2-y^2} \rightarrow d_{xy}$	19,870	18,200	x
$d_{yz} \rightarrow d_{xy}$	21,290	20,000	$x(y)$
$\sigma_L \rightarrow d_{xy}$	21,500	$\sim$26,600(sh)	
$\pi_L \rightarrow \pi_L^*$	$\begin{cases} 32,400 \\ 32,900 \end{cases}$	$\begin{cases} 32,300 \\ 34,600 \end{cases}$	
$\sigma_L \rightarrow d_{xy}$	42,350	40,000	
$d_{yz} \rightarrow \pi_L^*$	50,600	48,600	

[a] See ref. 78.

[b] Principal polarization, weaker polarization in parentheses.

The electron spin resonance spectrum of $Cu(DPM)_2$ doped into the isostructural diamagnetic $Ni(DPM)_2$ has been measured by Cotton and Wise (79). The values $g_\parallel = 2.244$ and $g_\perp = 2.051$ are in excellent agreement with those of Maki and McGarvey on $Cu(acac)_2$ (242).

10. $BaCuSi_4O_{10}$

Clark and Burns (67a) have determined the spectrum of $BaCuSi_4O_{10}$ in a mull, and by observation of the pleochroism have attributed this to the axial (E $\perp$ c) spectrum. The synthetic silicate $BaCuSi_4O_{10}$ has been shown by Pabst (295b) to be isostructural with $BaFeSi_4O_{10}$ (see p. 354). It has the pleochroic scheme $E \perp c$, blue; $E \parallel c$, pale-rose colorless.

The observed absorption peaks are attributed to electric dipole $d \rightarrow d$ transitions, allowed by a vibronic coupling mechanism. The possible allowed transitions and resulting assignments are given in Table LXXXVA.

TABLE LXXXVA

Vibronically Allowed Transitions of Cu^{2+} in $BaCuSi_4O_{10}$ (67a)

Polarization of electric vector	Transition	Normal modes allowing vibronic coupling	Band max., cm^{-1}	Intensity at band maximum
$E \perp c$	$^2B_{1g} \rightarrow {}^2B_{2g}$	E_u	12,900	0.16
	$^2B_{1g} \rightarrow {}^2E_g$	$A_{2u} + B_{2u}$	15,800	0.71
	$^2B_{1g} \rightarrow {}^2A_{1g}$	E_u	18,800	0.49
$E \parallel c$	$^2B_{1g} \rightarrow {}^2B_{2g}$	None		
	$^2B_{1g} \rightarrow {}^2E_g$	E_u		
	$^2B_{1g} \rightarrow {}^2A_{1g}$	B_{2u}		

It is assumed (67a) that a peak position corresponds to an energy $\Delta E_{el} + h\nu$ where ΔE_{el} is the difference in energy of the vibrationally unexcited states, and ν is the frequency of the vibration excited during the transition. The observations are corrected for the vibrational quanta by using the values (57c,408b): $\nu(E_u) = 1000 \pm 250$ cm^{-1}; $\nu(A_{2u}) \approx \nu(B_{2u}) = 500 \pm 250$ cm^{-1}. The results for the vibrationally unexcited states are then [$^2B_{1g}(d_{x^2-y^2})$ being taken as zero]: $^2B_{2g}(d_{xy})$ 11,900 cm^{-1}; $^2E_{2g}(d_{xz},d_{yz})$ 15,300 cm^{-1}; $^2A_{1g}(d_{z^2})$ 17,800 cm^{-1}.

11. Cu(oxine)$_2$

The crystal spectrum of anhydrous copper oxinate (copper 8-hydroxyquinolate), has been measured by Basu (31). The crystal structure has been reported by Kanamaru, Ogawa, and Nitta (194). It is monoclinic, space group $B2_1/a$, with eight molecules in a cell of dimensions $a_0 = 23.66$, $b_0 = 8.72$, $c_0 = 15.30$ Å, and $\beta = 117° 30'$. A copper atom is surrounded by two nitrogen (Cu—N = 1.92 and 1.90 Å) and two oxygen (Cu—O = 2.01 and 2.07 Å) atoms in a square–planar configuration. A weak dimeric unit of Cu$_2$(oxine)$_4$ with a Cu—Cu distance of 3.54 Å has been postulated, although magnetic measurements down to liquid air temperature do not show any antiferromagnetism as in copper acetate.

Bands were observed at $\sim$20,000, $\sim$16,000 (sh), and 10,000 cm^{-1}. The bands at 20,000 and 16,000 cm^{-1} have a strong temperature dependence.

12. Cu(en)$_2^{2+}$ Complexes

The polarized crystal spectra of bisethylenediamine complexes of copper(II) have been determined by Yamada and Tsuchida (437). Two bands are observed at $\sim$16,700 and $>$30,000 cm^{-1}. Preliminary ESR investigations of single crystals of Cu(en)$_2$Br$_2$H$_2$O have been reported by Reddy and Rajan (343). The Cu^{2+} ion is in sixfold coordination with the four nitrogen atoms of the two ethylenediamine groups in a plane and the Br and H$_2$O coordinated along the tetragonal axis perpendicular to this plane.

13. Cu(DMG)$_2$

Yamada and Tsuchida (429,431) have also determined the polarized spectrum of copper dimethylglyoxime, Cu(DMG)$_2$. Crystals of this compound are monoclinic, space group $P2_1/c, C_{2h}^5$, with a tetramolecular unit cell of dimensions $a_0 = 9.80$, $b_0 = 17.10$, $c_0 = 7.12$ Å, $\beta = 107° 20'$ at 20°C and $a_0 = 9.71$, $b_0 = 16.88$, $c_0 = 7.08$ Å, $\beta = 108° 26'$ at -140°C (133,134). There is no phase transition between these temperatures. The coordination of the copper is unusual. The four nitrogens are nearly equidistant from the metal atom, but lie in a plane just below making a very flat pyramid. Above the copper there is also an oxygen atom of another molecule, with Cu—O = 2.43 Å. Bands were observed at 18,000 cm^{-1} and 19,300 cm^{-1} for light with the electric vector polarized ∥ and ⊥ to the plane of the molecule (429).

Roos (346b) has calculated the energies of the excited states of $Cu(DMG)_2$ using a semiempirical open-shell SCF—MO method including configuration interaction. It is considered that the allowed transitions $^1B_{3u}^-$ and $^1B_{2u}^-$ may be responsible for some of the in-plane polarized absorption observed in the crystal. (This notation is explained in ref. 346b.) It was not found possible to classify the excited states into ligand-field and charge-transfer states.

A study (346b) of the spectral shifts occurring when axial ligands are added to the complex led to the conclusion that in the presence of an amine, a five-coordinated adduct with $Cu(DMG)_2$ is formed, while the complex forms a six-coordinated adduct with water.

14. $Cu[S_2CN(C_2H_5)_2]_2$

The ESR and crystal spectrum of copper diethyldithiocarbonate, $Cu[S_2CN(C_2H_5)_2]_2$, $[Cu(detc)_2]$ has been determined by Reddy and Srinivasan (344). The crystal structure of this compound has space group $P2_1/c, C_{2h}^5$ with a monoclinic unit cell containing four molecules and dimensions $a_0 = 9.907$, $b_0 = 10.627$, $c_0 = 16.591$ Å, $β = 113° 52'$ (20,43). Each copper atom is at the apex of a four-sided pyramid having sulfur atoms at the corners of the base; Cu—S = 2.30, 2.30, 2.31, and 2.33 Å, with a fifth at 2.86 Å. Zinc(detc)$_2$ is isomorphous with the copper complex having a cell of dimensions $a_0 = 10.015$, $b_0 = 10.661$, $c_0 = 16.357$ Å, and $β = 111° 58'$ (43,370).

A broad absorption band centered at 21,980 cm^{-1} was observed in the xy polarization which was assigned to $B_1 \rightarrow E$ in C_{4v}. ESR measurements on a diluted crystal in $Zn[S_2CN(C_2H_5)_2]_2$ gave $g_∥ = 2.1085$ and $g_⊥ = 2.023$. Absorption measurements at 90°K showed that although there was a small decrease in the width of the band, the intensity increased.

This is inconsistent with a vibronic intensity-gaining mechanism, and it is proposed that the intensity results from the presence of low-symmetry components in the ligand field.

15. $Cu(bipy)_3^{2+}$

The polarized crystal spectrum of the $Cu(bipy)_3^{2+}$ ion has been determined by Palmer and Piper (300) in the matrices $Cu(bipy)_3Br_26H_2O$,

TABLE LXXXVI
Band Maxima of Cu(bipy)$_3^{2+}$ Spectra (300)

Matrix	Spectrum[a]	Excited state	Band max., cm^{-1} 300°K	Band max., cm^{-1} 80°K	$\Delta\nu_{1/2}$, cm^{-1}	ϵ	I^b
Cu(bipy)$_3$Br$_2$6H$_2$O	Axial	^2E ($\perp$)	14,400	14,700	3,550	54.4	14.1
		$^2A_1$	6,400	6,400	4,100	28.2	18.5
Zn(bipy)$_3$Br$_2$6H$_2$O	Axial	^2E ($\perp$)	14,300	14,800			
		$^2A_1$	6,400	6,400			
Zn(bipy)$_3$SO$_4$7H$_2$O	σ	^2E ($\perp$)	14,400	14,800			
		$^2A_1$	6,500	5,800^c			
	π	^2E ($\parallel$)	14,200	14,500			

 ^a Polarization σ and π refers to electric vector $\perp$ C_3 and $\parallel$ C_3 axis.

[b] Integrated band intensity $I = (1/\nu_{max}) \int \epsilon d\nu$.

[c] Uncertainty of ± 300 cm^{-1} owing to strong OH and CH overtones in the thicker crystals.

Zn(bipy)$_3$Br$_2$6H$_2$O and Zn(bipy)$_3$SO$_4$7H$_2$O. The crystal structures of these compounds have been discussed under d^6 configurations, p. 356. The observed absorption bands are listed in Table LXXXVI while in Figure 87 is shown the orthoaxial spectra of Zn(Cu)(bipy)$_3$SO$_4$7H$_2$O at 80°K.

The axial spectrum of Cu(bipy)$_3^{2+}$ diluted in Zn(bipy)$_3$SO$_4$7H$_2$O is essentially the same as the orthoaxial spectrum with the electric vector polarized $\perp$ to the C_3 axis, Figure 87. However, with the electric vector polarized $\parallel$ to the C_3 axis, only a single band is observed at 14,200 cm^{-1}.

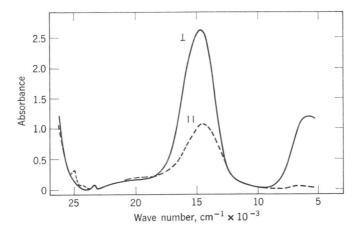

Fig. 87. Polarized orthoaxial spectra of Zn(Cu)(bipy)$_3$SO$_4$7H$_2$O at 80°K. Crystal thickness, 2.9 mm; approximately 15 mole % copper (300).

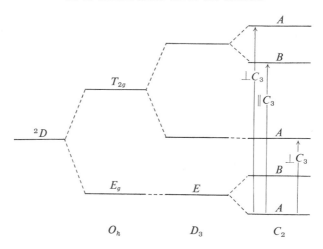

Fig. 88. *d*-orbital splitting diagram for $Cu(bipy)_3^{2+}$ assuming a predominantly trigonal ligand field (300).

The observed bands are explained on the basis of a trigonal distortion in the complex. The energy level diagram for this symmetry is given in Figure 88.

The temperature dependence of the band intensities is small, and it is assumed that electric dipole selection rules based on the crystal symmetry are effective. The band at 6400 cm^{-1} is $\perp$ polarized and is assigned to the $^2E \rightarrow {}^2A_1$ transition in D_3. The band at 14,300 cm^{-1} has intensity in both polarizations and is consistent with the assignment of $^2E \rightarrow {}^2E$. The small splitting (200–300 cm^{-1}) of the $\perp$ and $\parallel$ components of the ^2E band might be ascribed to a weak digonal field or may be vibronic in origin. An interpretation consistent with C_2 symmetry is also shown in Figure 88 (300). Dq is found to be 1170 cm^{-1} neglecting interaction between the two ^2E levels. Jørgensen (188) has proposed an energy scheme based on a large tetragonal distortion in the complex to explain the observed bands in the solution spectra of $Cu(bipy)_3^{2+}$ and $Cu(phen)_3^{2+}$. The splitting of the octahedral ^2E ground state is so large that its upper component is found among the relatively slightly split levels of $^2T_2$ at about 14,000 cm^{-1}. This model is however rejected by Palmer and Piper (300) as the 6400 cm^{-1} band would be assigned to a $A \leftrightarrow B$ transition in C_2 which is only $\perp C_2$ allowed (i.e., both $\perp$ and $\parallel C_3$ allowed) but it is observed only in polarization $\perp C_3$.

The large trigonal splitting gives K a value of -2600 cm^{-1} while for $Ni(bipy)_3^{2+}$ it was found to be only $+50$ cm^{-1}. This increased tetragonality could occur from three possible sources: (*1*) The much larger radius of the Cu^{2+} ion relative to Ni^{2+} may lead to considerable distortion of the six nitrogens from octahedral symmetry; (2) The greater size of the Cu^{2+} ion may allow solvation and ion pairing to occur along the trigonal axis; (3) The greater electronegativity of Cu^{2+} would lead to increased covalence in both σ and π bonds.

16. $Zn(Cu)(phenanthroline)_3(NO_3)_2 2H_2O$

The polarized crystal spectrum and ESR spectrum of copper(II) doped tris(phenanthroline) zinc(II) dihydrate has been determined by Kokoszka, Reimann, Allen, and Gordon (204b). The crystal structure of $(C_{12}H_8N_2)_3Zn(Cu)(NO_3)_2 2H_2O$ is not known and so no unambiguous assignment of energy levels may be made. Absorption bands were observed at 7000, 14,700, and 15,200 cm^{-1} at 77°K.

The two transitions observed near 7000 and 15,000 cm^{-1} are both assigned as $d \rightarrow d$ transitions. The 7000 cm^{-1} band arises from a Jahn-Teller splitting of the E_g ground state. The liquid superfine structure observed at 77°K on the 15,000 cm^{-1} band is considered consistent with the assignment of the ground state as primarily as $d_{x^2-y^2}$ and suggests an approximate tetragonal symmetry (204b). However, the observed polarization properties do not correspond to a simple tetragonal distortion in which the magnetic z axes correspond to the principal molecular axis and thus it is assumed that the magnetic and optical axes do not coincide. Palmer and Piper (300) found transitions near 7000 and 15,000 cm^{-1} when studying the $Cu(bipy)_3^{2+}$ complexes and interpreted them on a trigonal model, but this conflicts with the essentially tetragonal model proposed for the magnetic data.

17. Cu^{2+} Salicylaldehyde Complexes

Ferguson (113) has studied the crystal spectrum of bissalicylaldiminato Cu(II) and bis-N-methylsalicylaldiminato Cu(II). $Cu(Salim)_2$ crystallizes (375) in a monoclinic unit cell with space group $P2_1/c, C_{2h}^5$, dimensions $a_0 = 12.96, b_0 = 5.83, c_0 = 8.11$ Å, and $\beta = 95° 35'$ with two molecules per unit cell. It is isomorphous with the corresponding Ni(II) compound. The molecule is planar and has a *trans* configuration. $Cu(MeSalim)_2$ has space group $Ibam, D_{2h}^{26}$ and contains four molecules (213) with lattice dimensions $a_0 = 24.71, b_0 = 9.25$, and $c_0 = 6.66$ Å. The packing of the molecules is similar to that found for nickel bisdimethylglyoxime, $Ni(DMG)_2$ and involves chains of Cu atoms with a Cu—Cu separation of 3.33 Å. The well-developed face is the (100); the (001) face was obtained by sectioning with a microtome. The plane of the molecule lies exactly parallel to the (001) face, and the x and y axes each make an angle of $10°$ with the b and a crystal axes, respectively.

In Cu(MeSalim)$_2$ for light incident upon the (001) face, bands were observed at 15,000 cm^{-1} ($\|a$ and $\|b$) and 17,000 cm^{-1} ($\|b$). The observed polarization ratio a/b for the band at 17,000 cm^{-1} is $\sim 4:1$. This is only possible for axes bisecting metal ligand bond angles. For light incident on the (001) face, the two bands at 15,000 and 17,000 cm^{-1} are polarized completely in the molecular plane but the band at 20,500 cm^{-1} has a component normal to the molecular plane as well as in the plane.

The order of the five nondegenerate levels in D_{2h} symmetry is assumed to be in increasing order of energy, $^2B_{1g}(xy), ^2A_g(z^2), ^2A_g(x^2 - y^2)$, $^2B_{2g}(xz), ^2B_{3g}(yz)$. Four absorption bands are expected, but the ligand absorption in the near ultraviolet region overlaps that of the copper ion. The bands at 17,000 and 15,000 cm^{-1} have no out-of-plane component and are both assigned to $^2B_{1g} \rightarrow ^2A_g$. The assignment of the next excited state is uncertain owing to overlap with the ligand absorption, but it is most probably $^2B_{1g} \rightarrow ^2B_{2g}$ with the $3d_{xz}$ orbital singly occupied in the excited state. The other excited state $^2B_{3g}$ either lies at higher energy or the transition to it occurs weakly in the same region, which accounts for the polarization in the b direction.

In Cu(Salim)$_2$ the spectrum on the (100) face shows one band at 16,500 cm^{-1}, polarization ratio $b/c = 2.3$, but an extra band at 18,500 cm^{-1} is observed in the (001) face which is stronger in the a polarization.

These two bands are assigned to the two $^2B_{2g} \rightarrow ^2A_g$ transitions, the observed polarizations being in agreement with these assignments. The higher energy bands are not observed in this complex due to the overlap by ligand absorption.

18. Cu(Sucim)$_n$ Complexes

The polarized spectra of complexes with succinimide have been determined by Yamada and Miki (261,447). The diethylethylenediamine copper(II) complex was also investigated. The main absorption bands are given in Table LXXXVII, but as no crystal structure analysis is available, assignments on the basis of a tetragonal distortion are to be regarded as very tentative.

Both the assignments given for the higher energy bands in the compounds Cu(Sucim)$_4$4H$_2$O and Cu(Sucim)$_2$(iso-C$_3$H$_7$NH$_3$)$_2$ are compatible with the observed polarization and an unambiguous assignment is not regarded as possible. From a comparison of these spectra with those of other cupric complexes, it has been concluded that these compounds

TABLE LXXXVII

Observed Absorption Bands in a Series of Cu(II) Complexes with Succinimide

Compound	Band max., cm^{-1}	$\Delta\nu_{1/2}$, cm^{-1}	log α	Assignment	Ref.
K$_2$Cu(Sucim)$_4$6H$_2$O	20,400 xy	4,160	1.08	$d_{xy} \rightarrow d_{x^2-y^2}$	447
	~16,700 xy and z	3,670	0.40	$d_{z^2} \rightarrow d_{x^2-y^2}$	
Cu(Sucim)$_2$	20,870 xy	4,000	1.38	$\left.\begin{matrix}d_{xy} \rightarrow \\ d_{z^2} \rightarrow\end{matrix}\right\} d_{x^2-y^2}$	447
(isopropylamine)$_2$					
	19,700 z	3,670	0.99	$d_{z^2} \rightarrow d_{x^2-y^2}$	
Cu(Sucim)$_2$Py$_2$	20,670				261
	17,300				
Cu(Sucim)$_2$Py(H$_2$O)	19,300				261
	17,670				
Cu(deen)$_2$(ClO$_4$)$_2$	21,130	6,270	1.04	$\left.\begin{matrix}d_{xy} \\ d_{z^2}\end{matrix}\right\} \rightarrow d_{x^2-y^2}$	447
	20,000	4,530	0.70	$d_{z^2} \rightarrow d_{x^2-y^2}$	

contain Cu(II) in a square–planar coordination and the fifth and sixth coordination sites are left empty.

19. Other Cu(II) Complexes

The polarized crystal spectra of copper(II) complexes of benzoic and substituted benzoic acids have been determined by Yamada, Nishikawa, and Miki (448). The crystal structures of these compounds are at present unknown but the observed bands are listed in Table LXXXVIII.

The polarization of band I is the reverse of band II. It is concluded that the substituted cupric carboxylates consist of binulear molecules with a Cu—Cu bond except for Cu(p-OH benz)$_2$6H$_2$O which shows no band in the 26,500 cm^{-1} region, so the crystal contains no Cu—Cu bond.

The dichroism of the cupric salts of monocarboxylic fatty acids have been determined by Yamada, Nakahara, and Nakamura and Tsuchida

TABLE LXXXVIII

Absorption Maxima of Cupric Benzoate and Substituted Benzoates in the Crystalline State (448)

	Band max. I, cm^{-1}	Band max. II, cm^{-1}
Cu(benz)$_2$3H$_2$O	13,800	None
Cu(benz)$_2$C$_2$H$_5$OH	14,800	26,500
Cu(benz)$_2$(benz H)0.5H$_2$O	14,770	~26,700
Cu(o-CH$_3$benz)$_2$	15,000	~26,700
Cu(o-Cl benz)$_2$2H$_2$O	14,700	~26,300
Cu(m-OH benz)$_2$H$_2$O	15,000	~26,300
Cu(p-OH benz)$_2$6H$_2$O	13,800	None
Cu(o-NO$_2$benz)$_2$	14,800	Masked

(441,442). Compounds of the form $Cu(CH_3(CH_2)_xCO_2)_2$ show absorption bands at approximately 14,300 and 26,700 cm^{-1}, and are included in Table LXXXIX.

From the appearance of the band at 26,700 cm^{-1} and the reversal of the polarization compared to the 14,300 cm^{-1} band, it is concluded that the cupric compounds have a dimeric structure similar to that of copper acetate.

The polarized absorption spectra of the cupric salts of the halomonocarboxylic acids have been studied by Yamada, Nishikawa, and Tsuchida (398,444). Similar bands to those in the unsubstituted acids were observed.

TABLE LXXXIX
Absorption Maxima in Cupric Salts of Monocarboxylic Acids (441)

	Band max. I, cm^{-1}	Band max. II, cm^{-1}
$Cu(HCO_3)_2 4H_2O$	13,300	
$Cu(CH_3CO_2)_2 H_2O$	14,300	26,700
$Cu(C_2H_5CO_2)_2$	14,300	26,700
$Cu(C_2H_5CO_2)_2 H_2O$	14,300	26,700

XII. CONFIGURATIONS d^0 AND d^{10}

A. $3d^0$, d^{10}

As ions having the configurations d^0 and d^{10} do not have spectra characteristic of other transition metal ions they will not be discussed in detail here. A number of features of the posttransition ion spectra are discussed by McClure (230).

1. KMnO$_4$

The series of ions MnO_4^-, CrO_4^{2-}, VO_4^{3-} all have unoccupied d orbitals in the ground state. These were originally studied by Teltow (390, 391) using the host crystals $KClO_4$, K_2SO_4, and $Na_3PO_4 12H_2O$, respectively. The polarized absorption spectrum of $KMnO_4$ in $KClO_4$ has been recently remeasured at 4°K by Holt and Ballhausen (183). $KClO_4$ and $KMnO_4$ are isomorphous having space group $Pnma, V_h^{16}$. The lattice constants for $KClO_4$ are $a_0 = 8.834$, $b_0 = 5.650$, and $c_0 = 7.240$ Å in an orthorhombic unit cell containing four molecules (149,246). The cell constants for $KMnO_4$ are $a_0 = 9.908$, $b_0 = 5.730$, and $c_0 = 7.394$ Å (235,342). Holt and Ballhausen (183) obtain a spectrum which contains four regions. The first region from 18,000–23,000 cm^{-1} is characterized by a great deal of fine structure. The second region from 24,500–30,000 cm^{-1} shows a simple vibrational progression, thought to be due to vibrations belonging to the third system, superimposed upon a nearly structureless band. The third region from 30,000–38,000 cm^{-1} consists of a single progression in quanta of 750 cm^{-1}. The fourth region from 30,000 to 46,000 cm^{-1} is made up of a broad featureless absorption band.

An SCF molecular orbital calculation by Dahl and Ballhausen (83) yields the energy-level scheme shown in Figure 89.

Teltow's (390,391) measurements had been analyzed and interpreted by Wolfsberg and Helmholz (420) and by Ballhausen and Liehr (13,15) on the basis of two different energy schemes. The main qualitative difference between these schemes lies in the relative ordering of the first empty orbital. According to Wolfsberg and Helmholz, the t_2 level lies lowest, while according to the later work of Ballhausen and Liehr, the e level lay lowest. This is in accord with crystal field prediction for a tetrahedral complex. The assignments of Wolfsberg and Helmholz were made before the general application of crystal field theory to transition ion spectra. ESR studies by Carrington, Ingram, Lott, Schonland, and Symons (60) by Schonland (356), and by Carrington and Schonland (61) led to a revised analysis by Carrington and Symons (59) and by Carrington and Jørgensen (62) and of the vibrational progressions in the bands by Ballhausen (18). MO calculations were carried out by Fenske and Sweeney (108) and by Viste and Gray (409); they were later slightly amended with improved parameters (157).

Following the most recent calculations of Dahl and Ballhausen (83) Holt and Ballhausen (183) made the assignments of the bands as follows. The $^1A_1 \rightarrow {}^1T_1(t_1 \rightarrow 2e)$ and $^1A_1 \rightarrow {}^1T_1(3t_2 \rightarrow 2e)$ transitions are expected to resemble each other as they are both of the $\pi \rightarrow \pi^*$ type and are accordingly assigned to the bands at 2.3 and 4.0 eV, respectively. The band at 3.5 eV is very tentatively assigned to the $t_1 \rightarrow 2e$ transition while the 5.5 eV system is assigned to the promotion of an electron from the bonding $2a_1$ orbital to the antibonding $2e$ orbital. There are no recent crystal spectra for other tetrahedral oxyions of this type.

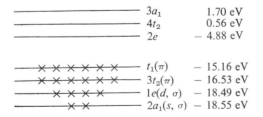

Fig. 89. Part of SCF—LCAO—MO energy level scheme for MnO_4^- (183).

2. $BaTiO_3$

The polarized light transmission of $BaTiO_3$ crystals has been studied by Casella and Keller (63). The absorption coefficients $\mu_{\parallel}$ and $\mu_{\perp}$ (to the c axis) were determined as a function of the incident photon energy and it was found that at the absorption edge $\mu_{\perp} > \mu_{\parallel}$.

The exciton spectra down to $4°K$ of a number of crystals including Cu_2O, $CuCl$, $CuBr$, CuI, AgI, HgI_2, PbI_2, $TlCl$, $TlBr$, and TlI have been studied by Nikitine and are summarized in reference 288.

REFERENCES

1. Allen, H. C. Jr., *J. Chem. Phys.*, *45*, 553 (1966).
1a. Andress, K. R., and C. Carpenter, *Z. Krist.*, *87*, 446 (1934).
1b. Anex, B. G., and A. V. Fratini, *J. Mol. Spectry.*, *14*, 1 (1964).
2. Anex, B. G., *Mol. Cryst.*, *1*, 1 (1966).
3. Anex B. G., M. E. Ross, and M. W. Hedgcock, *J. Chem. Phys.*, *46*, 1090 (1967).
4. Anex, B. G., and F. K. Krist, *J. Am. Chem. Soc.*, *89*, 6114 (1967).
5. Astbury, W. T., *Proc. Roy. Soc.* (*London*), *A112*, 448 (1926).
6. Atoji, M., J. W. Richardson, and R. E. Rundle, *J. Am. Chem. Soc.*, *79*, 3017 (1957).
7. Bacon, G. E., and W. E. Gardner, *Proc. Roy. Soc.* (*London*), *A246*, 78 (1958).
8. Bacon, G. E., and N. A. Curry, *Proc. Roy. Soc.* (*London*), *A266*, 95 (1962).
9. Ballhausen, C. J., and C. K. Jørgensen, *Kgl. Danske Videnskab. Selskab. Matfys. Medd.*, *29*, No. 14 (1955).
10. Ballhausen, C. J., and W. J. Moffitt, *J. Inorg. Nucl. Chem.*, *3*, 178 (1956).
11. Ballhausen, C. J., *Z. Physik. Chem.*, N.F., *11*, 205 (1957).
12. Ballhausen, C. J., *Z. Physik. Chem.*, N.F., *17*, 246 (1958).
13. Ballhausen, C. J., and A. D. Liehr, *J. Mol. Spectry.*, *2*, 342 (1958).
14. Ballhausen, C. J., and A. D. Liehr, *J. Am. Chem. Soc.*, *81*, 538 (1959).
15. Ballhausen, C. J., and A. D. Liehr, *J. Mol. Spectry.*, *4*, 190 (1960).
16. Ballhausen, C. J., and H. B. Gray, *Inorg. Chem.*, *1*, 111 (1962).
17. Ballhausen, C. J., *Introduction to Ligand Field Theory*, McGraw-Hill, New York, 1962.
18. Ballhausen, C. J., *Theoret. Chim. Acta*, *1*, 285 (1963).
19. Ballhausen, C. J., N. Bjerrum, R. Dingle, K. Eriks, and C. R. Hare, *Inorg. Chem.*, *4*, 514 (1965).
20. Bally, R., *Compt. Rend.*, *257*, 425 (1963).
21. Banks, C. V., and D. W. Barnum, *J. Am. Chem. Soc.*, *80*, 3579 (1958).
22. Banks, C. V., and D. W. Barnum, *J. Am. Chem. Soc.*, *80*, 4767 (1958).
23. Barnum, D. W., *J. Inorg. Nucl. Chem.*, *21*, 221 (1961).
24. Barnum, D. W., *J. Inorg. Nucl. Chem.*, *22*, 183 (1961).
25. Basch, H., and H. B. Gray, *Inorg. Chem.*, *6*, 365 (1967).
26. Basolo, F., C. J. Ballhausen, and J. Bjerrum, *Acta Chem. Scand.*, *9*, 810 (1955).
26a. Basu, G., R. L. Belford, and R. E. Dickerson, *Inorg. Chem.*, *1*, 438 (1962).
27. Basu, G., W. Yeranos, and R. L. Belford, *Inorg. Chem.*, *3*, 929 (1964).
28. Basu, G., G. M. Cook, and R. L. Belford, *Inorg. Chem.*, *3*, 1361 (1964).

29. Basu, G., and R. L. Belford, *J. Mol. Spectry.*, *17*, 167 (1965).
30. Basu, G., *Spectrochim. Acta*, *22*, 155 (1966).
31. Basu, G., *J. Inorg. Nucl. Chem.*, *29*, 1544 (1967).
32. Baur, W. H., *Naturwiss.*, *44*, 349 (1957).
33. Baur, W. H., *Acta Cryst.*, *11*, 488 (1958).
34. Baur, W. H., *Naturwiss.*, *49*, 464 (1962).
35. Beevers, C. A., and H. Lipson, *Proc. Roy. Soc. (London)*, *A146*, 570 (1934).
36. Belford, R. L., M. Calvin, and G. Belford, *J. Chem. Phys.*, *26*, 1165 (1957).
37. Belford, R. L., and G. G. Belford, *Theoret. Chim. Acta*, *3*, 465 (1965).
37a. Belford, R. L., and J. W. Carmichael, Jr., *J. Chem. Phys.*, *46*, 4515 (1967).
38. Bernal, I., and S. E. Harrison, *J. Chem. Phys.*, *34*, 102 (1961).
39. Bethe, H., *Ann. Physik* (5) *3*, 133 (1929).
40. Bjerrum, J., C. J. Ballhausen, and C. K. Jørgensen, *Acta Chem. Scand.*, *8*, 1275 (1954).
41. Bleaney, B., and K. D. Bowers, *Proc. Roy. Soc. (London)*, *A214*, 451 (1952).
42. Bokii, G. B., T. I. Malinovskii, and A. V. Ablov, *Kristallografiya*, *1*, 49 (1956).
43. Bonamico, M., G. Dessy, G. Mazzone, A. Mugnoli, A. Vaciago, and L. Zambonelli, *Atti Acad. Naz. Lincei, Rend., Classe Sci. Fis. Mat. Nat.*, *35*, 338 (1963).
44. Bonamico, M., G. Dessy, and A. Vaciago, *Theoret. Chim. Acta*, *7*, 367 (1967).
45. Bor, G. P., *J. Inorg. Nucl. Chem.*, *17*, 174 (1961).
46. Bosnich, B., *J. Am. Chem. Soc.*, *88*, 2606 (1966).
47. Boudreaux, E. A., *Inorg. Chem.*, *3*, 506 (1964).
48. Brasseur, H., and A. de Rassenfosse, *Z. Krist.*, *97*, 239 (1937).
49. Brasseur, H., and A. de Rassenfosse, *Bull. Soc. Franc. Mineral. Crist.*, *61*, 129 (1938).
50. Brasseur, H., *Bull. Classe Sci. Acad. Roy. Belg.*, *49*, 1028 (1963–1964).
51. Braterman, P. S., *Inorg. Chem.*, *2*, 448 (1963).
52. Brehler, B., *Z. Krist.*, *109*, 68 (1957).
53. Brode, W. R., *Proc. Roy. Soc. (London)*, *A118*, 286 (1928).
54. Brosset, C., and N. G. Vannerberg, *Nature*, *190*, 714 (1961).
55. Brumage, W. H., C. R. Quade, and C. C. Lin, *Phys. Rev.*, *131*, 949 (1963).
56. Brumage, W. H., and C. C. Lin, *Phys. Rev.*, *134*, 950 (1964).
57. Brumage, W. H., E. C. Segraves, and C. C. Lin, *J. Chem. Phys.*, *42*, 3326 (1965).
57a. Brunetière, F., *Compt. Rend.*, *255*, 3394 (1962).
57b. Burns, G., *Phys. Rev.*, *123*, 1634 (1961).
57c. Burns, R. G., *J. Sci. Instr.*, *43*, 58 (1966).
57d. Burns, R. G., M. G. Clark, and A. J. Stone, *Inorg. Chem.*, *5*, 1268 (1967).
58. Burstall, F. M., and R. S. Nyholm, *J. Chem. Soc.*, *1952*, 3570.
58a. Carlin, R. L., and I. M. Walker, *Chem. Commun.*, *1965*, 139.
58b. Carlin, R. L., and I. M. Walker, *J. Chem. Phys.*, *46*, 3921 (1967).
58c. Carlin, R. L., and E. G. Terezakis, *J. Chem. Phys.*, *47*, 4901 (1967).
58d. Carmichael, J. W., Jr., L. K. Steinrauf, and R. L. Belford, *J. Chem. Phys.*, *43*, 3959 (1965).
59. Carrington, A., and M. C. R. Symons, *J. Chem. Soc.*, *1960*, 889.
60. Carrington, A., D. J. E. Ingram, K. A. K. Lott, D. S. Schonland, and M. C. R. Symons, *Proc. Roy. Soc. (London)*, *A254*, 101 (1960).
61. Carrington, A., and D. S. Schonland, *Mol. Phys.*, *3*, 331 (1960).
62. Carrington, A., and C. K. Jørgensen, *Mol. Phys.*, *4*, 395 (1961).
63. Casella, R. C., and S. P. Keller, *Phys. Rev.*, *116*, 1469 (1959).

64. Cavalca, L., M. Nardelli, and A. Braibant, *Gazz. Chim. Ital.*, *86*, 942 (1956).
65. Chakravorty, A., and S. Basu, *J. Chem. Phys.*, *33*, 1266 (1960).
66. Chakravorty, A. S., *J. Chem. Phys.*, *39*, 1004 (1963).
67. Chatt, J., G. A. Gamlen, and L. E. Orgel, *J. Chem. Soc.*, *1958*, 486.
67a. Clark, M. G., and R. G. Burns, *J. Chem. Soc. (A)*, *1967*, 1034.
68. Clark, R. J. H., and T. M. Dunn, *J. Chem. Soc.*, *1963*, 1198.
69. Clark, R. J. H., T. M. Dunn, and C. F. Stoneman, unpublished data quoted in *Some Aspects of Crystal Field Theory*, T. M. Dunn, R. G. Pearson, and D. S. McClure, Eds., Harper, 1965, p. 52.
70. Colin, M. L., *Bull. Classe Sci. Acad. Roy. Belg.*, *49*, 973 (1963–1964).
71. Colin, M. L., *Bull. Soc. Roy. Sci. Liège*, *34*, 130 (1965).
72. Collins, R. J., and D. A. Kleinman, *J. Chem. Phys. Solids*, *11*, 190 (1959).
73. Condon, E. U., and G. H. Shortley, *The Theory of Atomic Spectra*, Cambridge University Press, London and New York, 1951.
74. Cotton, F. A., *Chemical Applications of Group Theory*, Interscience, New York, 1963.
75. Cotton, F. A., and G. Wilkinson, *Advanced Inorganic Chemistry*, 2nd ed., Interscience, New York, 1966.
76. Cotton, F. A., and J. J. Wise, *Inorg. Chem.*, *5*, 1200 (1966).
77. Cotton, F. A., and C. B. Harris, *Inorg. Chem.*, *6*, 369 (1967).
78. Cotton, F. A., C. B. Harris, and J. J. Wise, *Inorg. Chem.*, *6*, 909 (1967).
79. Cotton, F. A., and J. J. Wise, *Inorg. Chem.*, *6*, 915 (1967).
80. Cotton, F. A., and J. J. Wise, *Inorg. Chem.*, *6*, 917 (1967).
81. Crosby, G. A., W. G. Perkins, and D. M. Klassen, *J. Chem. Phys.*, *43*, 1498 (1965).
82. Dahl, L. F., private communication to T. S. Piper and R. L. Belford, ref. 326.
83. Dahl, J. P., and C. J. Ballhausen, in *Advances in Quantum Chemistry*, Vol. 4, P. O. Löwdin, Ed., Academic Press, New York, 1968, p. 173.
83a. Danon, J., and L. Iannarella, *J. Chem. Phys.*, *47*, 382 (1967).
84. Day, P., A. F. Orchard, A. J. Thompson, and R. J. P. Williams, *J. Chem. Phys.*, *42*, 1973 (1965).
85. Day, P., A. F. Orchard, A. J. Thompson, and R. J. P. Williams, *J. Chem. Phys.*, *43*, 3763 (1965).
86. Davydov, A. S., *J. Exptl. Theoret. Phys. (USSR)*, *18*, 210 (1948); *21*, 673 (1951).
87. Denning, R. G., *Chem. Commun.*, *1967*, 120.
88. Deutschbein, O., *Ann. Physik*, *14*, 712 (1932).
89. Dickinson, R. G., *J. Am. Chem. Soc.*, *44*, 2404 (1922).
90. Dijkgraaf, C., *Theoret. Chim. Acta*, *3*, 38 (1965).
91. Dijkgraaf, C., and J. P. G. Rousseau, *Spectrochim. Acta*, *23A*, 1267 (1967).
92. Dingle, R., *J. Mol. Spectry.*, *9*, 426 (1962).
93. Dingle, R., *Chem. Commun.*, *1965*, 304.
94. Dingle, R., *J. Mol. Spectry.*, *18*, 276 (1965).
95. Dingle, R., unpublished work. Ref. 94.
96. Dingle, R., *Inorg. Chem.*, *4*, 1287 (1965).
97. Dingle, R., *Acta Chem. Scand.*, *20*, 33 (1966).
98. Dingle, R., *J. Chem. Phys.*, *46*, 1 (1967).
99. Dingle, R., unpublished work, ref. 94.
100. Dodge, R. P., D. H. Templeton, and A. Zalkin, *J. Chem. Phys.*, *35*, 55 (1961).
101. Dunitz, J. D., *Acta Cryst.*, *10*, 307 (1957).

102. Dunn, T. M., in *Modern Coordination Chemistry*, J. Lewis and R. Wilkins, Eds., Interscience, New York, 1960, Chap. 4.
103. Ebsworth, E. A. V., and J. A. Weil, *J. Phys. Chem.*, *63*, 1890 (1959).
104. Englman, R., *Mol. Phys.*, *4*, 183 (1961).
105. Elliot, H., and B. J. Hathaway, *Spectrochim. Acta*, *21*, 1047 (1965).
106. Felsenfield, G., *Proc. Roy. Soc. (London)*, *A236*, 506 (1956).
107. Fenske, R. F., D. S. Martin, Jr., and K. Ruedenberg, *Inorg. Chem.*, *1*, 441 (1962).
108. Fenske, R. F., and C. C. Sweeney, *Inorg. Chem.*, *3*, 1105 (1964).
109. Ferguson, J., *J. Chem. Phys.*, *32*, 528 (1960).
110. Ferguson, J., *J. Chem. Phys.*, *32*, 533 (1960).
111. Ferguson, J., *J. Chem. Phys.*, *34*, 611 (1961).
112. Ferguson, J., *J. Chem. Phys.*, *34*, 1609 (1961).
113. Ferguson, J., *J. Chem. Phys.*, *35*, 1612 (1961).
114. Ferguson, J., K. Knox, and D. L. Wood, *J. Chem. Phys.*, *35*, 2236 (1961); *Erratum*, *37*, 193 (1962).
115. Ferguson, J., R. L. Belford, and T. S. Piper, *J. Chem. Phys.*, *37*, 1569 (1962).
116. Ferguson, J., H. J. Guggenheim, L. F. Johnson, and H. Kamimura, *J. Chem. Phys.*, *38*, 2579 (1963).
117. Ferguson, J., *J. Chem. Phys.*, *39*, 116 (1963).
118. Ferguson, J., D. L. Wood, and K. Knox, *J. Chem. Phys.*, *39*, 881 (1963).
119. Ferguson, J., *Rev. Pure Appl. Chem.*, *14*, 1 (1964).
120. Ferguson, J., H. J. Guggenheim, and D. L. Wood, *J. Chem. Phys.*, *40*, 822 (1964).
121. Ferguson, J., *J. Chem. Phys.*, *40*, 3406 (1964).
122. Ferguson, J., *Theoret. Chim. Acta*, *3*, 287 (1965).
122a. Ferguson, J., H. J. Guggenheim, and Y. Tanabe, *J. Appl. Phys.*, *36*, 1046 (1965).
123. Ferrari, A., and F. Giorgi, *Atti Accad. Naz. Lencei Rend. Classe Sci. Fiz. Mat. Nat.*, *9*, 1134 (1929).
124. Ferrari, A., and F. Giorgi, *Atti Accad. Naz. Lencei Rend. Classe Sci. Fiz. Mat. Nat.*, *10*, 522 (1929).
125. Figgis, B. N., and R. L. Martin, *J. Chem. Soc. 1956*, 3837.
126. Figgis, B. N., J. Lewis, and F. Mabbs, *J. Chem. Soc.*, *1960*, 2480.
127. Figgis, B. N., M. Gerloch, and R. Mason, *Acta Cryst.*, *17*, 506 (1964).
128. Foner, S., and W. Low, *Phys. Rev.*, *120*, 1585 (1960).
129. Forman, A., and L. E. Orgel, *Mol. Phys.*, *2*, 362 (1959).
130. Forster, L. S., and K. DeArmond, *J. Chem. Phys.*, *34*, 2193 (1961).
131. Forster, L. S., and C. J. Ballhausen, *Acta Chem. Scand.*, *16*, 1385 (1962).
132. Fox, M. R., P. L. Orioli, E. C. Lingafelter, and L. Sacconi, *Acta Cryst.*, *17*, 1159 (1964).
133. Frasson, E., R. Bardi, R. Zanetti, and M. Mamni, *Ann. Chim. (Rome)*, *48*, 1007 (1958).
134. Frasson, E., R. Bardi, and S. Bezzi, *Acta Cryst.*, *12*, 201 (1959).
135. Frasson, E., C. Panattoni, and L. Sacconi, *J. Phys. Chem.*, *63*, 1908 (1959).
135a. Frasson, E., and C. Panattoni, *Acta Cryst.*, *13*, 893 (1960).
136. Frossard, L., *Schweiz. Mineral. Petrog. Mitt.*, *36*, 1 (1956).
137. Furlani, C., and G. Morpurgo, *Theoret. Chim. Acta*, *1*, 102 (1963).
138. Furlani, C. E., F. Cervone, F. Calzona, and B. Baldanza, *Theoret. Chim. Acta*, *7*, 375 (1967).

139. Gabriel, J. R., D. F. Johnston, and M. J. D. Powell, *Proc. Roy. Soc.* (*London*), *A264*, 503 (1961).

139a. Geller, S., and D. P. Booth, *Z. Krist.*, *111*, 117 (1958).

139b. Geller, S., *Z. Krist.*, *114*, 148 (1960).

140. Geller, S., C. E. Miller, and R. G. Treuting, *Acta Cryst.*, *13*, 179 (1960).

141. Gersmann, H. R., and J. D. Swalen, *J. Chem. Phys.*, *36*, 3221 (1962).

142. Geschwind, S., and J. P. Remeika, *J. Appl. Phys.*, Supplement No. 1, *33*, 370 (1962).

143. Geschwind, S., P. Kisliuk, M. P. Klein, J. P. Remeika, and D. L. Wood, *Phys. Rev.*, *126*, 1684 (1962).

144. Gielessen, J., *Ann. Physik.*, *22*, 537 (1935).

145. Godychi, L. E., and R. E. Rundle, *Acta Cryst.*, *6*, 487 (1953).

146. Golling, E., *Ann. Physik*, *9*, 181 (1951).

147. Goode, D. H., *J. Chem. Phys.*, *43*, 2830 (1965).

148. Gossner, B., *Z. Krist.*, *38*, 50 (1904).

149. Gottfried, G., and C. Schusterius, *Z. Krist.*, *84*, 65 (1932).

150. Gladney, H. M., and J. D. Swalen, *J. Chem. Phys.*, *42*, 1999 (1965).

151. Gray, H. B., and C. R. Hare, *Inorg. Chem.*, *1*, 363 (1962).

152. Gray, H. B., and C. J. Ballhausen, *J. Am. Chem. Soc.*, *85*, 260 (1963).

153. Gray, H. B., and C. J. Ballhausen, *Molecular Orbital Theory*, Benjamin, New York, 1964.

154. Gray, H. B., and N. A. Beach, *J. Am. Chem. Soc.*, *85*, 2922 (1963).

155. Gray, H. B., P. T. Manoharan, J. Pearlman, and R. F. Riley, *Chem. Commun.*, *1965*, 62.

156. Gray, H. B., *Transition Metal Chem.*, *1*, 240 (1965).

157. Gray, H. B., *Coord. Chem. Rev.*, *1*, 2 (1966).

158. Greenberg, A. L., and G. H. Walden, *J. Chem. Phys.*, *8*, 645 (1940).

159. Griffith, J. S., *Theory of Transition Metal Ions*, Cambridge University Press, Cambridge, 1961.

160. Grum Grzhimailo, V., N. A. Brilliantov, and R. K. Sviridova, *Opt. i Spektroskopiya*, *6*, 238 (1959); *Opt. Specty. Engl. Trans.*, *6*, 152 (1959).

161. Grum Grzhimailo, V., N. A. Brilliantov, R. K. Sviridova, and A. S. Dzhamalova, *Optika i Spectroskopiya*, *6*, 240 (1959); *Opt. Spectry., Engl. Trans.*, *6*, 154 (1959).

162. Hansen, A. E., and C. J. Ballhausen, *Trans. Faraday Soc.*, *61*, 631 (1965).

163. Hare, C. R., and C. J. Ballhausen, *J. Chem. Phys.*, *40*, 788 (1964).

164. Hare, C. R., and C. J. Ballhausen, *J. Chem. Phys.*, *40*, 792 (1964).

165. Hartmann, H., and H. L. Schläfer, *Z. Physik.*, *Chem.*, *B197*, 116 (1951).

166. Hartmann, H., and H. L. Schläfer, *Z. Naturforsch.*, *6a*, 754 (1951).

167. Hartmann, H., and H. L. Schläfer, *Z. Naturforsch.*, *6a*, 760 (1951).

168. Hartmann, H., *Z. Naturforsch.*, *6a*, 781 (1951).

169. Hartmann, H., and H. L. Schläfer, *Angew. Chem.*, *66*, 768 (1954).

170. Hartmann, H., C. Furlani, and A. Bürger, *Z. Physik. Chem.*, *9*, 62 (1956).

171. Hartmann, H., and M. Müller, *Discussions Faraday Soc.*, *26*, 49 (1958).

172. Hassel, O., and J. R. Salvesen, *Z. Physik. Chem.*, *128*, 345 (1927).

173. Hatfield, W. E., *Inorg. Chem.*, *3*, 605 (1964).

174. Hatfield, W. E., and T. S. Piper, *Inorg. Chem.*, *3*, 841 (1964).

175. Hatfield, W. E., and T. S. Piper, *Inorg. Chem.*, *3*, 1295 (1964).

176. Haussühl, S., *Z. Krist.*, *116*, 371 (1961).

177. Hawkins, C. J., E. Larson, and I. Olson, *Acta Chem. Scand.*, *19*, 1915 (1965).

178. Heiland, G., E. Mollwo, and F. Stöckmann, *Solid State Phys.*, *8*, 195 (1959).
179. Helmholz, L., and R. F. Kruh, *J. Am. Chem. Soc.*, *74*, 1176 (1952).
180. Herpin, P., *Bull. Soc. Franc. Mineral. Crist.*, *81*, 245 (1958).
181. Hoard, J. L., and L. Goldstein, *J. Chem. Phys.*, *3*, 645 (1935).
182. Holmes, O. G., and D. S. McClure, *J. Chem. Phys.*, *26*, 1686 (1957).
183. Holt, S. L., and C. J. Ballhausen, *Theoret. Chim. Acta*, *7*, 313 (1967).
184. Hush, N. S., and G. C. Allen, in *Progress in Inorganic Chemistry*, Vol. 8, F. A. Cotton, Ed., Interscience, New York, 1967, p. 357.
185. Jacobs, G., and F. Speeke, *Acta Cryst.*, *8*, 67 (1954).
186. Jaeger, F. M., and J. A. Van Dijk, *Z. Anorg. Allgem. Chem.*, *227*, 273 (1936).
187. Jarrett, H. S., *J. Chem. Phys.*, *27*, 1298 (1957).
188. Jørgensen, C. K., *Acta Chem. Scand.*, *9*, 1362 (1955).
189. Jørgensen, C. K., Report to the 10th Solvay Conference, Brussels, 1956.
190. Jørgensen, C. K., *Acta Chem. Scand.*, *11*, 166 (1957).
191. Jørgensen, C. K., *Absorption Spectra and Chemical Bonding in Complexes*, Pergamon Press, New York, 1961.
192. Jørgensen, C. K., *Orbitals in Atoms and Molecules*, Academic Press, New York, 1962.
193. Jørgensen, C. K., *Chem. Phys., Letters*, *1*, 11 (1967).
194. Kanamaru, F., K. Ogawa, and I. Nitta, *Bull. Chem. Soc. (Japan)*, *36*, 422 (1963).
195. Karipides, A. G., and T. S. Piper, *Inorg. Chem.*, *1*, 970 (1962).
196. Karipides, A. G., Ph.D. thesis, University of Illinois, 1964.
197. Karipides, A. G., and T. S. Piper, *J. Chem. Phys.*, *40*, 674 (1964).
198. Keeling, R. O., *Acta Cryst.*, *10*, 209 (1957).
199. Kida, S., K. Nakamoto, and R. Tsuchida, *J. Chem. Soc. (Japan)*, *72*, 749 (1951).
200. Kida, S., J. Fujita, and K. Nakamoto, *Bull. Chem. Soc. (Japan)*, *31*, 79 (1958).
201. Kida, S., Y. Nakashima, Y. Morimoto, K. Niimi, and S. Yamada, *Bull. Chem. Soc. (Japan)*, *37*, 549 (1964).
202. King, W. R., and C. S. Garner, *J. Chem. Phys.*, *18*, 689 (1950).
203. Kobayashi, M., and R. Tsuchida, *J. Chem. Soc. (Japan)*, *60*, 769 (1939).
204. Koide, S., and M. H. L. Pryce, *Phil. Mag.*, *3*, 607 (1958).
204a. Kokoszka, G. F., H. C. Allen, Jr., and G. Gordon, *J. Chem. Phys.*, *46*, 3013 (1967).
204b. Kokoszka, G. F., C. W. Reimann, H. C. Allen, Jr., and G. Gordon, *Inorg. Chem.*, *6*, 1657 (1967).
205. Kondo, Y., *Sci. Light (Tokyo)*, *11*, 76 (1962).
205a. Krishnamurthy, R., W. B. Schaap, and J. R. Perumareddi, *Inorg. Chem.*, *6*, 1338 (1967).
206. Krishnan, R. S., *Proc. Indian Acad. Sci.*, *26A*, 450 (1947).
206a. La Mar, G. N., *Acta Chem. Scand.*, *20*, 1359 (1966).
207. Lambe, J., and C. Kikuchi, *Phys. Rev.*, *118*, 71 (1960).
208. Lambot, H., *Bull. Soc. Roy. Sci. Liège*, 12, 439 (1943).
209. Liehr, A. D., and C. J. Ballhausen, *Phys. Rev.*, *106*, 1161 (1957).
210. Liehr, A. D., and C. J. Ballhausen, *Ann. Phys. (N.Y.)*, *6*, 134 (1959).
211. Liehr, A. D., *J. Phys. Chem.*, *64*, 43 (1960).
212. Liehr, A. D., *J. Phys. Chem.*, *67*, 1314 (1963).
213. Lingafelter, E. C., G. L. Simmons, B. Morosin, C. Scheringer, and C. Frieburg, *Acta Cryst.*, *14*, 1222 (1961).
213a. Lingafelter, E. C., *Coord. Chem. Rev.*, *1*, 151 (1966).

214. Linz, A. Jr., and R. E. Newnham, *Phys. Rev.*, *123*, 500 (1961).
215. Lipson, H., *Proc. Roy. Soc. (London)*, *A151*, 347 (1935).
216. Lipson, H., and C. A. Beevers, *Proc. Roy. Soc. (London)*, *A148*, 664 (1935).
217. Lohr, L. L. Jr., and W. N. Lipscomb, *J. Chem. Phys.*, *38*, 1607 (1963).
218. Lohr, L. L. Jr., and W. N. Lipscomb, *Inorg. Chem.*, *2*, 911 (1963).
219. Longuet-Higgins, H. C., M. H. L. Pryce, and R. A. Sack, *Proc. Roy. Soc. (London)*, *A244*, 1 (1958).
220. Longuet-Higgins, H. C., and J. N. Murrell, *Proc. Physik. Soc.*, *A68*, 601 (1955).
221. Lopez-Castro, A., and M. R. Truter, *J. Chem. Soc.*, *1963*, 1309.
222. Low, W., *Z. Physik. Chem., N.F.*, *13*, 107 (1957).
223. Low, W., *Phys. Rev.*, *105*, 801 (1957).
224. Low, W., *Phys. Rev.*, *109*, 247 (1958).
225. Low, W., *Phys. Rev.*, *109*, 256 (1958).
226. Low, W., and G. Rosengarten, *J. Mol. Spectry.*, *12*, 319 (1964).
227. McCaffery, A. J., and S. F. Mason, *Trans. Faraday Soc.*, *59*, 1 (1963).
228. McClure, D. S., *J. Phys. Chem. Solids*, *3*, 311 (1957).
229. McClure, D. S., *Solid State Phys.*, *8*, 1 (1958).
230. McClure, D. S., *Solid State Phys.*, *9*, 399 (1959).
231. McClure, D. S., *Electronic Spectra of Molecules and Ions in Crystals*, Academic Press, New York, 1959.
232. McClure, D. S., in *Advances in the Chemistry of Coordination Compounds*, S. Kirschner, Ed., Macmillan, New York, 1961, p. 498.
233. McClure, D. S., *J. Chem. Phys.*, *36*, 2757 (1962); *Erratum*, *37*, 1571 (1962).
233a. McClure, D. S., *J. Chem. Phys.*, *38*, 2289 (1963).
234. McClure, D. S., *J. Chem. Phys.*, *39*, 2850 (1963).
235. McCrone, W. C., *Anal. Chem.*, *22*, 1459 (1950).
236. Macfarlane, R. M., *J. Chem. Phys.*, *39*, 3118 (1963).
237. Macfarlane, R. M., *J. Chem. Phys.*, *40*, 373 (1964).
238. Macfarlane, R. M., *J. Chem. Phys.*, *42*, 442 (1965).
239. McGarvey, B. R., *J. Chem. Phys.*, *38*, 388 (1963).
240. McGarvey, B. R., *Transition Metal Chem.*, *3*, 89 (1966).
241. Maki, G., *J. Chem. Phys.*, *28*, 651 (1958).
242. Maki, G., and B. R. McGarvey, *J. Chem. Phys.*, *29*, 31 (1958).
243. Maki, G., *J. Chem. Phys.*, *29*, 162 (1958).
244. Maki, G., *J. Chem. Phys.*, *29*, 1129 (1958).
245. Malatesta, L., *Gazz. Chim. Ital.*, *72*, 287 (1942).
246. Mani, N. V., *Proc. Indian Acad. Sci.*, *46A*, 143 (1957).
247. Manoharan, P. T., and W. C. Hamilton, *Inorg. Chem.*, *2*, 1043 (1963).
248. Manoharan, P. T., and H. B. Gray, *J. Am. Chem. Soc.*, *87*, 3340 (1965).
249. Manoharan, P. T., and H. B. Gray, *Inorg. Chem.*, *5*, 823 (1966).
250. Margerie, J., *Compt. Rend.*, *255*, 1598 (1962).
251. Martin, D. S., Jr., and C. A. Lenhardt, *Inorg. Chem.*, *3*, 1368 (1964).
252. Martin, D. S., Jr., M. A. Tucker, and A. J. Kassman, *Inorg. Chem.*, *4*, 1682 (1965).
253. Martin, D. S., Jr., J. G. Foss, M. E. McCarville, M. A. Tucker, and A. J. Kassman, *Inorg. Chem.*, *5*, 491 (1966).
254. Martin, D. S., Jr., M. A. Tucker, and A. J. Kassman, *Inorg. Chem.*, *5*, 1289 (1966).
255. Martin, D. S., Jr., *Coord. Chem. Rev.*, *1*, 39 (1966).

256. Martin, D. S., Jr., M. A. Tucker, and A. J. Kassman, *Coord. Chem. Rev.*, *1*, 44 (1966).
256a. Martin-Brunetière, F., and L. Couture, *Compt. Rend.*, *256*, 5327 (1963).
257. Mason, S. F., and B. J. Norman, *Chem. Commun.*, *1965*, 48.
258. Mathieson, A. McL., private communication to J. Ferguson, ref. 109.
258a. Mehra, A., and P. Venkateswarlu, *J. Chem. Phys.*, *47*, 2334 (1967).
259. Merritt, L. L., Jr., C. Guare, and A. E. Lessor, Jr., *Acta Cryst.*, *9*, 253 (1956).
260. Michel, R. E., private communication to H. A. Weakliem and D. S. McClure, ref. 412; ESR experiments on Co:ZnS gave a ground-state splitting of 1 cm^{-1}.
261. Miki, S., and S. Yamada, *Bull. Chem. Soc. (Japan)*, *37*, 1044 (1964).
262. Miller, J. R., *Proc. Chem. Soc.*, *1960*, 318.
263. Miller, J. R., *J. Chem. Soc.*, *1961*, 4452.
264. Miller, J. R., *J. Chem. Soc.*, *1965*, 713.
265. Mizuno, J., K. Ukei, and T. Sugawara, *J. Phys. Soc. (Japan)*, *14*, 383 (1959).
266. Moffitt, W., and C. J. Ballhausen, *Ann. Rev. Phys. Chem.*, *7*, 107 (1956).
267. Morgan, G. T., and H. D. K. Drew, *J. Chem. Soc.*, *1921*, 1058.
268. Morosin, B., and E. C. Lingafelter, *Acta Cryst.*, *12*, 611 (1959).
269. Morosin, B., and E. C. Lingafelter, *Acta Cryst.*, *12*, 744 (1959).
270. Morosin, B., and E. C. Lingafelter, *Acta Cryst.*, *13*, 807 (1960).
271. Morosin, B., and E. C. Lingafelter, *J. Phys. Chem.*, *65*, 50 (1961).
272. Morosin, B., and J. R. Brathovde, *Acta Cryst.*, *17*, 705 (1964).
273. Morosin, B., and K. Lawson, *J. Mol. Spectry.*, *12*, 98 (1964).
274. Morosin, B., and K. Lawson, *J. Mol. Spectry.*, *14*, 397 (1964).
275. Morosin, B., *Acta Cryst.*, *19*, 131 (1965).
276. Mortenson, O. S., *Acta Chem. Scand.*, *19*, 1500 (1965).
277. Mulliken, R. S., *Phys. Rev.*, *43*, 279 (1933).
278. Naiman, C. S., *Bull. Am. Phys. Soc.*, Ser II, *7*, 533 (1962).
279. Naiman, C. S., and A. Linz, *Polytech. Inst. Brooklyn, Microwave Res. Inst. Symp. Ser.*, *13*, 369, 383 (1963).
280. Nakahara, A., Y. Saito, and H. Kuroya, *Bull. Chem. Soc. (Japan)*, *25*, 331 (1952).
281. Nakamoto, K., Y. Morimoto, and A. E. Martell, *J. Am. Chem. Soc.*, *83*, 4533 (1961).
282. Nakatsu, K., Y. Saito, and H. Kuroya, *Bull. Chem. Soc. (Japan)*, *29*, 428 (1956).
283. Nakatsu, K., M. Shiro, Y. Saito, and H. Kuroya, *Bull. Chem. Soc. (Japan)*, *30*, 158 (1957).
284. Naray-Szabo, I., *Muegyet. Kozlemen.*, *1*, 30 (1947).
285. Natta, G., P. Garradini, and G. Allegra, *J. Polymer Sci.*, *51*, 399 (1961).
286. Neuhaus, A., and W. Schilly, *Fortschr. Mineral.*, *36*, 64 (1958).
287. Newnham, R. E., and Y. M. de Haan, *Z. Krist.*, *117*, 235 (1962).
288. Nikitine, S., *Phil. Mag. (8)*, *4*, 1 (1959).
289. Oi, S., Y. Komiyana, Y. Saito, and H. Kuroya, *Bull. Chem. Soc. (Japan)*, *32*, 263 (1959).
290. Oi, S., Y. Komiyana, and H. Kuroya, *Bull. Chem. Soc. (Japan)*, *33*, 354 (1960).
291. Okaziaki, A., and Y. Seumone, *J. Phys. Soc. (Japan)*, *16*, 671 (1961).
292. Orgel, L. E., and J. D. Dunitz, *Nature*, *179*, 462 (1957).
293. Orgel, L. E., *Nature*, *179*, 1348 (1957).
294. Ortolano, T. R., J. Selbin, and S. P. McGlynn, *J. Chem. Phys.*, *41*, 262 (1964).

295. Owen, J., *Proc. Roy. Soc.* (*London*), *A227*, 183 (1955).
295a. Pabst, A., *Am. Mineralogist*, *28*, 372 (1943).
295b. Pabst, A., *Acta Cryst.*, *12*, 733 (1959).
296. Le Paillier-Malecot, A., and L. Couture, *J. Chim. Phys.*, *62*, 359 (1965).
297. Le Paillier-Malecot, A., *Compt. Rend.*, *260*, 2777 (1965).
298. Le Paillier-Malecot, A., *Compt. Rend.*, *261*, 943 (1965).
299. Palma-Vittorelli, M. B., M. U. Palma, D. Palumbo, and F. Sgarlata, *Nuovo Cimento*, *3*, 718 (1956).
300. Palmer, R. A., and T. S. Piper, *Inorg. Chem.*, *5*, 864 (1966).
301. Palmer, R. A., G. D. Stucky and T. S. Piper, to be published.
302. Pappalardo, R., *Nuovo Cimento*, *6*, 392 (1957).
303. Pappalardo, R., *Phil. Mag.*, (*8*), *2*, 1397 (1957).
304. Pappalardo, R., *Nuovo Cimento*, *8*, 955 (1958).
305. Pappalardo, R., *Phil. Mag.*, (*8*), *4*, 219 (1959).
306. Pappalardo, R., *J. Chem. Phys.*, *31*, 1050 (1959).
307. Pappalardo, R., *J. Mol. Spectry.*, *6*, 554 (1961).
308. Pappalardo, R., and R. E. Dietz, *Phys. Rev.*, *123*, 1188 (1961).
309. Pappalardo, R., D. L. Wood, and R. C. Linares, Jr., *J. Chem. Phys.*, *35*, 1460 (1961).
310. Pappalardo, R., D. L. Wood, and R. C. Linares, Jr., *J. Chem. Phys.*, *35*, 2041 (1961).
311. Pariser, R., and R. G. Parr, *J. Chem. Phys.*, *21*, 466, 767 (1953).
312. Pauling, L., and S. B. Hendricks, *J. Am. Chem. Soc.*, *47*, 781 (1925).
313. Pauling, L., *Z. Krist.*, *72*, 482 (1930).
314. Pauling, L., and J. L. Hoard, *Z. Krist.*, *74*, 546 (1930).
315. Pelletier-Allard, N., *Compt. Rend.*, *252*, 3970 (1961).
316. Pelletier-Allard, N., *Compt. Rend.*, *258*, 1215 (1964).
317. Pelletier-Allard, N., *Compt. Rend.*, *259*, 2999 (1964).
318. Pelletier-Allard, N., *Compt. Rend.*, *260*, 2170 (1965).
319. Pelletier-Allard, N., *Compt. Rend.*, *261*, 1259 (1965).
320. Perummareddi, J. R., A. D. Liehr, and A. W. Adamson, *J. Am. Chem. Soc.*, *85*, 249 (1963).
320a. Perumareddi, J. R., *J. Phys. Chem.*, *71*, 3144 (1967).
320b. Perumareddi, J. R., *J. Phys. Chem.*, *71*, 3155 (1967).
321. Piper, T. S., and N. Koertage, *J. Chem. Phys.*, *32*, 559 (1960).
322. Piper, T. S., and R. L. Carlin, *J. Chem. Phys.*, *33*, 608 (1960).
323. Piper, T. S., and R. L. Carlin, *J. Chem. Phys.*, *33*, 1208 (1960).
324. Piper, T. S., *J. Chem. Phys.*, *35*, 1240 (1961); *Erratum*, *36*, 1089 (1962).
325. Piper, T. S., and R. L. Carlin, *J. Chem. Phys.*, *35*, 1809 (1961).
326. Piper, T. S., and R. L. Belford, *Mol. Phys.*, *5*, 169 (1962).
327. Piper, T. S., and A. G. Karipides, *Mol. Phys.*, *5*, 475 (1962).
328. Piper, T. S., *J. Chem. Phys.*, *36*, 2224 (1962).
329. Piper, T. S., and R. L. Carlin, *J. Chem. Phys.*, *36*, 3330 (1962).
330. Piper, T. S., and R. L. Carlin, *Inorg. Chem.*, *2*, 260 (1963).
331. Pople, J. A., *Trans. Faraday Soc.*, *49*, 1375 (1953).
332. Porai-Koshits, M. A., *Kristallografiya*, *1*, 291 (1956).
333. Porai-Koshits, M. A., and L. M. Dikareva, *Kristallografiya*, *4*, 650 (1959); *Soviet Phys. Cryst.* (*Engl. Trans.*), *4*, 611 (1959).
334. Powell, H. M., and A. F. Wells, *J. Chem. Soc.*, *1935*, 359.

335. Powell, M. J. D., J. R. Gabriel, and D. F. Johnston, *Phys. Rev. Letters*, 5, 145 (1960).
336. Pratt, G. W., and R. Coehlo, *Phys. Rev.*, 116, 281 (1959).
337. Prout, C. K., personal communication to P. Day et al., ref. 85.
338. Pryce, M. H. L., and U. Öpik, *Proc. Roy. Soc. (London)*, A238, 425 (1957).
339. Pryce, M. H. L., and W. A. Runciman, *Discussions Faraday Soc.*, 26, 34 (1958).
340. Pryce, M. H. L., G. Agnetta, T. Garofano, M. B. Palma-Vittorelli, and M. U. Palma, *Phil. Mag. (8)*, 10, 477 (1964).
341. Racah, G., *Phys. Rev.*, 62, 438 (1942).
342. Ramaseshan, S., K. Venkatesan, and N. V. Mani, *Proc. Indian. Acad. Sci.*, 46A, 95 (1957).
343. Reddy, T. R., and R. Rajan, *Proc. Nucl. Phys. Solid State Phys. Symp.*, Chandigarh, India, Pt.B., pp. 344–347, 1964.
344. Reddy, T. R., and R. Srinivasan, *J. Chem. Phys.*, 43, 1404 (1965).
345. Reimann, C. W., G. F. Kokoszka, and G. Gordon, *Inorg. Chem.*, 4, 1082 (1965).
345a. Reimann, C. W., G. F. Kokoszka, and H. C. Allen, Jr., *J. Rev. Natl. Bur. Std.*
345b. Remy, H., and G. Laves, *Ber.*, 66, 401 (1933).
346. Roof, R. B., *Acta Cryst.*, 9, 781 (1956).
346a. Roos, B., *Acta Chem. Scand.*, 20, 1673 (1966).
346b. Roos, B., *Acta Chem. Scand.*, 21, 1855 (1967).
347. Ross, I. G., *Trans. Faraday Soc.*, 55, 1058 (1959).
348. Ross, I. G., and J. Yates, *Trans. Faraday Soc.*, 55, 1064 (1959).
349. Runciman, W. A., private communication to R. M. Macfarlane, ref. 237.
350. Runciman, W. A., *Repts. Progr. Phys.*, 22, 30 (1958).
350a. Sabatini, A., *Inorg. Chem.*, 6, 1756 (1967).
351. Saito, Y., and H. Iwasaki, *Bull. Chem. Soc. (Japan)*, 35, 1131 (1962).
352. Santos, J. A., and H. Grime, *Z. Krist.*, 88, 136 (1934).
353. Sayre, E. V., K. Sancier, and S. Freed, *J. Chem. Phys.*, 23, 2060 (1955).
354. Schaeffer, C. E., *Chemistry of the Coordination Compounds*, Pergamon Press, London, 1958.
354a. Schafer, H. L., J. C. Morrow, and H. M. Smith, *J. Chem. Phys.*, 42, 504 (1965).
355. Schläfer, H. L., *Z. Phys. Chem.*, *N.F.*, 4, 116 (1955).
356. Schonland, D. S., *Proc. Roy. Soc. (London)*, A254, 111 (1960).
357. Schroeder, K. A., *J. Chem. Phys.*, 37, 1587 (1962).
358. Sears, D. R., *Dissertation Abstracts*, 19, 1225 (1958).
359. Seed, D. P., *Phil. Mag. (8)*, 7, 1371 (1962).
360. Seed, T. J., *J. Chem. Phys.*, 41, 1486 (1964).
361. Selbin, J., T. R. Ortolano, and F. J. Smith, *Inorg. Chem.*, 2, 1315 (1963).
362. Selbin, J., and T. R. Ortolano, *J. Inorg. Nucl. Chem.*, 26, 37 (1964).
362a. Selbin, J., *Coord. Chem. Rev.*, 1, 293 (1966).
362b. Selbin, J., G. Maus, and D. L. Johnson, *J. Inorg. Nucl. Chem.*, 29, 1735 (1967).
363. Sharnoff, M., *J. Chem. Phys.*, 41, 2203 (1964).
364. Sharnoff, M., *J. Chem. Phys.*, 42, 3383 (1965).
365. Sharnoff, M., and C. W. Reimann, *J. Chem. Phys.*, 43, 2993 (1965).
366. Sharnoff, M., and C. W. Reimann, *J. Chem. Phys.*, 46, 2634 (1967).
366a. Sharpe, A. G., and D. B. Wakefield, *J. Chem. Soc.*, 1957, 281.
367. Shenstone, A. G., *J. Opt. Soc. Am.*, 44, 749 (1954).
368. Shkolnikova, L. M., and E. A. Shugam, *Kristallografiya*, 5, 32 (1960), *Soviet Phys. Cryst. (Engl. Trans.)*, 5, 27 (1960).

369. Shkolnikova, L. M., and E. A. Shugam, *Zh. Strukt. Khim.*, *2*, 72 (1961).
370. Simonsen, S. H., and J. Wah Ho, *Acta Cryst.*, *6*, 430 (1953).
371. Singer, L. S., *J. Chem. Phys.*, *23*, 379 (1955).
372. Smith, G. P., and T. R. Griffiths, *J. Am. Chem. Soc.*, *85*, 4051 (1963).
373. Sone, K., *J. Am. Chem. Soc.*, *75*, 5207 (1953).
374. Spacu, P., M. Brezeanu and C. Lepadatu, *Studii, Cercatari. Chim. Bucharest*, *13*, 525 (1964); *Rev. Roumaine Chim.* (*Eng. Trans.*), *9*, 475 (1964).
374a. Stanko, J. A., Dissertation Abstr., *27*, 3838 (1967); Ph.D. thesis, University of Illinois.
374b. Steadman, J. P., and R. D. Willett, *Inorg. Chem.*, in press.
374c. Stemple, N., private communication to G. Burns, Ref. 57b.
374d. Stevenson, R., *Can. J. Phys.*, *43*, 1732 (1965).
374e. Stevenson, R., *Phys. Rev.*, *152*, 531 (1966).
375. Stewart, J. M., and E. C. Lingafelter, *Acta Cryst.*, *12*, 842 (1959).
376. Stewart, J. M., E. C. Lingafelter, and J. D. Breazedale, *Acta Cryst.*, *14*, 888 (1961).
377. Stout, J. W., and S. A. Reed, *J. Am. Chem. Soc.*, *76*, 5279 (1954).
378. Stout, J. W., *J. Chem. Phys.*, *31*, 709 (1959).
379. Stout, J. W., *J. Chem. Phys.*, *33*, 303 (1960).
380. Struckturbericht, *2*, 85 (1928–1932).
381. Sturge, M. D., *Phys. Rev.*, *130*, 639 (1963).
381a. Sturge, M. D., *J. Chem. Phys.*, *43*, 1826 (1965).
382. Sugano, S., and Y. Tanabe, *J. Phys. Soc.* (*Japan*), *13*, 880 (1958).
383. Sugano, S., and I. Tsujikawa, *J. Phys. Soc.* (*Japan*), *13*, 899 (1958).
384. Sugano, S., *J. Chem. Phys.*, *33*, 1883 (1960).
385. Sugano, S., and M. Peter, *Phys. Rev.*, *122*, 381 (1961).
386. Tanabe, Y., and S. Sugano, *J. Phys. Soc.* (*Japan*), *9*, 753 (1954).
387. Tanabe, Y., and S. Sugano, *J. Phys. Soc.* (*Japan*), *9*, 766 (1954).
388. Tanaka, N., N. Sato and J. Fujita, *Spectrochim. Acta*, *22*, 577 (1966).
389. Tanito, Y., Y. Saito, and H. Kuroya, *Bull. Chem. Soc.* (*Japan*), *25*, 328 (1952).
390. Teltow, J., *Z. Physik. Chem.*, *B40*, 397 (1938).
391. Teltow, J., *Z. Physik. Chem.*, *B43*, 198 (1939).
392. Theilacker, W., *Z. Anorg. Allgem. Chem.*, *234*, 161 (1937).
393. Tonnet, M. L., S. Yamada and I. G. Ross, *Trans. Faraday Soc.*, *60*, 840 (1964).
393a. Townsend, M. G., *J. Chem. Phys.*, *41*, 3149 (1964).
394. Trees, R. E., *Phys. Rev.*, *83*, 756 (1951).
395. Tsuchida, R., and M. Kobayashi, *Bull. Chem. Soc.* (*Japan*), *13*, 619 (1938).
396. Tsuchida, R., and M. Kobayashi, *J. Chem. Soc.* (*Japan*), *60*, 769 (1939).
397. Tsuchida, R., and M. Kobayashi, *J. Chem. Soc.* (*Japan*), *64*, 1268 (1943).
398. Tsuchida, R., and S. Yamada, *Nature*, *182*, 1230 (1958).
399. Tsujikawa, I., and L. Couture, *J. Phys. Radium*, *16*, 430 (1955).
400. Tsujikawa, I., and L. Couture, *Compt. Rend.*, *250*, 2013 (1960).
401. Vainshtein, B. K., *Dokl. Akad. Nauk. SSSR*, *83*, 227 (1952).
401a. Van Heuvelen, A., *J. Chem. Phys.*, *46*, 4903 (1967).
402. Van Niekerk, J. N., and F. R. L. Schoening, *Acta Cryst*, *4*, 35 (1951).
403. Van Niekerk, J. N., and F. R. L. Schoening, *Acta Cryst.*, *5*, 196 (1952).
404. Van Niekerk, J. N., and F. R. L. Schoening, *Acta Cryst.*, *5*, 499 (1952).
405. Van Niekerk, J. N., and F. R. L. Schoening, *Acta Cryst.*, *6*, 227 (1953).

406. Van Niekerk, J. N., F. R. L. Schoening, and J. F. de Wet, *Acta Cryst.*, 6, 501 (1953).
407. Van Vleck, J. H., *J. Phys. Chem.*, 41, 67 (1937).
408. Van Vleck, J. H., *J. Chem. Phys.*, 7, 72 (1939).
408a. Varfolomeeva, L. A., G. S. Zhdanov, and M. M. Umanskii, *Kristollagrafiya*, 3, 368 (1958); *Soviet Phys. Cryst.* (*Engl. Trans.*), 3, 369 (1958).
408b. Vedder, W., *Am. Mineralogist*, 49, 736 (1964) and references therein.
409. Viste, A., and H. B. Gray, *Inorg. Chem.*, 3, 1113 (1964).
409a. Vossos, P. H., D. R. Fitzwater, and R. E. Rundle, *Acta Cryst.*, 16, 1037 (1963).
409b. Walker, I. M., and R. L. Carlin, *J. Chem. Phys.*, 46, 3931 (1967).
410. Walsh, W. M., *Phys. Rev. Letters*, 4, 507 (1960).
411. Watanabe, H., *Operator Methods in Ligand-Field Theory*, Prentice-Hall, Englewood Cliffs, N.J., 1966.
412. Weakliem, H. A., and D. S. McClure, *J. Appl. Phys.*, Supplement No. 1, 33, 347 (1962).
413. Wentworth, R. A. D., and T. S. Piper, *J. Chem. Phys.*, 41, 3884 (1964).
414. Wentworth, R. A. D., *Chem. Commun.*, 1965, 532.
415. Wentworth, R. A. D., and T. S. Piper, *Inorg. Chem.*, 4, 202 (1965).
416. Wentworth, R. A. D., and T. S. Piper, *Inorg. Chem.*, 4, 709 (1965).
417. Wentworth, R. A. D., and T. S. Piper, *Inorg. Chem.*, 4, 1524 (1965).
418. Wentworth, R. A. D., *Inorg. Chem.*, 5, 496 (1966).
418a. Wickersheim, K., *J. Appl. Phys.*, 34, 1224 (1963).
418b. Willett, R. D., C. Dwiggens, R. F. Kruh, and R. E. Rundle, *J. Chem. Phys.*, 38, 2429 (1962).
418c. Willett, R. D., *J. Chem. Phys.*, 41, 2243 (1964).
418d. Willett, R. D., *J. Chem. Phys.*, 44, 39 (1965).
418e. Willett, R. D., and O. L. Liles, Jr., *Inorg. Chem.*, 6, 1666 (1967).
418f. Willett, R. D., O. L. Liles, Jr., and C. Michelson, *Inorg. Chem.*, 6, 1885 (1967).
418g. Willett, R. D., *J. Inorg. Nucl. Chem.*, 29, 2482 (1967).
418h. Measurements of Windsor and Wilson quoted by M. B. Walker and R. W. H. Stevenson, *Proc. Phys. Soc.* (*London*), 87, 35 (1966).
419. Wolf, H. C., *Solid State Phys.*, 9, 1 (1959).
420. Wolfsberg, M., and L. Helmholz, *J. Chem. Phys.*, 20, 837 (1952).
421. Wood, D. L., and J. P. Remeika, *J. Chem. Phys.*, 46, 3595 (1967).
421a. Wong, E. Y., *J. Chem. Phys.*, 32, 598 (1960).
421b. Wybourne, B. G., *J. Chem. Phys.*, 43, 4506 (1965).
422. Wyckoff, R. W. G., *Crystal Structures*, Vol. 1, 2nd ed., Interscience, New York, 1963.
423. Wyckoff, R. W. G., *Crystal Structures*, Vol. 3, 2nd ed., Interscience, New York, 1965.
424. Yamada, S., *Bull. Chem. Soc.* (*Japan*), 24, 125 (1951).
425. Yamada, S., *J. Am. Chem. Soc.*, 73, 1182 (1951).
426. Yamada, S., *J. Am. Chem. Soc.*, 73, 1579 (1951).
427. Yamada, S., and R. Tsuchida, *Bull. Chem. Soc.* (*Japan*), 25, 127 (1952).
428. Yamada, S., Y. Shimura, and R. Tsuchida, *Bull. Chem. Soc.* (*Japan*), 26, 72 (1953).
429. Yamada, S., and R. Tsuchida, *Bull. Chem. Soc.* (*Japan*), 26, 156 (1953).
430. Yamada, S., and R. Tsuchida, *Bull. Chem. Soc.* (*Japan*), 26, 489 (1953).
431. Yamada, S., and R. Tsuchida, *J. Am. Chem. Soc.*, 75, 6351 (1953).

432. Yamada, S., and R. Tsuchida, *Bull. Chem. Soc. (Japan)*, *27*, 156 (1954).
433. Yamada, S., and R. Tsuchida, *Bull. Chem. Soc. (Japan)*, *27*, 436 (1954).
434. Yamada, S., and R. Tsuchida, *J. Chem. Phys.*, *22*, 1273 (1954).
435. Yamada, S., A. Nakahara, Y. Shimura, and R. Tsuchida, *Bull. Chem. Soc. (Japan)*, *28*, 222 (1955).
436. Yamada, S., and R. Tsuchida, *Bull. Chem. Soc. (Japan)*, *28*, 664 (1955).
437. Yamada, S., and R. Tsuchida, *Bull. Chem. Soc. (Japan)*, *29*, 289 (1956).
438. Yamada, S., and R. Tsuchida, *Bull. Chem. Soc. (Japan)*, *29*, 421 (1956).
439. Yamada, S., and R. Tsuchida, *Bull. Chem. Soc. (Japan)*, *29*, 694 (1956).
440. Yamada, S., and R. Tsuchida, *Bull. Chem. Soc. (Japan)*, *29*, 894 (1956).
441. Yamada, S., S. Nakahara, and R. Tsuchida, *Bull. Chem. Soc. (Japan)*, *30*, 953 (1957).
442. Yamada, S., H. Nakamura, and R. Tsuchida, *Bull. Chem. Soc. (Japan)*, *31*, 303 (1958).
443. Yamada, S., and R. Tsuchida, *Bull. Chem. Soc. (Japan)*, *33*, 98 (1960).
444. Yamada, S., H. Nishikawa, and R. Tsuchida, *Bull. Chem. Soc. (Japan)*, *33*, 1278 (1960).
445. Yamada, S., paper presented at Proceedings of the International Symposium Molecular Structural and Spectroscopy, Tokyo, A403, 1962.
446. Yamada, S., *Bull. Chem. Soc. (Japan)*, *35*, 1427 (1962).
447. Yamada, S., and S. Miki, *Bull. Chem. Soc. (Japan)*, *36*, 680 (1963).
448. Yamada, S., H. Nishikawa, and S. Miki, *Bull. Chem. Soc. (Japan)*, *37*, 576 (1964).
449. Yamada, S., *Coord. Chem. Rev.*, *2*, 83 (1967).
450. Yamatera, H., *Bull. Chem. Soc. (Japan)*, *31*, 95 (1958).
451. Yeranos, W. A., *Dissertation Abstr.*, *25*, 6991 (1965).
452. Zahner, J. C., and H. G. Drickamer, *J. Chem. Phys.*, *33*, 1625 (1960).
453. Zdhanov, G. S., and Z. V. Zvonkhova, *Zh. Fiz. Khim.*, *24*, 1339 (1950).
454. Zverov, G. M., and A. M. Prokhorov, *Zh. Eksperim. i Teor. Fiz.*, *34*, 1023 (1958). *Soviet Phys., JETP (Eng. Trans.)*, *7*, 707 (1958).
455. Zverov, G. M., and A. M. Prokhorov, *Zh. Eksperim. i Teor. Fiz.*, *38*, 449 (1960); *Soviet Phys., JETP (Engl. Trans.)*, *11*, 330 (1960).
456. Zverov, G. M., and A. M. Prokhorov, *Zh. Eksperim. i Teor. Fiz.*, *40*, 1016 (1961); *Soviet Phys., JETP (Engl. Trans.)*, *13*, 714 (1961).

Author Index

Numbers in parentheses are reference numbers and show that an author's work is referred to although his name is not mentioned in the text. Numbers in *italics* indicate the pages on which the full references appear.

Subject Index

Progress in Inorganic Chemistry

CUMULATIVE INDEX, VOLUMES 1–10